INTRODUCTION TO GLASS SCIENCE

INTRODUCTION TO GLASS SCIENCE

Proceedings of a Tutorial Symposium held at the State University of New York, College of Ceramics at Alfred University, Alfred, New York, June 8-19, 1970

Edited by

L. D. Pye, H. J. Stevens, and W. C. LaCourse

Division of Engineering/Science
State University of New York
College of Ceramics at Alfred University
Alfred, New York

PLENUM PRESS · NEW YORK-LONDON · 1972

Library of Congress Catalog Card Number 72-76933

ISBN 978-1-4757-0330-6 ISBN 978-1-4757-0328-3 (eBook)
DOI 10.1007/978-1-4757-0328-3

Softcover reprint of the hardcover 1st edition 1972

A Division of Plenum Publishing Corporation
227 West 17th Street, New York, N. Y. 10011

United Kingdom edition published by Plenum Press, London
A Division of Plenum Publishing Company, Ltd.
Davis House (4th Floor), 8 Scrubs Lane, Harlesden, London, NW10 6SE, England

FOREWORD

Glass technologists are fascinated by glass; exploration as well as application of glass is expanding and the influx of documentation is bewildering. There were about 200 papers on just semiconduction in glasses in 1970 and one has to scan about 200 papers a month to sense the pulse of glass science. Yet there are many in industry and education in science or engineering who require or wish to have coherent, comprehensive and contemporary information on this exciting material "glass."

The Tutorial Symposium offered as an Introduction to Glass Science in Alfred represents an earnest attempt to fulfill this need. It has been designed to provide both broad and technical instruction for participants and readers who are not specialists. Glass is not only a material but a condition of matter: the vitreous state. The topic, therefore, is introduced by a careful consideration of the nature of glass, or the vitreous state. The universality of the vitreous state is now generally recognized: not just a few, but very many structures can be obtained without appreciable crystallization. There is no restricted family of structures characteristic of glass formation: as long as crystallization is avoided, every liquid will solidify to a non-crystalline substance. Structural analysis in each case is now to be postulated and has become increasingly successful. The Alfred "Introduction to Glass Science" offers a representative overview of methods and results. The most important techniques considered are X-ray analysis, infrared and Raman spectroscopy, electron and nuclear magnetic resonance, Mossbauer spectroscopy. Analysis alone is rarely successful in the case of glass structrues and will depend on ingenious models. The central body of the volume discusses some of the most important properties of frequently used glasses: refraction, dispersion, color, strength, fracture, viscosity, diffusion, electrical properties, surface properties, corrosive action.

Last, but not least, the new materials melted and shaped as glasses, but processed and used as crystallized ceramics are introduced: the glass ceramics. Phase separation and crystallization are treated as the fundamental phenomena of this new and spreading technique. The tutorial symposium concludes with hints for laboratory experimenters in glass melting.

This volume is a contribution of this institution to the limited amount of introductory texts on glass science.

N. J. Kreidl
Professor of Ceramic Engineering
University of Missouri (Rolla)

PREFACE

Although glass has been used as an artistic medium and industrial material for centuries it has been only in recent years that "Glass Science" has emerged as a field of study in its own right. As is typical of any new field, Glass Science is currently in a state of rapid development, both in terms of our fundamental understanding of this material and in its commercial exploitation. On the other hand, glasses are complex materials, a fact which has impeded progress in achieving this understanding and exploitation. Noting this, it is not unreasonable to speculate that, perhaps, the most dramatic advances in this field are yet to come.

"Introduction to Glass Science" is a logical outgrowth of the increasing interest in non-crystalline materials. The book is based on a tutorial symposium held on the campus of Alfred University in June 1970. The objectives of both the symposium and this text were to present, in a single source, an introduction to the broad field of Glass Science. The material covered encompasses fundamental aspects, from chemical bonding to mechanical properties, as well as technologically important subjects such as melting and glass-refractory interactions.

Considering the broad subject area covered, this book should be suitable for introductory courses in the vitreous state at the senior level. At the same time, however, most subjects are covered in sufficient depth to render the book useful to those students and researchers interested in a particular aspect of glass science.

The authors wish to point out their gratitude and indebtedness to a number of people who assisted in various capacities in the presentation of the symposium and the preparation of the material included herein. They are Dr. L. Miles, President of Alfred University, Dr. E. E. Mueller, Dean of the S.U.N.Y. College of Ceramics at Alfred University, Dr.'s H. Kay, C. H. Greene, R. C. Martin, Mr. R. Dodge, L. Hanks, L. Manfredo, S. Boysen, and R. Kelly.

The authors acknowledge the enormous contribution of Mr. C. Bloomquist for his skilled duplications of the manuscripts for distribution at the symposium, as well as his preparation of the figures and tables appearing in this book. Similarly, the fine effort extended by Mr. W. Earl in planning and executing the symposium is also pointed out. Typing of the manuscript by Mrs. D. Snowden, Mrs. J. Franum, and Mrs. D. Stevens is also appreciated. The financial assistance provided by the State of New York technical services program is gratefully acknowledged. Finally, we wish to thank the contributing authors and Plenum Press for agreeing to publish the proceedings.

L. D. Pye
H. J. Stevens
W. C. LaCourse

Alfred, N. Y.
May, 1972

CONTENTS

The Vitreous State

L. David Pye

SUNY College of Ceramics
at Alfred University
Alfred, N. Y.

AN OVERVIEW

Glass is a versatile and sometimes enigmatic substance. It functions equally well as a carrier of light, a protector of man and his inventions, an object of art, or an indispensable tool for the exploration of science. Depending upon composition, it can be made stronger than steel[1] or soluble in water; a detector of nuclear radiations[2] or the source of a powerful laser beam.[3,4] Certain compositions are known to possess either negative, zero or positive expansion coefficients. Still others are colorless or black, or even variable in their light transmissive properties.[5,6,7] Electrical conductivities can vary from 10^{-3} - 10^{-18} ohm^{-1} cm^{-1} with both electronic and ionic type mechanisms being clearly substantiated. Paramagnetic or diamagnetic glasses can also be made, depending again upon choice of composition. Given this enormous diversity of properties, it is clear as to why the use of glass might transcend many industries and fields of science. It is also found in nature. Terrestrial (obsidian), celestial (tektites) and more recently, lunar varieties have been identified.

By in large this diversity can be attributed to any or all of the following factors:

1) The composition of glasses may vary over wide limits and includes 60 of the chemical elements.

2) The properties of glass vary continuously with composition in a predictable manner. Among other things this means that unlike many crystalline substances, stoichiometry of a cooled glass is never a problem. In fact, that term is of little value in describing multicomponent glasses.

3) The internal energy of a glass is almost always greater than that corresponding to a crystalline phase of the same composition.

4) Glass is isotropic in its properties; it can be made extremely pure and homogeneous. It can also be fabricated into a variety of shapes and sizes.

Even without a complete and detailed knowledge of the structure of a cooled glass, the exploitation of the above factors is readily achieved by simply viewing glass as a substance somewhat continuous with its liquid state. This state, like all liquids, possesses a finite vapor pressure,[8,9] a boiling point, a temperature dependent viscosity, and can be regarded as a solution of oxides or elements.

While at this liquid state, or at somewhat lower temperatures where conversion to a solid may ensue, many chemical reactions such as decomposition, oxidation-reduction, precipitation and/or dissolution, are readily achieved.[10,11] For silicate systems, which are particularly viscous in their molten state, these reactions are very sensitive to time and temperature. Because these parameters are easily controlled, it is comparatively straight forward to start, halt, advance, or slow down these reactions as desired. Coupling this with a natural tendency for all glasses to revert to a lower energy crystalline state, gives rise to a highly desirable materials design capability. That is, as glassy solids convert to a crystalline phase(s), the properties of these solids will vary continously in a controllable manner from those associated with the parent glass to those characteristic of the crystalline phase.

All of these considerations constitute a powerful control feature for the fabrication of materials with specific properties within certain limits. In fact, the potential use of glass as a precursor for derivative materials represents an additional diversity unqiue in materials science.

THE NATURE OF GLASS

(A) General Considerations:

There have been, of course, several attempts to give a technological definition to glass. In chronological order these definitions are:

1) Glass is an amorphous, i.e., structureless solid - (early 20's based on a diffuse X-ray diffraction pattern of glass).

2) After Morey[12] (1938) - "A glass is an inorganic substance in a condition which is continuous with and analagous to the liquid state of that substance, but which, as the result of a reversible change in viscosity during cooling, has attained so high a degree of viscosity as to be for all practical purposes, rigid."

3) From A.S.T.M.[13] (1949) - "Glass is an inorganic product of fusion which has cooled to a rigid condition without crystallizing."

4) After Secrist and Mackenzie[14] (1960) - "Glass is a non-crystalline solid."

In a sense, the historical development of these definitions reflects an evolution of thought regarding the scope, growth and understanding of the field of glass science. For example, organic glasses, which behave almost identically to oxide glasses, were known to exist at least in the earlier 1930's[13] but were not considered technologically important enough to be included within a definition until at least 1960. Similarly, shock waves, neutron bombardment, and vapor deposition can be used to fabricate a glass which is sensibly identical to that made by a fusion method.[14] Hence, it is idle to exclude them from any definition of glass. It might be argued, however, that definitions two and three were by in large, intended to include only those glasses which are of high commercial value. While it is true that these definitions for the most part accomplish this, other glasses produced by vapor decomposition are becoming more and more technologically and commercially important.

There are two good examples of this:

a) An arsenic-selinide glass made by vapor phase deposition is a basis of high speed xerography.

b) Vitreous silica made by vapor phase hydrolysis is a highly desirable optical glass and finds many applications.

A precise and measurable atomic structure of glass, as ruled out by number one is a matter of perspective. Although everyone agrees that vitreous substances lack the long range atomic order characteristic of crystalline materials, there is rather convincing evidence to indicate that a short range order, over a few atomic distances, is indeed possible. A precise estimate of the extent of this order is an extraordinary difficult task and may in the end prove to be intractable. It is interesting to note here that Warren's* first work on glass, using X-ray diffraction techniques, was concerned with calculating the known diffuse X-ray pattern for vitreous silica.[15] In this calculation it was assumed that the interatomic distances and coordination(s) were the same as those found in crystalline forms of silica and that a high degree of order existed in the glass over a chosen atomic grouping up to about 13 Å in diameter. The agreement between the observed and calculated X-ray pattern was quite good.

Thus at this point, the negative definition (number four) appears to be the most satisfactory since it includes all materials which are non-crystalline regardless of their composition and mode of formation. This definition is also consistent with Mott's suggestion that all solids can be classified as either crystalline or amorphous.[16] It implicitly assumes, however, that one can always distinguish between the crystalline and non-crystalline state. This may not always be true. In fact, many "crystalline" substances are themselves highly disordered whenever their lattice defect concentration becomes excessive. Eckstein would extend this concept to infer that the vitreous state is describable in terms of its defect concentration.[17] That is, lattice distortions and relaxations almost always accompany point defects in crystalline solids. If the

* see Chapter 3

defect concentration is high, then the local areas affected by these distortions and relaxations can overlap and in so doing, cause a breakdown of long-range crystalline order. On this basis then, the difference between crystalline and non-crystalline solids may be a matter of degree rather than a fixed absolute.

In attempting to further distinguish between these solids, it cannot even be stated with certainty that all glassy substances show identical directional changes in their properties when compared to their crystalline counterparts. As shown in Table I, two simple glass forming compounds, SiO_2 and B_2O_3, show contradictory changes in expansion and density upon crystallization. That is, vitreous silica undergoes an increase in density and expansion coefficient when crystallized, whereas vitreous B_2O_3 shows a decrease in expansion coefficient while increasing in density.

All of the above suggest then, the great difficulty of attempting to broadly classify such a widely diverse substance as glass. Definition number four would appear to most nearly accomplish this.

A similar difficulty is encountered when trying to develop uniform theories which are applicable to all glass forming systems. This is not always possible. This uneven applicability is most obvious when contrasting simple glasses (i.e., single component) with complex ones (i.e. multicomponent).

Table I. Density and Expansion Coefficient of Vitreous and Crystalline SiO_2 or B_2O_3

Compound	Approximate Room Temperature Expansion Coefficient/°C	Density g/cm^3
SiO_2		
a) vitreous	7×10^{-7}	2.20
b) cristobalite	170×10^{-7}	2.32
B_2O_3		
a) vitreous	$150\text{-}160 \times 10^{-7}$	1.81
b) crystalline	$130\text{-}140 \times 10^{-7}$	2.46

(B) Energetics of Glass Formation:

As mentioned above, vapor phase hydrolysis and vapor deposition are important methods of fabricating commercial glasses. These techniques are discussed more thoroughly in Chapter 10.

One of the best ways to gain an understanding of the vitreous state is to examine in some detail the manner in which the vast majority of glasses are formed--namely by quenching their melts to room temperature. To this end it is instructive to compare the energetics of two widely different substances such as bismuth and GeO_2, which through this process, will give a crystalline and glassy solid respectively. In making this comparison, it is assumed that the respective melt temperatures can be followed with time during quenching.

For bismuth, it is found (see Fig. 1-a) that a temperature arrest occurs exactly at the melting point T_o of this compound. This constancy of temperature indicates an evolution of heat content ΔH; otherwise the temperature would continue to fall as before. The duration of this arrest t, is just that required to completely crystallize the melt. It represents the time interval for the bismuth atoms to form a periodic crystalline lattice and is a direct consequence of the phase rule.

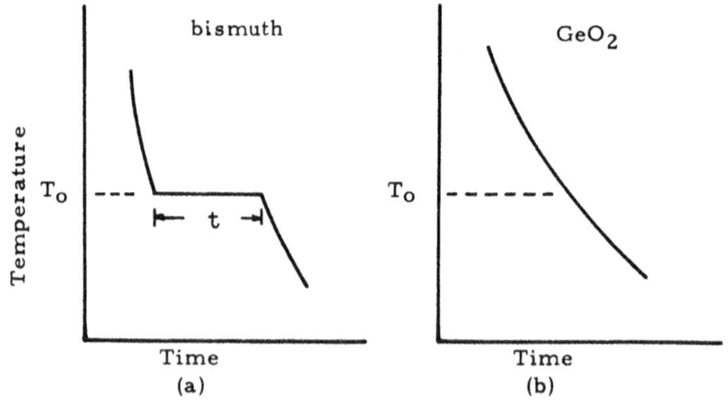

Fig. 1. Typical cooling curves for a) bismuth, b) GeO_2. T_o is different for each compound.

Since

$$\Delta S = \frac{\Delta H}{T_o} \tag{1}$$

and for this process ΔH is < 0, it is concluded that a reduction of entropy S, also occurs during this time interval. After crystallization, the once fluid melt is now solid.

In contrast to this behavior, GeO_2 (see Fig. 1-b) does not exhibit a temperature arrest when cooled through its melting point. This absence indicates that crystallization has not occurred, and subsequently the melt has not undergone a structural transformation to a more ordered crystalline phase. In this case, $\Delta H = 0$ at T_o, which requires that $\Delta S = 0$. For temperatures just below T_o, the still fluid melt of GeO_2 becomes metastable and behaves as a supercooled liquid. Continued cooling occasions a gradual decrease in heat content and entropy.

The metastability of a supercooled melt of GeO_2 indicates an equilibrium of its atomic structure with temperature. However, its internal energy is still higher than its crystalline state and that is why it is labeled metastable. A further decrease in temperature is accompanied by an increase in viscosity of the melt. Continued metastable equilibrium is possible as long as the viscosity is low enough to permit a structural rearrangement of the melt to occur on an atomic level. Eventually, as will be shown below, its metastability is destroyed; this substance becomes unstable and is no longer at structural equilibrium. Shortly thereafter, the once fluid melt converts to a solid whose internal energy is greater than its corresponding crystalline phase.

The departure from metastable equilibirum is sensitive to the rate of cooling employed and manifests itself in two distinct forms:

a) A failure to evolve energy in the form of heat consistent with the melt temperature.

b) A failure to equilibrate dimensionally with temperature.

These effects are of enormous importance in glass

science and will be discussed separately below.

1) Heat content of a glass:

Again consider cooling a melt to form a glass. Following Charles,[18] the heat content of this melt may be graphed against temperature (Fig. 2). It is seen that at just below the freezing point, the melt is at metastable equilibrium. That is for every temperature, there is a single heat content (which is higher than that corresponding to its crystalline state). Cooling to lower temperatures causes a sharp departure from linearity. The position of this departure is sensitive to the rate employed, lower rates giving rise to lower temperatures.

Reheating this glass (at a rate slightly different from that of cooling) will produce a characteristic DTA curve illustrated in Fig. 3.[19] During reheating, heat is evolved from the glass and transferred to its surroundings. This is followed by an absorption of heat by the glass. The magnitude of these effects are dependent upon the original cooling rates in the manner indicated. With the aid of Fig. 4, these peaks are explained as follows: both effects represent a departure from true metastable equilibrium. During reheating, the glass structure adjusts itself to its lowest energy level and in so doing, reduces its

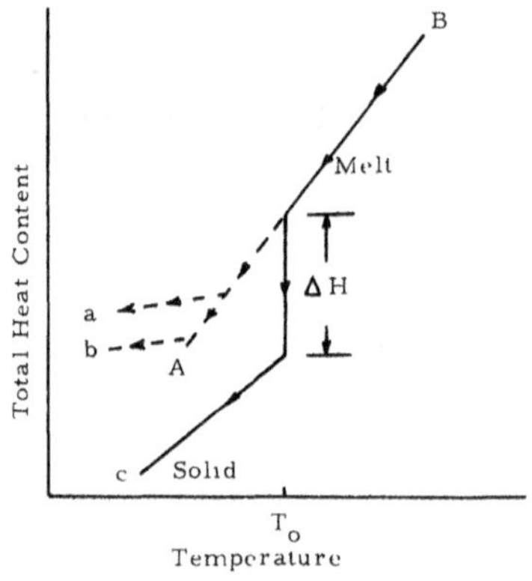

Fig. 2. Heat content of a glass forming substance when a) highly quenched, b) slowly cooled to form a glass or c) crystallized at its freezing point T_o. ΔH is the heat content evolved during crystallization.

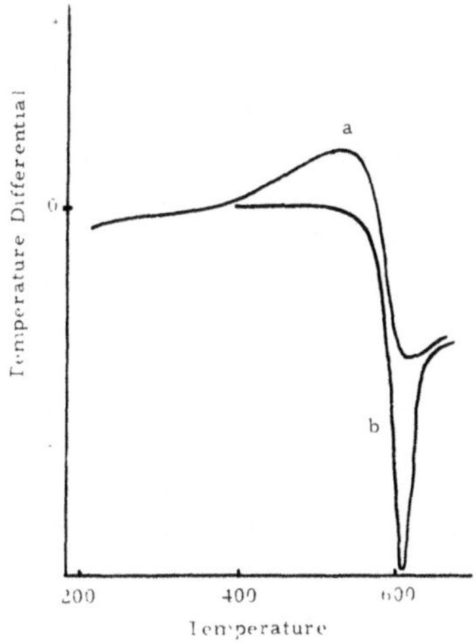

Fig. 3. Typical DTA curve of a) highly quenched glass, b) slowly cooled glass.[19]

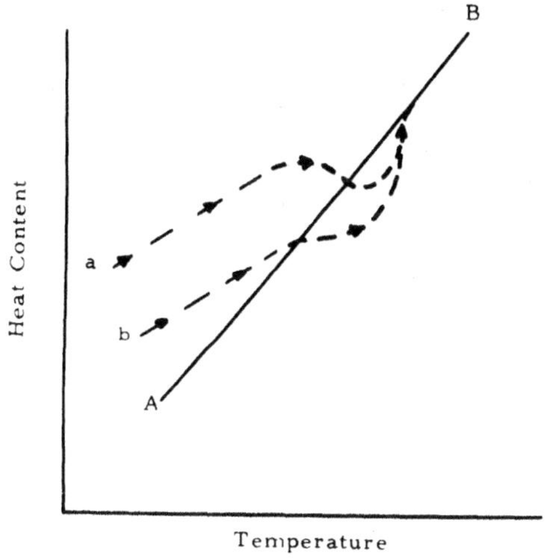

Fig. 4. Metastable, thermal equilibration behavior of a) highly quenched glass, b) slowly cooled or well annealed glass, upon reheating.

heat content to the equilibrium line A-B in Figs. 2 and 4. For highly quenched glasses this reduction in heat content is much larger than that for slowly cooled glasses. In the extreme case, very slow cooling of the original melt, or a long-term annealing of the residual glass will nearly eliminate the initial heat evolution as is indicated in Fig. 3.

In both cases, further heating after this initial heat evolution creates an additional sub-linear departure from equilibrium which persists until a temperature is reached where the rate of equilibration is fast enough to match the temperature rise. At this point, the melt returns to full metastable equilibrium. This equilibration is accompanied by a heat absorption which is larger for slow cooled or well annealed glasses than for highly quenched glasses.

2) Dimensional equilibration of glass forming melts:

Inasmuch as it is comparatively more difficult to measure the heat content of a solid, the glass scientist prefers to describe the above phenomena using density as a measure of equilibrium. This parameter can in turn be related to other properties of glass such as index of refraction, etc. From this viewpoint glass formation is depicted in Fig. 5. As before, departures from equilibrium manifest themselves at temperatures where the internal atomic structure cannot adjust itself in a sufficiently rapid manner. Again, the cooling rate will largely dictate the temperature at which disequilibrium occurs. This point is defined as T_g, the glass transition temperature. It is for these

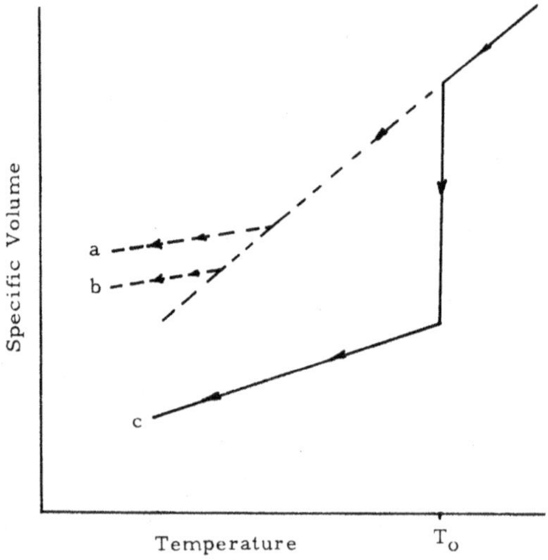

Fig. 5. Metastable and stable dimensional behavior of a glass forming substance when a) highly quenched, b) slowly cooled to form a glass, and c) crystallized at its melting point T_o.

reasons that the properties of a glass are slightly dependent upon its thermal history. More importantly, uneven cooling of a glass article gives rise to different residual densities throughout the article. This creates internal strains which may be as permanent as the glass article itself unless the article is annealed and allowed to cool very slowly. On the other hand, the controlled quenching of a glass, i.e., tempering, may induce permanent strains which can considerably increase the average breaking strength of a glass. (see Chapter 15)

a) the transformation range of a glass -

Occasionally, it is desirable to reheat glass and allow it to fully equilibrate at those temperatures where the above described departures from equilibrium may occur. After reheating, the glass can be quenched to room temperature and its density measured. There are two main points of interest here: 1) the rate of approach to equilibrium and 2) the range of temperatures where equilibration is measurable. The data of Ritland[20] in Fig. 6-a shows the equilibration of a highly quenched and well annealed glass of the same composition. The rate of approach to equilibrium is clearly faster for the quenched glass. This is consistent with the notion that such glasses have a much more open and therefore

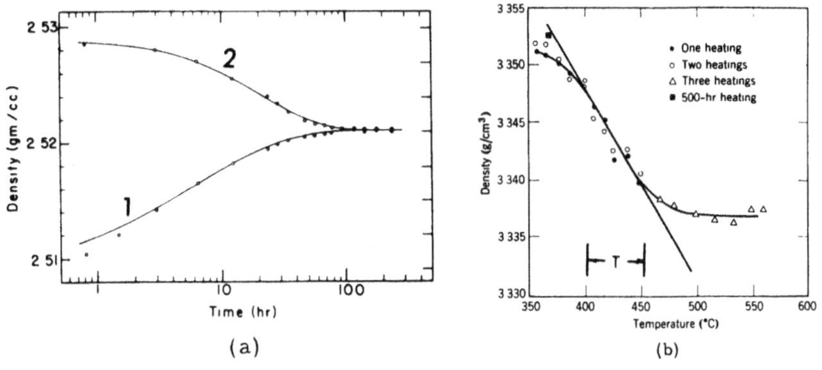

Fig. 6a. Equilibration behavior of (1) a highly quenched glass, (2) slowly cooled glass of the same composition.[20] The equilibrium density at this temperature (520°C) is about 2.5215 gm/cm^3. 6b. Typical S-shaped curve of a common glass showing its approximate transformation range T.[21]

easily alterable atomic structure than well annealed glasses which are more compact. This same experiment may be repeated at different temperatures and if the measured equilibrium densities are graphed against temperature, there results a characteristic S-shaped curve such as in Fig. 6-b. Vitreous silica shows an anomalous exception to this behavior.

From Fig. 6-b it is further apparent that the true equilibrium densities may not always be determined for all temperatures. At low values, the equilibration time is so long as to preclude its practical measurement. Conversely, at higher temperatures the density changes so rapidly with time, that during quenching, the density will actually adjust itself to a value corresponding to some lower temperature. This has created the concept of the transformation range of a glass, meaning that temperature interval in which the change in properties of a glass will occur at a measurable rate. A typical interval, T, is shown in Fig. 6-b.[21]

b) expansion characteristics of a glass -

From the above discussion, it is possible to understand the expansion characteristics of a glass. As before, the two extremes of a well annealed and quenched glass are used. Fig. 7 shows typical results. It is generally agreed that the linear, low temperature expansion is associated with changing interatomic distances and bond angles. At slightly higher temperatures, departures from linearity occur in a direction that is characteristic of the thermal history of the glass and is a consequence of the glass attempting to reach dimensional equilibrium as it enters its transformation range. As might be expected, quenched glasses with their more open atomic structure begin to show this effect at a lower temperature than that observed for well annealed glasses with their more compact structures. At higher temperatures, both glasses equilibrate with temperature and their expansion coefficients become sensibly coincident.

In this context, then, the expansion curve should be able to, more or less, delineate the transformation range of a glass. Fig. 8 indicates a typical analysis of an expansion curve showing the lower annealing limit T_ℓ, the transformation temperature T_g, the upper annealing limit T_u, the point of incipient deformation M_g, and the dilatometric softening point T_d. The viscosity at T_d is about 10^{11} - 10^{12}

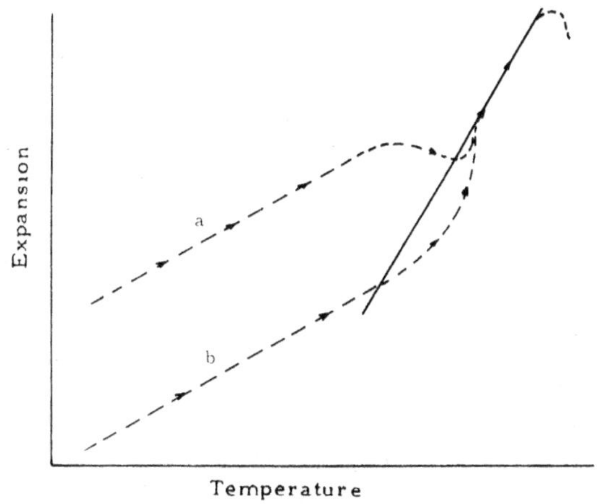

Fig. 7. Thermal expansion of a) highly quenched glass, b) slowly cooled glass of the same composition. The ordinate scale for 'a' has been displaced for clarity.

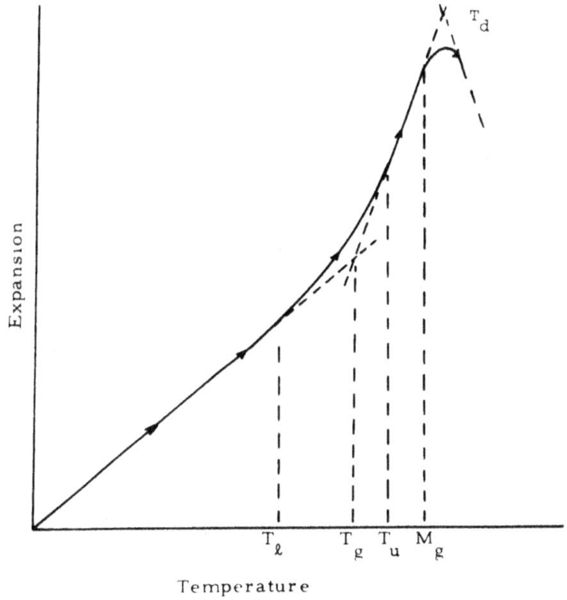

Fig. 8. Analysis of typical expansion curve, T_ℓ. T_u and M_g are points of tangency. T_g and T_d are points of intersection.

poise. In contrast, the viscosity associated with the Littleton softening point (which is defined as the temperature at which a uniform fiber of glass, 0.55 to 0.75 mm in diameter and 23.5 cm long, elongates under its own weight when the upper 10 cm of its length is heated in a specified furnace at the rate of approximately 5°C/min.) is about $10^{7.6}$ poise. (see Chapter 11)

c) the fictive temperature concept -

Clearly, the residual, room temperature density of a glass formed by cooling a melt is not necessarily the equilibrium density at room temperature. However, the true equilibrium density corresponding to any temperature can be estimated (particularly within the transformation range) as was shown in Fig. 5. From these observations, it occurred to Tool[22] that the structural state of a glass could be characterized by a temperature parameter T, meaning that temperature at which the density of a glass would not change if brought to that temperature, i.e., the equilibrium temperature corresponding to its actual density. On this basis the difference between the ambient temperature of a glass sample T, and its fictive temperature T, is a measure of departure from equilibrium. Consequently, if T = T, then the glass is at structural equilibrium. It is thus possible to speak of a glass having a high fictive temperature (meaning rapidly quenched) or a low fictive temperature (meaning slowly cooled). At room temperature, T is much greater than T for most common glasses.

In its broadest implications, this concept requires that all properties of a glass can be characterized by this single parameter T, regardless of its prior thermal history in arriving at T. Thus, if two glass samples of identical composition have the same density, then their other physical properties should also be the same. Careful measurements by Ritland,[23] however, showed that this is not generally true. As shown in Fig. 9, it was observed that the average thermal expansion coefficient varies markedly with index of refraction (density) depending upon whether or not the samples were soaked at a given temperature or cooled at a constant rate.

Hence it may be that the fictive temperature concept has severe limitations--and that there are subtle effects of thermal history on the physical properties of a glass beyond those outlined above. That is, each equilibrium density is not necessarily characteristic of a unique glass structure, but rather is a consequence of yet to be identified variables. Nonetheless, the formalism of the fictive temperature characterization of a glass is prevalent in the present day literature on glass science, and is therefore worthwhile noting as long as its limitations are recognized.

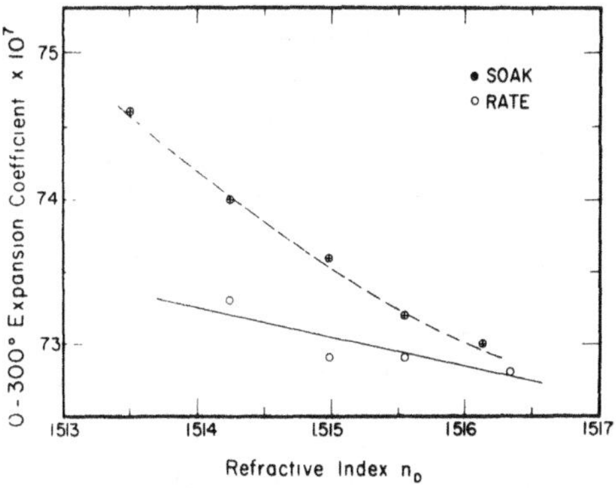

Fig. 9. The influence of thermal history on refractive index and expansion coefficient (0-300°C).[23] For the fictive temperature concept to hold, all points should fall on a smooth, continuous line.

d) effects of very high pressure on glass -

It is known that when subjected to external pressures, say up to 10 kiloatmospheres, most glasses will behave as a perfectly elastic substance. At much larger pressures, i.e., 100-200 kiloatmospheres, it is possible to induce permanent changes in the density of glass.[24,25,26] For example, the density of vitreous silica can be changed from 2.22 to 2.61 gm/cm3 when subjected to a pressure of 200 kiloatmospheres. Figs. 10 and 11 show results obtained by Bridgeman and Simon for pure vitreous silica, and also for several soda-silica glasses. These changes are sensibly permanent at room temperature but upon heating to a few hundred degrees centigrade, these changes fade with time, the glass soon reverting back to its original state. Fig. 12 indicates the density changes for a pure B_2O_3 glass with pressure. Unlike vitreous silica, there is no observable threshold pressure for this glass. Furthermore, vitreous B_2O_3 appears to exhibit an asymptotic density increase of about 6%. Therefore, the mechanisms of densification for these glasses may be quite different. Later work by Mackenzie and others indicated the importance of shear in properly interpreting the densification mechanism for all glasses.[26]

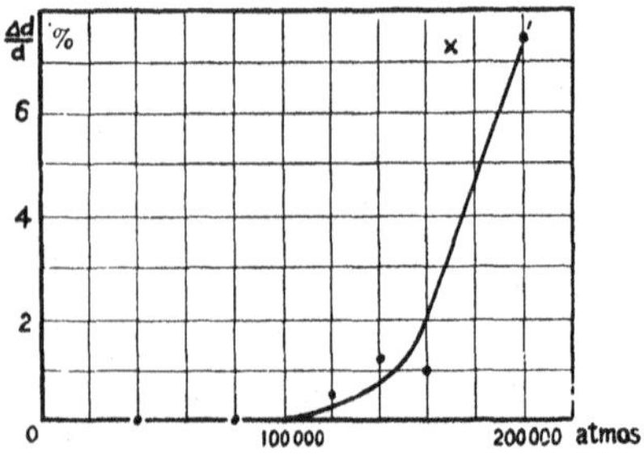

Fig. 10. The fractional increase of density, d, i.e., Δd/d for vitreous silica when subjected to various pressures.[24]

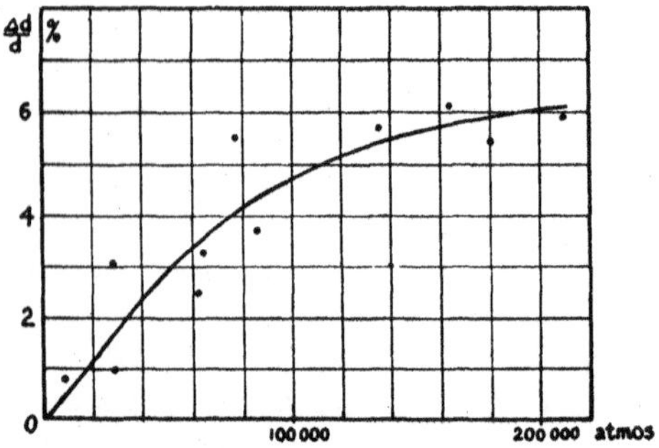

Fig. 11. The fractional increase of density, Δd/d, for several soda-silica glasses when subjected to various pressures.[24]

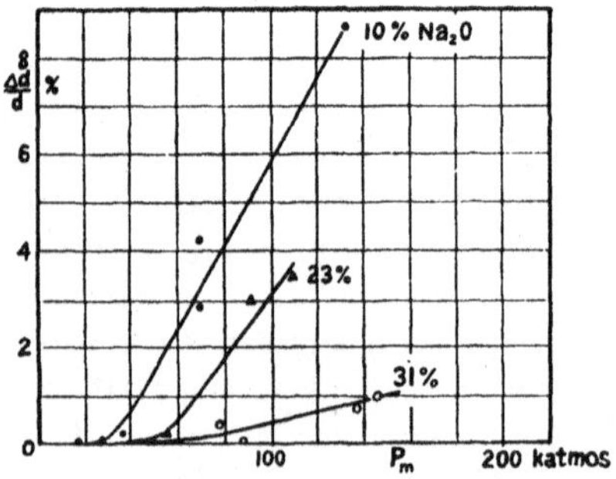

Fig. 12. Fractional increase of density, Δd/d, for pure B_2O_3 when subjected to various pressures.[24]

As might be expected, the observed densification is strongly dependent upon composition. This is clearly demonstrated in Fig. 11 and in Table II. Since an increase in alkali content decreases the observed densification and also since larger cations occasion a further decrease in densification, this is taken to mean that these cations play a very important role in bringing about a collapse of the glass structure. Specifically, it is thought that with an increase in the number of cations embedded in the interstices of a silica network, then a collapse of the glass structure is impeded. Larger ions should be more efficient in this process. This is observed. It is suggested here, however, that the threshold value for observing densification should increase with decreasing alkali cation size and content. More work is needed in this area.

Mackenzie further studied the densification of B_2O_3 and SiO_2 glasses at temperatures higher than T_g.[27] These studies showed that the annealing of a glass densified in the non-rigid state is related to the stabilization of these glasses towards their equilibrium density at the annealing temperature. Glasses densified in the rigid state, however, return to the density of the uncompressed state and not to the equilibrium value corresponding to the annealing temperature.

Table II. Relative Density Increase for Alkali Silicate Glasses[24]

Molar Percentage of Alkali	Cation Radius	P K(atmos)	$\Delta d/d^*$ (percent)
22 Li_2O	0.60	112	6.9
23 Na_2O	0.95	109	3.5
23 K_2O	1.33	128	1.2

*d = density

Δd = increased density increment

GLASS FORMATION

(A) General

The forming of a stable, homogenous glass by cooling a melt requires that the melt be brought through a temperature interval where crystallization may occur. The by-passing of crystallization and a reduction of the melt temperature to a point where it acts as a barrier to crystallization, are requisite to claiming that a vitreous state has been achieved. This is true irrespective of the melt composition; it may be composed of certain metals, oxides, nitrates or even combinations thereof (see Table III). The question of why only certain substances readily form glasses is a subject cardinal to all others in glass science. In spite of this, a full understanding of this phenomena has not yet been achieved. This stems from both the enormous range of substances which form glasses and also the subtle relationships which exist between several variables involved. Beyond stating that glass formation is neither related to periodic law nor phase equilibria, it is often difficult to generalize in this subject area. This is particularly evident when comparing the glass forming tendencies of simple systems with more complex ones.

Table III. Representative Glass Forming Substances

Substance	Melting Point (°K)	Heat of Fusion (Kcal/mole)	Entropy of Fusion (cal/°)
<u>Se</u>	491	1.2	2.5
Na	371	0.63	1.7
<u>SiO_2</u>	1980	1.3	0.7
<u>S</u>	394	0.29	0.75
LiCL	887	4.7	5.3
<u>B_2O_3</u>	733	5.3	7.2
<u>GeO_2</u>	1388	10.0	7.0
<u>$K_2O \cdot 4SiO_2$</u>	1073	11.7	10.9
Al_2O_3	2318	26.0	1.1

Glass formers are underlined

In keeping with the concept that glasses are non-crystalline solids, and bearing in mind that there is an inherent tendency for all melts to crystallize when cooled below their freezing points, it is clear that to examine glass formation, is in reality, to examine those factors which aid, impede or control this tendency to crystallize. This problem may be approached from either a crystal-chemistry or dynamic viewpoint. Both aspects are discussed below.

In a more general sense, it is important to realize that glass formation is neither a function of energy nor structural differences which might exist between the crystalline and vitreous states of a given material.[28] This may be seen by comparing the most direct indication of these differences; namely, the heats of fusion and entropy of a number of glass forming and non-glass forming materials. These parameters are listed in Table IV. Inspection of this data shows that there is no clear relationship to glass formation.

These results are somewhat surprising since it might be expected that large structural differences between a melt and its corresponding crystalline state would hinder crystallization and thus enhance glass formation. This is apparently true for some substances such as sulphur and phosphorous pentoxide. In contrast, vitreous silica closely resembles cristobalite, yet is easily fabricated by a variety of methods.

Table IV. Molar Heats of Fusion and Entropy for Several Substances

Elements	S, Se Bi, Ge (vapor deposited at low temperatures) Mixtures of As, Se, Te, S, In, Ge, etc.
Oxides	B_2O_3, GeO_2, As_2O_3, TeO_2 SiO_2, P_2O_5, V_2O_5 Mixtures of CaO-Al_2O_3, PbO-V_2O_5, TiO_2-K_2O, etc.
Salts	BeF_2, $ZnCl_2$ Mixtures of KNO_3-$CaNO_3$
Carbonates	$K_2Mg(CO_3)_2$ (at 1000 atmospheres)
Organic Compounds	Glycerol, Glucose Various high polymers.

However, the intrinsic melt structure relates directly to glass formation. This is illustrated by the observation that sulphur, just above its melting point, consists of ring-shaped molecules of the formula S_8. This melt will not yield a glass when quenched. To produce vitreous sulphur, it is necessary to heat the melt to a higher temperature where it undergoes a transition to a polymer-like chain.[28] This liquid, when quenched, will readily give a glass. Other examples of this behavior are found in certain borate and silicate systems.

Since a given glass will crystallize when properly heat treated, it at least seems certain that vitreous substances possess higher internal energies than their crystalline counterparts. Although a rigorous, thermodynamic proof of this statement is lacking,[29] it follows that should this energy difference be large, i.e., giving rise to very unstable supercooling, then crystallization should be enhanced, thus reducing the likelihood of glass formation. Yet, from the data of Table IV, it does not follow directly that the energy difference between the two states is always small.

Given this contradictory influence of structural and energy differences upon glass formation, it would seem that a better approach to this problem, as noted above, might be to consider those factors which bear a more direct relation to crystallization. Ideally, this would include a discussion of not only the mechanisms and activation energies associated with this process, but also the thermal energy available for them to be carried out.

(B) Dynamic Aspects of Glass Formation:

The complete crystallization of a melt has two prerequisites: nucleation and crystal growth. If the former is internal or intrinsic to the melt, it is termed homogenous. Otherwise, it is called extrinsic or heterogeneous, meaning that surface effects, impurities, or deliberate seeding of the melt dominate the nucleation process. The latter types are more commonly observed. Following nucleation, complete crystallization is dependent upon growth of the precipitating crystalline phase. Both of these prerequisites are time dependent and may occur, for example, in the time interval t depicted in Fig. 1. This dependence is a consequence of the need for atom migrations and/or spatial adjustments which are an integral part of the structural transformation of the disordered melt into an orderly, periodic atomic lattice.

Yet, the quenching of a melt to prevent crystallization is also a rate process, which in this case, is concerned with the extraction of heat or thermal energy necessary for the above noted atomic displacements to take place. From a viewpoint of glass formation, then, these rate processes, crystallization and quenching, are competitive. Also, they are not necessarily independent.

Fig. 13 gives a listing of the major independent variables associated with both processes and which can be controlled in the laboratory. These are: the type of quench; the melt temperature above ($+\Delta T$) or below ($-\Delta T$) the thermodynamic melting point from which the quench is to be made; and three others associated with the melt itself--volume, composition and thermal history. Interacting collectively, these variables determine the maximum rate of quenching for any substance and also its lowest rate of crystallization. Clearly, glass formation is favored by a high rate of quench and a low rate of crystallization.

For a given composition, the maximum quench rate obtainable depends upon a) the size of the melt to be quenched, b) the temperature from which the quench is to be made and c) the method of quenching used. The quenching rate obtained by pouring molten glass on a casting plate is probably at most 10 - 15°C/sec. Slightly larger rates are possible if small melts, i.e., 1 - 2 gms, are used or by dropping molten glass into water or mercury. Recently, however, the application of the techniques of splat cooling[30,31] to glass melting has made possible quenching rates of the order 10^6 - 10^7°C/sec., a marked improvement. The essential feature of this technique is to blow the melt away from a strip heater in such a way as to spread the melt in the form of a thin film on either a stationary or rotating surface which is usually maintained at a very low temperature. By using this method, even the oxides TeO_2, V_2O_5, WO_3 and MoO_3 can be quenched to the glassy state. Here, quenching from a temperature high above the liquidus temperature may actually facilitate glass formation. The higher temperature reduces the melt viscosity which in turn allows a larger spreading of the melt on a cold surface, and thus enhances the rate of heat extraction.

Within any system, an increase in quenching rate will enlarge the glass forming region in the manner shown in Fig. 14.[32] Conversely, an increase in melt volume may decrease the quenching rate adversely. In fact, it is often

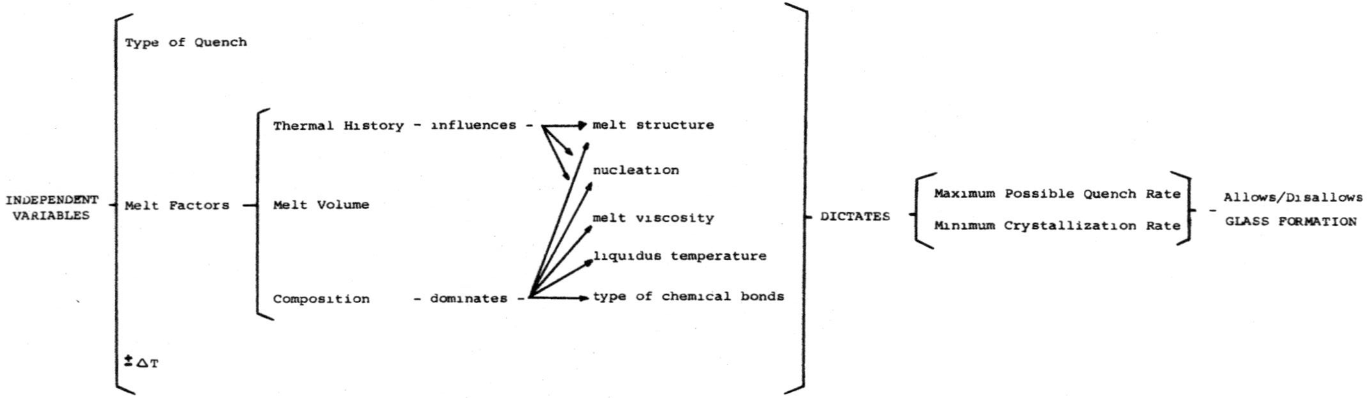

Fig. 13. A summary of the variables associated with glass formation and their interactions. The quantity ΔT refers to the temperature difference above ($+\Delta T$) or below ($-\Delta T$) the true melting point of the substance from which the quench will be attempted.

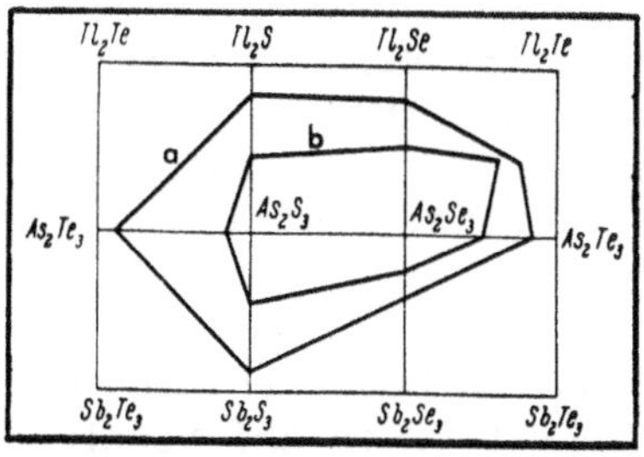

Fig. 14. Expansion of glass forming region by increasing the quenching rate. Outer polygon (a) highly quenched, inner polygon (b) slowly cooled.[32]

observed that although certain glass forming compositions can be made into a stable glass whenever the melt size is 50 - 100 gms, a scaling to 25 - 30 kilograms gives rise to a melt which is prone to partial crystallization. This results directly from the greater difficulty in extracting heat from the larger melt and thus, in this case, provides additional time for crystallization to occur. Consequently, the stabilization of glasses which are to be made in large melts must often be achieved through compositional adjustments to the base glass. Thus, in outlining the glass forming region of any system, the melt volume and temperature, and the type and rate of quench should be clearly specified.

(C) Crystal Growth:

The viscosity of many liquids is given by Fulcher's equation

$$\eta = A \exp \left[\frac{B}{T-T_o}\right] \tag{2}$$

where A, B, T_o are constants, T is the melt temperature. For pure SiO_2 or GeO_2, T_o is nearly zero while B is quite large and nearly equal to the order of the chemical bond energy in these compounds.[29] Subsequently, their viscosity shows a near Arrhenius type temperature dependence. For many organic liquids, B is smaller and T_o is a sizable fraction of their thermodynamic melting point T_m. For these cases, their melt viscosity is lower at T_m and increases substantially upon supercooling. All substances which can be quenched to form a glass show large increases of viscosity with falling temperature at or below their melting points.

It is known that the rate of crystallization R, of a melt is nearly represented by

$$R = \frac{K\Delta T}{\eta} \tag{3}$$

where K is a material and model related constant, ΔT the temperature difference between the melting point T_m and the actual melt temperature T . From this expression it is expected that melts with high viscosities should show small rates of crystallization and hence tend towards glass formation. The data in Table V illustrate this relationship and its converse.[33]

Similar to crystallization, the rate of nucleation is also inversely proportional to melt viscosity. It is possible that many viscous melts can be quenched to a glassy state simply because they fail to nucleate. From either viewpoint, melt viscosity is a major parameter in studying glass formation.

Table V. Viscosity of Several Glass Forming and Non-Glass Forming Substances at Their Melting Point[33]

	Melting Point (°C)	Viscosity (poise)
Glass Formers		
BeF_2	540	$> 10^6$
B_2O_3	460	10^5
GeO_2	1150	10^7
SiO_2	1710	10^7
Non-Glass Formers		
LiCL	613	0.02
$CdBr_2$	567	0.03
Na	98	0.01
Zn	420	0.03
Fe	1535	0.07

(D) Crystal Chemistry Approach:

It is clear that the energy term 'B' in Equation (2) has a strong influence on viscosity of oxide melts and thus directly affects the propensity of these melts towards glass formation. The magnitude of 'B' depends on 1) the intrinsic melt structure and 2) the type and strength of the chemical bonds of the constituent atoms in the melt. Owing perhaps to the difficulty of precisely estimating the contribution to B from melt structure, several authors have sought direct relationships between glass formation and chemical bonding. Stanworth, for example, using Pauling's electronegativity criteria, observed that the chemical bonds of all the common glass forming oxides possess a high degree of covalency.[34] This effect is shown in Fig. 15. A reduced atom mobility, associated with the directional character of the covalent bond, is often cited as the underlying basis for this trend. However, the highly ionic compound BeF_2 is readily quenched to a vitreous state. Indeed, upon categorizing all solids by their predominate bond type, i.e., ionic, covalent, metallic, hydrogen or Van der Waals, it is found that each category contains a glass forming substance. Smekal suggests that a mixture of bond type is conducive to glass formation.[35]

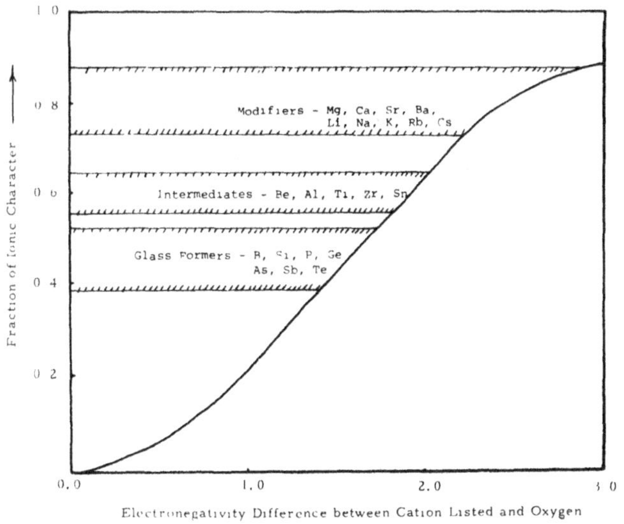

Figure 15. The influence of chemical bonding on glass formation. Covalency favors glass formation.

A much more satisfying correlation is found in associating bond strength with glass formation. This correlation rests on the assumption that during crystallization of the melt, the chemical bonds of the constituent atoms must continuously break and reform as the disordered melt converts to a periodic atomic lattice. It follows that the stronger these bonds, the greater will be the tendency to resist crystallization and thus promote glass formation. This approach was first suggested by Sun[36] in 1947 who was able to demonstrate that such a correlation exists. To do this, it was necessary to estimate the total energy required to dissociate an oxide MO_x, into its gaseous constituent atoms. This energy, when divided by the most probably coordination number of M, gives the single bond strength of the M-O bond. A partial summary of Sun's data is given in Table VI. These and other considerations have led to the division of oxides in Table VI and Fig. 15 into those that exhibit glass formation, those that promote glass formation (called intermediates), and those that only alter the properties of the resulting glasses (called modifiers).

Rawson[37,38] further recognized that it is important to consider not only the individual bond strengths, but also the thermal energy available to rupture these bonds so that crystallization of the melt may proceed. The maximum available thermal energy for this process is given by the melting point of a substance. Thus, the data of Table VI can be normalized by dividing the calculated bond strength of each oxide by its melting point. Clearly, a high bond strength and low melting temperature is more conducive to a glass formation than the reverse condition. Of all the oxides this ratio is largest for B_2O_3, and may explain the extreme difficulty encountered when attempting to crystallize this compound.

This approach also illustrates a working concept long known to the glass scientist, namely, to enhance glass formation, suppress the thermodynamic melting point. An extreme example of this is illustrated by what Rawson has termed the "liquidus temperature" effect. Here, two non-glass forming compounds, i.e., CaO and Al_2O_3, will readily form a eutectic mixture which can be quenched to the glassy state. This phenomena is shown in Fig. 16. Other compound pairs such as $PbO-V_2O_5$, show similar behavior.

Table VI. Relationship of Glass Formation to Dissociation Energy as Suggested by Sun[36]

M in MO_x	Dissociation Energy per MO_x (Kcal/mole)	Coordination Number	Single Bond Strength (Kcal/mole)
Glass Formers			
B	356	3	119
Si	424	4	106
Ge	431	4	108
Al	317-402	4	79-101
B	356	4	89
P	442	4	88-111
Intermediates			
Ti	435	6	73
Zn	144	2	72
Pb	145	2	73
Al	317-402	6	53-67
Be	250	8	63
Modifiers			
Mg	222	6	37
Ba	260	8	33
CaO	254	8	32
Li	144	4	36
Na	120	6	20
K	115	9	13
Rb	115	10	12
Cs	114	12	10

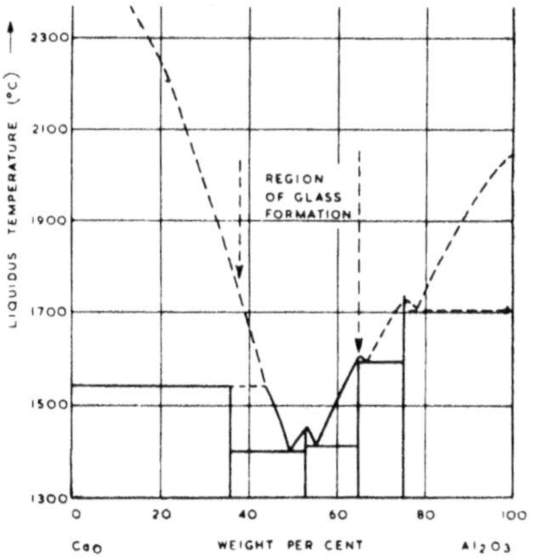

Fig. 16. Region of glass formation in the system $CaO-Al_2O_3$, illustrating Rawson's "liquidus temperature effect".[37]

RECAPITULATION

Thus, there are many facets of glass science--the structure of glass, its optical, mechanical, electrical and thermal properties. All of these need further attention and are worthy of additional research. It is difficult to develop absolutes in this field; what seems appropriate for one glass forming system may be inapplicable to others. It is somewhat ironic that the exploitation of glass as an engineering material has been achieved in spite of the difficulty in fully understanding it. It has been suggested that this exploitation has been made possible because of the great complexity of glass.[39] The present situation is this: there are a great many theories to explain the physical properties of solids. Many are simplified by an assumed periodic atomic lattice. Their application to non-crystalline solids is more difficult, but not always impossible. This, coupled with the enormous diversity of glass properties, has served as a powerful stimulus for research into vitreous materials. It is confidently expected that future discoveries will cast new light on our present thinking about this substance and at the same time, further broaden its applications in science and society.

REFERENCES

1. Charles, R. J., "A Review of Glass Strength", Progress in Ceramic Science, Vol. 1, 1961, Edited by J. E. Burke.
2. Kreidl, N. J., Blair, G. E., "A System of Megaroentgen Glass Dosimetry", Nucleonics, 45 [1] 56-60 (1956).
3. Pearson, A. D., Peterson, G. E., "Energy Exchange Processes and Laser Oscillation in Glasses", Proceedings 7th Inter. Congr. Glass.
4. Snitzer, E., "Glass Lasers", Glass Industry, Vol. 48, 498-503 (1967).
5. Cohen, A. J., "Mechanisms in Variable Transmission Silicate Glasses", Proceedings 7th Inter. Congr. Glass.
6. Armistead, W. H., Stookey, S. D., "Photochromic Silicate Glasses Sensitized by Silver Halides", Science, Vol. 144, 150-154 (1964).
7. Smith, G. P., "Photochromic Silver Halide Glasses", Proceedings 7th Inter. Congr. Glass.
8. Bradley, C. A., "Fume or Vapors from Molten Glass", Bulletin of the Am. Ceram. Soc., 23 [10] 379-381(1944).

9. Barlow, D. F., "Volatilization of Fluorides, Borates and Arsenic from Glass", Proceedings 7th Inter. Congr. Glass.
10. Weyl, W., Pincus, A. G., "The Role of Gases in Glass: I. Oxygen", Glass Industry, 19, 255 (1938).
11. Stookey, S. D., Maurer, R. D., "Catalyzed Crystallization of Glass--Theory and Practice", Progress in Ceramic Science, Vol. 2 (1962).
12. Morey, G. W., The Properties of Glass, 1st edition Reinhold Publishing Corp., New York, p. 28 (1938).
13. Stanworth, J. E., Physical Properties of Glass, Clarendon Press, Oxford, chap. 1 (1950).
14. Secrist, D. R., Mackenzie, J. D., Modern Aspects of the Vitreous State, Vol. 3, chap. 6, Butterworths, London (1960).
15. Warren, B. E., "X-ray Diffraction of Vitreous Silica", Z. Krist, 86, 349-358 (1933).
16. Mott, N. F., "The Solid State", Scientific American, 217 [3] 80-89 (1967).
17. Eckstein, B., "Vitreous States", Mat. Res. Bull., 3, 199-208 (1968).
18. Charles, R. J., "The Nature of Glass", Scientific American, 217 [3] 127-136 (1967).
19. Tool, A. Q., Eichlin, C. G., "Variations in Glass Caused by Heat Treatment", J. Am. Ceram. Soc., 8 [1] 1-17 (1925).
20. Ritland, H. N., "Density Phenomena in the Transformation Range of a Borosilicate Crown Glass", J. Am. Ceram. Soc., 37 [8] 370-78 (1954).
21. Tool, A. Q., Hill, E. E., "On the Constitution and Density of Glasses", J. Soc. Glass Tech., 9, 185 (1925).
22. Tool, A. Q., "Relation Between Inelastic Deformability and Thermal Expansion of Glasses in its Annealing Range", J. Am. Ceram. Soc., 29 [9] 240-53 (1946).
23. Ritland, H. N., "Limitations of the Fictive Temperature Concept", J. Am. Ceram. Soc., 39 [12] 403-406 (1956).
24. Bridgeman, P. W., Simon, I., "Effect of Very High Pressure on Glass", J. Appl. Phys., 24 [4] 405-13 (1953).
25. Cohen, H. M., Roy, R., "Effects of Ultra High Pressures on Glass", J. Am. Ceram. Soc., 44 [10] 523-24 (1961).
26. Mackenzie, J. D., "High Pressure Effects on Oxide Glasses: I, Densification in Rigid State", J. Am. Ceram. Soc., 46 [10] 461-70 (1963).
27. Mackenzie, J. D., "High Pressure Effects on Oxide Glasses: III, Densification in Nonrigid State", J. Am. Ceram. Soc., 47 [2] 76-80 (1964).

28. Weyl, W., Marboe, E. C., The Constitution of Glasses: A Dynamic Interpretation, Vol. I, chapter 13, 219, Interscience Publishers (1964).
29. Turnbull, D., "Under What Conditions Can a Glass be Formed?", Contemp. Phy., 10 [5] 473-488 (1969).
30. Sarjeant, P. T., Roy, R., "New Glassy and Polymorphic Oxide Phases Using Rapid Quenching Techniques", J. Am. Ceram. Soc., 50 [10] 500-503 (1967).
31. Sarjeant, P. T., Roy, R., "A New Approach to the Prediction of Glass Formation", Mat. Res. Bull., 3, 265-80 (1968).
32. Goriunova, V. A., Kolomiets, B. T., and Shilo, V. P., "Vitreous Semiconductors", Zuhr. Tekhn. Fiz., 28, 981 (1958).
33. Mackenzie, J. D., in Modern Aspects of the Vitreous State, Vol. I, J. D. Mackenzie, ed., pp. 188-218, Butterworths, London.
34. Stanworth, J. E., "The Ionic Structure of Glass", J. Soc. Glass. Tech., 30, 54-64T (1946).
35. Smekal, A. G., "On the Structure of Glass", J. Soc. Glass Tech., 35, 411-420T (1951).
36. Sun, H. K., "Fundamental Condition of Glass Formation", J. Am. Ceram. Soc., 30, 277-81 (1947).
37. Rawson, H., "The Relationship Between Liquidus Temperature Bond Strength and Glass Formation", Proceedings IV Inter. Congr. Glass, Paris (1956).
38. Rawson, H., Inorganic Glass Forming Systems, Academic Press, p. 26 (1967).
39. Stookey, S. D., "Glass Chemistry as I Saw it", Chem. Technol., 1, 458-465 (1971).

Chemical Bonding In Non-Crystalline Solids

Roger Eiss

Oregon Graduate Center

INTRODUCTION

Since the readers of this chapter include both chemists and physicists who are interested in better understanding the field of glass technology, a serious problem arises in relating a body of fundamental information to an audience with differing levels of understanding of structural chemistry. The chemist who has taken a good graduate course in inorganic chemistry will find few new concepts or insights into the chemistry of glasses; this chapter can only serve to present material he has previously seen in a new context. On the other hand a physicist whose formal study of chemistry was concluded in his freshman year may find the chapter to be little more than an outline of a body of material that will not be mastered without further reading in standard texts on the subject (some of which are listed in the references.) The chapter is directed primarily to those who have studied most of the concepts presented but have lost a sense of familiarity with the material because of a lack of opportunity to use it in daily applications, and to provide a practical context that will help to keep it from being lost again.

The primary short-range forces that bond neighboring atoms in the condensed phases are essentially the same in all materials, whether liquid or solid, crystalline or non-crystalline. The observed bond lengths and bond angles between an atom and its nearest neighbors (its first coordination sphere) are generally thought to be similar in a glass or a crystal of a given chemical composition. It is the presence or absence of long-range order that is responsible for determining which of these states are observed. Attractive (and repulsive) forces exist between "non-bonded" atoms* in near proximity, though these forces are considerably weaker than those between "bonded" atoms.

In the melt these forces are not sufficiently strong to overcome the thermal motion of the atoms and the degree of ordering beyond the first coordination sphere is extremely limited. In an ideal crystal solidification has proceeded in a manner permitting the "non-bonding" forces to establish a lowest-energy arrangement of atoms, with complete long-range order. A glass has been cooled from a melt so quickly that these "non-bonding" forces are established before the atoms have attained a minimum energy arrangement and further re-arrangement is prevented by the lack of sufficient thermal energy to break these weak forces. In fact the extremes of (1) a perfect crystal or (2) a glass with no long-range order seldom (if ever) are observed, and there is a continuum of possibilities between these limits.

In this chapter the major types of chemical bonds observed in solid materials will be briefly reviewed and some examples of the relation between the nature of bonding and the properties of non-crystalline materials will be presented. Emphasis will be placed on interactions between nearest neighbors; the question of longer-range order and the structure of the vitreous state, as well as such structure-related phenomena as spectral and electrical properties, will be presented in following chapters.

---

* In this sense the term "bonding" implies formation of a strong chemical bond of one of the three classical types (covalent, ionic, metallic) described later in this chapter.

INTRODUCTION TO CHEMICAL BONDING

1. The Electronic Structure of Atoms and Ions

An electron which is bound to a nucleus of an atom or ion behaves essentially like a wave and consequently can be described in terms of wave mechanics,[1] with the following consequences:

a) Only certain discrete values of energy are allowed, with a transfer from one energy state to another being permitted only when accompanied by the absorption or emission of an amount of energy equal to the energy difference between the two electronic states. The energy transfer can be thermal (vibrational) or optical.

b) The state of an electron can be described by an algebraic expression called a wave function, Ψ , having the property that the value of Ψ^2 at any point in space is a measure of the probability that the electron will be at that point at any given time.

c) The energy of an electron, the direction of its spin, and the shape and orientation of the wave function Ψ are all determined by four variable integers (called quantum numbers) such that the energy state of an electron in an atom or ion is determined when these four quantum numbers are specified. Further, the Pauli exclusion principle states that in a given atom no two electrons can have the same set of four quantum numbers.

The symmetry of the wave function is determined by the angular momenum wave number ℓ . Figure 1 shows electron density contour maps for wave functions with $\ell = 0,1,2$ (called s,p and d orbitals respectively) which are the orbitals normally occupied by the electrons of all atoms excepting the rare earth elements and the actinides. The s orbital has spherical symmetry and can be occupied by two electrons, in which case the Pauli exclusion principle requires that their directions of spin are opposite. There are three equivalent suborbitals in each p orbital, mutually perpendicular in direction and each capable of containing two electrons for a total of six electrons in the orbital. Note that each p suborbital has a "nodal plane" on which the wave function and thus the probability

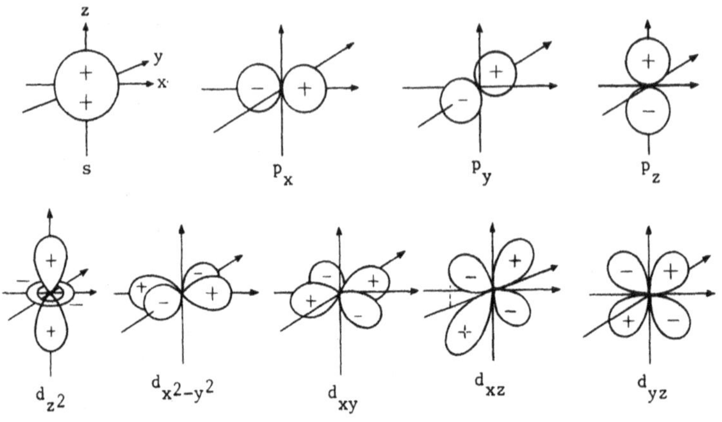

Figure 1. Shape and Orientation of Electronic Suborbitals in a Free Atom. The surfaces represent a contour of constant electron density in the atom.

of finding the electron is zero. The five d suborbitals are also equivalent, though this is not immediately obvious because the $d_z{}^2$ suborbital is the result of a linear combination of two other suborbitals having the symmetry of the other four d suborbitals. The d orbital can contain ten electrons.

An atom or ion can be considered as having an "inert gas core" (or an inner core of electrons having a filled p orbital, identical to that for one of the inert gas elements), very stable and non-reactive, and a number of "valence electrons" that determine its chemical properties. A glance at the long form of the periodic table (Figure 2) shows that the column furthest left contains a series of elements (the alkali metals) having a single s electron in the valence shell, and the alkaline earth elements (the next column to the right) have an s^2 configuration (the superscript indicates two s electrons, or a filled s orbital). For the heavier elements, starting with the third row, there follows the transition metal elements, for which there is a partially or totally filled d orbital, the ten columns corresponding to the ten electrons that can be accomodated in a d orbital. In all rows except the first (which contains only hydrogen and helium) there follows

1 H																	2 He
3 Li	4 Be											5 B	6 C	7 N	8 O	9 F	10 Ne
11 Na	12 Mg											13 Al	14 Si	15 P	16 S	17 Cl	18 Ar
19 K	20 Ca	21 Sc	22 Ti	23 V	24 Cr	25 Mn	26 Fe	27 Co	28 Ni	29 Cu	30 Zn	31 Ga	32 Ge	33 As	34 Se	35 Br	36 Kr
37 Rb	38 Sr	39 Y	40 Zr	41 Nb	42 Mo	43 Tc	44 Ru	45 Rh	46 Pd	47 Ag	48 Cd	49 In	50 Sn	51 Sb	52 Te	53 I	54 Xe
55 Cs	56 Ba	[a] 57 La	72 Hf	73 Ta	74 W	75 Re	76 Os	77 Ir	78 Pt	79 Au	80 Hg	81 Tl	82 Pb	83 Bi	84 Po	85 At	86 Rn
87 Fr	88 Ra	[b] 89 Ac															
			[a] Lanthanide series	58 Ce	59 Pr	60 Nd	61 Pm	62 Sm	63 Eu	64 Gd	65 Tb	66 Dy	67 Ho	68 Er	69 Tm	70 Yb	71 Lu
			[b] Actinide Series	90 Th	91 Pa	92 U	93 Np	94 Pu	95 Am	96 Cm	97 Bk	98 Cf	99 Es	100 Fm	101 Md	102 No	103 Mv

Figure 2. The Periodic Table of the Elements (Long Form).

six columns of elements having electrons in the highest p orbital. The last two rows contain an additional series of elements, the rare earths(or lanthanides) and the actinides, which contain a partly filled f orbital. Since these elements are important in only a limited number of glasses for special applications and since their chemical behavior can be understood using the principles we will develop for the elements having only s, p and d electrons we will not be concerned with the chemistry of these two groups.

2. Chemical Bonding [1a,b,c]

It has been stated, without explanation, that chemical properties are related to the configuration of the outermost partly-filled electron shell (or "valence" shell). In this section we will see why this is the case, showing how the valence electrons of neighboring atoms interact to form chemical bonds. There are three types of strong chemical bonds formed: ionic (or electrostatic), covalent and metallic. We will next briefly examine each of these types and give some examples of compounds in which each is predominant.

a) The Ionic Bond

The inert gas configurations are exceptionally stable and have a very low tendency to participate in chemical bonding; the discovery of stable chemical compounds involving the inert gases was made in 1962 and their known chemistry is very limited. The high stability of these electronic configurations provides a very strong force favoring the filling of one or two vacancies to form an ion having the inert gas shell, such as F^- or O^{-2}. Conversely the s electrons just beyond an inert gas shell are rather weakly held and can be lost to form positive ions such as K^+ or Sr^{+2} without a large expenditure of energy. In Table I the ionization potentials (the energy required to remove an electron from a neutral atom) and the electron affinities (the energy released when an electron is added to a neutral atom) are listed for a number of elements.[2]

Comparison of these quantities shows that in no case is it possible to remove an electron from one isolated atom and add it to a second without external energy being

Table I

Element	Ionization Potential (kcal/mole)	Electron Affinity (kcal/mole)
F	403.3	83.5
Cl	300.3	87.3
Br	274.6	82.0
I	242.2	75.7
H	315.0	17.8
Li	125.8	<0
Na	120.0	<0
K	101.6	<0
Rb	97.8	<0
Cs	91.3	<0

applied to the system. In an ionic crystal, however, a great deal of energy is provided by the electrostatic attraction between each positive ion (cation) and the negative ions (anions) that are close enough to show electrostatic interaction. For NaCl, for example, the energy provided by this effect is 181.3 kcal/mole.[3] It is this electrostatic attraction which is believed to be the predominant bonding force in classical ionic crystals, such as KF or Na_2O.

The structures expected would be those which minimize contact between ions with like charge and maximize contacts between ions of opposite charge. This is the case for the purely ionic crystals, with the particular lattice type dependent on the respective charges of positive and negative ions (since these determine the ratio of the number of anions to the number of cations) and on the ratio of the radii of the anions and the cations.

Because of the high mobility of ions in an ionic lattice at temperatures near the melting point it is extremely difficult to form glasses from ionic materials, and glass technologists are generally not concerned with ionic salts in the pure form. All commercial oxide glasses contain appreciable amounts of these ionic oxides and addition of the ionic oxides to the glass-forming oxides (e.g. SiO_2, B_2O_3 and P_2O_5) have very significant effects on the physical properties of the glasses produced, as will be discussed later in more detail.

b) The Metallic Bond

A simple metal (such as sodium) can be considered to be composed of metal ions (in their normal valence state) and electrons. The metal ions are usually arranged in one of three packing arrangements: body centered cubic, face centered cubic or hexagonal closest packed. The lower energy electronic orbitals (those filled with electrons in the metal ion) remain associated with the atomic nucleus; the higher electronic orbitals of each metal ion combine to form a large number of delocalized orbitals, lying so close together they are called the valence band, extending throughout the metal. The valence electrons are in these delocalized orbitals and are free to move throughout the structure without remaining associated with any particular metal ion.

The electrostatic repulsion between the metal ions is counteracted by the negative charges of the electrons in the valence band,although the metal ion radius in sodium metal is approximately twice as great as the ionic radius of sodium in its ionic salts. The total energy of the metal, compared with the energy of the same number of individual metal atoms, is a measure of the strength of the metallic bond. The high mobility of the electrons in the solid metal (and even in the melt) gives rise to many of the characteristic properties of metals, including the high electrical and thermal conductivity.

The metallic bond is of practical importance to glass technologists only at the interface of a glass and a metal, where there is a very narrow region over which the bonding is changed from predominantly metallic to predominantly covalent and ionic. This case has been presented in detail in a review of the subject.[4]

c) The Covalent Bond

In all commercial glasses there is a high degree of covalent bonding. The factors that permit the existence of a covalent bond and that determine its strength are complex and will not be presented here (see Reference 1). We will say only that a covalent bond is the result of the sharing of an electron pair (or sometimes a single electron) between two (or more) atoms and can be formed only when there is an overlap of the wave functions of the atoms

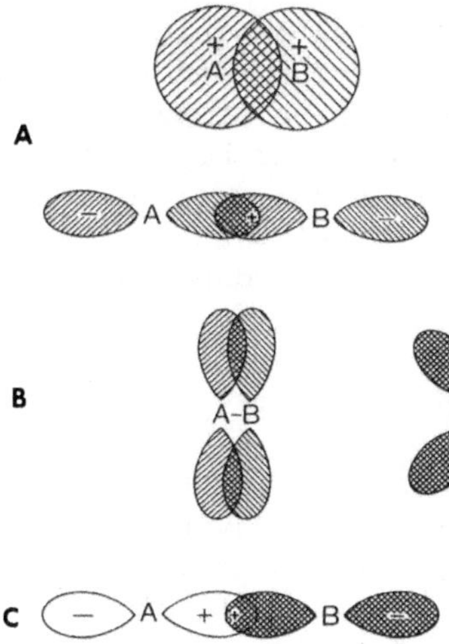

Figure 3. Examples of the Combination of Atomic Wave Functions to Form Covalent Bonds. A. Sigma Bonding. B. Pi Bonding. C. Charge-Transfer Complexes Singly cross-hatched areas represent suborbitals occupied by a single electron Doubly cross-hateched areas represent suborbitals occupied by an electron pair.

involved. The signs of the bonding wave functions must be the same and the strength of the bond is related to the extent of the overlap of the wave functions.

Figure 3 shows some representative examples of covalent bonding: a sigma bond has appreciable electron density between the two nuclei and is the normal single-bond generally found in covalent molecules. The hydrogen and halogen (fluorine, chlorine, bromine and iodine) gaseous diatomic molecules are examples of molecules containing simple covalent sigma bonds. Diamond is a network of carbon atoms in which each carbon atom is connected to four others by simple sigma bonds. A pi bond (Figure 3b) has a nodal plane containing the two nuclei and thus no electron density along the internuclear axis. It is most commonly found in compounds having sigma bonds and contributes to a "multiple bond" character. In ethylene ($H_2C{=}CH_2$), for example, there is a carbon-carbon double bond consisting of a sigma and a pi bond, while in acetylene ($HC{\equiv}CH$) there is a sigma and <u>two</u> pi bonds to give a triple bond. Multiple bonds tend to be shorter and stronger than corresponding single bonds between the same pair of atoms. The second example of pi-bonding in

Figure 3 shows a filled metal d-orbital interacting with an empty p-orbital of an atom bonded to the metal. This provides an example of a common situation where the electrons involved in forming a bond both come from one of the atoms involved in the bond; when all the electrons involved in forming a chemical bond come from one of the atoms the compound is called a "donor-acceptor complex".

The covalent bonds in a complex molecule need not be localized between single atoms in the molecule; as in the case with metals, bonds may extend over numbers of atoms or even over the entire crystal. Graphite is an example of a covalently-bonded material with very high electrical conductivity owing to electrons in orbitals that extend throughout the crystal lattice. More often, as in benzene or the nitrate ion the delocalization extends only for a limited number of atoms.

Accepting this highly simplified picture of covalent bond formation, consider the hypothetical reaction of a carbon atom with four hydrogen atoms to form a molecule of methane, CH_4. The ground state of the hydrogen is $1s^1$; that for carbon is $[He]2s^22p^2$. Since this represents only two unpaired electrons, there must be a promotion of one of the s electrons to the p orbital in order to permit the formation of four bonds. Presumably the energy in forming two additional C-H bonds will more than compensate for that required to cause the promotion $[He]2s^22p^2 \rightarrow [He]2s^12p^3$.

If the bonding in CH_4 resulted from simple overlap of the 1s orbital in the four hydrogen atoms with a carbon 2s orbital and three carbon 2p orbitals, the C-H bonds involving the p orbitals would be expected to be approximately mutually perpendicular, since the p suborbitals are mutually perpendicular and only an arrangement of hydrogens very close to this would provide sufficient overlap for bonding. The hydrogen atom bonding to the 2s orbital could be in any direction, since s orbitals are spherically symmetric, and thus would be directed away from the other three hydrogens by steric factors (Figure 4). In fact, methane is rigorously tetrahedral, with all four hydrogen atoms equivalent. The explanation of this equivalence lies in the concept of hybridization of electron orbitals, which we shall examine next.

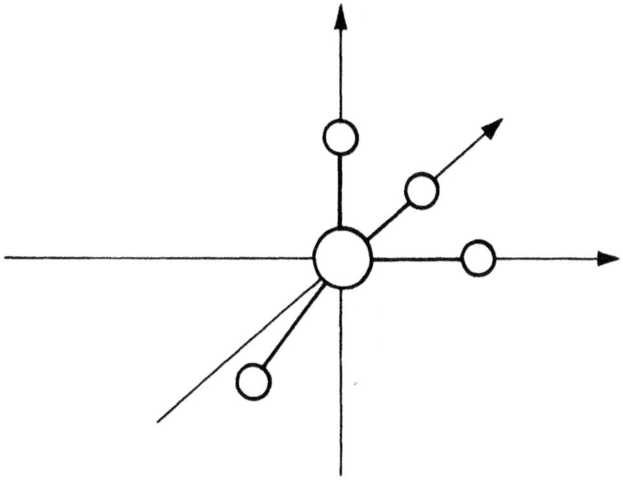

Figure 4. Hypothetical Structure of a Molecule of Methane (CH_4) in which the Carbon Atom Forms Bonds Using Pure s and p Orbitals.

Mercury has the ground state electronic configuration $[Xe]5d^{10}6s^2$, and in a covalent compound such as $Hg(CN)_2$ it is assumed that the two bonds are formed after promotion of one of the 6s electrons to the 6p orbital. Figure 5 shows one of the p suborbitals (p_z) and the s orbital. While the bonding could involve these orbitals as they exist in a free atom, the lower portion of the figure shows two new orbitals that can be constructed by taking linear combinations of the s and p_z orbitals. The new wave equations, expressed as linear combinations of the original wave functions, are as follows:

$$\Psi_1 = \frac{1}{\sqrt{2}} (s + p_z)$$

$$\Psi_2 = \frac{1}{\sqrt{2}} (s - p_z)$$

Such orbitals, called hybrid orbitals, can be favored over the original pure orbitals because the extent of the orbital in the direction of the bond is increased by forming the hybrid. Since, as we have pointed out, the bond strength is related to the degree of overlap of the wave functions

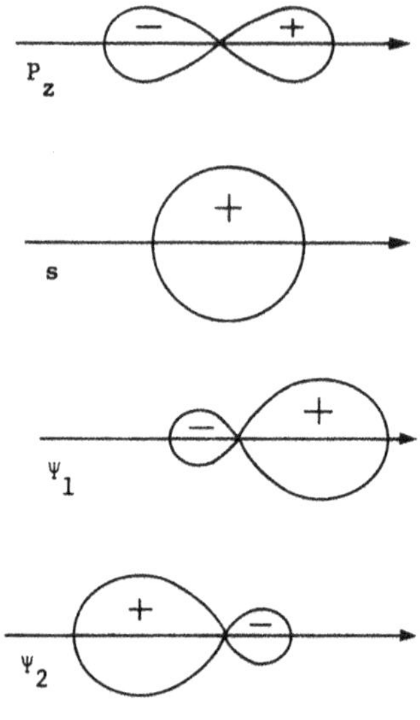

Figure 5. Hybridization of s and p_z Orbitals to Give sp Hybrid Orbitals.

of the two bonding atoms, the increased extent of the hybrid orbital in the direction of the bond provides an increase in bond energy sufficient to compensate for the energy required to form the hybrid bonds. In the case of $Hg(CN)_2$, or any other molecule having a central atom with bonding orbitals Ψ_1 and Ψ_2 (called sp hybridization) the hybridized bonding scheme would lead to prediction of a linear molecule involving two equivalent bonds.

Figure 6 shows the situation for sp^2 hybridization, or hybrid bonds made up of one s and two p orbitals. As shown, the three resulting hybrid bonds are directed to the corners of a plane equilateral triangle and the bonds are again equivalent. The equations[5] for the hybrid orbitals are as follows:

$$\Psi_1 = \frac{1}{\sqrt{3}} s + \frac{2}{\sqrt{6}} p_x$$

$$\Psi_2 = \frac{1}{\sqrt{3}} s - \frac{1}{\sqrt{6}} p_x + \frac{1}{\sqrt{2}} p_y$$

$$\Psi_3 = \frac{1}{\sqrt{3}} s - \frac{1}{\sqrt{6}} p_x - \frac{1}{\sqrt{2}} p_y$$

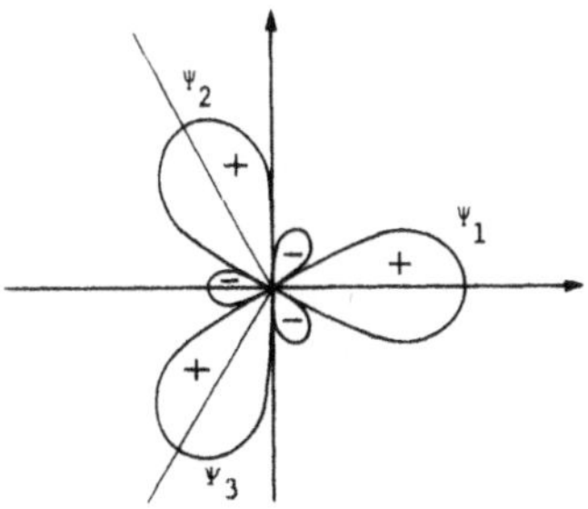

Figure 6. sp^2 Hybrid Orbitals.

For the sp^3 case observed in methane the four equivalent bonds are directed toward the corners of a regular tetrahedron and have the following equations:

$$\Psi_1 = 1/2\ (s + p_x - p_y + p_z)$$

$$\Psi_2 = 1/2\ (s - p_x + p_y + p_z)$$

$$\Psi_3 = 1/2\ (s - p_x - p_y - p_z)$$

$$\Psi_4 = 1/2\ (s + p_x + p_y - p_z)$$

Figure 7 shows the overall energetic requirements for the formation of CH_4 from graphite and hydrogen gas,[6] showing how the formation of four C-H bonds provides an amount of energy sufficient to compensate for the energy required for the dissociation of the hydrogen molecule, vaporization of the graphite, promotion from the s^2p^2 to the sp^3 state and the formation of the sp^3 hybrid orbitals (indicated by the term V_4 in the Figure). The fact that H_2 does not react with graphite at room temperature but requires considerably higher temperature for initiation of the reaction is an indication that the energy barrier for the reaction is reasonably large.

Hybrid orbitals involving d electrons are also commonly found in the transition metals. Figure 8 shows the most common hybridization schemes observed in these cases and the orientation of the resulting bonds.

We conclude the introductory section on covalent bonding with a final example combining many of the concepts that have been summarized in this section, that of benzene. Each carbon atom is bonded to two other carbon atoms and

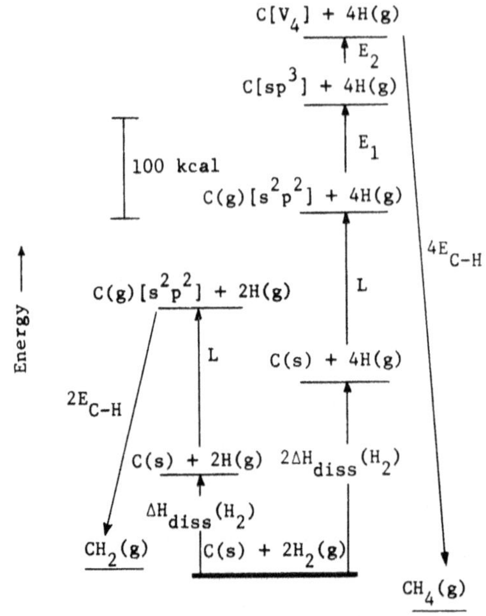

Figure 7. Energy Requirements for the Formation of CH_4 from Graphite and Hydrogen Gas.

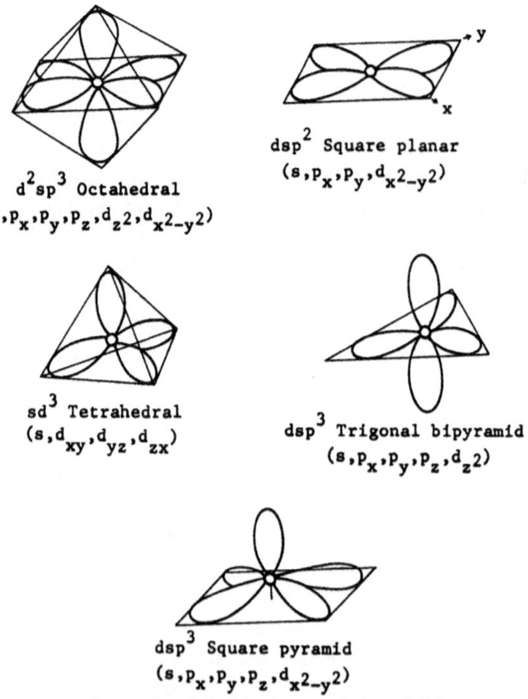

Figure 8. Hybrid Orbitals Involving d Electrons

to a hydrogen atom, involving three sigma bonds which are hybridized in the sp^2 (trigonal planar) form. If we assume the molecule lies in the xy plane, the orbitals involved in the hybridized sigma bonds are s, p_x and p_y. There remains on each carbon atom a single electron and a single p suborbital, p_z, which are not involved in sigma bonding and which are available for the formation of a set of delocalized π - orbitals that extend over the entire molecule (Figure 9). Three such pi orbitals,[7] each capable of holding a pair of electrons, are shown and can accommodate the remaining six electrons in delocalized multiple bonding. In agreement with this interpretation we find all bond angles about the carbon atoms to be 120° and all carbon-carbon bond lengths to be intermediate between a single and a double-bond length.

d) Intermediate Types of Strong Bonds

We have shown some examples of relatively pure types of bonds: The metallic bond in sodium metal, the ionic bond in KF and the covalent bond in diamond. Most chemical compounds are bonded by a combination of these forces. Metallic copper has a significant degree of covalent bonding

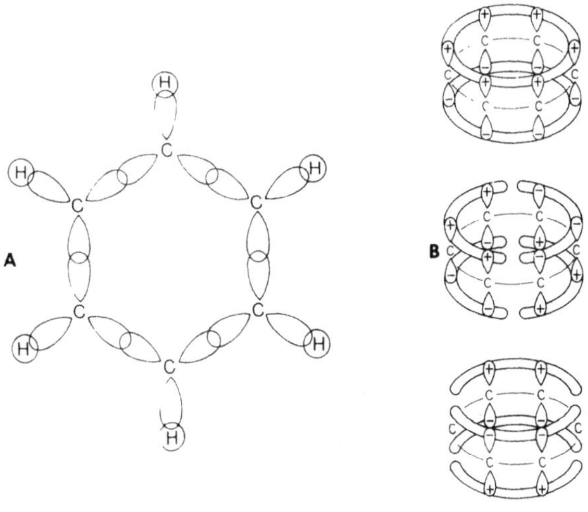

Figure 9. Bonding Orbitals in Molecular Benzenes.
A. Sigma Bonds (Localized). B. Pi Bonds (Delocalized).

while many metal alloys have both ionic and covalent contributions to the total bonding scheme. The electron pair may not spend full time on one of the members of a bonded pair of atoms (as in an ionic compound) or be equally shared between the atoms (as in a covalent compound); even in H_2, where the bond is predominantly covalent, approximately 5% of the total bonding energy can be attributed to electrostatic attraction of the type $\bar{H}:\overset{+}{H}$, where the electron pair becomes momentarily associated with one of the atoms.[8] When the atoms are dissimilar, so the attraction for the electron pair is unequal, the effect will be more pronounced.

The ability of an element to attract electrons is called its electronegativity. A number of formulas have been proposed for the calculation of a quantitative measure of electronegativity, one of which[9] will be used in this material: $x = 0.359\ Z^*/r^2 + 0.744$, where Z^* is the effective nuclear charge (the total nuclear charge minus the amount "shielded" by the electrons already surrounding the atom) and r is the mean distance of the incoming electron from the nucleus. Table II lists the electronegativity of a number of elements as calculated by this formula; the precise values differ if another formula is used, but the relative magnitudes are not greatly different. Values of electronegativity are sensitive to the oxidation state and the hybridization of electronic orbitals, since both of these can be expected to influence the interaction between an atom (or ion) and incoming electron.

Table II. Allred-Rochow Values[9] for Electronegativity of Some Elements

H	2.20												
Li	0.97	Be	1.47	B	2.01	C	2.50	N	3.07	O	3.50	F	4.10
Na	1.01	Mg	1.23	Al	1.47	Si	1.74	P	2.06	S	2.44	Cl	2.83
K	0.91	Ca	1.04	Ga	1.82	Ge	2.02	As	2.20	Se	2.48	Br	2.74
Rb	0.89	Si	0.99	In	1.49	Sn	1.72	Sb	1.82	Te	2.01	I	2.21
Cs	0.86	Ba	0.97	Tl	1.44	Pb	1.55	Bi	1.67				

When the electronegativity values for two atoms that form a chemical bond are identical or closely similar (e.g. H-H, Cl-Cl, As-H, H-I) the bond should be predominantly covalent with a very small degree of ionic character. Bonds between atoms having large differences in electronegativities (e.g. $K^{+}F^{-}$ or $Ca^{+2}O^{-2}$) should be very highly ionic owing to the tendency of the electron pair to reside exclusively in the vicinity of the more electronegative atom. It is generally agreed that between these extremes there is a continuous transition from highly covalent to highly ionic bond character, but all attempts to assign quantitative values based on electronegativity differences have been quite unsuccessful.[10] Equations calculated for one series of related compounds (e.g. hydrogen halides) are unsuccessful when applied to a different series with errors of up to 45% of the total bond strength depending on the method used for calculation. Therefore, without attempting to assign quantitative values we shall be content with the simple "rule of thumb" that the degree of ionic character of a bond tends to increase as the difference in electronegativities of the atoms participating in the bond increases.

e) Weaker Types of Bonds

While most solids are predominatly bonded by one or a combination of the three types of bonding schemes described above, there are a number of types of weaker bonds, generally electrostatic in origin, which often have a significant influence on the properties of materials. Most of these, including interactions between ions and dipoles, between pairs of dipoles, between ions or dipoles and induced dipoles or even between pairs of induced dipoles are of only secondary importance in normal glasses and will not be considered further here.

One final type of relatively weak bonding that is of importance in some glasses (especially at the surface of oxide glasses) is hydrogen bonding. It is essentially an electrostatic bond between the positive end of the $-\overset{-}{O}-\overset{+}{H}$ dipole and a neighboring oxygen atom and is stronger than the average dipole-dipole interaction for two reasons: (1) the unusually small size of the polarized hydrogen atom and (2) the unusually high electronegativity of the oxygen atom (second only to fluorine, which also forms hydrogen bonds) leading to a high charge to radius ratio for the hydrogen-bonded proton. Since the surface of oxide glasses contains negatively charged oxygen ions, it is a favorable site for the formation of a hydrogen bond with any available dipolar hydrogen compound, most typically water. The strength of the hydrogen bond is weak, usually 5-12 kcal/mole, compared with 30 to 100 kcal/mole for normal covalent chemical bonds.

THE CHEMISTRY OF OXIDE GLASSES

1. Silica and the Silicates

While silicon, having the ground state electronic configuration [Ne]$3s^2 3p^2$ is formally analagous to carbon, the differences between the chemistry of carbon and silicon are more pronounced than the similarities. Silicon, like carbon, tends to form sp^3hybrid orbitals that favor a tetrahedral arrangement of bonding atoms around the silicon atom. However the difference in the electronegativity (electron attracting power) and radius of the atoms, and the availability of an empty silicon 3d orbital at moderate energy above the valence orbitals give rise to the following important differences in chemical behavior:

a) Silicon shows less tendency to form bonds with atoms of its own kind; the Si-Si bond in Si_2H_6 is only about half the strength of the C-C bond in C_2H_6.

b) Silicon rarely participates in the formation of multiple bonds, so there is no silicon analog of the common carbonyl group, O=C< , and the common dioxide, SiO_2, forms an "infinite" three-dimensional network consisting of Si-O single bonds rather than an analog of the simple multiple-bonded O=C=O molecule. Both the size of the central atom and inability of silicon to participate in multiple bonding help to explain the simplest silicate ion being the orthosilicate, SiO_4^{-4}, rather than an analog of the carbonate ion, CO_3^{-2}.

c) The C-H bond is much stronger than the Si-H bond, so that silanes (compounds containing Si-H bonds) are much less stable than hydrocarbons; however the Si-O bond is appreciably stronger than the C-O bond, leading to a very extensive series of silicate compounds.

d) The increased difference in electronegativity between the atoms forming the Si-O bond, compared to those of the C-O bond, suggests a greatly increased degree of ionic character in the former. It is convenient to assume that the basic directional forces that determine the bond angles in the oxide glass-formers are covalent in nature with the ionic contribution arising from unequal attraction for the shared electron pair, though a reasonable case can also be made for assuming the materials to be essentially ionic in nature with the covalent contribution arising as a deformation (or polarization) of the electron cloud. The true state is probably a combination of these; at room temperature the covalent contribution is important and greatly influences the arrangement of nearest atomic neighbors in the glass (a factor that will be discussed later). As the temperature rises thermal energy is available for breaking these covalent bonds until, in the melt, the ionic contribution will clearly be dominant.

Silica, SiO_2 can be found in a number of crystalline forms, the most common three being quartz, tridymite and cristobalite, each having silicon atoms that are tetrahedrally bonded to four oxygen atoms and each oxygen atom in turn bonded to two silicon atoms to maintain the overall Si:O ratio of 1/2. While the silicon atom is believed to maintain an O-Si-O angle very close to the theoretical value of 109.4^{o}, the Si-O-Si angles are believed to vary considerably. For example various crystalline minerals containing the pyrosilicate anion, $Si_2O_7^{-6}$ (two SiO_4 tetrahedra sharing a common apex), have Si-O-Si angles ranging from 131 to 180°. This variation can be explained by assuming that the hybridization[11] of the oxygen atom can vary in a continuous fashion from pure p^2 (for which a bond angle of 90° would be predicted) to sp hybridization (predicted value 180°).

The three crystalline forms of silica, shown in Figure 10, differ with respect to the arrangement of the SiO_4 tetrahedra in the unit cell and conversion from one form to another is possible only by breaking Si-O bonds and reforming them in a different arrangement. This process has a high energy of activation, explaining the ability of the higher temperature forms to exist indefinately at temperatures below their transition temperature (both can be found in nature). Each of the three forms also exist in α and β modifications, with only minor changes in the

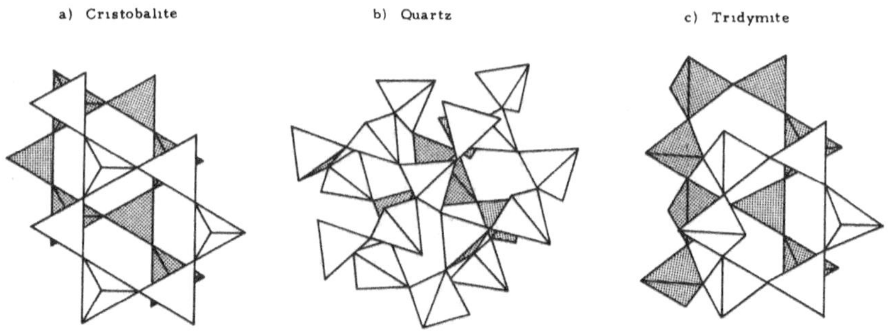

Figure 10. Crystalline Forms of Silica. Silicon atoms are located at the centers of the tetrahedra; oxygen atoms at the corners of the tetrahedra. The shaded tetrahedra are below the plane of the diagram; unshaded tetrahedra are on or above this plane.

arrangements of tetrahedra and no Si-O bond breaking required for the transition, resulting in the much higher rate of transition observed and the lower temperature at which the $\alpha \rightarrow \beta$ transition occurs.

In the non-crystalline state (both liquid and solid) the SiO_4 tetrahedron is believed to remain the basic structural unit, but with high numbers of disordered sites (such as silicon atoms surrounded by three, or even five, oxygen atoms or oxygen atoms having a coordination number of one or three). No long-range order in the arrangement of the tetrahedra has been observed, and will be considered, for our purposes, as a random arrangement of the tetrahedral SiO_4 units. Increasing the temperature lowers the viscosity of the non-crystaline SiO_2 for two reasons:

(1) there is increased thermal energy available for rehybridization of the Si-O-Si bond, permitting an increased degree of twisting and flexing in the network without actually breaking any covalent bonds. This is the dominant contribution at lower temperatures.

(2) at higher temperatures the thermal energy is

sufficient to break covalent bonds, leaving only electrostatic bonding forces. In this case the electron pair would be expected to remain associated with the oxygen atom, owing to its considerably higher electronegativity:

$$\geq Si-O-Si \leq \rightleftharpoons \geq Si^{+} + :\ddot{O}^{-}-Si \leq$$

As the temperature increases the process of conversion from predominantly covalent to predominantly ionic bonding can be expected to continue, resulting in a further decrease in the viscosity of the melt.

Addition of an ionic oxide (e.g. Na_2O) to the SiO_2 melt provides an excess of O^{-2} ions that can react with the available $^{+}SiO_3$ centers or can directly attack a Si-O-Si bond to cause a further breakdown of the network structure:

$$\geq Si^{+} + O^{-2} \rightarrow \geq Si-O^{-}$$

$$\geq Si-O-Si \leq + O^{-2} \rightarrow 2 \geq Si-O^{-}$$

The cations, meanwhile, are accommodated in the structure in a manner dictated by pure electrostatic forces; i.e. contacts between like charges are minimized and those between opposite charges are optimized. In general this suggests that they tend to remain in the neighborhood of the singly charged non-bridging oxygen atoms because of the higher charge in these regions, though they also interact with the partial negative charges associated with the bridging oxygen atoms. As the process continues with the addition of larger amounts of ionic oxides there is a considerable loosening of the network structure as physically observed by the decreased viscosity and the lower transition temperature for SiO_2 glass to which moderate amounts of ionic oxides have been added (the softening point of the glass decreases from 1500°C for pure SiO_2 glass to 400-800°C for many commercial silicate glasses).

It can be seen that continued addition of ionic oxides will give rise to progressive breakdown in the long-range chain nature of the SiO_4 tetrahedra until, in the limit, all oxygen atoms are in non-bridging positions (that is, all silicon atoms are present as SiO_4^{-4} ions) and addition of further oxygen ions can only give rise to "free" $O^{=}$

ions in the liquid. Long before this point is reached the ease of glass formation is destroyed and the liquid, like that of a normal ionic material, crystallizes readily on cooling. Between the extremes of an "infinite" three-dimensional network having all oxygens shared and the orthosilicate complex ion, with no shared oxygens, there is an infinite number of intermediate possibilities, some of which have been identified in crystalline samples by means of x-ray diffraction techniques. Figure 11 shows a number of the more common. These complex ions, plus other less regular intermediates are probably present in glasses, the relative quantities of the varying types depending, among other factors, on the atomic ratio of Si:O, which ranges from 1/2 for pure SiO_2 to 1/4 and less for the orthosilicate-oxide melts. In general a high proportion of chain-and network-type complex ions are considered important for a high tendency to glass formation. The central silicon atom, like others to be discussed in this section, are considered to be "network formers", while the metal oxides which serve to break up the network are called "network modifiers". There is no clear line of demarkation between these extreme types and those oxides capable of satisfying either role, depending on the composition of the melt, are known as "intermediates".

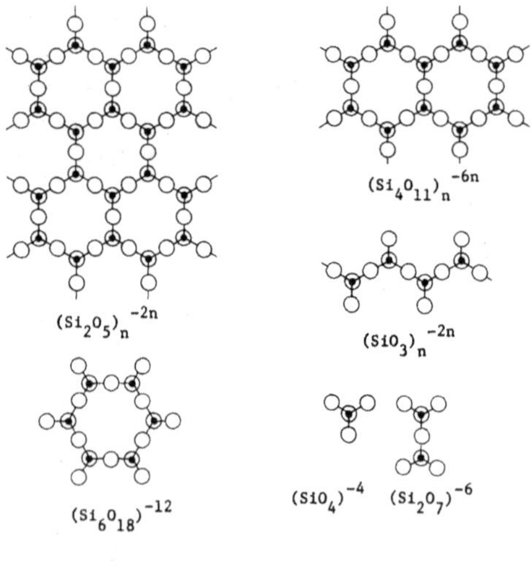

Figure 11. Some Silicate Anions Observed in Crystalline Minerals.

2. GeO_2 and the Germinate Glasses

Differences between the chemical behavior of germanium and silicon are less pronounced than those between silicon and carbon; those that exist can be explained by differences in atomic size and in energetics of electron orbitals. The two forms of germania, GeO_2, found in nature provide an example of these differences. One form, crystobalite-like in structure, involves the tetrahedral GeO_4 basic structural unit; the other, a rutile-like crystal form, contains GeO_6 octahedra. A series of germinate anions, with tetrahedral structural units similar to those listed for the silicates has been reported, along with others involving octahedral coordination of oxygen atoms about germanium.

In view of the electronegativity difference between germanium and oxygen, which is significant but not extreme, both covalent and ionic forces should contribute to the total bond strength. Each of these contributions can be expected to contribute to an increasing tendency to octahedral coordination: (1) For a purely ionic crystal the ideal radius ratio of a tetrahedral coordination of anions about a central cation lies in the range $2.44 < r^-/r^+ < 4.44$; for octahedral coordination the range is $1.37 < r^-/r^+ < 2.44$. If the ratio exceeds this range mutual contact between negatively charged anions decreases the energy of the system. A ratio below this range is unstable with respect to a higher coordination arrangement having an increased number of contacts between ions of opposite charge. For $O^{=}/Si^{+4}$ the ratio is 3.4, clearly in the range for tetrahedral coordination. The $O^{=}/Ge^{+4}$ ratio is 2.6, very close to the theoretical value of 2.44 for which the two structures should have equal stability. (2) The energy gap between the highest filled p orbitals of a covalently bonded silicon atom and the lowest empty d orbitals is such that only a limited number of atoms are able to form bonds strong enough to compensate for the activation energy required to involve two of these orbitals in an octahedral d^2sp^3 hybridized system. For germanium these lowest vacant orbitals lie at energies which are much closer to those of the highest filled orbitals; the energy for d^2sp^3 hybridization is much lower and is observed in a number of compounds.

In pure glassy GeO_2 the structural units are thought to be primarily tetrahedral (equilibrium crystallization of the melt gives the cristobalite crystal form), but upon addition of alkali oxides there is a competition between breaking of bridging bonds between tetrahedra (as in silicate glasses) and conversion from the tetrahedral to the octahedral structural unit with no bridge breaking. Both processes are believed to occur, so the fusion point and viscosity of the melt are lowered by the addition of ionic oxides to pure GeO_2 but to a much lesser extent than observed for SiO_2.

3. P_2O_5 and Phosphate Glasses

In P_2O_5 and the phosphate glasses the basic structural unit is again the tetrahedral PO_4 group. Even in the pure oxide, however, the P:O ratio is 2/5, less than the 1/2 observed for a structure having all oxygens bridging, so that one-fourth of the oxygen atoms must be terminal. The thermodynamically stable form of P_2O_5 at room temperature consists of discrete P_4O_{10} molecules, in which each phosphorous atom is surrounded by three bridging and one terminal oxygen atom (Figure 12). With a moderate amount of simple electron bookkeeping the reader should be able to satisfy himself that each phosphorous atom and each oxygen atom is surrounded with an inert gas octet of electrons. The P-O bond is appreciably more covalent than the Si-O bond, owing to the higher electronegativity of P, but still has a very appreciable degree of ionic char-

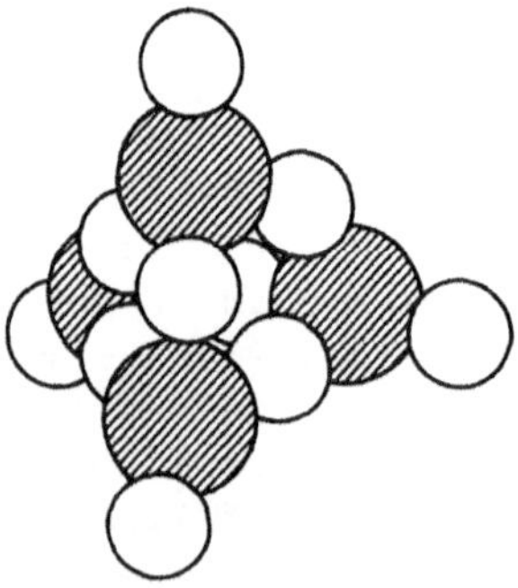

Figure 12. Structure of the P_4O_{10} Molecule.

acter. Further, unlike the terminal Si-O bonds in silicate glasses the terminal P-O bonds are significantly shorter[12] than the bridging bonds (1.39 Å vs. 1.62 Å), indicative of a partial multiple bonding character, made possible by an overlap of filled oxygen non-bonding orbitals and empty phosphorous d orbitals.

When the stable rhombohedral form of P_4O_{10} is subjected to prolonged heating at temperatures in excess of 400°C in a closed system a gradual breaking and reforming of P-O bonds leads to a melt having an extended network of PO_4 tetrahedra. Depending on the rate of cooling the melt can form a glass or one of two orthorhombic crystal forms. In contrast with the rhombohedral (molecular) form, which sublimes at 360° at atmospheric pressure, the two orthorhombic forms and the glass have low volatility and, when molten, a high viscosity. Addition of increasing amounts of ionic oxides to the molten network provide a set of complex anions completely analagous to the silicate glasses having a Si:O ratio less than 2/5, ultimately resulting in an ionic melt having as anions a mixture of tetrahedral PO_4^{-3} ions and O^{-2} ions.

While phosphorous is capable of 5- and 6- coordination, as in PF_5 and PF_6^-, these higher coordination numbers do not appear to be important in oxide systems and thus are not known to exist in phosphate glasses. It must be assumed that the d orbitals of the phosphorus atom can be more effectively used in forming multiple bonds to oxygen than in forming complex ions with higher coordination numbers. The partial multiple bonding character is shared among all non-bridged oxygen atoms, as observed for x-ray determinations of many bond lengths in crystalline materials (the P-O distances in the PO_4^{-3} ion, for example, are all equal[13] to 1.54 Å, compared to 1.39 Å for the non-bridged oxygen atom and 1.62 Å for the bridged oxygen atoms in the P_4O_{10} molecule.

4. As_2O_3 and the Oxides of Trivalent Arsenic

While not of great importance in commercial glass manufacture, As_2O_3 is the only glass-forming oxide permitting us to assess the relative importance of covalent and ionic factors in determining the shape of the fundamental structural unit in oxide glasses. The tetrahedron of SiO_4^{-4} and PO_4^{-3} and the planar triangle of B_2O_3 (see below) would be expected on the basis of either ionic or covalent arguments. The shift of coordination about germanium from tetrahedral to octahedral on adding ionic oxides to GeO_2 could be due to rehybridization of the valence electrons from sp^3 (tetrahedral) to d^2sp^3 (octahedral), using the low-lying vacant d orbitals on germanium, or could simply be a consequence of the anion-cation radius ratio which for GeO_2 is on the borderline between 4- and 6- fold coordination.

Crystalline As_2O_3, however, has a trigonal pyramidal AsO_3 building unit which can only be a result of tetrahedral sp^3 coordination of the eight valence electrons about arsenic (plus a filled d shell which is assumed to have spherical symmetry and to lie below the s and p electrons), with three of the tetrahedral positions occupied by oxygen atoms and the fourth by a single pair of non-bonded electrons. The O-As-O angle is approximately 100°, somewhat less[14] than the idealized 109.4° for a tetrahedron, in agreement with the common observation that an unshared electron pair tends to occupy more space in the coordination sphere than does a bonded atom (see reference 1a, Chapter 15). An ionic bonding scheme would predict a planar triangle of oxygen ions about the central arsenic ion, possibly somewhat distorted by the unshared electron pair, and certainly could not explain the rather classis sp^3 hybridization coordination observed.

It can be assumed, then, that in glassy As_2O_3 there is a network of AsO_3 trigonal pyramids with bridging oxygen atoms and that addition of ionic oxides tends to break up the network by breaking the As-O-As bridges and introducing non-bridging terminal oxygen ions to the structure. This model should not be used to argue that the ionic character of the M-O bonds in glasses is low, for the covalent contribution could be quite small and still be the factor determining the arrangement of atoms within the inner coordination sphere of the central atom.

5. B_2O_3 and Borates

The electronegativity of boron is very close to that for phosphorus or germanium and the covalent bond nature of the B-O bond is therefore intermediate between that for Si-O and that for As^{III}-O. Unlike the oxides discussed thus far, however, the B:O ratio of the pure oxide exceeds 1/2 and a tetrahedral unit with singly-bridging oxygen atoms is no longer possible.[15] Studies using nuclear magnetic resonance techniques have indicated that glasssy B_2O_3 consists of planar BO_3 units which are randomly arranged in a three-dimensional network by sharing all three oxygen atoms with adjacent BO_3 units. The planar BO_3 unit presumably involved sp^2 hybridization, with the third p orbital being vacant and extending in a direction perpendicular to the BO_3 plane. This vacant orbital accepts electron density from the unpaired electrons of the oxygen atoms, forming a partial double bond as evidenced by a shortening of the B-O bond from a normal single bond value of approximately 1.50Å to about 1.35Å.

Upon addition of ionic oxides to a B_2O_3 melt the following network modifications may be postulated: (1) B-O-B bonds may be broken by oxygen anions to form non-bridging terminal oxygen atoms, in a manner completely analogous to the breakdown of the silica network of tetrahedra; (2) a filled orbital on an oxygen anion may overlap with an empty p orbital on a boron atom, resulting in a change of hybridization of the boron atom to the sp^3 tetrahedral arrangement, resulting in a BO_4 tetrahedron with three bridging and one terminal oxygen atoms; (3) an oxygen atom may contribute electron pairs for <u>two</u> BO_3 units resulting in the change in coordination of two borons from sp^2 to sp^3 hybridization and resulting in <u>no</u> terminal oxygen atoms. In fact, all of these processes are believed to occur, with the predominant process determined by the mole fraction of ionic oxide present. Successive addition of ionic oxides to B_2O_3 have been shown to result in the following changes in the average coordination number of boron:[16] (1) from pure B_2O_3 to 30 mole percent alkali oxide all oxide ions are consumed in the process of increasing the coordination number of boron atoms from 3 to 4; (2) between 30 and 50 mole percent the proportion of tetrahedra and planar triangles remains relatively constant, and (3) above 50 mole percent the process reverses and the planar triangle rapidly becomes favored until at 70 mole percent alkali oxide all boron atoms again have the coordination number 3.

NON-OXIDE GLASSES

While the discussion has concerned glasses formed from network-forming oxides, it should be noted that glass formation is not limited to these materials. If the rate of solidification (particularly from a vapor) is extreme nearly all covalently-bonded materials can be obtained in a glassy state. More practically the requirements for a tendency to glass formation are similar to those in the oxide systems: (1) a degree of flexibility in the arrangement of second or third nearest neighbors and (2) a fundamental structural unit that is extended in at least one dimension. Glasses of elementary sulfur and selenium, as well as mixtures of many of the group Va and VIa elements (phosphorus, arsenic, antimony, sulfur, selenium and tellurium) can readily be obtained by rapid cooling from the melt, as can a number of organic compounds, generally those involving long chain molecules and/or hydrogen bond formation. A brief introduction to these systems is available elsewhere[1b] and will not be repeated here.

SUMMARY

This chapter has presented a brief overview of some basic principles of structural inorganic chemistry and their applications in understanding the behavior of simple oxide glasses (glass-forming oxides and network modifiers). It is hoped that this model will provide a basis for other contributers to elaborate upon and, in areas where the degree of oversimplification necessarily has been excessive, to correct.

REFERENCES

1. For further details the reader is referred to standard texts, such as (in order of increasing depth of the material covered):
   a) F. A. Cotton and G. Wilkinson, "Advanced Inorganic Chemistry", Second Edition, Interscience, New York (1966).
   b) H. Krebs, "Inorganic Crystal Chemistry", McGraw-Hill, London (1968).
   c) L. Pauling, "The Nature of the Chemical Bond", Third Edition, Cornell University Press, Ithaca, New York (1960).
   d) L. Harris and A. L. Loeb, "Introduction to Wave Mechanics", McGraw-Hill, New York (1963).
   e) L. Pauling and E. B. Wilson, Jr., "Introduction to Quantum Mechanics", McGraw-Hill, New York (1935).
2. Reference 1c, p. 96.

3. L. Helmholz and J. E. Mayer, J. Chem. Phys. 2 245 (1934).
4. J. A. Pask, from "Modern Aspects of the Vitreous State, Vol. 3", J. D. MacKenzie, ed., p. 1, Butterworth, Washington (1964).
5. A description of the method used for the calculation of the coefficients of hybrid orbitals can be found in F. A. Cotton, "Chemical Applications of Group Theory", Interscience, New York (1963).
6. Reference la, p. 68.
7. Three other orbitals can be constructed but are of higher energy and so are not normally occupied by the electron pairs. The derivation of these orbitals is given in Reference 5.
8. Reference lc, p. 25.
9. A. L. Allred and E. G. Rochow, J. Inorg. Nucl. Chem. 5, 264, 269 (1958).
10. See, for example, A. F. Wells, "Structural Inorganic Chemistry", Third Edition, Oxford Press, London (1962), pp. 32-35.
11. D. W. J. Cruickshank, H. Lynton and G. A. Barclay, Acta Cryst. 15, 491 (1962).
12. H. C. J. deDecker and C. H. MacGillavry, Rec. Trav. Chim. 60, 153 (1941).
13. G. E. R. Schulze, Z. physik, Chem. B24, 215 (1934).
14. A. J. Frueh, Amer. Min. 36, 833 (1951).
15. For a discussion of possible three-dimensional networks of the compound A_mB_n, for various m:n ratios, see A. F. Wells, Structural Inorganic Chemistry, Oxford University Press, London (1962). Chapter IV.
16. P. J. Bray and J. G. O'Keefe, Phys. Chem. Glasses 4, 37 (1963).

The Structure of Glass

Harold T. Smyth

School of Ceramics
Rutgers University
The State University of New Jersey

INTRODUCTION

The search for the truth about the structure of glass is very much like the search for the truth about religion. The search itself is very much influenced by what one would like the truth to be. It is difficult to find an impartial discussion of the different ideas on the structure of glass since anyone who knows enough to write such a discussion knows enough to have formed his own ideas and these keep creeping in. This chapter is going to be no exception.

This chapter is written to introduce the student in the field of glass to some of the ideas in the structure of glass and methods whereby the structure can be studied, but is mainly a description of the mathematical methods worked out by the author and his students at Rutgers University to describe the structure of simple glasses and relate this structure to physical properties such as density, expansion, viscosity, surface tension and miscellaneous mechanical and electrical properties and how they vary with temperature and composition.

USE OF X-RAY DIFFRACTION

X-ray diffraction has, of course, long been recognized as a powerful tool in the study of crystal structures. The regularly repeating unit cell pattern is of necessity associated with uniformly spaced planes of atoms. The diffraction pattern of X-rays scattered from these planes consists of sharp lines or spots, depending on the method used. These spots or lines occur when the wavelength of the X-rays, angle of incidence and spacing of the planes are all just right to give constructive interference. The sharpness arises from the very large number of such evenly spaced planes just as the resolving power of a diffraction grating depends on the number of rulings. When the individual crystals are very small and the number of planes in any set of parallel planes is correspondingly small the lines are broadened. The broadening of the lines can be used to obtain measures of the crystal size.

Since glasses do not have regularly repeating unit cells there are no uniformly spaced sets of planes and consequently no sharp diffraction pattern such as a single crystal or a powder of randomly oriented crystals would produce. It is found, however, that a glass of given composition, prepared in a specified manner, gives a perfectly reproducible diffraction pattern which is characteristic of this particular glass. If the intensity of scattered radiation is plotted against $\frac{\sin\theta}{\lambda}$ where λ is the wavelength of the X-rays the same glass always gives the same pattern regardless of the wavelength of the X-rays. The pattern usually consists of one or more very broad peaks.

These diffraction patterns, since they are reproducible for each glass, should carry information on the atomic or molecular arrangements in the glass. This turns out to be the case and there are a number of ways of extracting this information.

One way of going about this is to make a guess as to some of the essential features of the structure and calculate a theoretical diffraction curve. This was done in some of the earlier work by B. E. Warren[1] and his associates who, following Zachariasen[2] postulated that each silicon was surrounded by four oxygens and each oxygen was bridging between two silicons. Although no crystallinity is assumed there are certain pairs of atoms such as Si-O, Si-Si and O-O nearest neighbors and several second or third nearest neighbor pairs which repeat with fixed distances throughout the structure. These pairs, occurring with all orientations in the material as a whole are sufficient to give rise to a calculable diffraction with one or more peaks. Warren computed such a pattern for vitreous silica in reasonable agreement with the experimentally obtained pattern. While such agreement is encouraging there is always the possibility that a number of variations of the original postulate might have given just as good agreement.

Warren and Krutter[3] showed that it is possible to calculate directly and uniquely from the diffraction curve a radial distribution function. This essentially plots the average density of X-ray scattering material around each atom in the structure as a function of the distance from the center of the atom. Where a certain atom-atom distance is frequently repeated there will be a peak in the radial distribution function. If the interatomic distance becomes random after a small number of nearest neighbor separations the curve blends into a smooth parabola such as would be obtained for a continuum of uniform density around each atom. The positions of the peaks give the average values of the interatomic distances and the areas under the peaks give a measure of the number of atoms of one kind surrounding an atom of another kind. This process of determination of radial distribution functions is the source for our information concerning the structures of simple glasses. For the mathematical details and for refinements of the method the reader is referred to the chapter on "Scattering by Non-crystalline Forms of Matter" in the book "X-Ray Diffraction" by B. E. Warren.[4] Some of the applications of this method to simple glasses will now be reviewed.

The radial distribution curve for vitreous silica shows that, on the average, each silicon is surrounded by four oxygens and each oxygen is bridging between two silicons. Warren's original work indicated that the silicon-silicon distance was almost exactly twice the silicon-oxygen distance implying that the silicon-oxygen-silicon angles were 180 degrees. Repeating the work some twenty years later Zarzycki showed that this angle was considerably smaller than this. Repetition by Mozzi and Warren has confirmed this. Their refined methods were able to indicate the spread existing in this quantity. They indicate that 144 degrees is the most probable value.

The addition of soda to silica introduces a new peak at 2.35A which increases in height and area with soda concentration. This is interpreted as a Na-O distance which is consistent with the ionic radii of these two ions and also with Na-O distances in crystals. The area under the peak suggests that on the average there are about five oxygens around each sodium. The sharpness of the peak contradicts some suggestions that the silicons and the oxygens form by themselves a network containing interstices of more or less random size into which the alkali ions fit as well as they can. This would lead to a much more variable Na-O distance than is actually found. It would probably be much more correct to think of the network as forming around the alkali and other modifying ions in a way very much related to the size of these ions. This concept is a very important part of the model which will be presented in this chapter. Warren's work on potash-silica glasses suggests the same kind of picture with the average co-ordination of the potassiums somewhat higher than that of the sodiums.

Investigation of B_2O_3 glass shows that each boron is surrounded by three oxygens forming an equilateral triangle in the plane with the boron and each oxygen is bridging between two borons. However, it is very difficult to build a model of such a structure having any rigidity and it is possible that there are some B-O bonds of a different length and character giving additional stability to the structure. Later work by Warren, incorporating suggestions by Krogh-Moe, has suggested the existence of planar boroxyl groups as essential structural elements. These do not, however, do much to add rigidity to the structure.

RANDOM NETWORK

The X-ray results have suggested the phrase "random network" to describe the structure of a glass as contrasted to the structure of a crystal. The suggestion is that the immediate neighbors of any cation or anion are very much the same as they would be in a crystal of the same composition, but that a certain randomness enters in describing how the co-ordination polyhedra are connected together.

Some years ago Tilton[5] suggested that it would be perfectly possible to have groups in vitreous silica consisting of dodecahedra of five membered silicon-oxygen rings. In fact, it is easy to build up a unit of twelve such dodecahedra around a central one without distorting the oxygen-silicon-oxygen angles very much from the ideal tetrahedral angle of 109° 28'. Any further extension of this unit very seriously distorts the angles. The density of one of these groups of dodecahedra is considerably less than that of vitreous silica and an interstitial or grain boundary structure with high density has to be postulated. The idea is not quite so readily adjustable to describe the structures of other glass formers.

Eckstein[6] has approached the problem by starting with a crystal of corresponding composition and increasing the concentration of defects until the deformed regions around the defects overlap. This implies an almost continuous passage from the structure of a perfect single crystal to the structure of the corresponding glass.

The diffraction patterns of soda-silica glasses from zero soda to 50 mole percent soda can be described very well as superpositions of a pure vitreous silica diffraction curve and a vitreous Na_2OSiO_2 curve properly weighted. This and other evidence has suggested to some of the Russian workers the co-existence of very small regions of two quite different compositions. The existence of these microheterogeneities is considered an essential part of the glass structure in the opinion of these investigators. Although these regions are not necessarily crystalline the theory is sometimes referred to as the revised crystallite theory to distinguish it from earlier theories in which there was an implication of crystallinity. Although these approaches do not form the basis of the method to be described it very often happens that two apparently quite distinct ways of going about the problem turn out to be different ways of saying the same thing.

DESCRIPTION OF THE MODEL

The model to be described will be referred to as the Rutgers model. It is formed as follows. A reasonable guess is made of the volume that one mole of the glass will occupy. All of the anions, if of more than one kind are mixed together and randomly distributed through this space so that no two are closer together than the sum of their radii. The cations are mixed together and randomly distributed in among the anions, again with the restriction that no two ions are closer together than the sum of their radii.

Each cation is then free to reach out to any anion whose center lies within a sphere whose radius is twice the normal cation-anion distance for this pair of ions and form a bond with this anion, but not necessarily with all of the anions within such a sphere of influence. The forming of a band will cause a slight local readjustment of positions to give approximately the correct cation-anion distance. This, of course, depends to some extent on the number of anions attached to this cation. All of the cations and anions may then form bonds in the same way but do it in such a way as to make the free energy a minimum or at least a relative minimum. The fact that there may be a regular crystalline arrangement that would make the free energy lower still does not enter into the immediate calculation, but would, of course, come into a study of whether the glass was thermodynamically stable in the presence of this crystal.

The exact definition of the size of the sphere of influence just used is not yet firmly established and will not be until many more comparisons between experimental and theoretical values of many physical properties have been made.

The calculation of free energy G will depend on calculation of the internal energy E, the entropy S and the volume V of one mole of the glass being studied. The methods and assumptions necessary to do this will now be studied.

All of the ions will be divided into species. A species will be all of those ions of one kind which are bonded to the same number of ions of each other kind. For instance, one species would be all of those silicons bonded each to four oxygens and another would be all of those oxygens bonded to one silicon and two sodiums.

The state of the material is described by specifying how many ions of each species are present in each mole of the glass.

The energy of the material is calculated by assigning an energy (cals. per mole) to each species present. This has been discussed in a previous publication by Smyth and Leipold.[7]

The volume is calculated by assigning a volume (cm^3 per mole) to each species. This has been discussed in the same publication. The use of volume calculations enters into calculations of molar volume and would come into calculations of compressibility at elevated temperatures. This latter application has not yet been attempted.

The entropy is calculated by computing the number of different ways of connecting the atoms together to give the same state. Boltzmann's constant k times the natural logarithm of this number is taken as the entropy. At the present state of development vibrational energy and vibrational entropy are not considered. This later introduction will probably not give rise to any insurmountable difficulties. The methods of applying these principles can best be demonstrated by specific examples.

In several of the following sections there is involved a certain amount of detailed but not very advanced mathematics which may act as a mental stumbling block on a first reading. An attempt will therefore be made to arrange the text to include the basic principles to be read at a first reading. The detailed derivation will be so marked that it can be omitted the first time through. The details are, however, a very essential part of the whole development and some plausibility is lost by omitting them.

VITREOUS SILICA

It is assumed that in one mole of vitreous silica there are N_3 moles of silicon connected to 3 oxygens and having energy U_3 cals per mole. N_4 moles of silicon connected to 4 oxygens and having energy U_4 cals per mole. N_1'' oxygens connected to 1 silicon and having energy U_1'' cals per mole. N_2'' oxygens connected to 2 silicons and having energy U_2'' cals per mole.

There is a certain amount of personal judgment used in setting up each of these problems. Certain species are omitted which seem unlikely. If later this appears not to be justified it is easy to include these species and the only purpose in omitting them is to simplify the first approach.

The internal energy, E, will then be equal to
$N_3U_3 + N_4U_4 + N''_1 U''_1 + N''_2 U''_2$
This will be in cals per mole of SiO_2.

The calculation is then carried through to choose N_3, N_4, N''_1 and N''_2 so that the free energy will be as low as possible. Since the entropy calculations are somewhat involved the result will be written down for later use followed by the details of the calculation.

It turns out that $N_3 = N''_1$ and the equation for either is given by:

$$\frac{N_3^{\,2}}{(1-N_3)\;(2-N_3)} = \frac{16(S+N_3-N_4)}{(S-2)\;(S-3)} \exp -\left[\frac{U_3-U_4+U''_1-U''_2}{RT}\right]$$

where S equals the number of oxygens in the sphere of influence of a silicon.

DERIVATION (This may be omitted in a first reading).

There is one mole of silicons so that:

$$N_3 + N_4 = 1$$

There are two moles of oxygens so that:

$$N''_1 + N''_2 = 2$$

The bonds from silica to oxygen must be the same as the number of bonds from the oxygens to silicons so that:

$$3N_3 + 4N_4 = N''_1 + 2N''_2$$

These are three equations for four unknowns so that only one of the four N's is independent. If it is known all of the others are known. For instance

$$N_4 = 1 - N_3$$
$$N''_1 = N_3$$
$$N''_2 = 2 - N_3$$

if N_3 is considered the independent variable.

In calculating the entropy there are $\frac{N!}{N_3N!\ N_4N!}$

ways of selecting which silicas will be connected to three oxygens and which to four.

For each of these there are $\frac{2N!}{N_1''\ N!\ N_2''\ N!}$ ways of selecting which oxygens will be connected to one silicon and which to two. The contribution to the entropy which these choices make will be

$$k \ln \left[\frac{N!}{N_3N!\ N_4N!} \quad \frac{2N!}{N_1''\ N!\ N_2''\ N!} \right]$$

By using the Stirling approximation for very large numbers such that if M is a very large number,

$$\ln M! = M \ln M - M,$$

this can be reduced to

$$R \left[- N_3 \ln N_3 - N_4 \ln N_4 + 2\ln 2 - N_1'' \ln N_1'' - N_2'' \ln N_2'' \right]$$

Having made the choice as to which silicon and which oxygens belong to each species one has to calculate how many ways there are to connect them together so as to give the specified distribution.

Because of the finite size of the sphere of influence each silicon on the average has a small number S of oxygens in its sphere of influence. Although S is certainly not always an integer and varies from one silicon to another it is treated as an integer and each silicon is assumed to have the same number S of oxygens in its sphere of influence. Similarly each oxygen will have S/2 silicons in its sphere of influence.

It can then be said that each silicon has S bond sites around it, not all of which may be occupied by a bond joining this silicon to one of the surrounding oxygens. In one mole of silica there are S moles of bond sites and there are $(3N_3 + 4N_4)$ bonds. The problem is to find how many ways there are of putting $(3N_3 + 4N_4)$ bonds into SN bond sites so that a particular N_3 moles of silicons are connected to 3 oxygens, a particular N_4 moles are connected to 4 oxygens, a particular N_1'' moles of oxygen are connected to one silicon and a particular N_2'' moles of oxygens are connected to two oxygens.

The number of ways of putting $(3N_3 + 4N_4)N$ bonds into SN bond sites is

$$^{SN}C_{(3N_3 + 4N_4)N} = \frac{SN!}{(3N_3 + 4N_4)N! \; (S-3N_3 - 4N_4)N!}$$

or since $N_4 = 1 - N_3$

$$\frac{SN!}{(4-N_3)N! \; (S-4 + N_3)N!}$$

If this has been done randomly the probability that a bond site is occupied is $\frac{4 - N_3}{S}$ and that it is unoccupied is $\frac{S - 4 + N_3}{S}$

The probability that a given silicon is bonded to 3 oxygens is obtained by first calculating the number of ways a silicon can choose three oxygens out of S which is $^SC_3$.

The probability that one specific site is occupied is $\frac{4 - N_3}{S}$ and that a particular three are occupied is $\left(\frac{4 - N_3}{S}\right)^3$

and that the other (S - 3) are vacant is $\left(\frac{S - 4 + N_3}{S}\right)^{S-3}$

Thus, the probability that a given silicon is bonded to 3 out of the S oxygens around it is

$$^SC_3\left(\frac{4 - N_3}{S}\right)^3 \left(\frac{S - 4 + N_3}{S}\right)^{S-3}$$

This, incidentally, is one term in the binomial expansion of

$$\left[\left(\frac{4 - N_3}{S}\right) + \left(\frac{S - 4 + N_3}{S}\right)\right]^S$$

which is of course 1. The probability is therefore a true fractional probability.

Similarly the probability that a silicon is connected to 4 oxygens is

$$^{S}C_4 \left(\frac{4 - N_3}{S}\right)^4 \left(\frac{S - 4 + N_3}{S}\right)^{S-4}$$

and that an oxygen is connected to one silica is

$$^{S/2}C_1 \left(\frac{4 - N_3}{S}\right) \left(\frac{S - 4 + N_3}{S}\right)^{S/2 - 1}$$

and that an oxygen is connected to two silicons is

$$^{S/2}C_2 \left(\frac{4 - N_3}{S}\right)^2 \left(\frac{S - 4 + N_3}{S}\right)^{S/2 - 2}$$

The probability that a specific N_3 moles of silica are each connected to 3 oxygens is

$$^{S}C_3 \left[\left(\frac{4 - N_3}{S}\right) \left(\frac{S - 4 + N_3}{S}\right)^{S-3} \right]^{N_3 N}$$

with similar expressions for each of the other species. Once it is decided which silicons and which oxygens belong to which species the total ways of connecting them together is

$$\frac{SN!}{(4-N_3)N!\ (S-4+N_3)N!} \left[^{S}C_3 \left(\frac{4-N_3}{S}\right)^3 \left(\frac{S-4+N_3}{S}\right)^{S-3} \right]^{N_3 N}$$

$$\left[^{S}C_4 \left(\frac{4-N_3}{S}\right)^4 \left(\frac{S-4+N_3}{S}\right)^{S-4} \right]^{N_4 N} \left[^{S/2}C_1 \left(\frac{4-N_3}{S}\right) \left(\frac{S-4+N_3}{S}\right)^{S/2-1} \right]^{N_1'' N}$$

$$\left[^{S/2}C_2 \left(\frac{4-N_3}{S}\right)^2 \left(\frac{S-4+N_3}{S}\right)^{S/2-2} \right]^{N_2'' N}$$

and using the Stirling's approximation this reduces to

$$R \Big[N_3 \ln {}^{S}C_3 + N_4 \ln {}^{S}C_4 + N_1'' \ {}^{S/2}C_1 + N_2'' \ {}^{S/2}C_2$$

$$+ (4-N_3)\ln (4-N_3) + (S-4+N_3)\ln (S-4+N_3) - S\ln S \Big] .$$

and to get the total structural entropy there must be added to this:

$$R \left[- N_3 \ln N_3 - N_4 \ln N_4 + 2\ln 2 - N_1'' \ln N_1'' - N_2'' \ln N_2'' \right]$$

which is the entropy arising from the number of silicons and oxygens which belong to which species.

The free energy G can then be written (for zero pressure)

$$G = N_3U_3 + N_4U_4 + N_1''U_1'' + N_2''U_2'' - RT\Big[- N_3\ln N_3 - N_4\ln N_4 + 2\ln 2 - N_1''\ln N_1'' - N_2''\ln N_2'' + N_3\ln {}^{S}C_3 + N_4\ln {}^{S}C_4 + N_1''\ln {}^{S/2}C_1 + N_2''\ln {}^{S/2}C_2 + (4-N_3)\ln(4-N_3) + (S-4+N_3)\ln(S-4+N_3) - S\ln S\Big].$$

The derivative of this with respect to N_3 the only independent variable must be zero for minimum G. Hence

$$U_3-U_4+U_1''-U_2''-RT\Big[-\ln N_3 + \ln N_4 - \ln N_1'' + \ln N_2'' + \ln {}^{S}C_3 - \ln {}^{S}C_4 - \ln {}^{S/2}C_1 - \ln {}^{S/2}C_2 - \ln(4-N_3) + \ln(S-4+N_3)\Big] = 0$$

or

$$\frac{U_3-U_4+U_1''-U_2''}{RT} = \ln \frac{(1-N_3)(2-N_3)}{N_3^2}\,\frac{4}{S-3}\,\frac{4}{S-2}\,\frac{S-4\,N_3}{4-N_3}$$

or

$$\frac{N_3^2\,(S-2)(S-3)(4-N_3)}{16(1-N_3)(2-N_3)(S-4+N_3)} = \exp -\left[\frac{U_3-U_4+U_1''-U_2''}{RT}\right]$$

The quantity $U_3-U_4+U_1''-U_2''$ is the energy required to break a bond connecting a four coordinated silica to a bridging oxygen and creating in its place a three coordinated silica and a non-bridging oxygen.

KINETIC APPROACH

The method just used is useful because it permits calculation of the free energy which is important particularly in binary and ternary glasses in studying immiscibility and liquidus shapes.

In studying rate processes such as viscous flow a different method of calculating the equilibrium conditions is desirable and this calculation goes as follows.

The bonds going out from the silicons finish up in the

oxygens. The N_4 moles of 4 coordinated silicons send out $4N_4$ moles of bonds which are divided among the bridging and non-bridging oxygens in the ratio of the number of bonds going out from each of these species. The bridging oxygens have $2N_2''$ moles of bonds and the non-bridging oxygens have N_1'' moles of bonds so that there are

$$\frac{4N_4 2N_2''}{N_1''+2N_2''}$$

moles of bonds from 4 coordinated silicons to bridging oxygens. This can be written:

$$\frac{8(1-N_3)(2-N_3)}{(4-N_3)}$$

If the energy barrier is to be overcome when one of these breaks is U the number breaking per second will be

$$\frac{8(1-N_3)(2-N_3)}{(4-N_3)} \nu \exp -\frac{U}{RT}$$

where ν is some kind of an average frequency of vibration of the network.

The opposite process may take place when a 3 coordinated silicon finds a non-bridging oxygen within its sphere of influence. There are $N_3(S-3)$ moles of such vacant bond sites which involve bridging and non-bridging oxygens in the ratio of the vacant bond sites associated with each of these species. The N_1'' non-bridging oxygens have N_1'' (S/2 - 1) and the N_2'' bridging oxygens N_2'' (S/2 - 2) so that the number of vacant bond sites between a 3 coordinated silica and a non-bridging oxygen is

$$\frac{N_3(S-3)\ N_1''\ (S/2 - 1)}{N_1''(S/2 - 1) + N_1''(S/2 - 2)}$$

which comes out to be

$$\frac{N_3{}^2(S - 2)(S - 3)}{2(S + N_3 - 4)}$$

If the barrier to be overcome to bring these together is

U^1 then the number of such bond reformations taking place per second is

$$\frac{N_3^2(S-2)(S-3)}{2(S+N_3-4)} \nu \exp - \frac{U^1}{RT}$$

If the rate of breaking and the rate of reforming are equal then

$$\frac{8(1-N_3)(2-N_3)}{4-N_3} \nu \exp - \frac{U}{RT} = \frac{N_3^2(S-2)(S-3)}{2(S+N_3-4)} \nu \exp - \frac{U^1}{RT}$$

Remembering that $U-U^1 = U_3-U_4+U_1''-U_2''$ this relation comes out to be exactly the same as that derived by minimizing the free energy.

END OF DERIVATION

If, as is usually the case, $N_3 \ll 1$

$$N_3^2 = \frac{8(S-4)}{(S-2)(S-3)} \exp - \frac{U_3-U_4+U_1''-U_2''}{RT}$$

or

$$N_3 = \sqrt{\frac{8(S-4)}{(S-2)(S-3)}} \exp - \frac{U_3-U_4+U_1''-U_2''}{2RT}$$

If each time a bond from a 4 coordinated silicon to a bridging oxygen breaks there is created a volume Vo. then the excess molar volume (over and above the volume of vitreous silica without the defects) is N_3Vo and this is given by

$$N_3Vo = \sqrt{\frac{8(S-4)}{(S-2)(S-3)}} V_o \exp - \frac{U_3-U_4+U_1''-U_2''}{2RT}$$

and the natural logarithm of N_3Vo plotted against 1/T should give a straight line. Such a curve is plotted in Figure 1 from the data of Bacon[8] et al., and is a good straight line for four points out of the five. The fifth point, which is the highest temperature point, may indicate the presence of a new species of silicon connected to two oxygens. Each one of these would require two non-bridging oxygens, and since these are probably the highest volume species this might explain the high volume associated with

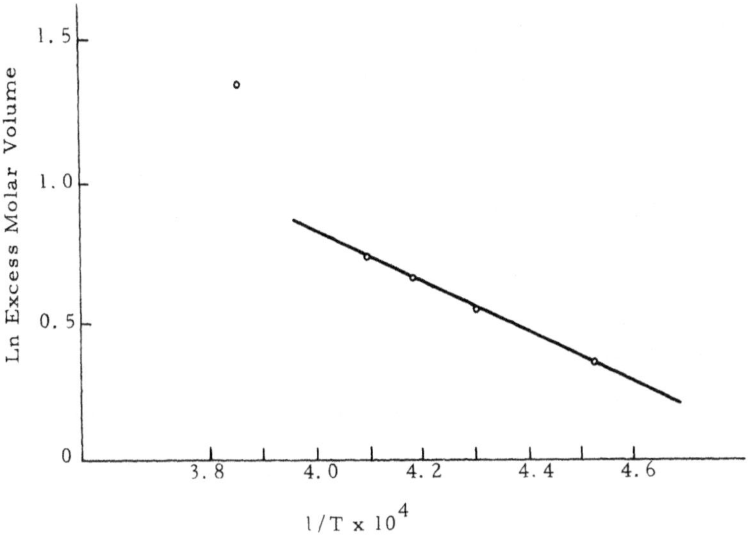

Figure 1. Plot of Ln excess molar volume vs reciprocal temperature for vitreous silica.

the high temperature point.

The slope of the straight line indicates that $U_3-U_4+U_1''-U_2''$ is about 36 k cals per mole. Work is currently being carried on to see if the introduction of different species entering at high temperature may account for the high temperature point on the experimental data having an excessively high volume.

SODA SILICA GLASSES

The method of handling a binary glass is not in principle different from that for a single oxide glass but will be more complicated to handle. The formula of the glass is taken to be xNa_2O $(1-x)$ SiO_2.

The species considered will be silicons connected to one or two oxygens, sodiums connected to anywhere from one to six oxygens, oxygens connected to one or two silicons and anywhere from zero to six sodiums as well. The choice of six as an upper limit is not important since it is very easy to extend the solutions obtained to any higher number.

The notation for the number of the various species and their energies are shown in Table I.

DERIVATION (May be omitted in first reading).

The following relations exist among the N's

$$N_3+N_4 = (1-x)$$

$$N_1'+N_2'+N_3'+N_4'+N_5'+N_6' = 2x$$

$$N_{10}''+N_{11}''+N_{12}''+N_{13}''+N_{14}''+N_{15}''+N_{16}''+N_{20}''+N_{21}''+N_{22}''+N_{23}''+N_{24}''+N_{25}''+N_{26}'' = 2-x$$

$$(N_{10}''+N_{11}''+N_{12}''+N_{13}''+N_{14}''+N_{15}''+N_{16}'') +2 (N_{20}''+N_{21}''+N_{22}''+N_{23}''+N_{24}''+N_{25}''+N_{26}'')$$

$$= 3N_3 + 4N_4 = (1-x)\ l_s.$$

$$N_{11}''+2N_{12}''+3N_{13}''+4N_{14}''+5N_{15}''+6N_{16}''+N_{21}''+2N_{22}''+3N_{23}''+4N_{24}''+5N_{25}''+6N_{26}''$$

$$= N_1'+2N_2'+3N_3'+4N_4'+5N_5'+6N_6' = 2xl_n$$

In these equations l_s is introduced and this is the average coordination of a silicon. Also l_n is the average coordination of a sodium. Since there are five relations it means that five of the N's are not independent and if the five dependent variable are chosen to be N_3, N_1', N_{11}'', N_{20}'', N_{10}'' these can be solved for to give

$$N_3 = (1-x) - N_4$$

$$N_1' = 2x - N_2'-N_3'-N_4'-N_5'-N_6'$$

$$N_{11}'' = 2x + N_2'+2N_3'+3N_4'+4N_5'+5N_6'-2N_{12}''-3N_{13}''-4N_{14}''-5N_{15}''-6N_{10}'' -N_{21}''$$

$$-2N_{22}''-3N_{23}''-4N_{24}''-5N_{25}''-6N_{26}''$$

$$N_{20}'' = 1-4x + N_4-N_{21}''-N_{22}''-N_{23}''-N_{24}''-N_{25}''-N_{26}''$$

$$N_{10}'' = 1-x-N_4-N_2'-2N_3'-3N_4'-4N_5'-5N_6'+N_{12}''+2N_{13}''+3N_{14}''+4N_{15}''+5N_{16}''$$

$$+N_{21}''+2N_{22}''+3N_{23}''+4N_{25}''+6N_{26}''$$

Table I.

Ion	Moles	Neighbors Si	Na	O	Energy
Si	N_3			3	U_3
Si	N_4			4	U_4
Na	N'_1			1	U'_1
Na	N'_2			2	U'_2
Na	N'_3			3	U'_3
Na	N'_4			4	U'_4
Na	N'_5			5	U'_5
Na	N'_6			6	U'_6
O	N''_{10}	1	0		U''_{10}
O	N''_{11}	1	1		U''_{11}
O	N''_{12}	1	2		U''_{12}
O	N''_{13}	1	3		U''_{13}
O	N''_{14}	1	4		U''_{14}
O	N''_{15}	1	5		U''_{15}
O	N''_{16}	1	6		U''_{16}
O	N''_{20}	2	0		U''_{20}
O	N''_{21}	2	1		U''_{21}
O	N''_{22}	2	2		U''_{22}
O	N''_{23}	2	3		U''_{23}
O	N''_{24}	2	4		U''_{24}
O	N''_{25}	2	5		U''_{25}
O	N''_{26}	2	6		U''_{26}

Also

$$(1-x)l_s = 3(1-x)+N_4$$

$$2xl_n = 2x+N_2'+2N_3'+3N_4'+4N_5'+5N_6'$$

There are $(1-x)\,l_s$ moles of Si-O bonds in $(1-x)\,S$ bond sites. The probability of any bond site being occupied is $\frac{l_s}{S}$ and of being vacant is $\frac{S-l_s}{S}$

If the sphere of influence of a sodium contains n oxygens the sphere of influence of an oxygen will contain $\frac{2xn}{2-x}$ sodiums.

If the sphere of influence of a silicon contains S oxygens the sphere of influence of an oxygen will contain $\frac{S(1-x)}{(2-x)}$ silicons.

There are $2xl_nN$ Na-O bonds in $2xnN$ bond sites. The entropy S will be

$$S=k\ln\left[\frac{(1+x)N!}{N_3N!N_4N!N_1'N!N_2N!N_3'N!\text{etc.}}\;\frac{2N}{N_{10}''N!N_{11}''N!N_{12}''N!\text{etc.}N_{20}N!N_{21}N!\text{etc.}}\right.$$

$${}^{2xnN}C_{2xl_nN}\quad{}^{(1-x)SN}C_{(1-x)l_sN}\left[{}^{S}C_3\left(\frac{l_s}{S}\right)^3\left(\frac{S-l_s}{S}\right)^{S-3}\right]^{N_3N}$$

$$\left[{}^{S}C_4\left(\frac{l_s}{S}\right)^4\left(\frac{S-l_s}{S}\right)^{S-4}\right]^{N_4N}\left[{}^{n}C_1\left(\frac{l_n}{n}\right)\left(\frac{n-l_n}{n}\right)^{n-1}\right]^{N_1'N}$$

$$\left[{}^{n}C_2\left(\frac{l_n}{n}\right)^2\left(\frac{n-l_n}{n}\right)^{n-2}\right]^{N_2'N}\quad\text{etc.}$$

$$\left[{}^{\frac{S(1-x)}{2-x}}C_1\left(\frac{l_s}{S}\right)\left(\frac{S-l_s}{S}\right)^{\left(\frac{S(1-x)}{2-x}-1\right)}\right]^{(N_{10}''+N_{11}''+N_{12}''+N_{13}''+N_{14}''+N_{15}''+N_{16}'')N}$$

$$\left[{}^{\frac{S(1-x)}{2-x}}C_{2} \left(\frac{1_s}{S}\right)^{2} \left(\frac{S-1_s}{S}\right)^{\frac{S(1-x)}{2-x}-2} \right]^{(N_{20}''+N_{21}''+N_{22}''+N_{23}''+N_{24}''+N_{25}''+N_{26}'')N}$$

$$\left[{}^{\frac{2xn}{2-n}}C_{o} \left(\frac{S-1_n}{S}\right)^{\frac{2xn}{2-x}} \right]^{(N_{10}''+N_{20}'')N} \left[{}^{\frac{2xn}{2-n}}C_{1} \left(\frac{1_n}{n}\right) \left(\frac{S-1_n}{n}\right)^{\frac{2xn}{2-x}-1} \right]^{(N_{11}''+N_{21}'')N}$$

$$\left[{}^{\frac{2xn}{2-x}}C_{2} \left(\frac{1_n}{n}\right)^{2} \left(\frac{S-1_n}{n}\right)^{\frac{2xn}{2-x}-2} \right]^{(N_{12}''+N_{22}'')N} \left[\;\right] \left[\;\right] \Bigg]$$

Using these relations and using a little manipulations the free energy G is given by

$$G = N_3U_3+N_4U_4+N_1'U_1'+N_2'U_2'+N_3'U_3'+N_4'U_4'+N_5'U_5'+N_6'U_6'$$

$$+N_{10}''U_{10}''+N_{11}''U_{11}''+N_{12}''U_{12}''+N_{13}''U_{13}''+N_{14}''U_{14}''+N_{15}''U_{15}''+N_{16}''U_{16}''$$

$$+N_{20}''U_{20}''+N_{21}''U_{21}''+N_{22}''U_{22}''+N_{23}''U_{23}''+N_{24}''U_{24}''+N_{25}''U_{25}''+N_{26}''U_{26}''$$

$$-RT \Big[(1+x)\ln(1+x) - N_3\ln N_3-N_4\ln N_4-N_1'\ln N_1'$$

$$-N_2'\ln N_2'-N_3'\ln N_3'-N_4'\ln N_4'-N_5'\ln N_5'-N_6'\ln N_6'+2\ln 2$$

$$-N_{10}''\ln N_{10}''-N_{11}''\ln N_{11}''-N_{12}''\ln N_{12}''-N_{13}''\ln N_{13}''-N_{14}''\ln N_{14}''-N_{15}''\ln N_{15}''$$

$$-N_{16}''\ln N_{16}''-N_{20}''\ln N_{20}''-N_{21}''\ln N_{21}''-N_{22}''\ln N_{22}''-N_{23}''\ln N_{23}''-N_{24}''\ln N_{24}''$$

$$-N_{25}''\ln N_{25}''-N_{26}''\ln N_{26}''$$

$$+N_3\ln {}^{S}C_3+N_4\ln {}^{S}C_4+N_1'\ln {}^{n}C_1+N_2'\ln {}^{n}C_2+N_3'\ln {}^{n}C_3+N_4'\ln {}^{n}C_4$$

$$+N_5'\ln {}^{n}C_5+N_6'\ln {}^{n}C_6+(N_{10}''+N_{11}''+N_{12}''+N_{13}''+N_{14}''+N_{15}''+N_{16}'') \; {}^{\frac{S(1-x)}{2-x}}C_1$$

$$+(N_{20}''+N_{21}''+N_{22}''+N_{23}''+N_{24}''+N_{25}''+N_{26}'')\ {}^{\frac{S(1-x)}{2-x}}C_2$$

$$+(N_{10}''+N_{20}'')\ln {}^{\frac{2xn}{2-x}}C_o+(N_{11}''+N_{21}'')\ln {}^{\frac{2xn}{2-x}}C_1+(N_{12}''+N_{22}'')\ln {}^{\frac{2xn}{2-x}}C_2$$

$$+(N_{13}''+N_{23}'')\ln {}^{\frac{2xn}{2-x}}C_3+(N_{14}''+N_{24}'')\ln {}^{\frac{2xn}{2-x}}C_4+(N_{15}''+N_{25}'')\ln {}^{\frac{2xn}{2-x}}C_5$$

$$+(N_{16}''+N_{26}'')\ln {}^{\frac{2xn}{2-x}}C_6$$

$$+(1-x)l_s \ln l_s+(1-x)(S-l_s)\ln(S-l_s)-(1-x)S\ln S$$

$$+2xl_n \ln l_n+2x(n-l_n)\ln(n-l_n)-2x\ n\ln n \Big]$$

For the free energy G to be a minimum the partial derivative of G with respect to each of the independent variables must be a minimum so that

$$\frac{\partial G}{\partial N_4} = -U_3+U_4+U_{20}''-U_{10}''-RT\Big[\ln N_3-\ln N_4-\ln N_{20}''+\ln N_{10}''$$

$$-\ln {}^3C_3+\ln {}^SC_4+\ln {}^{\frac{S(1-x)}{2-x}}C_2-\ln {}^{\frac{S(1-x)}{2-x}}C_1+\ln l_S-\ln(S-l_S)\Big] = 0$$

$$\frac{\partial G}{\partial N_2'} = -\ U_1'+U_2'-U_{10}''+U_{11}''-RT\Big[\ln N_1'-\ln N_2'+\ln N_{10}''-\ln N_{11}''$$

$$-\ln {}^nC_1+\ln {}^nC_2-\ln {}^{\frac{2xn}{2-x}}C_o+\ln {}^{\frac{2xn}{2-x}}C_1+\ln l_n-\ln(n-l_n)\Big] = 0$$

$$\frac{\partial G}{\partial N_3'} = -\ U_1'+U_3'-2U_{10}''+2U_{11}''-RT\Big[\ln N_1'-\ln N_3'+2\ln N_{10}''-2\ln N_{11}''$$

$$-\ln{}^{n}C_1+\ln{}^{n}C_3-2\ln{}^{\frac{2xn}{2-x}}C_o+2\ln{}^{\frac{2xn}{2-x}}C_1+2\ln l_n-2\ln(n-l_n)\Big] = 0$$

$$\frac{\partial G}{\partial N_4'} = -U_1'+U_4'-3U_{10}''+3U_{11}''-RT\Big[\ln N_1'-\ln N_4'+3\ln N_{10}''-3\ln N_{11}''$$

$$-\ln{}^{n}C_1+\ln{}^{n}C_4-3\ln{}^{\frac{2xn}{2-x}}C_o+3\ln{}^{\frac{2xn}{2-x}}C_1+3\ln l_n-3\ln(1-l_n)\Big] = 0$$

etc. up to $\frac{\partial G}{\partial N_6'}$

$$\frac{\partial G}{\partial N_{12}''} = U_{10}''-2U_{11}''+U_{12}''-RT\Big[-\ln N_{10}''+2\ln N_{11}''-\ln N_{12}''$$

$$+\ln{}^{\frac{2xn}{2-n}}C_o-2\ln{}^{\frac{2xn}{2-x}}C_1+\ln{}^{\frac{2xn}{2-x}}C_2\Big] = 0$$

$$\frac{\partial G}{\partial N_{13}''} = 2U_{10}''-3U_{11}''+U_{13}''-RT\Big[-2\ln N_{10}''+3\ln N_{11}''-\ln N_{13}''$$

$$+2\ln{}^{\frac{2xn}{2-x}}C_o-3\ln{}^{\frac{2xn}{2-x}}C_1+\ln{}^{\frac{2xn}{2-x}}C_3\Big] = 0$$

etc. up to $\frac{\partial G}{\partial N_{16}''}$

$$\frac{\partial G}{\partial N_{21}''} = U_{10}''-U_{11}''-U_{20}''+U_{21}''-RT\Big[-\ln N_{10}''+\ln N_{11}''+\ln N_{20}''-\ln N_{21}''$$

$$+\ln{}^{\frac{2xn}{2-x}}C_o-\ln{}^{\frac{2xn}{2-x}}C_1-\ln{}^{\frac{2xn}{2-x}}C_o+\ln{}^{\frac{2xn}{2-x}}C_1\Big] = 0$$

$$\frac{\partial G}{\partial N_{22}''} = 2U_{10}''-2U_{11}''-U_{20}''+U_{22}''-RT\Big[-2\ln N_{10}''+2\ln N_{11}''+\ln N_{20}''-\ln N_{22}''$$

$$+2\ln{}^{\frac{2xn}{2-x}}C_o-2\ln{}^{\frac{2xn}{2-x}}C_1-\ln{}^{\frac{2xn}{2-x}}C_o+\ln{}^{\frac{2xn}{2-x}}C_2\Big] = 0$$

$$\frac{\partial G}{\partial N''_{23}} = 3U''_{10}-3U''_{11}-U''_{20}+U''_{23}-RT\left[-3\ln N''_{10}+3\ln N''_{11}+\ln N''_{20}-\ln N''_{23}\right.$$

$$\left.+3\ln {}^{\frac{2xn}{2-x}}C_o-3\ln {}^{\frac{2xn}{2-x}}C_1-\ln {}^{\frac{2xn}{2-x}}C_o+\ln {}^{\frac{2xn}{2-x}}C_3\right] = 0$$

etc. up to $\frac{\partial G}{\partial N''_{26}} = 0$

With a little rearranging these can be written

$$-U'_1+U'_2-RT\ln\left[\frac{N'_1}{N'_2}\,\frac{N-1}{2}\,\frac{{}^1{}_n}{n-1_n}\right] = -U'_2+U'_3-RT\ln\left[\frac{N'_2}{N'_3}\,\frac{N-2}{3}\,\frac{{}^1{}_n}{n-1_n}\right]$$

$$= -\,U'_3+U'_4-RT\ln\left[\frac{N'_3}{N'_4}\,\frac{N-3}{4}\,\frac{{}^1{}_n}{n-1_n}\right] = -U'_4+U'_5-RT\ln\left[\frac{N'_4}{N'_5}\,\frac{N-4}{5}\,\frac{{}^1{}_n}{n-1_n}\right]$$

$$= -\,U'_5+U'_6-RT\ln\left[\frac{N'_5}{N'_6}\,\frac{N-5}{6}\,\frac{{}^1{}_n}{n-1_n}\right]$$

$$= U''_{10}-U''_{11}-RT\ln\left[\frac{N''_{11}}{N''_{10}}\,\frac{1}{\frac{2xn}{2-x}}\right] = U''_{11}-U''_{12}-RT\ln\left[\frac{N''_{12}}{N''_{11}}\,\frac{2}{\frac{2xn}{2-x}-1}\right]$$

$$=U''_{12}-U''_{13}-RT\ln\left[\frac{N''_{13}}{N''_{12}}\,\frac{3}{\frac{2xn}{2-x}-2}\right] =U''_{13}-U''_{14}RT\ln\left[\frac{N''_{14}}{N''_{13}}\,\frac{4}{\frac{2xn}{2-x}-3}\right]$$

$$=U''_{14}-U''_{15}-RT\ln\left[\frac{N''_{15}}{N''_{14}}\,\frac{5}{\frac{2xn}{2-x}-4}\right] = U''_{15}-U''_{16}-RT\ln\left[\frac{N''_{16}}{N''_{15}}\,\frac{6}{\frac{2xn}{2-x}-5}\right]$$

$$=U''_{20}-U''_{21}-RT\ln\left[\frac{N''_{21}}{N''_{20}}\,\frac{1}{\frac{2xn}{2-x}}\right] = U''_{21}-U''_{22}-RT\ln\left[\frac{N''_{22}}{N''_{21}}\,\frac{2}{\frac{2xn}{2-x}-1}\right]$$

$$=U''_{22}-U''_{23}-RT\ln\left[\frac{N''_{23}}{N''_{22}}\,\frac{3}{\frac{2xn}{2-x}-2}\right] = U''_{23}-U''_{24}-RT\ln\left[\frac{N''_{24}}{N''_{23}}\,\frac{4}{\frac{2xn}{2-x}-3}\right]$$

$$=U_{24}''-U_{25}''-RT\ \ln\left[\frac{N_{25}''}{N_{24}''}\,\frac{5}{\frac{2xn}{2-x}}-4\right]=U_{25}''-U_{26}''-RT\ \ln\left[\frac{N_{26}''}{N_{25}''}\,\frac{6}{\frac{2xn}{2-x}}-5\right]$$

For purposes of computation each one of these is set equal to an arbitrary number A.

Also

$$-U_3+U_4-RT\ \ln\left[\frac{N_3}{N_4}\,\frac{S-3}{4}\,\frac{1_s}{S-1_s}\right]=-U_{20}''+U_{10}-RT\ \ln\left[\frac{N_{20}''}{N_{10}''}\,\frac{2}{\frac{s(1-x)}{2-x}}-1\right]$$

Each of these is set equal to B

Having assumed values for the U's and for S and N one of the ways which has been used to get numerical solutions is to take arbitrary values of A and B and from these equations to calculate the ratios of the N's. The absolute values are then calculated to make the silicons, sodiums and oxygens add up to the correct amount. The number of bonds from silica to oxygen then has to be the same as the number of bonds from oxygen to silicon so that

$$3N_3+4N_4 = (N_{10}''+N_{11}''+N_{12}+N_{13}''+N_{14}''+N_{15}''+N_{16}'')$$

$$+2(N_{20}''+N_{21}''+N_{22}''+N_{23}''+N_{24}''+N_{25}''+N_{26}'')$$

and the equivalent relation for the sodiums and oxygens is

$$N_1'+2N_2'+3N_3'+4N_4'+5N_5'+6N_6' = N_{11}''+2N_{12}''+3N_{13}''+4N_{14}''+5N_{15}'' + 6N_{16}''$$

$$+N_{21}''+2N_{22}''+3N_{23}''+4N_{24}''+5N_{25}''+6N_{26}''$$

If these are not satisfied A and B are adjusted to get a better fit and this is continued until all of the relations are satisfied. On an electronic computer such calculations can be programmed and ran with no great difficulty.

END OF DERIVATION

Even without any calculation there are some indications of how this approach may be useful. When the alkali is quite low there are very few sodiums on the sphere of influence of the oxygens. This means that the non-bridging oxygens may be attached to possibly only one sodium at the most and therefore the sodiums cannot themselves have as many non-bridging oxygens as they normally need to make the internal energy as low as possible. As soda is increased it becomes easier to form these low energy species. This results in a dimple in the free energy curve with the resulting possibility of immiscibility. Calculations in several binary silicates have shown that the free energy curves close to the silica side of the diagram do indeed show the shap that would give rise to immiscibility.

The inclusion of bridging oxygens with some sodiums attached is an important part of the treatment. When the bond between a four coordinated silica and a bridging oxygen breaks the energy barrier to be overcome is about 140 k cals per mole as will be seen from viscosity measurements discussed in the next section. If, however, the oxygen is already attached to one or more sodiums it will be easier to break. It is probably this more than anything else that allows modifying oxides like soda to lower the softening point as effectively as they do.

VISCOUS FLOW

As long as a shearing stress is applied to a glass there is a continuing change of shape. This implies a changing of neighbors and this would imply the breaking and reforming of bonds from 4-coordinated silicons to bridging oxygens. The calculations in the previous sections indicate that these processes are taking place in vitreous silica in the absence of stress. The only thing that stress does is to give some direction to the process. One way of expressing this, which leads to Newtonian flow would be as follows. For simplication it is assumed that the stress is a simple tension in the z direction leading to a strain pattern such that those bonds so oriented that they make an angle θ with the z direction have an elongation ξ_θ.

The assumption is then made that the energy barrier

U that must be overcome when a bond oriented in the θ direction breaks is decreased by an amount $\beta_1\xi_\theta$ when this bond is stretched to a state ε_θ and that the energy barrier to be overcome when a 3-coordinated silica and a non-bridging oxygen come together to form a new bond in direction θ is increased by an amount $\beta_2\varepsilon_\theta$.

This means that as long as the orientation of the Si-O bonds is random there are more bonds breaking than reforming for some values of θ and the other way around for these values of θ for which ε_θ is negative. At first glance this implies a contradiction since the excess of breaking ones reforming in some direction is going to disturb the randomness of orientation. In fact, it can be shown that it would disturb it so much that flow would stop. However, there are randomizing processes at work. When the bond from a 4-coordinated silica to a bridging oxygen breaks it changes the character of the other three bonds to the silica and they are essentially destroyed as far as their tendency to break goes. Similarly when a new bond is formed three other new ones are created. This essentially keeps the orientation random in spite of preferential breaking in some directions.

It can be shown that this leads to an expression for the viscosity of η of the form

$$\eta = \frac{C}{T} \exp \frac{U}{RT}$$

where the U is the barrier to be overcome in breaking a bond. Viscosity measurements place this around 140 k cals per mole. The constant C is made up of elastic constants, β_1 & β_2, and the volume created when a bond breaks.

THERMAL EXPANSION

It has been shown[9] that those vibrations in a network which are such that their frequency decreases as the network expands make a negative contribution to thermal expansion and those whose frequency increases as the network expands make a negative contribution. To the extent that one can associate modes of vibration with vibrations of individual atoms the transverse vibrations of bridging oxygens if they lie in a straight line between two silicons would lead to negative expansion thereby accounting for the very low expansion of vitreous silica and the complicated negative expansion pattern of some titania-silica glasses. If the oxygen is not in the line which is the same as saying the Si-O-Si angle is less than 180° there is still a contribution to negative expansion. One of the modes will be a rotation of the oxygen around the line of centers of the two silicons and this leads to negative expansion. The vibration parallel to the line of centers also leads to negative expansion but the vibration perpendicular to the line of centers this time leads to a positive expansion. Most of the structures, crystalline or vitreous, showing negative expansion have a high concentration of bridging oxygens or some very similar structural unit.

DERIVATION (May be omitted in first reading).

It is assumed that the glass has a number of modes of vibration which is three times the number of atoms present. Also the glass has a potential energy E_s which is a function of the volume of the glass and that this has a minimum where the volume V has a value V_o and close to Vo, E_s can be expressed as:

$$E_s = E_o + \gamma\ (V-V_o)^2$$

Since the compressibility β at absolute zero is given by

$$\left(\frac{d^2E}{dV^2}\right)_{V=V_o} = \frac{1}{V_o\beta} = 2\gamma$$

$$\gamma = \frac{1}{2\beta V_o}$$

To calculate the thermodynamic properties it is convenient to calculate a partition function Q which for this model is given by:

$$Q = \exp\left(\frac{-E_s}{RT}\right)\prod_i \frac{\exp\left(\frac{-h\nu_i}{2kT}\right)}{1-\exp\left(\frac{-h\nu_i}{kT}\right)}$$

where ν_i is the frequency of one of the modes of vibration and the product is carried out in all the modes of vibration.

$$\ln Q = \frac{-Es}{kT} - \sum_{o} \left[\frac{h\nu_i}{2kT} + \ln \left(1-\exp \frac{-h\nu_i}{kT}\right) \right]$$

The free energy F is given by

$$F = - KT\ln Q$$

$$F = E_s = \sum_{o} \left[\frac{h\nu_i}{2} + KT\ln \left(1-\exp \frac{-h\nu_i}{kT}\right) \right]$$

This is a function of V since E_s is a function of V and each frequency ν_i is also a function of V. The glass will take that volume at any constant temperature T which makes $\frac{dF}{dV} = 0$

$$\text{or} \quad \frac{dE_s}{dv} + \sum_{o} \left[\frac{h}{2} \frac{d\nu_i}{dt} + KT \frac{\frac{h}{kT} \frac{d\nu_i}{dt} \exp \frac{-h\nu_i}{kT}}{1-\exp \frac{-h\nu_i}{kT}} \right] = 0$$

$$2\gamma \; (V-V_o) + \sum \left[\frac{h}{2} + \frac{h}{\exp \left(\frac{h\nu_i}{kT}\right) -1} \right] = 0$$

$$V = V_o - \frac{1}{2\gamma} \sum h \left[\frac{1}{2} + \frac{1}{\exp \left(\frac{h\nu_i}{kT}\right) -1} \right] \frac{d\nu i}{dV}$$

$$\text{or } V = V_o - V_o \; \beta \; h \sum \left[\frac{1}{2} + \frac{1}{\exp \left(\frac{h\nu_i}{KT}\right) -1} \right] \frac{d\nu_i}{dV}$$

where $T = 0$

$$V = V_o \left[1 - \beta h \frac{1}{2} \sum \frac{d\nu_i}{dV} \right]$$

The zero point energy does not make any contribution to the change of volume with temperature but has a slight

For any given value of ν_i the expression

$$\frac{1}{\exp \frac{h\nu_i}{KT} - 1}$$

has a value of zero where T=0 and also a slope of zero. A plot of V against temperature has then zero slope at T=0.

The function

$$\frac{1}{\exp \frac{h\nu_i}{KT} - 1}$$

has an increasing slope as T increases from zero but gradually merges into a straight line with slope $\frac{k}{h\nu_i}$.

Ideally if the curve of thermal expansion were known to a sufficient precision it would be possible to work backwards from this curve to a curve giving the contribution to the change in volume of all of those modes having frequencies between ν and $\nu + d\nu$. However, with the of precision of most expansion measurements a number of different distributions will permit an accurate fit between calculated and experimental curves. About the best that can be done is to pick four values of ν or of $\frac{h\nu}{k}$ and piece the experimental curve together out of these. It is found that using $\frac{h\nu}{k} = 120^o$, 250°, 500°, and 1000° all of the complicated expansion curves in the TiO_2-SiO_2 system can be quite accurately fitted.

This suggests the possibility, although it has not yet

been tried, of assigning to each species of ion, in the sense that species has been used in the earlier sections, a coefficient for each of these values of $\frac{h\nu}{k}$ and attempting to put their contributions additionally together to predict expansion curves. This will probably not be possible until many more convincing calculations have been made of species populations in simple glasses and also until many more low temperature data are available in such glasses.

END OF DERIVATION

INTERNAL FRICTION, CONDUCTION AND DIFFUSION

One of the characteristics of a good glass forming ion will be a very strong preference for one particular number of neighbors, probably about four according to Zachariasen's rules. The other extreme will be an ion, such as an alkali, for which the energy differences for different numbers of oxygen neighbors are quite small. Such an ion will settle for a number of neighbors which might be considerably smaller than the number of oxygens which would geometrically fit around it and any process involving the breaking of a sodium-oxygen bond could rather easily take place.

When a stress, such as a tension on one direction is applied to a glass it could cause a breaking of those Na-O bonds nearly parallel to the tension and a reforming of those nearly perpendicular to it. Such a process would naturally give rise to the large internal friction peak in soda containing glasses. The activation energy is a little less than 20 k cals per mole.

The application of an electric field could in the same way cause the ultimate breaking of those Na-O bonds which prevent the migration of the sodiums in the direction of the field and the forming of new ones in the opposite direction. The controlling process is the breaking of the Na-O bonds and the activation energy for such ionic conductivity is the same order of magnitude as that for large internal friction peaks. The same can be said of diffusion of sodium ions.

There is a large amount of work to be done in expressing these ideas more formally and with detailed mathematical rigor but no essential difficulties are foreseen.

DENSITIES AND MOLAR VOLUMES

If it is assumed that in a soda-silica glass every sodium had within its sphere of influence enough non-bridging oxygens and every non-bridging oxygen had enough sodiums that each ion could pick out the pattern or distribution of patterns that would make the free energy a minimum , then it is easy to show mathematically that under these conditions the ratios of the number of sodium species and of the non-bridging oxygen species would be independent of the amount of soda present. Under these circumstances the molar volume of a glass whose formula is written $xNa_2O \cdot SiO_2$ would be a linear function of x and the slope would be the partial molar volume of Na_2O. The same would apply for potash-silica glasses. In Figure 2 such curves are plotted for soda-silica and potash silica glasses and the curves, with slight systematic deviations, are not far from being straight lines. The soda curve has a slope of about 18 cm^3 per mole and the potash curve about 32 cm^3 per mole. The sphere of influence

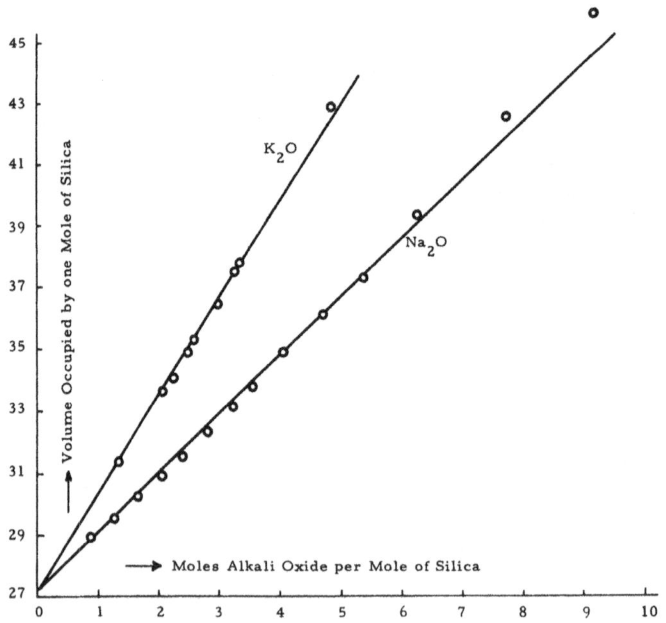

Figure 2. Partial molar volume vs. mole percent alkali oxide.

could very well be responsible for the slight deviation from strict linearity. At very low soda the sodiums must choose some bridging oxygens as neighbors and since these are a high volume species this lowers the molar volume. As the soda increases there comes a time when each sodium has within its reach all of the non-bridging oxygens it needs and a higher percentage of bridging oxygens with no sodiums occurs sending the volume above the line. Any such argument will need to be backed up with rigorous calculations. These have not yet been carried out.

There is one system where the molar volume curve is an excellent straight line even to values of x well over 1.0, in fact as far as the experimental data are available. This is the system $xB_2O_3\ SiO_2$.

Not only is the curve a straight line but the slope is, within experimental error, equal to the molar volume of B_2O_3 itself almost suggesting that the B_2O_3 and SiO_2 each maintain some kind of identity. It is not necessary, however, to make any use of such a microheterogeneity and the glasses in this system can be described in terms of the random network concept with the sphere of influence. First of all, pure B_2O_3 will be considered.

It is assumed that every oxygen is bridging between two borons, but that each boron is bonded to 2, 3 or 4 oxygens. The derivation from here is along the same general lines as those of the other systems considered, and may be omitted in a first reading. Some essentially new points are taken up at the end of the derivation and illustrate principles not yet presented.

DERIVATION (May be omitted in first reading).

For pure B_2O_3 it is assumed that one mole contains:
N_2' moles of boron attached to 2 oxygens and having energy U_2';
N_3' moles of boron attached to 3 oxygens and having energy U_3';
N_4' moles of boron attached to 4 oxygens and having energy U_4'.
All of the oxygens are bridging with energy U_{O2}''.

The N's are not all independent but satisfy the relations

$$N_2' + N_3' + N_4' = 2$$

$$2N_2' + 3N_3' + 4N_4' = 6$$

which leaves only one independent variable which may be

taken as N'_3 in which case

$$N'_2 = 1 - \frac{1}{2} N'_3$$

$$N'_4 = 1 - \frac{1}{2} N'_3$$

If the sphere of influence of a boron contains b oxygens then the sphere of influence of an oxygen contains $\frac{2b}{3}$ borons. Omitting some of the steps in setting up the entropy the free energy G comes out to be

$$G = N'_2U'_2 + N'_3U'_3 + N'_4U'_4 + 3U''_{O2}$$

$$-RT \left[2\ln 2 - N'_2\ln N'_2 - N_3\ln N'_3 - N'_4\ln N'_4 + N_2\ln {}^bC_2 + N'_3\ln {}^bC_3 \right.$$

$$\left. + N'_4\ln {}^bC_4 + 3\ln {}^{\frac{2b}{3}}C_2 + 2\left[3\ln 3 + (b-3)\ln(b-3) - b\ln b\right]\right]$$

$$\frac{\partial G}{\partial N'_3} = -\frac{1}{2}U'_2 + U'_3 - \frac{1}{2}U'_4 - RT\left[\frac{1}{2}\ln N'_2 - \ln N'_3 + \frac{1}{2}\ln N'_4\right.$$

$$\left. -\frac{1}{2}\ln {}^bC_2 + \ln {}^bC_3 - \frac{1}{2}\ln {}^bC_4\right] = 0$$

from which

$$\frac{N'_2}{N'_3}\ \frac{(b-2)}{3}\exp\left(\frac{U'_2 - U'_3}{RT}\right) = \frac{N'_3}{N'_4}\ \frac{b-3}{4}\exp\left(\frac{U'_3 - U'_4}{RT}\right)$$

This equation with the two relations already connecting N'_2 N'_3 and N'_4 is sufficient to determine all three.

In the glass $xB_2O_3 \cdot SiO_2$ it is assumed that each silicon has four oxygens and that each oxygen bridges between two silicons, a silicon and a boron, or two borons. There are N''_{20} moles of oxygens bridging between two silicons and having energy U''_{20}.

N''_{11} moles of oxygen bridging between a silicon and a boron and having energy U''_{11}.

N_{02}'' moles of oxygen bridging between two borons and having energy U_{22}''.

Then

$$N_2' + N_3' + N_4' = 2x$$

$$N_{20}'' + N_{11}'' + N_{02}'' = 2 + 3x$$

$$N_{11}'' + 2\,N_{02} = 2n_2' + 3N_3' + 4N_4'$$

$$2N_{20}'' + N_{11}'' = 4$$

If N_{11}'' and N_3' are taken as the independent variables

$$N_2' = x - \frac{1}{2}N_3' \qquad N_4' = x - \frac{1}{2}N_3'$$

$$N_{02}'' = 3x - \frac{1}{2}N_{11}'' \qquad N_{20}'' = 2 - \frac{1}{2}N_{11}''$$

If the sphere of influence of a silicon contains S oxygens the sphere of influence of an oxygen contains

$$\frac{S}{2 + 3x}$$

silicons. If the sphere of influence of a boron contains b oxygens the sphere of influence of an oxygen contains $\frac{2\ x\ b}{2 + 3x}$ borons.

The free energy G then comes out to be

$$G = N_2'U_2' + N_3'U_3' + N_4'U_4' + N_{20}''U_{20}'' + N_{11}''U_{11}'' + N_{02}''U_{02}''$$

$$-RT\left[-N_2'\ln N_2' - N_3'\ln N_3' - N_4'\ln N_4' - N_{20}''\ln N_{20}'' - N_{11}''\ln N_{11}'' - N_{02}''\ln N_{02}''\right.$$

$$\left. + 4\ln 4 - (s-4)\ln(s-4) - S\ln S + 2x\left[3\ln 3 + (b-3)\ln(b-3) - b\ln b\right]\right.$$

$$+N_2'\ln{}^bC_2+N_3'\ln{}^bC_3\ln{}^bC_3+N_4'\ln{}^bC_4+N_{20}''\ln^{\frac{S}{2+3x}}C_2+N_{11}''\ln^{\frac{S}{2+3x}}C_1$$

$$+N_{02}''\ln^{\frac{S}{2+3x}}C_o\ +N_{20}''\ln^{\frac{2xb}{2+3x}}C_o + N_{02}''\ln^{\frac{2xb}{2+3x}}C_2 + N_{11}''\ln^{\frac{2xb}{2+3x}}C_1\Bigg]$$

from which

$$\frac{\partial G}{\partial N_3'} = -\frac{1}{2}U_2'+U_3'-\frac{1}{2}U_4' - RT\Bigg[\ \frac{1}{2}\ln N_2' - \ln N_3'+\frac{1}{2}\ln N_4'$$

$$-\frac{1}{2}\ln{}^bC_2+\ln{}^bC_3 - \frac{1}{2}\ln{}^bC_4\ \Bigg] = 0$$

$$\frac{\partial G}{\partial N_{11}''} = -\frac{1}{2}U_{20}''+U_{11}'' - \frac{1}{2}U_{02}'' - RT\Bigg[\ \frac{1}{2}\ln N_{20}''-\ln N_{11}''+\frac{1}{2}\ln N_{02}''$$

$$-\frac{1}{2}\ln^{\frac{S}{2+3x}}C_2 + \ln^{\frac{S}{2+3x}}C_1 - \frac{1}{2}\ln^{\frac{S}{2+3x}}C_o - \frac{1}{2}\ln^{\frac{2xb}{2+3x}}C_o$$

$$+\ln^{\frac{2xb}{2+3x}}C_1 - \frac{1}{2}\ln^{\frac{2xb}{2+3x}}C_2\ \Bigg] = 0$$

END OF DERIVATION

There have been derived two equations for the two unknowns N_3' and N_{11}''. The first of these is exactly the same equation as for pure B_2O_3 which means that $N_2' : N_3' : N_4'$ is the same. This depends on b, the number of oxygens in the sphere of influence of a boron being independent of composition in this system which has not been shown but is probably not far from the truth. All of this means that the borons occupy exactly the same volume per mole in the binary glass as they do in pure B_2O_3. This volume will be denoted by V_B.

Then if V_{si} is the volume occupied by the silicon atoms, V_{20}'' the volume of an oxygen bridging between two silicas, V_{11}'', the volume of an oxygen bridging between a silicon and a boron, and V_{02}'' the volume of an oxygen bridging between two borons the molar volume V_m will be given by

$$V_m = 2xV_B + V_{si} + N_{20}''V_{20}'' + N_{11}''V_{11}'' + N_{02}''V_{02}''$$

At this point a new assumption is made and this is that V_{11}'' is the arithmetic mean between V_{20}'' and V_{02}'' or

$$V_{11}'' = \frac{1}{2}(V_{20}'' + V_{02}'')$$

Then $V_m = 2xV_B + V_{si} + N_{20}''V_{20}'' + \frac{1}{2}N_{11}''(V_{20}''+V_{02}'')+N_{02}''V_{02}$

$$= 2xV_B + V_{si} + V_{20}''(N_{20}''+\frac{1}{2}N_{11}'') + V_{02}''(N_{02}'' + \frac{1}{2}N_{11}'')$$

But it has been shown that

$$N_{20}'' + \frac{1}{2}N_{11}'' = 2$$

$$N_{02}'' + \frac{1}{2}N_{11}'' = 3x$$

Hence $V_m = 2xV_B + V_{si} + 2V_{20}'' + 3xV_{02}''$

$$= V_{si} + 2V_{20}'' + 2x(2V_B + 3V_{02}'')$$

$$= V_{SiO_2} + x(V_{B_2O_3})$$

which is the linear relationship shown by the measurements.

SURFACE TENSION

For an ion lying on the surface, the sphere of influence is cut in two. It may then be impossible for a silicon to be attached to four oxygens and it will be connected to two or possibly three oxygens. This would result in a high concentration of high energy species at or near the surface resulting in an energy per unit area in the surface layer. The method of applying the sphere of influence model to this problem is as follows.

Consider first the case of a single oxide such as SiO_2 or B_2O_3. The assumption will be made, rightly or wrongly, that the composition of each very thin layer close to the surface can be expressed by the formula SiO_2. The correctness of this assumption is not an essential part of the theory but it makes the first approach much simpler. Let there be N moles of the liquid in a cylindrical crucible whose area of cross section is A. The crucible is of such a material that the angle of contact with the liquid is 90°.

The skin layer, of unspecified thickness will be treated as a surface "phase" as it is commonly handled in the physical chemistry textbooks. Then this surface phase will be broken up into thin layers of the order of magnitude of 0.1A in thickness for computer calculations. However, the single surface phase is a good way to start the problem. The surface phase will be formed by taking n moles per cm^2 out of the bulk and spreading it on the surface.

Let F^b be the free energy per mole of the bulk material and F^σ be the free energy per mole of the surface phase. The forming of the surface phase will remove nA moles from the bulk thereby changing the free energy of the bulk material to

$$(N - nA)F^b$$

The free energy of the whole system is

$$(N - nA)F^b + nAF^\sigma$$

and n has to be chosen so that this is a minimum.

By the formation of the surface the free energy has been increased by an amount

$$A(nF^\sigma - nF^b)$$

The quantity $(nF^\sigma - nF^b)$ is what is commonly called the surface free energy(ergs per cm^3) and can be shown to be numerically equal to the surface tension in dynes per cm. The bulk free energy F^b is a constant which can be computed by applying the sphere of influence theory to the bulk material without any consideration of surface structure. The quantity F^σ is very much a function of n. To minimize the free energy of the whole system

$$\frac{d}{dn}\left[(N-nA)F^b + nAF^\sigma\right] = 0$$

or $\frac{d(nF^\sigma)}{dn} - F^b = 0$

This, of course, is a way of saying that the bulk material is in thermodynamic equilibrium with the surface layer in the sense that an infinitesimal amount of material taken from the bulk and put into the surface will not change the free energy of the whole system.

Instead of using a single layer of unspecified thickness the whole surface region is now broken up into thin layers as mentioned before and each layer will be so constructed that this same relation holds with the same value of F^b in the relation. Each layer will be in equilibrium with the bulk liquid and also with the vapor if it is a closed system. The existence of the surface layer does not in any way modify the relations between the bulk liquid phase and the vapor phase. This whole treatment depends on the material being a liquid where each layer is free to adjust its structure to satisfy the above relation. This is in general not true of a crystal face.

To start the program some assumptions of somewhat questionable validity will be invoked. The first is that the energy of an ion bonded to some number of ions of the opposite kind will be the same close to the surface as in the bulk interior. It is possible that the fact that an ion is in or close to the surface might necessitate different geometrical patterns from what would exist in the bulk. Such differences are for the moment ignored.

It is assumed that the bonds going out from the silicons in any layer exactly match in number the bonds going out from the oxygens. Such a balance must be maintained for the whole surface region, but it is not immediately obvious that it holds for each individual layer.

The volumes assigned to any ion in the bulk is obtained by erecting planes which perpendicularly bisect the straight lines from the ion to each of the neighbors to which it is bonded. If an ion is within a bond length from the surface, the volume is modified by including among its neighbors an image ion which is the image of the ion in the plane of the surface. This creates a polyhedron face in the surface plane. A large part of the polyhedron around an ion in any layer lies in neighboring layers and the layer itself contains silicons from the polyhedron of ions in neighboring layers. It is assumed that these balance. This would certainly be true in the homogeneous bulk material. Close to the surface the density gradient may modify this sharing but this will be ignored.

The doubtfulness of some of these assumptions is tolerated because at the moment information probably does not exist to do very much better. The only way such information can be obtained is by how much these assumptions have to be modified to fit experimental measurements. The plausibility of this approach remains to be tested.

Accepting these assumptions there are then certain relations that exist connecting the concentration of species in each layer. Assume, in the case of silica, that in any layer of thickness Δ there are.

M_2 moles per unit volume of silicons joined to two oxygens having energy U_2 and volume V_2 cm^3 per mole.

M_3 moles per unit volume of silicons joined to three oxygens having energy U_3 and volume V_3 cm^3 per mole

M_4 moles per unit volume of silica joined to four oxygens having energy U_4 and volume V_4 cm^3 per mole.

M_1'' moles per unit volume of oxygens joined to one silica having energy U_1'' and volume V_1'' cm^3 per mole.

M_2'' moles per unit volume of oxygens joined to two silica having energy U_2'' and volume V_2'' cm^3 per mole.

Then

$$2(M_2 + M_3 + M_4) = M_1'' + M_2''$$

$$2M_2 + 3M_3 + 4M_4 = M_1'' + 2M_2''$$

$$M_2V_2 + M_3V_3 + M_4V_4 + M_1''V_1'' + M_2''V_2'' = 1$$

This leaves only two independent variables. The method of attack will be as follows. It is assumed that all of the liquid of the bulk structure on one side of a plane is suddenly removed. Then the outside layer of thickness Δ is removed. The "sphere" of influence of ions on this outside layer is now a hemisphere. By maintaining the bulk structure in each of the layers except the surface layer of thickness Δ, the number of oxygens in the sphere of influence of a silicon in the surface layer and the number of silicons in the sphere of influence of an oxygen in the surface layer are computed. Using these numbers, values of M_2, M_3, M_4, M_1'' and M_2'' are calculated so that

$$nF^{\sigma} - nF^{b}$$

is a minimum.

The quantity nF^{σ} will simply be the free energy of the surface layer per unit area. The internal energy will be

$$M_2U_2 + M_3U_3 + M_4U_4 + M_1''U_1'' + M_2''U_2''$$

and the entropy per unit area of the layer of thickness Δ, will be calculated essentially as was done in the bulk calculations except that the spheres of influence will contain fewer neighbors.

The quantity N in the term nF^{b} will be given by

$$N = (M_2 + M_3 + M_4) = 1/2\ (M_1'' + M_2'')$$

F^{b} will already have been computed for the bulk material. In this way a new surface layer will be put in with minimum free energy. Keeping this new surface layer, the second layer of thickness, Δ will be removed and having calculated the neighbors for each silica and oxygen a new layer will be put in with minimum free energy. This will be continued until the composition of the new layer matches the bulk. The process will be repeated until further repetition does not change the structure. The values of $nF^{\sigma} - nF^{b}$ will be calculated for each layer and summed to give the surface free energy or surface tension.

For binary glasses the layers close to the surface will not necessarily have the same composition as the bulk and the quantity to be minimized will involve the partial molar free energies of the two components on the bulk glass but otherwise the calculations will parallel the corresponding calculations for structure of the bulk phase of the corresponding binary composition.

REFERENCES

1. Warren, B. E. & Biscoe, J., J. Am. Ceram. Soc., 21, 29, 1938.
2. Zachariesen, W. H., J. Am. Chem. Soc., 54, 3841, 1932.
3. Warren, B. E., Krutter, H., Morningstar, O., J. Am. Ceram. Soc., 19, 202, 1936.
4. Warren, B. E., X-ray Diffraction, Addison-Wesley, 1969.
5. Tilton, L. W., J. Res. Nat. Bull. Stds., 59, 139, 1957.
6. Eckstein, B., Glas. Ber. 36, 323, 1963.
7. Smyth, H. T., Leipold, M. H., Tech. Report No. 32-1042, Jet Propulsion Lab., April 15, 1967.
8. Bacon, J. F., Hasapis, A. A., Wholley, J. W., Phys. Chem. Glass 1, 90, 1960.
9. Smyth, H. T., J. Am. Ceram. Soc., 38, 140, 1955.

The Infrared and Raman Spectra of Glasses

Robert A. Condrate Sr.

College of Ceramics
Alfred University
Alfred, N.Y. 14802

INTRODUCTION

Over the last several decades, many investigators have found that the analysis of vibrational spectra of glasses can be applied to the solution of both industrial and basic research problems. In an applied industrial study, the vibrational spectra can be used to identify glasses or specific groups which are present in glasses. However, in a theoretical study, information can be obtained concerning the fundamental structural nature of glasses on which their external properties are dependent.

The structural geometry in crystals is clearly defined and can be described by the symmetry properties and the geometrical parameters of the unit cell. For these types of solids, x-ray crystallography can be used to completely determine the positions of all atoms in the crystalline lattices. However, the lack of long-range order in glasses prevents the use of such methods in studying their structures. Techniques have been developed in which one determines the radial distribution functions of electron densities about specific atoms. The results of this type of investigation are somewhat limited because the symmetry and extent of short range ordering can not be adequately resolved. Analysis of vibrational spectra, however, can provide this type of structural information when a thorough theoretical analysis of the data is carried out. Another important point to consider is that diffraction methods cannot distinguish the presence of minor concentrations of defect centers whose geometries may be hidden in the range of variation of the bond distances and angles found for the bulk of the normal structural units. In contrast, vibrational spectra can be readily used to identify the presence of defect groups within a glass. For example, R. Hanna[1] has inferred the presence of three-coordinated silicon groups in vitreous silica by means of infrared spectra. Both infrared and Raman spectroscopy have also been used to identify low-concentration impurities such as water, hydroxyl ion, carbonate ions, etc. in glass. The identification of these impurities in glasses is one of the most significant contributions of vibrational spectroscopy to the study of glasses.

In this review, we will discuss the theoretical origin of vibrational spectra, the differences between infrared and Raman spectra, and the effect of environment on the vibrational modes of specific molecular groups. The use of infrared and Raman spectra for the identification of specific groups within glasses will be illustrated with the spectra of different borate and silicate glasses. Finally, several special applications of vibrational spectroscopy to glass problems will be considered.

THEORETICAL ORIGINS OF VIBRATIONAL SPECTRA

Vibrational spectra originated from the vibrations of nuclei constituting a molecular group, and are observed as infrared and Raman spectra in the wave number region between approximately 10^4 cm $^{-1}$ to 10 cm $^{-1}$. The wave number values of vibrational modes observed in spectra are determined by the masses of the constituent atoms, the geometry of the atomic arrangement and the interatomic forces within the group of atoms that is being investigated.

During a given vibration, the atoms will behave as though classical springs are connecting them together. Each group of atoms has its own specific set of characteristic vibrational frequencies. These complex motions of atoms can be resolved into a set of vibrational modes in which all atoms move harmonically in phase with each other. To describe these motions, one must introduce the concept of the normal mode. This type of coordinate change can be simply described by specific linear combinations of changes in bond lengths and interbond angles. There are four types of coordinate change; bond stretching, in-plane angle deformation, torsion and out-of-plane bending. Figure 1 illustrates the four normal modes for the carbon dioxide molecule. As can be seen there are the two C = O stretching modes (the symmetric and the antisymmetric modes) and the two degenerate O = C = O bending modes. It may be noted that specific linear combinations of coordinates belong to specific characteristic wave numbers. This inter-relationship of specific groups of atoms vibrating at specific wave number values forms the basis of group frequencies, which is important in the interpretation of vibrational spectra for many materials.

Although vibrational spectra are observed experimentally as infrared or Raman spectra, the physical origins of these two types of spectra are different. Infrared spectra originate from the vibrational energy level transitions of a molecule in the electronic ground state. Thus, if a molecule is placed in an electromagnetic field, energy transfer from the field to the molecule occurs when Bohr's frequency relation is satisfied. This is strictly an absorption process and infrared spectra are usually observed as absorption spectra. For absorption to occur, the probability of transition between the two stationary states

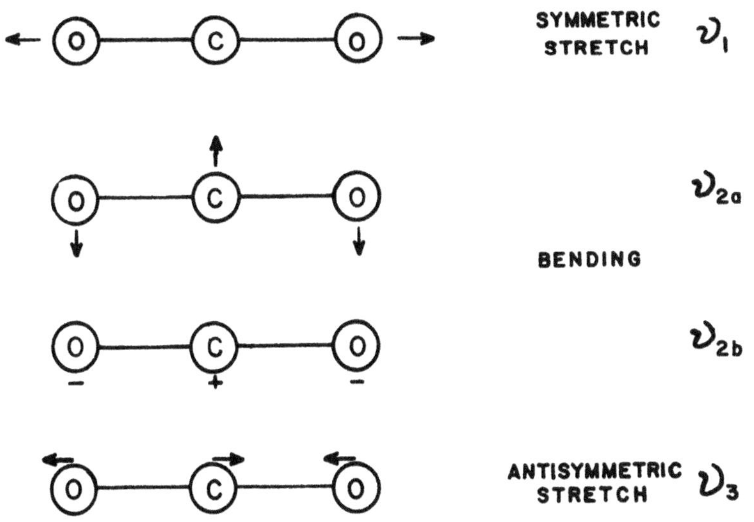

Figure 1. The normal modes of carbon dioxide.

involved must be non-zero. This corresponds to the classical condition that the vibrational mode must be associated with an oscillating dipole moment. Absorption of energy occurs if the frequency of this dipole moment is the same as that of the electromagnetic field with which it interacts. Quantum mechanically, one must have a non-zero transition moment.

$$\int \psi_{V'} \, \mu \, \psi_{V''} \, dT \qquad (1)$$

in which μ is the dipole operator, and $\psi_{V'}$ and $\psi_{V''}$ are wave functions of the ground and excited vibrational states. Looking at the normal modes of carbon dioxide, one can see that the antisymmetric C = O stretching mode and the two degenerate bending modes are infrared active because they involve changes in dipole moment during the vibrational modes. The symmetric C = O stretching mode is inactive in the infrared spectra because no dipole moment change occurs during the stretching mode.

In the Raman effect, photons interact with the material's molecules and are scattered with higher or lower energies than the incident exciting photons. The energy of the scattered photon changes by quantized amounts from that of the incident photon, corresponding to energy differences of the energy levels of the molecule. In the infrared spectra, absolute wave numbers are measured whereas in the Raman spectra, shifts in wave numbers are measured. Both infrared absorption and the Raman effect depend upon the same energy levels but their selection rules,(i.e., whether transitions between these energy levels are allowed or not) and band intensities differ. Therefore, the data obtained by these two different methods are not necessarily the same and are complementary in nature.

When a light quantum of frequency ν_0, and energy $h\nu_0$, collides with a molecule in a stationary state, E' or E" in Figure 2, the molecules energy is increased to $h\nu_0 + E''$ or $h\nu_0 + E'$, where h is Plank's constant. If the molecule does not possess a stationary state with this energy, the quantum is scattered in any direction and the molecule either returns to its original state or to a different state. If it returns to its original state, no energy is either taken from or given to the light quantum and its frequency remains unaltered. This is called Rayleigh scattering. However, if the molecule returns to a

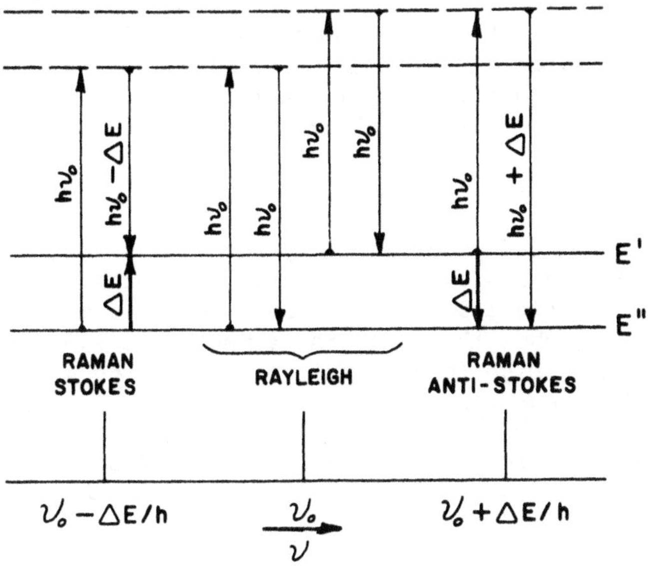

Figure 2. Quantum theory of Rayleigh and Raman scattering.

different stationary state, it has either taken energy from or given energy to the light quantum, with the change in frequency given by:

$$\frac{E' - E''}{h} = \frac{\Delta E}{h} \quad \text{or} \quad \frac{E'' - E'}{h} = -\frac{\Delta E}{h} \qquad (2)$$

depending on the initial state being E' or E". Thus, the frequencies $\nu_o + \Delta E/h$ and $\nu_o - \Delta E/h$ appear in the scattered light in addition to the unaltered frequency ν_o. This is the so called Raman effect. The Raman lines displaced to lower frequencies are called Stokes lines, and the lines displaced to higher frequencies are called anti-Stokes lines. For solids, both vibrational and electronic transitions can yield Raman lines. The Stokes lines for these transitions are found to be more intense than the anti-Stokes lines, since the concentration of molecules in the various states follow Boltzmann's distribution law. At the usual temperatures of observation, the populations of excited levels are less than that of the ground level.

Not all fundamental vibrational modes are observed

in the Raman spectra. In order that a band be Raman active, a change in polarizability must occur during the vibrational transition. Quantum mechanically, one must have a non-zero polarizability term,

$$\int \psi_{v'} \, \alpha \, \psi_{v''} \, d\tau \qquad (3)$$

in which α is the polarizability operator and $\psi_{v'}$ and $\psi_{v''}$ are the ground and excited vibrational states wave function. A quick investigation of the normal modes of carbon dioxide will show that only the symmetric C = O stretching mode can give rise to a polarizability change and be active in the Raman spectra.

THE EFFECT OF SYMMETRY AND SOLID-STATE ENVIRONMENT ON VIBRATIONAL SPECTRA

(A) Symmetry Properties of Isolated Molecular Groups or Ions.

Symmetry plays an important role in the structure of molecules: some structures are highly symmetric, some possess little symmetry, and many have no symmetry at all. The way in which vibrational modes of molecules behave with respect to these symmetry properties will decide their activity in the Raman and infrared spectra. From the standpoint of vibrational spectroscopy, molecules can be conveniently classified according to the following symmetry elements:

1) The identity, I, a trivial symmetry element which corresponds to leaving the molecule in its original configuration and is introduced for purposes of mathematical group theory.

2) The center of inversion designated by i.

3) The p-fold axis of rotation designated by C_p, for which rotation through $2\pi/p$ produces an orientation of the molecule which is indistinguishable from the original orientation.

4) Planes of symmetry designated by σ followed by subscripts, h, v, or d to indicate whether the plane is a horizontal, vertical or diagonal plane.

5) A p-fold rotation-reflection axis designated by S_p for which a rotation through $2\pi/p$ followed by a reflection through a plane perpendicular to the axis produces an orientation of the molecule indistinguishable from the original orientation.

Dipole moment and polarizability changes for the various vibrational modes can only occur if the modes vibrate in specific directions with respect to the symmetry elements present for the given group of atoms. For example, the normal modes of carbon dioxide, which change the sign of their direction of oscillation by inversion through a center of symmetry will be infrared active. For each of these cases, one of the dipole moment changes in either the X-, Y- or Z- directions which change unsymmetrically with respect to the center of symmetry will be present, causing the mode to be infrared active. The presence of a dipole moment for a vibration that oscillates unsymmetrically with respect to a center of symmetry is illustrated by the antisymmetric C = O stretching mode:

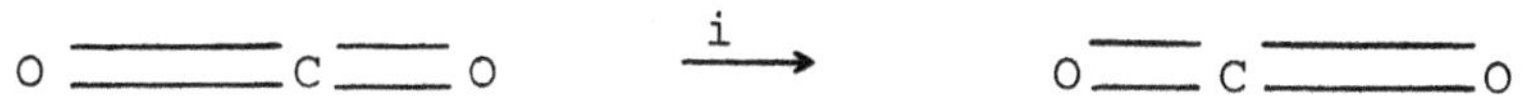

For this mode, a dipole moment change occurs along the chemical bonds. In contrast, only modes that behave symmetrically with respect to a center of inversion will possess polorizability changes and will be Raman active. The symmetrical C = O stretching mode is an example of this type of vibration. In general, for any polyatomic molecule possessing a center of symmetry , a similar set of rules hold and are referred to as "the mutual exclusion principle".

The activity of the stretching and bending modes of carbon dioxide in the infrared and Raman spectra gives some indication of the importance of symmetry with respect to the selection rules for vibrational modes. In the case of more complex polyatomic molecules, it is difficult to determine the different possible normal modes and their spectral activities in a simple way. The vibrational spectra for these molecules must be analyzed systematically on the basis of symmetry using group theory. Since we are interested only in the spectral predictions for different arrangements of atoms, we will delete any discussion concerning the methods used by group theory, and only discuss the results of the theoretical analysis. The reader will find more thorough discussions of the group-theoretical methods elsewhere.[2-4]

Group-theoretical concepts can be applied to a tetrahedral pentatomic molecule or ion of the form XY_4, which is assumed to be isolated from its environment. A tetrahedral molecule possesses the symmetry elements shown in Figure 3. The associated symmetry operations comprise the point group T_d. For the XY_4 -group in this point group, group-theoretical analysis reveals that there are only four bands that will be observed in the vibrational spectra: a symmetrical stretching mode ($\tilde{\nu}_1$), a doubly degenerate bending mode ($\tilde{\nu}_2$), and a triply degenerate stretching mode ($\tilde{\nu}_3$) and a triply degenerate bending mode ($\tilde{\nu}_4$). Whereas all of these modes are Raman active, only $\tilde{\nu}_3$ and $\tilde{\nu}_4$ are infrared active. XO_4 - ions is solution behave as isolated tetrahedral groups. Therefore, the predictions made by group theory should be observed in their vibrational spectra. Table I illustrates the vibrational

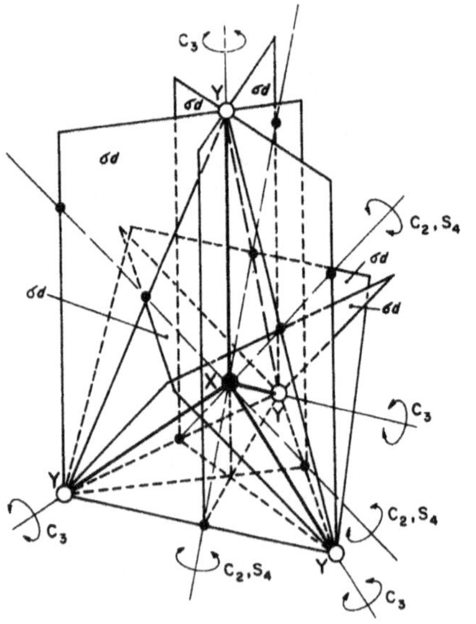

Figure 3. Symmetry elements of an XY_4 molecule having tetrahedral symmetry.

Table I. The Raman Fundamentals of Several XO_4 -Ions in Aqueous Solution (cm^{-1})[5]

Molecule	$\tilde{\nu}_1$	$\tilde{\nu}_2$	$\tilde{\nu}_3$	$\tilde{\nu}_4$
$[SiO_4]^{2-}$	800	500	1050	625
$[PO_4]^{3-}$	970	358	1080	500
$[SO_4]^{2-}$	983	450	1105	611
$[ClO_4]^{-}$	935	460	1050-1170	630
$[SeO_4]^{2-}$	833	335	875	432

bands observed in the Raman spectra for a series of XO_4 -ions in aqueous solution. The observed fundamentals totally agree with the predictions. Two higher wavenumber bands were observed in the Raman spectra which could be assigned to stretching modes and two lower wavenumber bands were observed which could be assigned to bending modes. The infrared fundamentals were not observed because water strongly absorbs infrared light in this region of the spectra.

(B) Site Groups in Crystals

In the previous section, we have assumed that the vibrating molecules were isolated from interacting strongly with other molecules or ions in their vicinity. In crystals, however, molecules are systematically arranged in close proximity to each other so that intermolecular interactions will have a significant effect on the vibrational modes. Two methods are available to take into account the effects of the crystalline lattice: site-groups and factor-group analysis.

In the site group approximation, one assumes that the coupling between the vibrations of similar molecules in the lattice are small and that the atoms surrounding a given molecule yield an environment of fixed symmetry. The new set of symmetry elements which these molecules possess is called a site group and is dependent upon the local symmetry of the crystalline environment around the center of gravity of the molecules. An acceptable site group must be a sub-group of the space group of the lattice and the molecule's point group. Using Halford's tables, the site group of a molecule can be found immediately if the space group of the crystal, the point group of the molecule and the number of molecules in the unit cell are known. Usually, the site symmetry is lower than the molecular symmetry in the free state. The selection rules may be relaxed in the crystalline state so that bands forbidden in the free state are observed, and the degenerate vibrations may split up in the crystalline state.

The application of site symmetry is nicely illustrated by CO_3^{2-} - ion in the aragonite lattice. If the CO_3^{2-} -ion behaves as an isolated ion, it would possess D_{3h} -symmetry and 4 fundamental bands would be observed, of which one would only be Raman active, one would be only infrared active and two would be both Raman and infrared active. Table II lists the observed fundamentals in the infrared and Raman spectra. Contrary to the isolated-ion approximation, all bands are observed in the infrared and Raman spectra. If we consider the lattice environment about the CO_3^{2-} -ion in the aragonite lattice, the site group for this ion will be C_s instead of D_{3h}. With this symmetry group, we get a new set of selection rules for the fundamental modes. In this case, we find that all modes will now be both infrared and Raman active. Furthermore, $\tilde{\nu}_3$ and $\tilde{\nu}_4$ should split into two bands. These predictions are consistent with the observed spectra. Slight differences in wave number values may be observed between the Raman and infrared bands. These differences arise from the neglect of lattice interactions that are not considered by the site group model.

Table II. The Fundamental Wave Numbers of the Carbonate Ion in the Aragonite Lattice (cm^{-1})[6]

	$\tilde{\nu}_1$	$\tilde{\nu}_2$	$\tilde{\nu}_3$	$\tilde{\nu}_4$
Infrared	1080	866	1492 1504	706 711
Raman	1084	852	1460	704

(C) Factor Groups in Crystals

The site-group analysis has proved to be a valuable and easily applied technique for the determination of the vibrational behavior of many solids. However, in some cases this method fails to give a satisfactory explanation of the observed spectra. The assumption of only static-field effects and complete neglect of possible intermolecular interactions is the cause of the occasional breakdown of the site-group method.

The most effective method of treating the vibrational behavior of crystals is the factor-group method.[7] The factor group is defined as the group formed by the symmetry elements existing in the smallest possible unit cell (Bravais cell). A factor group is isomorphous with one of the thirty-two crystallographic point groups. Since the wavelength of infrared radiation is large with respect to the unit cell, the vibrations of several unit cells are effectively moving in phase. Thus, of all the possible vibrations, only vibrations of one unit cell appear as distinct modes.

One striking example where site group analysis is insufficient to interpret vibrational spectra exists in the case of Fluorapatite, $Ca_{10}(PO_4)_6F_2$.[8] If site group analysis was adequate, the site group of the phosphate ions would be C_s and nine fundamentals should be observed coincidentally in the infrared and Raman spectra. Table III lists the fundamentals observed. There are more than the predicted number of fundamentals in the Raman spectra. Also, coincidence between the infrared and Raman wavenumbers does not occur. These results are contrary to the site group predictions. The site group, as we have said previously, assumes a static environment about the phosphate ion. If we assume that the 6 phosphate ions in the unit cell couple their vibrational motions together, we can consider that these ions interact by the factor group, C_{6h}. The number of predicted bands in the infrared and Raman spectra, and noncoincidence of bands in the Raman and infrared spectra from this factor group model agree with the observed results.[8]

(D) Localized Models in Glass Networks

The basic problem with glasses is that the long-range order or symmetry of a crystalline lattice does not exist.

Table III. Fundamental Wave Numbers for the Phosphate Ion of Fluorapatite (cm^{-1}).[8]

	Infrared	Raman
ν_1	965	965
ν_2	473	455
		447
		431
ν_3	1096	1080
	1076	1061
	1046	1051
		1040
		1032
ν_4	602	615
	575	605
	569	590
		580

One way to minimize this problem is to assume that short-range symmetry plays the most significant role in influencing the vibrational spectra and that the effective vibrating group of atoms can be localized to a fundamental structural unit. This type of approximation is the glass analogue of the site group approximation in crystals. This approach has been carried out by many investigations of silicate and borate glasses. SiO_4 and BO_3 units were used to analyze spectra of these glasses. The simplest applied models were models in which the basic units were considered isolated and unattached to the rest of the network.[9,10] However, the free unit model does not adequately approximate the true unit because the influence of the unit connecting with the rest of the lattice is neglected. Wadua and Balloomal,[11] and Su et. al.[12] have minimized this problem by using the bound or constrained model for vitreous silica shown in Figure 4. The effect of the lattice on the SiO_4 - unit was assumed to be rigid and was accounted for by fixed walls or infinitely-weighted masses attached to the extremeties of the tetrahedral unit. With this model, the high frequency modes will be adequately interpreted. However, for infrared modes such as those at 95 and 200 cm^{-1} can not be explained

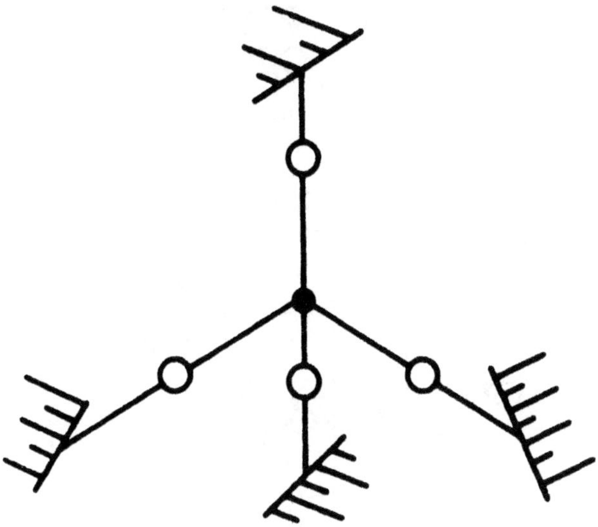

Figure 4. Constrained SiO_4 unit.

directly by this bound model. These modes involve coupled motion of several attached basic units.

(E) Lattice Models in Glass Networks

Recently, Bock and Su[13] have introduced a new structural model to be used for vibrational analysis as an improvement over the simple bound localized model. They assumed that the average vibrating structural unit was the structural unit of α - quartz. (Figure 5). This type unit can be considered as the glass analogue of a factor group unit in crystals. The model for vitreous silica appears to be reasonable on the basis of Zarzycki and Prod' Homme's X-ray analysis of vitreous silica.[14] The fundamental wavenumbers calculated by theoretical analysis with the above structural unit agree well with the observed values.[13] The major drawback with this so-called lattice model is in the choice of structural unit. For example, quartz-like, trydimite-like or cristobalite-like structural units can be chosen for vitreous silica. The final choice is decided by the observation of which structural unit will give the best calculated fit with observed vibrational wavenumbers.

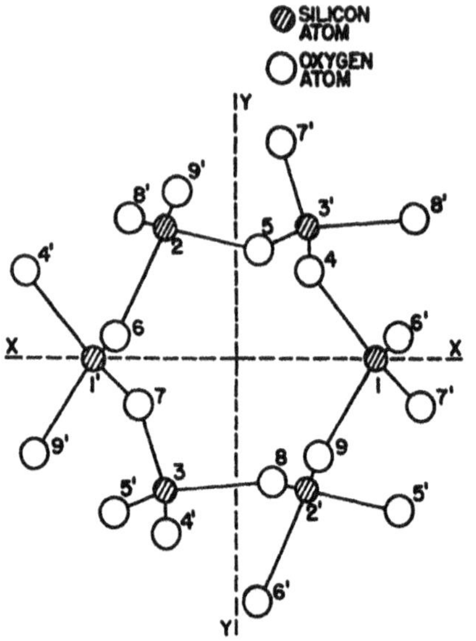

Figure 5. Structural unit for lattice model of vitreous silica.[13]

(F) Statistical Models in Glass Networks

One important development in the vibrational analyses of glasses has been the application of the statistical model of Bell and Dean[15] to the infrared spectra of vitreous silica. They constructed several models of vitreous silica consisting of about 500 atoms using a random arrangement of SiO_4 - tetrahedra and calculated the vibrational spectra. The wavenumber distributions showed distinct peaks which interrelate with the observed infrared peaks. However, as the authors related, a strict comparison of this wavenumber distribution cannot be made without further analysis of the dipole nature of the vibrational modes. Future improvement of this statistical approach will probably lead to the best theoretical method for analysis of the vibrational modes of glass networks.

THE EMPIRICAL INTERPRETATION OF THE VIBRATIONAL SPECTRA OF SEVERAL GLASSES

Empirical analysis of the vibrational spectra of glasses can give in many instances useful information concerning the arrangement of atoms in glasses. The empirical method involves investigating a large series of similar materials to assign certain vibrational bands to specific groups of atoms within the materials. Although this approach is simple and straight forward, care has to be taken because in some instances ambiguous results may be obtained even in the high wavenumber region. The principle of the empirical method rests on the idea of a group vibration. That is, the concept that some of the vibrations of a particular group of atoms in a molecule or lattice are relatively independent of motions of the rest of the atoms. In this section, we will consider several vibrational modes in selective glasses, which appear to behave as group vibrations. A rather thorough investigation of the spectra of a series of silicate, borate and borosilicate glasses has been carried out by Jellyman and Procter.[16]

(A) Silicate Glasses

The infrared reflection spectra of vitreous silica along with those of a series of sodium silicate glasses are shown in Figure 6. The peak at 1125 cm^{-1} for vitreous silica is assigned to the Si-O stretching mode associated

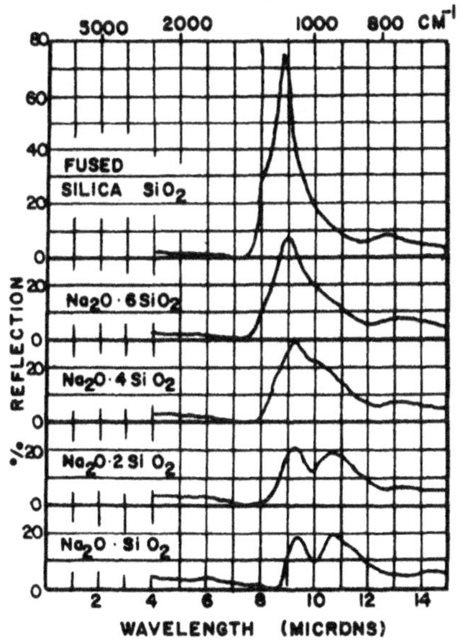

Figure 6. The infrared reflection spectra of various sodium silicate glasses.[16]

with Si-O-Si linkages. This mode moves to lower wavenumbers*and decreases in intensity as the concentration of Na_2O is increased. This band also becomes broader, indicating the addition of Na_2O increases the disorder within the structure. A secondary band occurs in the region 740-800 cm^{-1} which can also be assigned to a linked Si-O stretching mode. A third major peak at ca. 950 cm^{-1} increases in intensity as the concentration of Na_2O increases. This band can be assigned to stretching modes of $Si-O^-$ and $^-O-Si-O^-$ groups. Substitution of potassium, lithium and alkaline earth oxides for sodium oxide in silicate glasses produce similar effects on the spectra. Differences in these reflection spectra are mainly related to the intensities of various vibrational bands.

(B) Borate Glasses

The infrared reflection spectra for a series of sodium borate glasses along with fused basic oxide are shown in Figure 7. The principal peaks at 1280 cm^{-1} for fused basic oxide is assigned to a B-O stretching mode associated with B-O-B linkages in which both boxon atoms are triangularly linked. A band ca. 710 cm^{-1} is also assigned as a subsidiary band associated with this group. These peaks become less intense and move to lower wavenumbers as the concentration of Na_2O is increased. Jellyman and Procter,[16] also noted a band at 1350-1400 cm^{-1} and a band at ca. 1050 cm^{-1} in the Borate glasses that will increase in intensity as the concentration of Na_2O is initially increased. They have attributed these bands to B-O stretching modes involving B-O-B linkages associated with a triangularly coordinated boron atom and a tetrahedrally coordinated boron atom. The limiting composition at which a tetrahedrally coordinated boron may be linked to four BO_3 triangles is $Na_2O{\cdot}5B_2O_3$. This is roughly the composition at which the 1270 cm^{-1} band and the 1350-1400 cm^{-1} band become equal in intensity, and no further increase in the intensity of the 1350-1400 cm^{-1} band occurs with further

* a wavenumber is defined as the reciprical of its wavelength, and has dimensions of cm^{-1}.

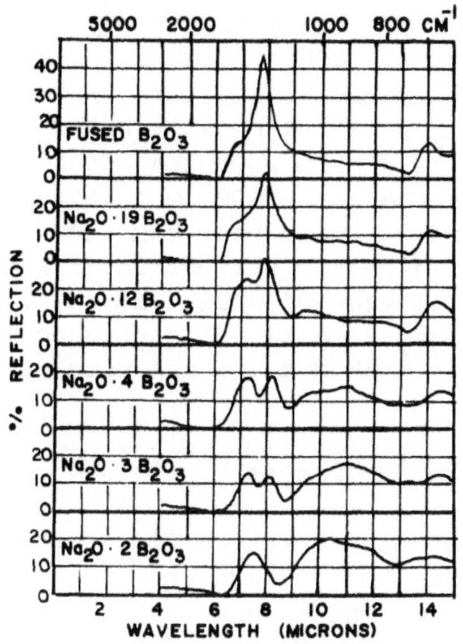

Figure 7. The infrared reflection spectra of various sodium borate glasses.[16]

addition of Na_2O. At higher concentrations of Na_2O, the band located at 850 to 1000 cm^{-1} is assigned to groups such as

$$B — O^- \quad \text{or} \quad B — O — B — O^-$$

At these concentrations, those groups will form. The changes which take place in lithium, potassium, and the alkaline earth borate glasses follow the same trends as is observed for sodium borate glasses. Differences are mainly in the relative intensities and the shapes of bands.

Recently Krogh-Moe[17] has shown by infrared and NMR spectroscopy that the total number of tetrahedrally coordinated boron atoms will increase if one adds up to 33 1/3 mole % alkali oxide to the glass.

(C) Borosilicate Glasses

The infrared reflection spectra for a series of different compositions of sodium borosilicate glass are shown in Figures 8 and 9. At higher concentrations of SiO_2, the band, at ca. 1100 cm^{-1} attributed to the Si-O-Si linkage can be observed. At lower concentrations, a band in this region has to be attributed to a B-O-Si linkage in which the B atom is tetrahedrally coordinated. A band is noted in the 1320-1400 cm^{-1} region at low concentration of Na_2O. This band has been assigned to a Si-O-B linkage in which the B atoms is triangularly coordinated. The absence of a peak at ca. 1250 cm^{-1} for composition in which there was an excess of SiO_2 over B_2O_3 indicated that no B-O-B linkage were present in these glasses.

(D) Chalcogenide Glasses

Oxide glasses such as fused silica cannot be used as optical materials in the infrared region because their major vibrational absorption peaks appear in that spectral region. However, non-oxide, chalcogenide glasses can transmit infrared radiation over a wide spectral range. Strong absorption bands do not occur in the middle infrared region (ca 1500-700 cm^{-1}) in such glasses since they are compounds of atoms which are much heavier than oxygen (i.e. sulfur, selenium and tellurium)

The effect of heavier atoms on absorption bands can be seen from equation (4) which is valid for the vibrational stretching mode of a diatomic molecule, A-B.

$$1/\lambda = \nu = \frac{1}{2\pi c} \left(\frac{f}{\mu}\right)^{1/2} \qquad (4)$$

where ν is the wavenumber in cm^{-1}, c is the speed of light, and f is the force constant of the bond. μ is the reduced mass of molecule A-B and is given by

$$\mu = \frac{m_a\, m_b}{m_a + m_b} \qquad (5)$$

where m_a and m_b are the atomic weights of A and B respectively.

For a hypothetical Si-O molecule μ is 10.2 m_p (m_p is the mass of a proton) while for As-Se and As-Te μ is 38 m_p and 47 m_p respectively. Force constants for the

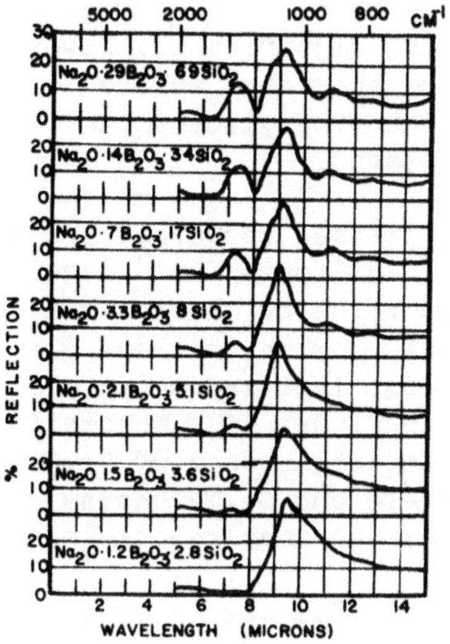

Figure 8. The infrared reflection spectra for various sodium borosilicate glasses with constant B_2O_3/SiO_2 molar ratio.[16]

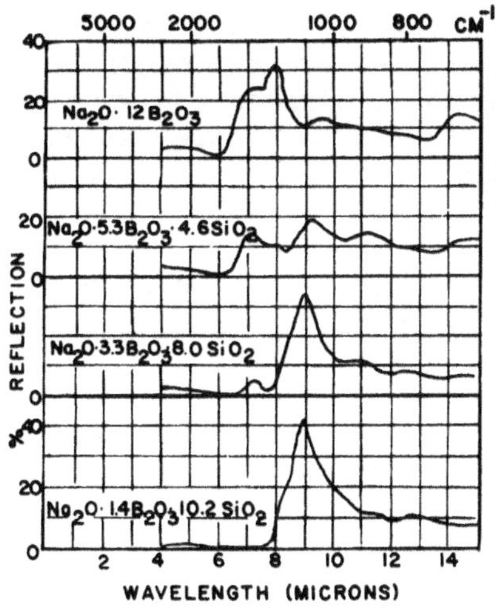

Figure 9. The infrared reflection spectra of various sodium borosilicate glasses with low Na_2O content.[16]

latter bonds are also significantly lower and infrared absorption bands are therefore shifted to significantly lower frequencies, and the glasses can be transparent well into the middle infrared.

The non-oxide glasses, however, are weaker and softer than oxide glasses. Hilton et al[18], have tried to improve these physical properties by obtaining a general understanding of chemical bonding in these glasses using infrared spectroscopy. The far infrared spectra of several of these glasses are shown in Figure 10. Theoretical analysis of such data shows that in the case of Si-As-Te and Ge-As-Te glasses, zig-zag Si-Te and Ge-Te chains occur. Figure 11 illustrates the chain parameters determined for Si-Te.

For As_2S_3 a pyramidal arrangement of As-S bands occurs in which many As-S molecules are trapped in a sulfur network. Similarly for P-S glasses, small P-S molecules are formed and dispersed in a sulfur matrix. Such information may be useful in the future for understanding how the addition of certain elements will perturb the structure of these glasses, and in turn alters their physical properties.

Hilton and Jones[19] have also used infrared spectroscopy to identify traces of various metal oxide impurities found in chalcogenide glasses, and have been able to identify oxides such as Cu_2O, Fe_2O_3, MgO and SiO_2 from infrared spectra. These impurities are the major sources of absorption found in these glasses for wavelengths less than 15 microns. Identification of these oxides is useful for their elimination which improves the infrared transmitting properties of the glasses.

Jerger[20] has investigated As-Te-Se glasses and has been able to identify an As_2O_3 impurity by a band occuring ca. 790 cm^{-1} in the infrared spectra.

SPECIAL APPLICATIONS OF VIBRATIONAL SPECTROSCOPY TO GLASS STUDIES

(A) Detection of Water in Glass

It is now widely accepted that infrared spectroscopy provides an excellent tool for the qualitative and quantitative determination of the water content in glasses.

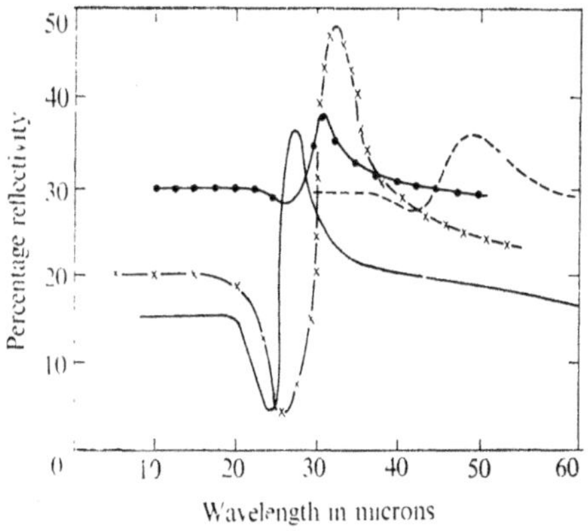

Figure 10. The far infrared reflection spectra of several non-oxide chalcogenide glasses (after Hilton et al.[18]).

————	Ge_2S_3	●-●-●-●	$Ge_{15}P_{15}Te_{70}$
X-X-X-X	As_2S_3	--------	$Si_{10}As_{10}Te_{80}$

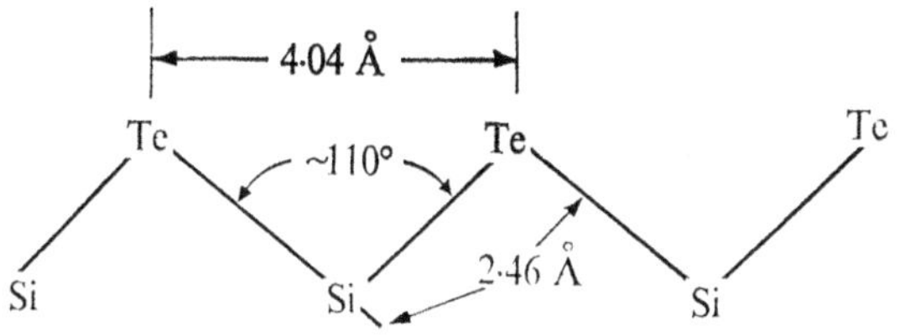

Figure 11. Chain model for the Si-Te system (after Hilton et al.[18]).

Silica or boric oxide containing glasses absorb infrared radiation strongly in the spectral region 2.7-3.0 microns (ca. 3350-3700 cm^{-1}). In 1933, H. P. Gage suggested that these absorption peaks were related to the water content in the glasses, but Anna Harrison[21] presented the first conclusive evidence that the absorption peak which can be observed at 2.7 microns was related to the amount of water present in silica glass. She also noted a band at 2.85 microns for boric oxide glasses which she related to the amount of water absorbed in these glasses. Figure 12 illustrates the transmission spectra for vitreous silica prepared by several different methods. Vitreous silica containing less water exhibited less intense peaks.

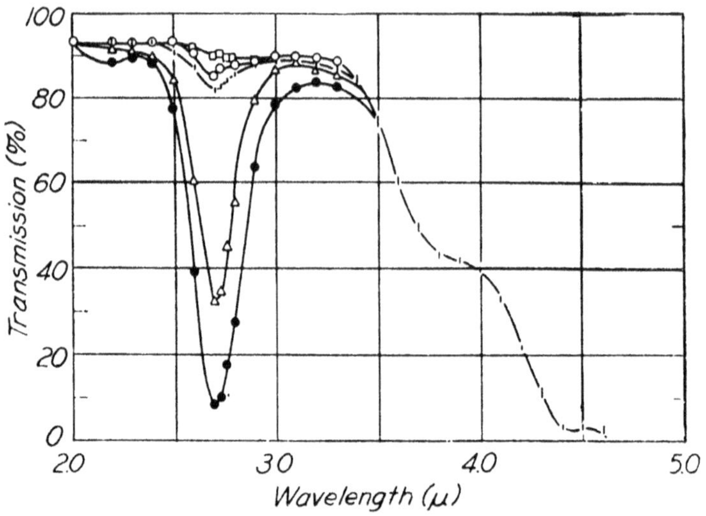

Figure 12. Transmission spectra of vitreous silica prepared by several different methods: Bottom- fused silica prepared from $SiCl_4$; 2nd- Neider fused silica; 3rd- General Electric fused silica; Top two- vitreosil.[21]

Adams[22] has explained the bands of fused silica as originating from OH groups bonded to one of the network-forming cations. This interpretation is consistent with the spectrum observed when D_2O is substituted for H_2O. Adams further points out that when other components are present in the glass, such as alkali oxides, hydrogen bonding may occur with the OH groups leading to an increased wavelength and broadening of the OH bands.

Sholze[23] thinks that OH groups from water occur in voids present in the glass network. He has described a method by which one can determine the concentration of water absorbed (present in) by the glass. Sholze relates band intensities to the concentration of OH groups through the Beer-Lambert equation,

$$\log \frac{T_o}{T} = \varepsilon c d$$

where c is the concentration of H_2O in moles/liter of glass, d is the thickness of the sample in cm., T_o/T is the transmissivity, (T_o is the transmission at the band position when no water is present.) and ε is the extinction coefficient for water in the glass and is constant for low concentrations of water. Franz[24] has noted that similar relations hold for water in boric oxide containing glasses. Therefore, infrared spectroscopy provides a relatively simple method for measuring the concentration of water absorbed by glass.

(B) Surface Studies on Porous Glasses

Because of the unique property of porous glass to scatter less light than either powders or pressed discs, this glass has been used to study the infrared spectra of many adsorbents on its surface. These glasses can be prepared as plates which can be as thin as 0.1 mm. Corning code 7930 porous glass possesses the chemical composition of about 96% SiO_2, 3% B_2O_3 and 1% other components. The surface area available on this porous glass is usually about 200 m^2/gm with an average pore diameter of 40 Å. The presence of boric oxide in this glass introduces a surface group that is not available on materials such as silica gel. Porous glass possesses many boron sites of the Lewis-acid type on which adsorption can occur. Three different infrared studies with porous glass will be illustrated:

1. Dehydration and Hydration on Porous Glass.[25]

The O-H stretching modes of O-H groups found on porous glass surface are located in the 3000-4000 cm^{-1} region of the infrared spectra. The spectra of porous glass after various hydration and dehydration treatments are shown in Figure 13. The isolated Hydroxyl group stretching mode appears at 3749 cm^{-1}. The band at 3665 has been assigned to the stretching mode of hydrogen bonded adjacent silanol groups.

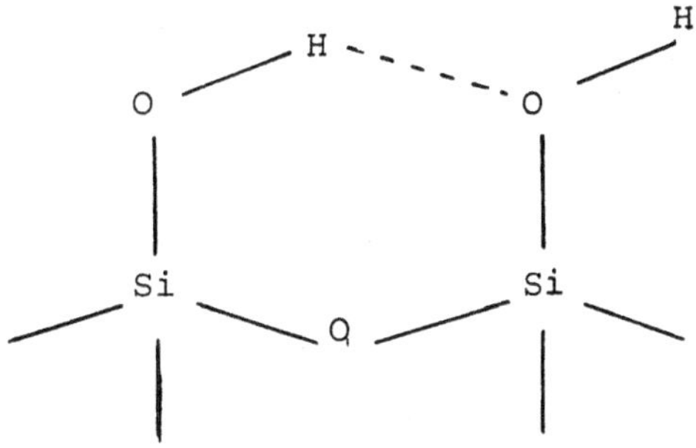

The band at 3450 cm^{-1} has been assigned to the O-H stretching mode of adsorbed water,

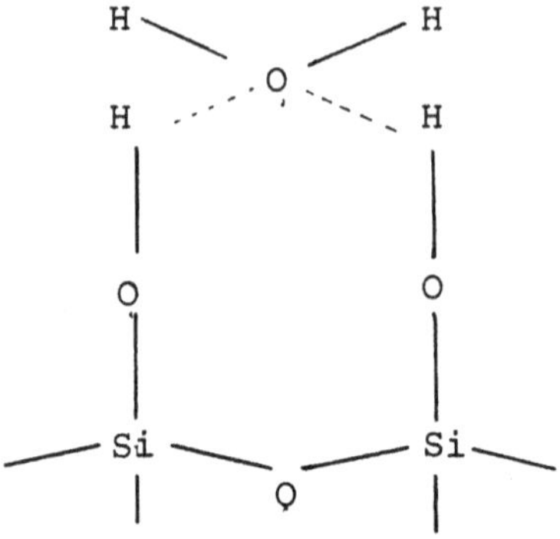

The results in Figure 13 indicate that all three types of hydroxyl groups are present after evacuating at 100°C. Evacuation at 300°C gives a spectra which shows that the adsorbed water has been removed. The two types of silanol O-H bands remain. Above 540°C, Evacuation removes the adsorbed water and also causes dehydration of adjacent silanol groups, eliminating the peak at 3665°C. Evacuation and heating at 650°C brings about the growth of a new peak at 3700 cm^{-1}. Many Russian investigators noted this band for porous glass samples. Heating porous glass above 500°C causes boric acid whiskers to migrate onto the glass surface. The band at 3700 cm^{-1} can be associated with the O-H stretching mode of a hydroxyl group attached to a boron atom.

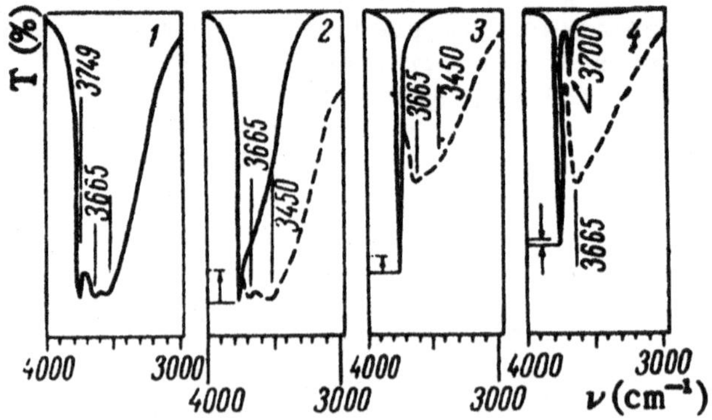

Figure 13. The infrared spectra of porous glass after heating in vacuum at 100° (graph 1), at 540° (graph 3), and at 650° (graph 4). Dashed lines indicate the spectra aft er the samples stand in water vapor at room temperature. The verticle arrows indicate the change in the intensity of the O-H stretching mode of the free O-H group after all treatments.[25]

The spectra of rehydrated porous glasses are also shown in Figure 13. Water vapor exposure of the completely dehydrated samples brings about the reappearance of the peak at 3665 cm^{-1}, indicating the reformation of adjacent silanol groups. The reappearance of the band at 3450 cm^{-1} indicates molecular water has been also physically absorbed again.

2. Ammonia on Porous Glass.[26]

The infrared spectra (3000-3800 cm^{-1}) of porous glass samples on which ammonia was adsorbed in different amounts both at 20°C and 150°C are shown in Figure 14. The modes at 3730 and 3665 cm^{-1} are assigned to O-H stretching modes discussed earlier. The bands observed at 3400 and 3320 cm^{-1} are assigned to the asymmetric and symmetric stretching modes of physically adsorbed ammonia. Interaction with the surface hydroxyl groups is indicated by the lowering of the O-H stretching modes with addition of ammonia to the sample. As can be seen from Table IV, the assignment of the two bands to physically adsorbed ammonia is realistic because of the close agreement of their wave number values with those of ammonia in carbon tetrachloride solution. Also, these peaks disappear with either heat treatment or evacuation and the hydroxyl bands return to their original intensity.

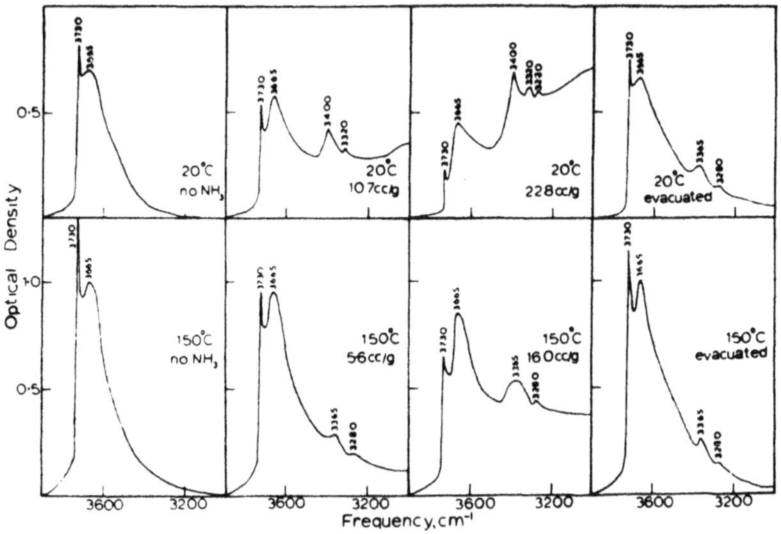

Figure 14. The infrared spectra of ammonia adsorbed on porous glass.[26]

Table IV. The Infrared Fundamentals for N H Groups for Various Molecules (cm^{-1}).[26]

Molecule	Phase	Asymmetric	Symmetric
NH_3	gas	3443	3337
NH_3	dissolved in CCl_4	3417	3315
$F_3B{:}NH_3$	solid	3340	3240
$-NH_2$	chemiadsorbed on glass	3520	3445
NH_4Cl	solid	3140	

Additional bands are observed at 3365 and 3280 cm^{-1}. The two bands are not influenced by low temperature heating, and will only be eliminated when the samples are heated above 200°C. These two bands will also appear when ammonia is adsorbed at 150°C. Physical adsorption of ammonia should not occur at these temperatures. The data in Table IV suggest the nature of the new ammonia complex on the glass. The wavenumber values observed for the ammonia - glass sample are very close to the N-H stretching modes observed for the $F_3B{:}NH_3$ complex. This seems to indicate that ammonia molecules can coordinate to Lewis-acid-type sites on the glass surface. Hair and Chapman[27] suggest that the ammonia forms the following type of complex with boron groups:

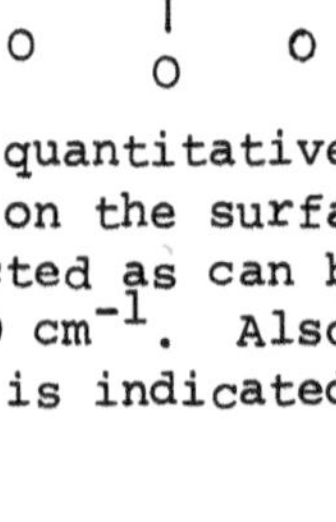

Using infrared spectra quantitatively, they suggest that the ratio B/Si was 1:3 on the surface. No ammonium groups (NH_4^+) were detected as can be noted by the fact that no band occurs ca. 3150 cm^{-1}. Also, ammonia was not chemically adsorbed as is indicated by the absence of a band above 3500 cm^{-1}.

3. Effect of Adsorption on Lengths of Porous Glass Rods.

An outstanding example of the application of infrared spectroscopy for the explanation of adsorption phenomena is illustrated by a study carried out by Folman and Yates.[28] In their investigation, they were able to interpret the infrared spectra in such a way that they could explain changes in lengths of porous glass rods during an adsorption process on the basis of structural changes. They noted that adsorption of certain compounds caused an initial contraction of porous glass rods. Thermodynamic arguments predict only expansion of the rods with adsorption occuring. The results obtained by them with acetone are illustrated in Figure 15.

Folman and Yates turned to infrared spectra to explain these contractions. Investigating the spectra of glass samples contracted by adsorption, they noted that interaction did take place between the surface hydroxyl groups and the adsorbate. Table V illustrates the correlation that is observed between the shift in O-H stretching vibrations versus the amount of contraction observed with an interferometer. It can be seen that the O-H mode decreases its wavenumber value and the amount of contraction decreases

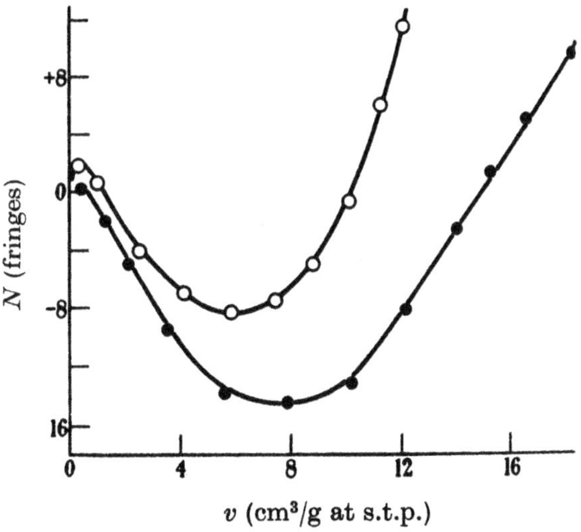

Figure 15. Length changes of porous glass rod with adsorption of acetone at 25° and 75° C.[28]

Table V. A Comparison Between the Shifts in OH Wave Numbers ($\Delta\nu$) and the Contractions Produced.[28]

Gas	Temp., °C	Average $\Delta\nu$ value, cm^{-1}	Contraction A-B (fringes)
CH_3Cl	24	110	0.6
SO_2	24	115	0.6
$(CH_3)_2CO$	25	330	15.0
	75	305	10.5
	135	270	--
NH_3	25	820	25.0
	75	750	21.0
	100	710	17.5
	150	640	10.2

with increasing temperature. The shift of the O-H mode with the addition of adsorbate indicates that Hydrogen banding is occuring. Methyl chloride, sulfur dioxide, acetone and ammonia all form this type of bond with O-H groups. The formation of a Hydrogen bond involves a change in electron distribution. A charge-transfer process occurs in which charge is transferred from the proton acceptor to the proton donor. The oxygen of O-H group involved in the hydrogen bonding becomes more negative. As a result, the hydrogen bond between adjacent O-H groups become stronger and a decrease between oxygen atoms can occur. The decrease in O...O distances will cause an overall contraction in the solid.

This interpretation explains other experimental observations. As the electron donating properties of the adsorbate increase, this should cause the O-H stretching mode to decrease its wavenumber value as is observed in Table V. Also, the amounts of contraction follow the same trend. The temperature dependence observed follows the predictions suggested by hydrogen bonding. Increasing the temperature weakens the hydrogen bond, causing smaller contractions to be observed.

(C) The Infrared Spectra of Heated Glasses

The infrared spectra of glasses are of importance to glass production because they can influence both the cooling rates of glasses and transfer of heat in glass tank furnaces. In order to provide fundamental data to develop expressions relating to such characteristics as radiation heat flow. A number of investigators have measured the infrared transmission spectra of different glasses at high temperatures.[29,30]

Grove and Jellyman[29] have studied a number of iron-containing glasses at temperatures up to 1400°C. Their spectrophotometer arrangement is shown in Figure 16. The sample was placed in a rectangular platinum-rhodium cell with fused-silica windows so that transmitted radiation can pass directly through the sample. A Perkin-Elmer model 12-C was also used to replace the monochromatic-detecter system in the Figure.

Glasses containing no iron show very little change in the region 3800-16500 cm^{-1} with change in temperature. Iron-containing absorbed in the region at room temperature

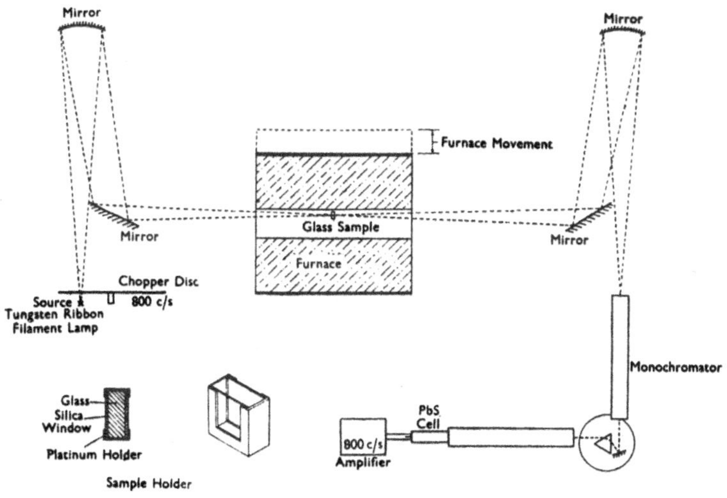

Figure 16. The spectrophotometer arrangement for measuring the transmission spectra of hot glasses.[29]

with a peak at ca. 9000 cm^{-1}. This peak was attributed to the iron in the ferrous state and decreases in intensity as the glass becomes more oxidized. As the glass is heated up to 1000°C, the intensity of the peak decrease continuously. Heating the sample above 1000°C, the intensity begins to increase again. These intensity changes are most pronounced for the oxidized glass. Figure 17 illustrates the spectra observed for oxidized iron-containing glass at different temperatures. The changes in absorption that take place suggest that iron prefers to exist in the ferric state at higher temperature. Strengthening this hypothesis is the observation that the ultraviolet bond attributed to iron in the ferric state intensifies with heating.

Another intensity high temperature study was carried out by Neuroth on vitreous silica.[31] He studied the infrared spectra in the region 900-1400 cm^{-1} for temperatures between 18° to 1270°C. Figure 18 illustrates the spectra obtained. As the temperature increases the absorption bond lowers in wavenumber value, and broadens. These characteristics can be used as a measure of disorder within the glass at different temperatures.[32]

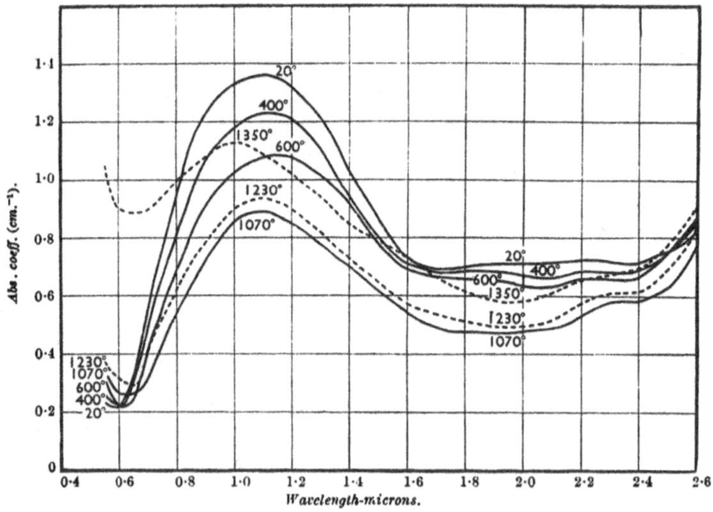

Figure 17. The infrared spectra of iron-containing glass prepared under reducing conditions.[29]

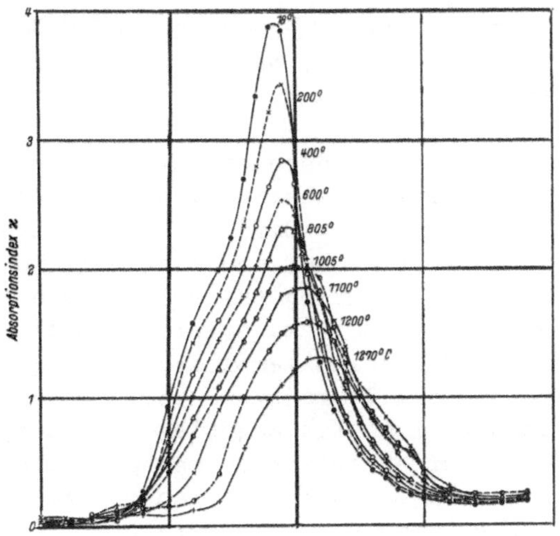

Figure 18. The infrared spectra of vitreous silica at several temperatures.[31]

(D) Gaseous Products from Molten Glasses

The application of the classical methods of spectroscopy and electron diffraction to the determination of structures of the vapor species of refractories and glass materials has been severly limited by the chemical, intensity and other difficulties associated with investigations at high temperatures. Several techniques are available for the investigation of vapors of ceramics with infrared spectroscopy that can give this type of information.

The conventional method for observing the infrared spectra of vapor species in equilibrium with materials at high temperatures are usually difficult. It involves heating the material in a tube furnace until the concentration of the vapor species are high enough to be observed with transmission spectra. The chief disadvantages are that the vapor might react with the furnace walls, complications might arise in the spectra from the appearance of hot bands or else sufficient concentrations of the vapor species of interest cannot be obtained to observe their spectra. Also, at sufficiently high temperature, the tube furnace walls begin to emit black body radiation in the infrared spectral region, making the measurement of good quality spectra difficult.

The application of Knudsen effusion* along with matrix isolation offers a relatively simple approach that, also, eliminates many of the difficulties of the tube furnace method.[33] This method involves heating the sample in a Knudsen cell and trapping the emitted species in solid matrices of relatively inert materials such as argon at low temperatures. The solid sample to be studied is prepared by spraying a small amount of the sample vapor along with a large amount of matrix material onto a cold plate. This plate is made of an infrared transparent material such as CsI which can be used for transmission studies, and is cooled to extremely low temperatures in order to freeze the gases onto the plate. Usually a cryotyp unit is used to cool down the plate.

Sommer et. al., have successfully used this method to study the vaporization of B_2O_3.[34] Figure 19 illustrates the infrared spectrum of a dilute concentration of B_2O_3

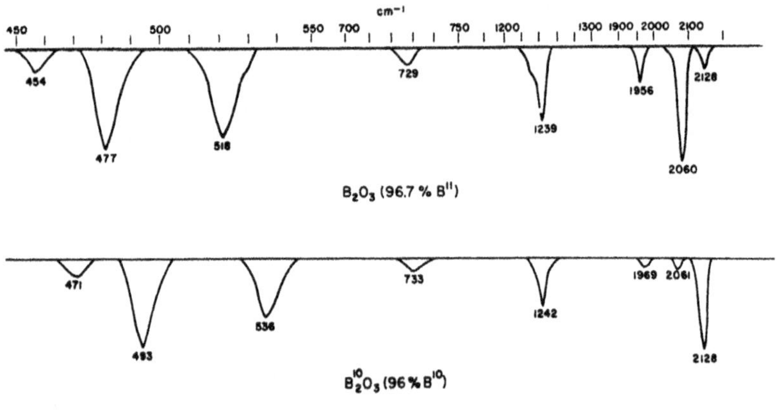

Figure 19. The infrared spectra of dilute concentrations of B_2O_3 in argon matrices.[34]

*a high temperature point source

trapped in an argon matrix. The analysis of the observed infrared spectra indicates that B_2O_3 in the gas phase is V-shaped with C_{2v} - symmetry. For this model, eight infrared fundamentals are expected. The limiting linear molecule would have only 4 infrared fundamentals. Band assignments for the B^{10} - and B^{11} - containing molecules are assigned in Table VI. B=O stretching modes occur near 2000 cm^{-1}, B-O stretching modes between 700-1400 cm^{-1} with the symmetric mode having the lower wavenumber and the bending modes below 600 cm^{-1}. Theoretical analysis using normal coordinate analysis and isotopic substitution indicates that the apex angle of the B_2O_3 molecules is 150°. Recently, Anderson and Ogden have observed the matrix isolated infrared spectra of the vaporous products from silica.[35] The spectra for Si-O and its polymeric companions, Si_2O_2 and Si_3O_3 have been observed and identified.

(E) The Infrared Spectra of Tri-coordinated Silicon Defects in Vitreous Silica.

Many investigators have indicated the possibility of tri-coordinated silicon defects in vitreous silica.[36] It is through that ionization of these centers by means of x- or γ- radiation results in a center that shows characteristic optical absorption in the ultraviolet spectra from 2150-2200 Å. Weeks and Lell have identified a band in the

Table VI. Fundamentals for B_2O_3 .[34]

Vibration	$B_2{}^{10}O_3$	$B_2{}^{11}O_3$
$V_1(A_1)$	2128	2060
$V_2(A_1)$	733	729
$V_3(A_1)$	536	518
$V_4(A_1)$	172 or 260	172 or 259
$V_5(A_2)$	476	456
$V_6(B_1)$	2128	2060
$V_7(B_1)$	1242	1239
$V_8(B_1)$	471	454
$V_9(B_2)$	493	477

EPR spectra corresponding to an E_1'-center.*

Groups of silicons coordinated with three oxygens should show fundamental modes differing from those of tetrahedrally coordinated systems. Hanna[37] has investigated the infrared spectrum of high purity, water free vitreous silica and found a broad weak band at 200 cm^{-1}. On the basis of normal coordinate analysis, he has assigned this band to the out-of-plane bending mode for tri-coordinated silicon groups.[1] If this assignment is correct, the band should appear weakly in the infrared spectra both because of the low concentration of centers and because the groups involved probably possess the lower C_{3v} -symmetry rather than D_{3h} -symmetry with the silicon atoms slightly out of the plane of the three oxygen atoms. Hanna, also, noted that the 200 cm^{-1} band was not present in $Na_2O \cdot xSiO_2$ glasses where x = 3, 4, or 5. He explained this observation by suggesting that Na_2O saturates the tri-coordinating silicon group removing this group and its related vibrational fundamentals.

---

* See R. A. Weeks - This book.

REFERENCES

1. R. K. Hanna, J. Am. Ceram. Soc., 48, 595 (1965).
2. A. G. Meister and F. F. Cleveland, Am. J. Phys. 11, 239 (1943).
3. F. A. Cotton, Chemical Applications of Group Theory. Interscience Publishers, New York, 1966.
4. E. B. Wilson, Jr., J. C. Decius and P. C. Cross, Molecular Vibrations. McGraw-Hill Book Company, Inc. New York, 1955.
5. K. Nakamoto, Infrared Spectra of Inorganic and Coordination Compounds. John Wiley and Sons, Inc., New York, 1963.
6. S. Bhagavantam and T. VenKatarayudu, Proc. Indian Acad. Sci. 9A, 224 (1939).
7. D. F. Hornig, J. Chem. Phys. 14, 1063 (1948).
8. S. R. Levitt and R. A. Condrate, Amer. Mineral. 55, 1562 (1970).
9. H. Moore and P. W. McMillan, J. Soc. Glass Technol. 40, 66T (1956).

10. R. V. Adams and R. W. Douglas, J. Soc. Glass Tech. 43, 147T (1959).
11. W. Wadua and W. S. Balloomal, J. Phys. Chem. Glasses, 9, 115 (1968).
12. G. J. Su, N. F. Borrelli and A. R. Miller, J. Phys. Chem. Glasses, 3, 167 (1962).
13. J. Bock and G. J. Su, J. Am. Cer. Soc., 53, 69 (1970).
14. J. Zarzycki and L. Prod'Homme, Glass Ind., 39, 587(1958).
15. R. J. Bell and P. Dean, Localized Excitons in Solids, Proc. 1st Int. Conf. (1967).
16. P. E. Jellyman and J. P. Procter, J. Soc. Glass Tech. 39, 173T (1955).
17. J. Krogh-Moe, Phys Chem. Glasses, 3, 101 (1962).
18. A. R. Hilton, C. E. Jones, R. D. Dobrott, H. M. Klein, A. M. Bryany, and T. D. George, Phys. Chem. Glasses, 7, 116 (1966).
19. A. R. Hilton and C. E. Jones, Phys. Chem. Glasses, 7, 112 (1966).
20. J. Jerger, Semiannual Technical Report, 1 June 1963-30 Nov. 1963, SCA-43600-Sr-1 (Dec. 31, 1963).
21. A. J. Harrison, J. Am. Cer. Soc., 30, 362 (1947).
22. R. V. Adams, Phys. Chem. Glasses, 2, 39 (1961).
23. H.Sholze, Glastech. Ber., 32, 142 (1959).
24. H. Franz, Glastech. Ber., 38, 54 (1965).
25. A. N. Siderov, Opt. Spectry. USSR, 8, 424 (1960).
26. N. W. Cant and L. H. Little, Can. J. Chem., 42, 802 (1964).
27. I. D. Chapman and M. L. Hair, Trans. Faraday Soc., 61, 1507 (1965).
28. M. Folman and D. J. C. Yates, Proc. Roy. Soc.(London) A246, 32 (1958).
29. F. J. Grove and P. E. Jellyman, J. Soc. Glass Tech., 39, 3T (1955).
30. N. Neuroth, Glastech. Ber., 25, 242 (1952).
31. N. Neuroth, Glastech. Ber., 28, 411 (1955).
32. I. Simon, Modern Aspects of the Vitreous State, Vol. 1, Edited by J. D. Mackenzie, Butterworth, London, 1960.
33. M.J. Linevsky, J. Chem. Phys., 34, 587 (1961).
34. A. Sommer, D. White, M. J. Linevsky and D. E. Mann, J. Chem. Phys., 38, 87 (1963).
35. J. S. Anderson and J. S. Ogden, J. Chem. Phys., 51, 4189 (1969).
36. G. Hetherington, K. H. Jack and M. W. Ramsay, Phys. Chem. Glasses, 6, 6 (1965).
37. R. K. Hanna, J. Phys. Chem., 69, 3846 (1965).

The Uses of Electron and Nuclear Magnetic Resonance and Nuclear Resonance Fluorescence In Studies of Glass

R. A. Weeks

Solid State Division
Oak Ridge National Laboratory
Oak Ridge, Tennessee 37830

INTRODUCTION

A legitimate question which one might ask about the title of this exposition is: what does "resonance" mean? In searching for an answer it is reasonable to consider the precedents for its usage. In Schiff's book on Quantum Mechanics, the term is used in two ways: (1) The scattering of a particle by a potential is said to be "resonance scattering" when certain relations between the scattered particle and the potential from which it is scattered obtain.[1] (2) A property of two classical oscillators which are in resonance (same unperturbed frequency) gives rise to two normal modes whose frequencies are lower and higher than the unperturbed frequency.[1] This characteristic of interacting harmonic oscillations in quantum mechanics provides a basis for the theory of homopolar binding in molecules. Pauling and Wilson note that it arises whenever a system contains two or more identical particles.[2] Another use of the term is found in the description of events which occur when an atom decays from an excited state and the emitted photon passes through a gas of the same kind of atoms being absorbed in the process and raising another atom to the same excited state as the one from which it was emitted. This process, called resonance fluorescence,[3] couples two identical oscillations by an electromagnetic field. Thus, the precedents for the use of the term "resonance" permit its application to the Mossbauer effect but not to the transitions between magnetic states of electrons or nucleons which have been called "magnetic resonance". Since the forces maintaining customs are diverse and subtle, no attempt will be made here to alter these and transitions between electron and nuclear magnetic states will be called electron and nuclear magnetic resonance, and the resonance fluorescence transitions observed in the Mossbauer effect will be called Mossbauer transitions.

In order that the value of these spectroscopic phenomena for investigating the glassy state might be estimated, it will be useful to consider the range of energies over which transitions can be observed in real experiments. Table I lists these energies for the three types of phenomena which are considered here and, for comparison, energies for two other phenomena widely used in the study of many solids. An important factor determining the usefulness of a particular spectroscopic method is the ratio of transition energy, E, to the range of energy, ΔE, over which the transition occurs. If this ratio is large, then it is usually possible to detect the dependence of transition energies upon the configurations of atoms surrounding the point in a solid at which the absorption or emission of a photon occurs. This ratio is very large for those favorable cases in which Mossbauer

Table I. Energies and Line Widths of Various Transitions

Method	Experimental Range of Transition Energies, E (electron volts)	Line Widths, ΔE* (electron volts) Single crystal	$\frac{E}{\Delta E}$
NMR	10^{-8} to 10^{-6}	10^{-14} to 10^{-8}	10^{6} to 10^{2}
EPR	10^{-5} to 10^{-3}	10^{-9} to $>10^{-3}$	10^{4} to 1
Mössbauer	10^{2} to 10^{5}	10^{-10} to 10^{-4}	10^{12} to 10^{9}
Optical Absorption Bands in a Solid	0.1 to 10	0.1 to 1	10 to 1
Zero Phonon Lines	0.1 to 10	$\sim 10^{-3}$	10^{4} to 10^{2}

*This range of widths for NMR and EPR encompasses, approximately, the observed range for solids.

transitions have been observed and much smaller for EPR and NMR transitions. In many cases in which NMR, EPR, or Mossbauer transitions occur solid state effects are such as to broaden the range of energies for a particular transition to such an extent that they cannot be detected. These solid state effects when of a form permitting the detection of transitions are the ones from which deductions can be made about the chemical and physical structure of glass in the vicinity of the resonating entity. Three principal classes of such effects are: (1) interactions of the resonating entity with the electric field produced at the entity by all the electric charges of the solid. Practically these can usually be limited, in the absence of electronic conduction, to those associated with the entity and with nearest and next nearest neighbor atoms. A point charge approximation[4] can frequently be made, particularly when the bonding forces between atoms are ionic in character. Most glasses of technological significance frequently have compositions in which the bonding between some of the ions is covalent and hence the directional character of these forces must be considered in evaluating electric field effects in such solids. (2) vibrations of atoms due to thermal effects. These vibrations modulate the electric field and if the interaction between the resonating entity and the electric field is strong then transition energies are broadened. These effects may be reduced by lowering the temperature of the solid. The temperature dependence of these resonance transitions can furnish information about some of the thermal properties of the solid. (3) magnetic interactions. In those materials in which the resonating entities have a low concentration and are randomly distributed these interactions are negligible. However, at high concentrations these interactions may be larger than other interactions and resonance transitions will be determined by them. Interactions between paramagnetic electrons and nuclei with net magnetic moments are particularly important in EPR experiments. Many details of structure can be obtained from these interactions. Each of these interactions will be considered in more detail as needed.

Extensive expositions of the theory and experimental results of these three types of resonance transitions are extant.[4,5,6] Hence, any extended discussion here would serve no useful function. EPR and NMR are essentially the same phenomenon and experimental procedures are similar.

The Mossbauer effect, or resonance fluorescence, is quite a different phenomenon, and the experimental procedures for observing it differ markedly from those for EPR and NMR. Briefly, the experimental apparatus for EPR and NMR experiments consists of a source of coherent electromagnetic radiation, a resonant structure in which the sample is placed and in which the electromagnetic radiation is polarized, a magnetic field in which the resonant system and sample is placed, and a detector of the radiation. The resonant system is usually oriented with its r.f. (radio frequency) magnetic field component perpendicular to the direction of the laboratory field. The separations of the magnetic spin states of a magnetic complex in a material are proportional to the applied field (in a first order approximation) if they are randomly distributed in a host material and if their concentration is small (small is $<1\%$). For a spectrometer operating at a particular frequency, ν, the magnetic field, H, may be increased until the energy difference of the spin states $E=h\nu$ ($h\equiv$Plancks constant) at which value of the field energy will be absorbed from the r.f. radiation. It has been pointed out by Burns[7] that the fine structure EPR Hamiltonian has the same mathematical form as the quadrupole Hamiltonian used in NMR.

The Mossbauer effect also requires a source of electromagnetic radiation. This radiation is that emitted by a nucleus in an excited state, e.g. ^{57}Co, when it decays to its ground state. A sample containing suitable atoms whose nuclei are in their ground state, e.g. ^{57}Fe, and which have an excited state whose energy with respect to the ground state is almost equal to the energy of the emitted radiation absorbs the radiation. A suitable detector then records the change in radiation intensity when the sample is placed between the source and detector. The energy of the absorber excited state with respect to the source radiation is altered by various solid state effects. These may be of such magnitude that a small relative velocity ($\lesssim 1$ cm sec^{-1}) of absorber with respect to the source will change the absorption. Thus by measuring absorption as a function of velocity an absorption curve will be determined. Hence the last piece of experimental apparatus for a study of Mossbauer transitions is a means of moving either the absorber or source with small variable relative velocities. Frauenfelder[6] notes that if hum exists in the velocity drive with an amplitude of the order of 10^{-8} cm the resonance of ^{67}Zn will not be detectable.

ELECTRON PARAMAGNETIC RESONANCE

Glasses of widespread technical value are made principally from the oxides of B, Si, Mg, Ca, Al, P, Na, and K. Elements such as the transition group, Ti through Ni, and Cu are usually present as impurities and are occasionally deliberately added. When formed under oxidizing conditions, such glasses are usually transparent to the visible region of the electromagnetic spectrum and diamagnetic. If members of the transition group of elements are present as impurities, they will contribute a small paramagnetic component to the dominant diamagnetic susceptibility. Frequently this component can be detected in an EPR experiment, but little else will be observed. If such a glass is irradiated with photons with energies greater than a few eV, the transparency of the glass will be altered and additional EPR transitions will be observed. In a single component glass formed from compounds such as SiO_2, GeO_2, or B_2O_3, these additional EPR transitions are frequently related to impurities which are normally diamagnetic. In many cases these impurities have valence states which differ from those of the principal components of the glass. Even when their principal valence state is the same, their size (ionic radius is a measure of this size) differs from those of the principal components. In either case there is a potential well generated by the impurity which is attractive to either the electrons excited by the radiation or to the state from which the electron was excited (this state is called a hole since it is an empty electronic state in an otherwise filled band of states). Hence, a fraction of these electrons and holes are trapped at these places in the glass. When this happens, the electronic states associated with these points may contain an odd number of electrons, frequently a single electron. In a magnetic field the spin degeneracy is removed and EPR transitions can be induced. (Frequently impurities can be detected by this technique which are not detectable by chemical analysis or optical spectroscopy).

This condition is exemplified in Fig. 1, in which the energy of the transition, the fraction of electrons in the ground state and in the excited state, and the population difference between the ground state and the excited state are given. For the usual conditions of an EPR measurement,

$$\mathcal{H} = g\beta\vec{H}\cdot\vec{S}$$

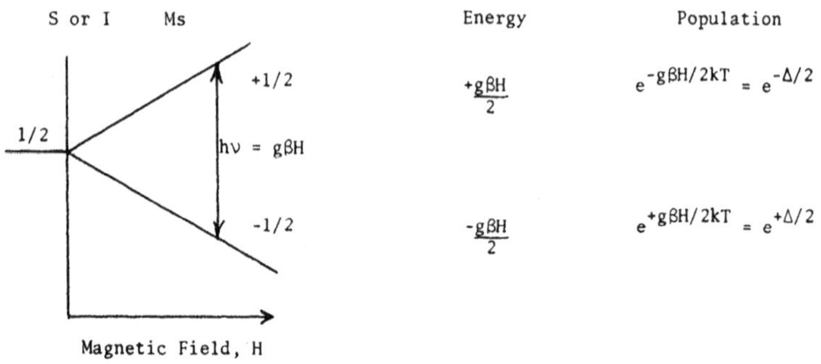

$$\text{Signal Intensity} \propto \frac{e^{+\Delta/2} - e^{-\Delta/2}}{e^{+\Delta/2} + e^{-\Delta/2}}$$

$$= \tanh \Delta/2$$

Figure 1. Effect of a Magnetic Field Upon Electron or Nuclear Spin States for which S = 1/2 or I = 1/2.

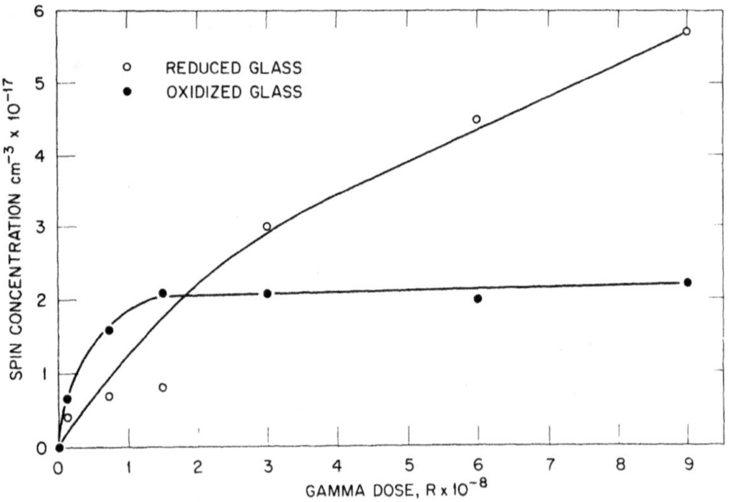

Figure 2. Concentration of Spins in the Superimposed β and γ Complexes as a Function of Gamma Ray Dose.[11] These paramagnetic centers are produced in pure GeO_2 glass and the β center is attributed to resonance transitions of O_2^- molecules in the form of peroxy radicals.

$\nu \sim 10$ GHz ($H \sim 3500$ oersteds) and room temperature the ratio of electrons in the excited state (excited by thermal vibrations) to those in the ground state is $\sim 10^{-3}$. The signal intensity is proportional to this difference and since the ratio increases with decreasing temperature, measurements at low temperature are, in some cases, desirable.

EPR transitions associated with impurities either normally paramagnetic (elements from the transition group for example) or altered to a paramagnetic state by photon irradiation will be called extrinsic. Another class of paramagnetic states, which will be called intrinsic, are due to various electronic states of the principal components. Some of these may be generated by fusing a glass under reducing conditions, under high oxygen pressure or by thermal treatments. These and others may be formed by irradiation with energetic photons or particles such as electrons, protons, or neutrons. In the alkali halides energetic photons have a high probability of displacing ions[8], while displacements by photons have a very low probability in many oxides.[9] In oxides energetic particles will displace ions. Threshold energies for displacement are of the order of 50 eV[9,10] and hence adequate momentum transfer is achieved for electron energies >0.4 Mev.

For an investigation of those structures in a glass which are a function of conditions of formation and in which paramagnetic states can be generated photon irradiation is probably desirable. In borate, silicate, germanate, and phosphate glasses, the ratio of the concentration of paramagnetic states to dose is high initially and then tapers off to a very small value. This effect is illustrated in Fig. 2 for an intrinsic paramagnetic state in an oxidized germanate glass[11] which has been called the β center by Purcell and Weeks. The maximum concentration corresponds to only 10^{-5} mole fraction, i.e. only one GeO_2 molecule in 10^5 is involved in this particular paramagnetic state. It is obvious that there are difficulties in relating such a small concentration to other properties of the glass. However, it should be noted that this concentration of states may have an optical absorption band in the visible region of the electromagnetic spectrum whose intensity would be sufficient to make the glass opaque in this region of the spectrum. Such concentrations may also alter dielectric properties by a significant factor. Thus an understanding of phenomena which involve a small fraction of the atoms in a glass can be crucial to controlling optical and dielectric properties important in technical use of such a glass.

The other curve in Fig. 2 shows the way in which the ratio of concentration to dose is altered by making the glass under reducing conditions. The ratio is constant for doses $>3 \times 10^8$R up to a dose of 9×10^8R in contrast to the oxidized glass in which the ratio decreased for doses $>3 \times 10^8$R. Irradiations to higher dose were not carried out in this case. Does this difference between the oxidized and reduced glass arise from a minor perturbation of the glass structure or is it indeed due to structural effects which are related to some bulk properties, to optical absorption bands, or to some other effect such as dielectric relaxation processes?

Establishing such relations is one of the difficult aspects of research on glasses. Studies of crystalline solids are in many respects simpler because of their periodic structure and the concomitant symmetry relations which can be utilized in determining the atomic surroundings of such an electronic state. Insight into the structure of these electronic states in glasses can be gained from investigations of crystalline forms of glass-forming compounds. This relation between a paramagnetic state in one of the crystalline forms of a compound and its glass form can be shown in the case of the E_1' center[12] in α-quartz and SiO_2 glass. In single crystals of α-quartz doped with Ge gamma-ray irradiation produces E_1' type centers at oxygen vacancies with Ge in one of the two nearest neighbor sites.[13] A similar paramagnetic state has been observed in irradiated hexagonal (α-quartz form) and glassy GeO_2[11,14] crystalline and glassy P_2O_5[15], and in alkali borate glasses.[16] In the hexagonal form of SiO_2 and GeO_2, in all the crystalline phases of P_2O_5, and in the alkali borate glasses the metal ion has four nearest neighbor oxygen ions and the oxygen two nearest neighbor metal ions(Me). (The fraction of fourfold coordinated B ions in alkali borate glasses is a function of composition[17,18].)

In all of these cases the angle formed by the metal ion-vacancy-metal ion is less than 180°. One consequence of this lack of colinearity and the highly directional character of the metal ion electron states (sp hybrids) through which the bonds are formed is that upon removal of an oxygen ion (by displacement or as a result of charge compensations required by incidental or deliberately added impurities) and subsequent trapping of an electron at the vacancy the electron density is not equally distributed between the two metal ions adjacent to the vacancy at any given time. This asymmetry in electron density is revealed by the interaction between the electron and the nuclear magnetic moment of the metal ion upon which its density is largest. Table II lists isotopes with their spins and natural abundances which have been observed through their interactions with an E' type center. Interaction between magnetic nuclei and the magnetic electrons splits the electron resonance transition into 2I + 1

Table II. Some Isotopes with Nuclear Magnetic Moments

Isotope	(Spin) I	Natural Abundance in %
	a) Elements Used to Study E' Centers	
^{10}B	3	18.83
^{11}B	3/2	81.17
^{29}Si	1/2	4.70
^{31}P	1/2	100
^{73}Ge	9/2	7.61
	b) Other Elements of Interest in NMR	
^{2}H	1	0.016
^{7}Li	3/2	92.57
^{9}Be	3/2	100
^{17}O	5/2	.04
^{23}Na	3/2	100
^{27}Al	5/2	100
^{51}V	7/2	99.75
^{85}Rb	5/2	72.5
^{133}Cs	7/2	100

transitions[4] of equal intensity. For the E' type centers in alkali borate glasses 18.83% will be split into seven and 81.17% will be split into four transitions, in SiO_2 glasses 4.70% exhibit two transitions, in P_2O_5 glasses 100% exhibit two transitions, and in GeO_2 glasses 7.6% exhibit 10 transitions. Although E' type centers have been detected in glasses fused from oxides of these elements investigations of the E' type paramagnetic states in crystal systems have been largely made on hexagonal SiO_2 and GeO_2.

The intrinsic nature of these defects has been confirmed by their increase in concentration with increasing time of irradiation in a nuclear reactor in which the high flux of energetic neutrons displaces ions and the intense gamma ray photon flux excites electrons. Figure 3a shows the concentration of the E' center in a hexagonal single crystal of SiO_2, and in four commercial varieties of SiO_2 glass as a function of the integrated flux of energetic

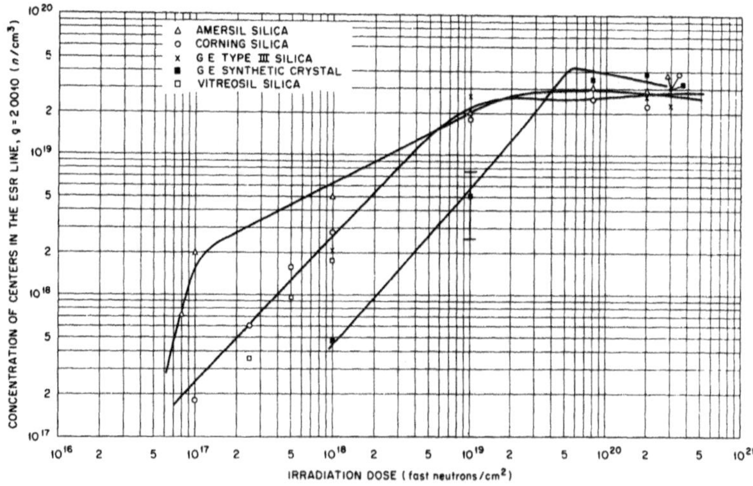

Figure 3. Concentration of Paramagnetic Centers as a Function of Fast Neutron Dose in Several Silica Glasses and a Synthetic Crystal.

a) Concentration of E_1' Centers Determined from ESR Measurements.

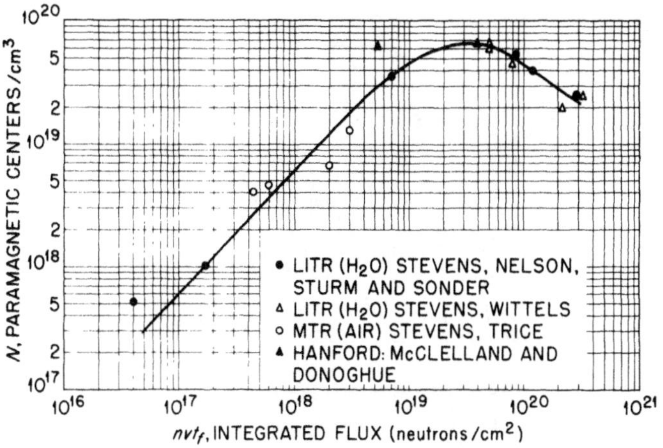

Fast-Neutron-Bombardment-Introduced Paramagnetic Centers in Natural Quartz.

b) Total Concentration of Paramagnetic Centers Determined from Magnetic Susceptibility Measurements. It has been assumed that all the paramagnetic centers have a spin state S = 1/2.[19]

neutrons (hereafter called "dose"), and in Fig. 3b[19] the total concentration of paramagnetic states, as determined from a measurement of magnetic susceptibility, is also shown as a function of dose. The number of E' centers produced in the glasses per incident neutron is approximately four times greater than in the single crystal at doses $\leq 10^{19}$ neutrons cm^{-2}. For doses $<3 \times 10^{19}$ neutrons cm^{-2} the concentrations ($\sim 2 \times 10^{19}$ centers cm^{-3}) are effectively the same and independent of dose. This effect is also evident in the total concentration as a function of dose (Fig. 3b). After a dose $>10^{20}$ neutrons cm^{-2} the crystalline order has been destroyed and the sample is in an amorphous state with a density (2.26) ∿12% less than that of α-quartz. The glass samples for the same dose have a density ∿3% greater than initially and equal to that of the irradiated crystal.[20,21] Annealing at 1000°C for several hours reduces the density of both materials to 2.20,[22] the initial density of the glasses. The E' centers disappear at lower temperatures as illustrated in Fig. 4. E' type centers disappear at much lower temperatures, ∿350°C, when produced only by gamma-ray irradiation.[23]

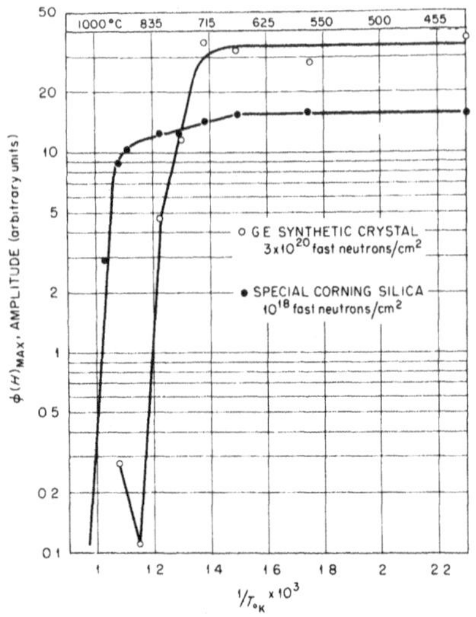

Figure 4. Intensity of E′ Center Resonance in Neutron Irradiated Crystalline and Glassy SiO_2 as a Function of Five-Minute Isochronal Anneals. The "Special Corning Silica" had a low concentration ($< 10^{18}$ cm^{-3}) of hydroxyl molecule ions.

The onset of the transformation of the crystal to a glass occurs only after a neutron dose $>3 \times 10^{19}$ fast neutrons cm^{-2} or after the number of oxygen vacancies as determined from the concentration of E' centers exceeds 10^{19} cm^{-3}. Guinier[21] has proposed mechanisms for this transformation in which the strains induced in the α-quartz structure by the vacancies and displaced ions induce transformations of the α structure to high temperature phases in local regions. The orientation of the transformed regions is essentially random. In this case a small concentration of defects, ∿0.1% of the crystal sites are altered, produce significant alterations in some of the bulk properties of the material, e.g.,density, thermal conductivity, and crystalline order.

Another aspect of these effects is the difference in the temperatures at which paramagnetic states of the E' center disappear in neutron irradiated material and in gamma-ray irradiated material. It was pointed out above that the probability of displacing ions by gamma-ray irradiation is orders of magnitude smaller than is the probability of displacing ions by energetic particles, e.g., neutrons. Hence the production of the paramagnetic E' state by photon irradiation takes place by exciting electrons which are then trapped at oxygen vacancies already present in the glass or crystal. (There is evidence that in some glasses displacement processes may occur[11,15].) Hence, annealing of the state requires thermal excitation of the trapped electron and its return to the site from which it was excited. Thermal energies required for this process are usually small. The higher thermal energies required for annealing the paramagnetic state in neutron irradiated glasses and crystals can be explained in the following way. The oxygen ion displaced by energetic neutrons is trapped at a site for which the thermal excitation energy is much higher than for the excitation of E' electrons. The E' electrons are derived from the displaced O^{--} ion leaving an O^{-} ion trapped. During a thermal treatment the E' electrons, although excited at the same energies as in photon irradiated crystals, are repulsed by electrostatic forces from the O^{-} ions from which they were initially excited and hence must return to the oxygen vacancy site. Annealing takes place only when thermal energies sufficient for excitation of the O^{-} ion are reached and these are higher than those for excitation of the E' electrons. The O^{-} ion then returns to the oxygen vacancy site.

These proposed models for radiation induced processes in SiO_2 glass and crystals are in reasonable agreement with observed effects. However, they may not be of general applicability to glasses of other compounds. Figure 5 shows the concentration of the E' type center in high purity germanate glass when fused under oxidizing and reducing conditions. The concentration in the oxidized glass is a linear function of dose and does not exhibit the initial rapid increase shown in Fig. 2 for another paramagnetic state in this glass. The reduced glass has a greatly enhanced ratio of concentration to dose. The effect upon the ratio of concentration to dose of the E' type center when an alkali oxide is added to the melt is shown in Fig. 6. Oxygen added to the glass in this way has altered the ratio of concentration to dose with the ratio decreasing after an initial rapid increase and also the maximum concentration is much less. This dependence of concentration upon dose is expected if no displacements are produced. Since the effects of photon irradiation on the pure glass whether oxidized or reduced are quite different from the alkali germanate glasses they may be due to displacement

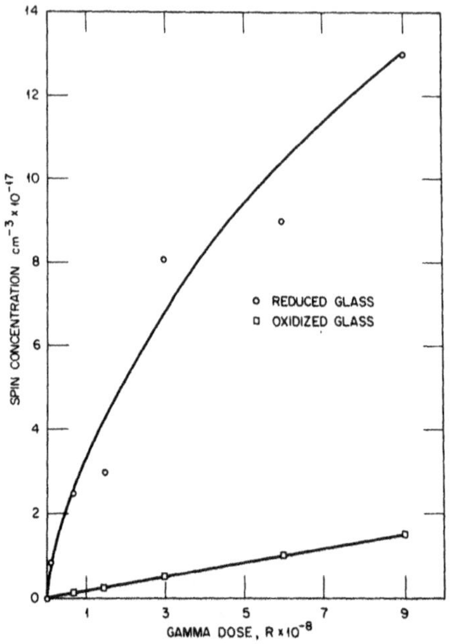

Figure 5. Spin Concentration of E′ Type Centers in GeO_2 Glasses as a Function of Gamma Ray Dose. The dose rate was $1.5 \times 10^6 R\ hr^{-1}$ in a ^{137}Cs source.[11]

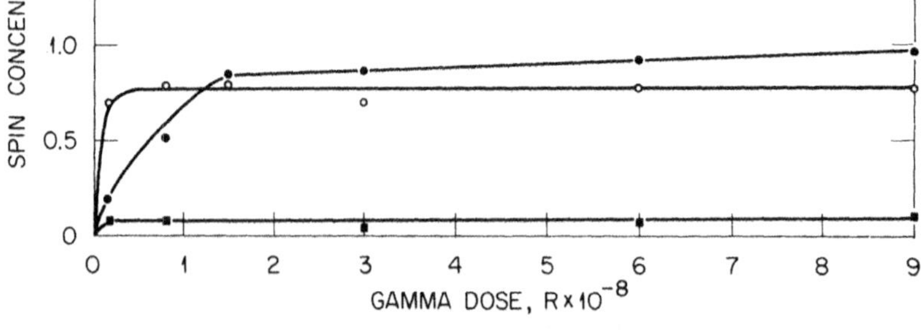

Figure 6. Spin Concentration of E′ Type Centers in Alkali Germanate Glasses as a Function of Dose. (T. Purcell, Ph. D. Thesis, The University of Tennessee Graduate School, August 1968.)

processes.

The preceding discussion has shown some of the ways in which an intrinsic paramagnetic state can be related to other processes and structures in glasses. Inferences about the structure of the glass in the vicinity of the E' center can also be drawn from the magnetic properties. Some characteristics of the electronic state of the electron can be inferred from the interaction of the electron magnetic dipole with nuclear magnetic dipoles in the vicinity of the electron. In all of the glasses considered here the electronic wave functions which describe the electron states are s and p functions (B^o: $2s^2 2p^1$, Si^o: $3s^2 3p^2$, Ge^o: $4s^2 4p^2$, P^o: $3s^2 3p^3$, O^o: $2s^2 2p^4$). As a first approximation the atomic configuration of the site of the E' center may be taken as a MeO_3 molecule ion, hence $(BO_3)^{4-}$, $(SiO_3)^{3-}$, $(GeO_3)^{3-}$ and $(PO_3)^{2-}$. The Me ions have a s-p hybrid orbital not involved in forming bonding and anti-bonding orbital states with the oxygen ions, and whose energy is greater than the bonding and less than the anti-bonding orbitals. It is this orbital in which the paramagnetic electron is found. The angle between the Me-O bonds and this nonbonding s-p hybrid orbital as well as the fraction of the electronic density which can be attributed to this orbital will determine the hyperfine interaction with those Me nuclei which have magnetic moments. Table III, part of which was assembled by Griscom,[16] gives these estimates along with the principal g-values. Most of the electronic density is in the nonbonding s-p hybrid. Hyperfine interactions with other nearby nuclei in the case of the E' (Si) and E' (Ge) centers observed in single crystals show that small additional admixtures of wave functions from other ions may account for part of the remaining electronic density. In addition polarization of the core electrons of the Me ion by the E' electron will alter the isotropic part of the hyperfine interaction thereby changing the estimate of the s part of the wave function.

Another paramagnetic state apparently associated with oxygen vacancies has a spin state $S = 1$[11,24,13] with separation between the singlet state and the triplet state of $\sim 0.04\ cm^{-1}$ with the singlet state lowest. That the singlet state is lowest is shown by measurements at 4^oK. Resonance transitions within the triplet state are not detected at this temperature. The model proposed for this defect is an oxygen vacancy with two trapped electrons

Table III. Electronic State of E' Type Centers

Material	p-State Density	Hybridization Ratio, p/S	Fraction of Electron Density in s-p Hybrid	g_x	g_y	g_z	Ref.
Alkali Borate glass	0.45	3	0.6	2.001	?		16
	0.85	5.7	1.0				
SiO_2 glass and α-quartz	0.63	2.6	0.87	2.0018	2.0006	2.0003	13
Ge in α-quartz	0.80	1.3	≤ 1.0	2.002	1.9950	1.9939	13
GeO_2 α-phase crystal and glass				2.002	1.9947		14
Phosphate glass	0.6	≤ 1.5	1.0	(2.00)			15
Alkali phosphate glasses	0.6	≤ 3	0.8				15

creating a defect with zero charge. One of the electrons is thermally excited between 78°K and 300°K since the state is observed only when gamma ray irradiations and measurements are carried out at 78°K. However, if a crystal is neutron irradiated at ∿150°K the S = 1 states are stable for a period of a day at room temperature, and several types are detected.[25]

Hydrogen, as an impurity, is common to almost all natural and synthetic α-quartz and to silicate glasses. A fraction of the E' type centers present initially in crystals of α-quartz have in their vicinities hydrogen ions. The nearby hydrogen ion perturbs the resonance spectrum of the E' center, (labeled E_2') by a hyperfine interaction between the E' electron and the nucleus of the hydrogen ion, the proton, which has a nuclear spin I = 1/2. One example of this interaction is shown in Fig. 7. The magnitude of the hyperfine interaction, assuming a simple electron magnetic dipole-nuclear magnetic dipole interaction, is determined by the separation of the two dipoles. In this case the separation is ∿3 Å,[26] which is also, approximately, the Si-Si separation. At other E' sites the hydrogen nucleus and the E' electron are separated by ∿1.5 Å[27]. This center has been labeled E_4' [23].

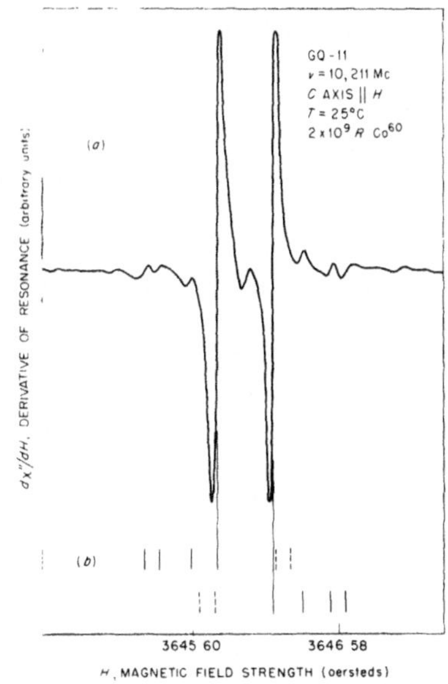

Figure 7. Spectrum of E_2' Centers in α-Quartz for the Magnetic Field Parallel to the Three-Fold Crystal Axis. The weak absorptions on either side of the two central lines are probably due to hyperfine interactions of the E_2' centers with nearby ^{29}Si nuclei.[26] The lines in b) indicate the probable positions of the hyperfine lines.

These E' type centers do give an indication of some of the ways in which hydrogen ions are incorporated into α-quartz. Because of the small magnitude of these hyperfine interactions and the averaging effect which occurs in a glass it is not possible to resolve these types of E' centers from E_1' centers in silica glass by the usual type of resonance experiment. Because the concentration of oxygen vacancies and of hydrogen is usually higher in silica glass than in α-quartz E_2' and E_4' type centers are expected to occur in silica glasses.

In those α-quartz crystals in which these defects have been detected the total hydrogen concentration is at least an order of magnitude greater than the total concentration of all the E' type centers. Hence most of the hydrogen is incorporated into these crystals in some other manner.

Although E_1' type centers have been observed in several other oxide glasses and their crystalline forms, E_2', and E_4' type centers have not been detected. Hence it may be that these particular vacancy-impurity associations are found only in α-quartz and, possibly, in silica glass.

Since oxygen vacancies are produced in oxide glasses by irradiation with energetic particles, interstitial oxygen ions must necessarily be present if they do not diffuse to the surface. There are no data showing diffusion to the surface hence these ions must be in interstitial sites. There are resonance absorptions that have been attributed to various forms of interstitial oxygen ions[11,28] and to other forms of paramagnetic oxygen[29,30]. One group of compounds, the peroxyborates, which are, in their crystalline forms, diagmagnetic become amorphous and intensely paramagnetic upon heating.[31]

These paramagnetic forms of oxygen are O^- and molecule ions such as O_2^- and O_3^-. The O^- ion may occur as an interstitial,[11] as a bridging ion between three- and four-fold coordinated boron or between a substitutional impurity such as Al and the nearest neighbor Si in a silicate glass, or as a nonbridging oxygen ion. Generally only a small fraction of the total concentration of such bridging and nonbridging ions became paramagnetic after an appropriate treatment such as irradiation with x-rays or gamma-rays. One reason for this small fraction is that trapping sites for electrons excited from these oxygen ions have a low concentration. Hence the concentration of paramagnetic forms of oxygen is limited by the concentration of electron trapping sites. Another reason for not detecting a paramagnetic state of a non-bridging oxygen ion is that most of these ions are in electric fields of axial symmetry.[28] Resonance transitions will not be detected when an O^- ion is in an electric field of cubic symmetry and in some cases for fields of axial symmetry. This possibility could be checked by a measurement of the temperature dependence of the magnetic susceptibility.

Normally diamagnetic impurities may also become paramagnetic. Some of these have been discussed in a review paper by Lell, et al.[22] Elements of the various transition groups (Ti--Cu, Zn--Pd, and Ce---Lu) are usually paramagnetic[4] and when present in a glass may exhibit resonance transitions. The most easily detected are the iron group of elements. The valence states of these ions which are usually observed are Ti^{3+}, (d^1), V^{4+} (d^1), Cr^{3+} (d^3), Mn^{2+} (d^5), Fe^{3+} (d^5), Ni^{2+} (d^8), and Cu^{2+} (d^9).[33] Kim and Bray[32] showed that low temperature gamma ray irradiations and measurements without warmup were necessary to observe the Ti^{3+} state in titanium silicate glasses. Purcell and Weeks[34] showed that these conditions were also necessary for the production of singly charged oxygen vacancies and interstitial Ti^{3+} in rutile single crystals. With respect to Fe^{3+} there is considerable controversy about the oxygen coordinations. Resonance and optical absorption measurements have not been sufficient to resolve the question. In the discussion of measurements of the Mossbauer transitions this problem will be considered again. The interaction of the Fe^{3+} electrons with the electric fields at the iron sites is frequently larger than the photon energies of available spectrometers and the spectra in glasses are not easily interpreted in terms of particular electric field symmetries. The other d^5 ion, Mn^{2+}, has a smaller interaction for a given electric field[4] and measurements are easier. The 100% abundant ^{57}Mn isotope also produces a six line hyperfine structure which is easily identified.

Mn^{2+} has been used as a probe of the electric field with which it interacts. This interaction has been used to detect the alterations of structure as a function of vitrification or devitrification. Chatelain and Weeks[35] observed such effects in $ZnCl_2$:Mn. Several stages in the process of devitrification from a glass produced by fusion were detected while the vitrification process resulting from neutron irradiation proceeded through different stages. Griscom and Griscom[36] have also investigated the spectrum of Mn^{2+} in lithium borate glasses and compounds. There are numerous other excellent studies of the transition elements in glasses too numerous to list here.

NUCLEAR MAGNETIC RESONANCE

Since the ratio of the mass of the electron to the mass of a nucleon, e.g., a proton, is $\sim 5 \times 10^{-4}$ and splitting of spin states by a magnetic field is proportional to the gyro-magnetic ratio of a particle, the splitting of nuclear magnetic states will be $\sim 5 \times 10^{-4}$ smaller than the splitting of electron magnetic states. Thus, the population difference between the two nuclear magnetic states will be $\sim 5 \times 10^{-4}$ that of the electron states. Absorption of energy from a photon flux is proportional to this difference in population and hence NMR signals are proportionately weaker than EPR signals for the same number of spin states. In addition to this basic physical difference NMR spectrometers have resonant systems whose Q's are ∿100 times smaller than those for EPR spectrometers. It is evident then, that NMR spectrometers are $\sim 10^6$ times less sensitive than EPR spectrometers. Because of this low sensitivity samples usually have concentrations of nuclei of interest which are greater than a few atomic percent and sample volumes of a few cubic centimeters. Resonance transitions of isotopes with a low natural abundance are difficult to detect without isotopic enrichment of compounds containing them. It is now possible to obtain from the Oak Ridge National Laboratory many elements highly enriched in specified isotopes.

Interactions between electric fields at the site of a nucleus and the nucleus are one of the primary concerns of NMR studies of glasses. This field is a function of electric charges and their distribution around the nucleus whose NMR is to be measured. Hence, the electrons of the atom in which the nucleus is located and those on nearest neighbor ions have the largest effect. The distribution of these charges are determined by the nature of the bonds between the atoms which in turn is determined by their chemical states. For elements whose nuclei have a nuclear spin (for example ^{2}H, ^{29}Si, ^{31}P), $I = 1/2$, the splitting of the two spin states is shifted by these effects and is called a chemical shift. If the electric field at the nucleus has a non-vanishing gradient determined by the symmetry of surrounding charges, this shift will be a function of the orientation of the field with respect to the applied laboratory field. In a single crystal a single symmetrical absorption line will be observed which will be observed at different magnetic fields as the crystal is rotated in the field. In a powder or glass of the single crystal the absorption will be asymmetric and have a width uncompassed by the maximum and minimum fields at which transitions could be observed in the crystal. The calculation of powder patterns from single crystal data and the converse extraction of important parameters from the spectrum of a glass have been dealt with in detail.[37]

If the nuclei have $I \geq 1$, the nuclear levels are split by a magnetic field into $(2I + 1)$ states and in an electric field of cubic symmetry the $(2I + 1)$ states are equally spaced and only one transition is observed (chemical shifts can then be measured to determine some of the characteristics of the bonding to surrounding atoms). For an electric field of lower symmetry the interaction between the field and quadrupole moment of the nucleus shifts the energies of the states with the separations of the $m_I = \pm 1/2$ being unchanged to first order. Second order terms in the interaction energy do perturb this separation which is then a function of the angle between the principal axes of the electric field and the applied laboratory field. Transitions between the other levels are dependent upon this angle to first order.[7] In a glass the electric field and its gradient at each nuclear site differs from that at the next site because of the non-periodicity of the glass structure. These perturbations may produce such a distribution of energies for those states whose energies are determined by first order effects that only the $m_I = 1/2 \longleftrightarrow m_I = -1/2$ transition is detected. In this case deductions about the distribution of ions about the nuclei whose NMR is being measured are made from second order perturbations of the $m_I = 1/2 \longleftrightarrow m_I = -1/2$ transition. Cohen and Reif[38] have considered in detail the effects of perturbations in single crystals which arise from crystal strains. The perturbations of electric fields in glasses by the randomness of structure is analogous to the effects of strain produced by point defects in crystals.[39] The structure of the $m_I = 1/2 \longleftrightarrow m_I = -1/2$ transition for various ratios of frequency and quadrupole coupling constants, i.e., the strength of the interaction between the electric field gradient and the quadrupole moment for the asymmetry parameter η which is a measure of the deviation of the electric field from axial symmetry and for chemical shifts have been analyzed in detail by Baugher et al.[40] These authors also relate their calculated powder patterns to the spectrum of ^{51}V in three metavanadate compounds.

One of the most extensive applications of nuclear resonance techniques to an investigation of the structure of glassy materials has been carried out by Bray and his collaborators. Bray has given summaries of a portion of this work.[18,41] Investigations of the structure of borate glasses have shown that the borate glass anomaly[42] is not related to any discontinuity in the transformation of three coordinated borons to four coordinated borons with increasing additions of alkali oxide to the glass.[18] In these cases the spectrum, the asymmetric $m_I = 1/2 \longleftrightarrow m_I = -1/2$ transition, of ^{11}B (samples are enriched in ^{11}B) surrounded by three oxygens can be resolved from the almost symmetric transition for those nuclei surrounded by four oxygen ions.

Curiously, in anhydrous sodium and potassium metaborates, in which a large fraction of the B ions are four coordinated, this transition is not detected.[43] One non-bridging oxygen has been shown to increase the asymmetry parameter of three coordinated ^{11}B from $\eta < 0.2$ to $\eta = 0.55 \pm 0.01$. The spectrum of four coordinated borons with one or two non-bridging oxygen ions has not been identified. Griscom's analysis of an EPR spectrum in borate glasses has shown that a small fraction of the boron sites have the boron forming the peak of a pyramid with the base formed by the three oxygens. The concentration of paramagnetic centers on such sites was not large, and the concentration of borons, determined from the EPR measurement, is certainly not enough to detect in an NMR measurement. The possibility does exist that the total concentration of such configurations is much larger or might be enhanced by an appropriate thermal treatment, or for some particular glass composition. Its presence does show that the deduction of structure from an analysis of NMR spectra is not exclusive of other atomic structures which may be important determinants of glass properties.

If a particular group of ions in a solid such as Li^+ are in motion and the motion is of a random walk type, and if the motion is sufficiently rapid, dipole-dipole and quadrupole interactions average to zero, and symmetric, isotropic absorption is observed. This motional narrowing[44] occurs when the reorientation rate or jumping frequency, is greater than the rigid lattice line width (expressed as a frequency). Bishop and Bray[45] have observed this effect in the spectrum of glassy and crystalline lithium borates. On the basis of the temperature dependence of this motional narrowing, activation energies were calculated from the Arrhenius equation, giving $0.12 \text{ eV} \leq E_a \text{ (glass)} \leq 0.22$. The value of E_a was shown to be dependent upon the glass composition, while for the compounds $Li_2O \cdot 4B_2O_3$ and $3Li_2O \cdot 2B_2O_3$, E_a was 0.14 and 0.43 eV respectively. Similar effects were observed in a lithium-silicate glass. They were not observed in potassium-borate or silicate glasses.

These activation energies are approximately three times smaller than those determined from ionic conductivity experiments carried out at higher temperatures. Bishop and Bray suggest that the ion motion which is observed in their experiments occurs between shallow potential wills, while that observed in ionic conductivity measurements occurs between much deeper potential wells. Since motional narrowing effects were also observed in crystalline compounds and activation energies were of a similar magnitude to those in the glasses it is of interest to know whether the crystalline compounds also exhibit the same type of ionic conductivity as the glasses. An investigation of the conductivity mechanisms in such crystalline compounds would certainly help in understanding the conductivity mechanisms in glasses fused from the compounds.

Chemical shifts of nuclei with $I = 1/2$ can be interpreted in terms of the degree of ionicity and covalency of the chemical bonds with which the atom containing the nucleus is bound in a compound or glass.[46] These shifts for ^{207}Pb ($I = 1/2$) have been measured as a function of composition in lead silicate glasses and in lead borate glasses. The chemical shift in lead silicate glasses with low PbO concentration up to $PbO.SiO_2$ is characteristic of the shift in the compound $PbO.SiO_2$. For compositions with higher PbO content the shift decreases indicating an increasing covalency in the Pb-O bonding. In lead borate glasses the shift decreases monotonically with increasing PbO. The shift in the glasses at compositions corresponding to crystalline compounds in the lead borate system is approximately the same and decreases towards the chemical shift of orthorhombic PbO. Bray[18] suggests that the resonance for a glass of any composition is an unresolved mixture of the resonances from the lead-oxygen configurations which are present in the crystalline compounds. This condition may exist in many other glasses in which the atomic structure on a local atomic scale is similar to that present in the compounds from which the glasses are formed. An investigation of the chemical shift of ^{57}Fe in glasses would certainly help towards resolving the controversies about the coordination of Fe.

Another recent application of NMR techniques has been an investigation of lunar material returned by the Apollo flights. These materials are principally silicate minerals of which a major component is the plagioclase group of minerals. Most of the Al in lunar material is in this mineral. Fig. 8 shows the resonance of ^{27}Al of one specimen of lunar material and of a terrestrial mineral whose quadrupole coupling constant and asymmetry parameters are of the same magnitude as those of the lunar material.[47] The ^{27}Al

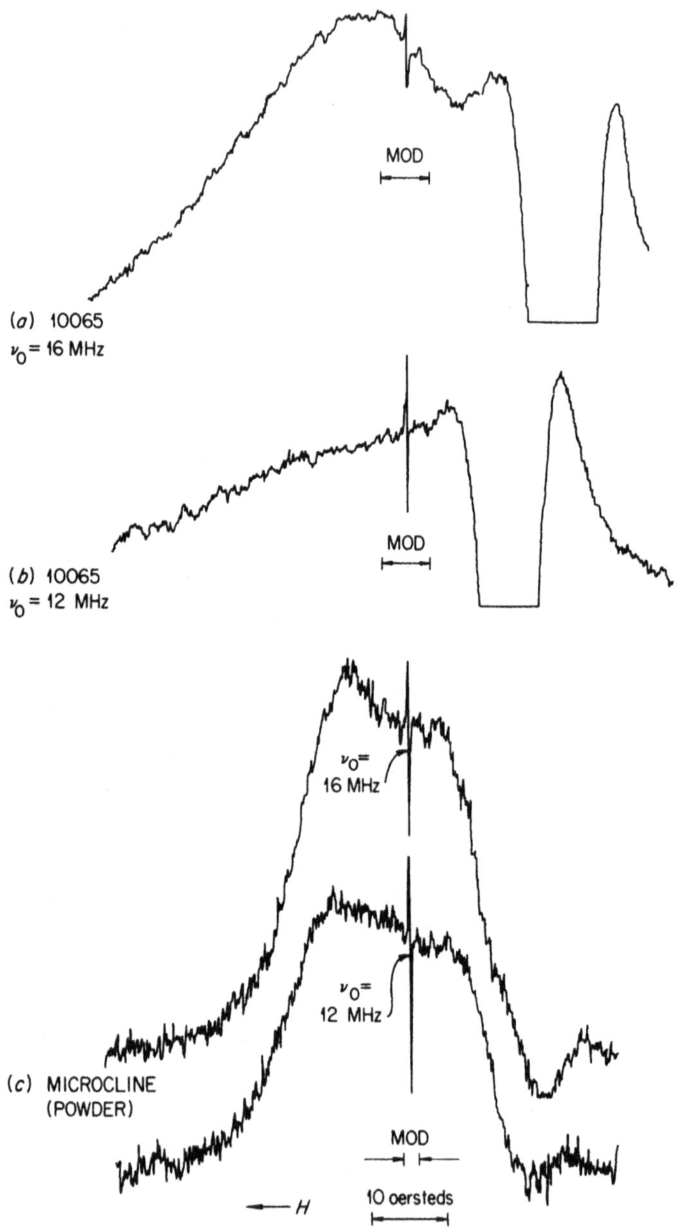

Figure 8. Nuclear magnetic resonance spectra of ^{27}Al in a specimen of lunar rock 10065 (a breccia) and in terrestrial microcline at two frequencies.[47] The dispersion mode was used for the recording.

resonance is attributed to the $m_I = 1/2 \longleftrightarrow m_I = -1/2$ transition whose structure is determined by second order effects. Using a gaussian distribution of quadrupole coupling constants and asymmetry parameters and adding a small local field of ~5 gauss, a computer calculated absorption curve fits the observed curve quite well. The small local field may be due to magnetic fields of ferrimagnetic particles or to an anomalous chemical shift.

These brief comments on the use of NMR of various nuclei to probe the structure of glasses have covered only a few of the applications. These probes of glass structure have penetrated to a microscopic scale not accessible to x-ray diffraction techniques and to most other techniques used for the study of atomic structure of solids. Other applications have been the determination of the effects of hydrogen ions on the structure of borate glasses[48] and minerals.[49] They have also been an important adjunct to structure determinations in crystalline compounds.[38] There is an important parameter which has not been thus far considered in glass research and that is the spin-relaxation time. This is the time required for 1/e of the spins excited to the higher energy state to return to the lower energy state. These times are frequently very long for NMR and it is then necessary to provide a magnetic coupling between the nuclear spin states and the thermal vibrations of the lattice in order that this time be reduced to permit observation of the absorption of energy. This coupling is usually to a paramagnetic ion such as Fe^{3+} or Mn^{2+} to which the nuclear spins can magnetically couple their energy. This energy is then transferred to the thermal vibrations much more rapidly than through the nuclear spin-structure vibration coupling because of the more rapid spin-lattice relaxation of the paramagnetic ions. However, the direct coupling to the temperature dependent thermal vibrations can be measured as a function of temperature. Since this coupling is determined by the atomic structure, it is another parameter specifying the structure.

MOSSBAUER SPECTROSCOPY

An excellent review article by Kurkjian[50] has recently been published and much of the following is based on this article . The article contains a good bibliography.

Many of the elements in common glass forming-compounds

do not have isotopes which can be used for Mossbauer spectroscopy, and Mossbauer transitions have not yet been detected for some isotopes which are of interest in these materials. Frauenfelder[6] lists nuclides of interest in Mossbauer experiments. Table IV lists a selection of these which may be of interest in studies of glasses. Imaginative experiments using isotopes not listed in this table should be considered. The parameters which are determined from Mossbauer experiments are:

1. ratio of recoil-free to recoil emissions and absorptions. This ratio should be as large as possible for a large Mossbauer effect. Since it is determined by the amplitude of atomic vibrations and these are a function of the ratio of sample temperature to the sample Debye temperature this ratio should be as small as possible. This effect is clearly demonstrated by experiments on the two crystal phases of GeO_2. The recoil-free fraction in the tetragonal phase is five times smaller than in the hexagonal phase for the same sample temperature.[51]

Table IV. Isotopes of Interest for Mossbauer Spectroscopy in Glasses

Isotope	Gamma-ray Energy (Kev) E_ν	Half-life of Excited State $T^{1/2}$ (sec)	Q E_ν/h	$\frac{lm^2}{T^{1/2}}$
^{40}K	29.4	3.9(-9)[2]	2.5(11)	
^{57}Fe	14.4	1.0(-7)	3.2(12)	
^{61}Ni	71	5.2(-9)	8.1(11)	
^{67}Zn	93	9.4(-6)	1.9(15)	
^{73}Ge	13.5	3.1(-6)	1.4(14)	
^{107}Ag	93	44.3	9.0(12)	
^{119}Sn	24	1.9(-8)	1.0(12)	
^{125}Te	35	1.6(-9)	1.2(11)	
^{133}Cs	81	6.0(-9)	1.1(12)	
^{238}U	44.7	2.3(-10)	2.2(10)	

Most rare earth elements have Mössbauer isotopes

The numbers in parentheses are the exponents of powers of 10, e.g., $(-9) = 10^{-9}$.

2. hyperfine interactions. If there are oriented magnetic fields, e.g., those from ferri-, or ferromagnetic states, interacting with the Mossbauer nucleus these will perturb the nuclear states, removing their degeneracies. Both the ground state and the excited state will experience this splitting. Assuming that such internal fields exist only in the absorber material, there will be (2I 'ground' + 1) + (2I 'excited' + 1) levels. In the case of ^{57}Fe there will be six sublevels, I(ground) = 1/2 + I (excited) = 3/2. Transitions between these sublevels may be resolved if magnetic domains are randomly oriented.

3. chemical or isomer shift. This interaction is similar to the NMR chemical shift. It is a function of s electron density at the nucleus and hence is determined in part by the nature of bonding between the Mossbauer ion and nearest neighbor ions.

4. quadrupole interactions. These are the same as those affecting NMR transitions. Whereas these are zero for I(ground) = 1/2 (^{57}Fe), they are not for the excited state I(excited) = 3/2 which is split into a $\pm$ 1/2 and a $\pm$3/2 state. Two transitions will be observed. It is possible to deduce atomic structure from these interactions. These deductions could not be made from an NMR experiment on the same nucleus. There are relativistic effects but these will not be discussed.

EPR spectra of Fe^{3+} in various glasses could be due to four-fold or six-fold coordination. Electric fields with symmetry lower than axial can result in either case. In these low symmetry fields in which the d state degeneracies are completely removed, it is not possible to unambiguously calculate g-values in the absence of a specific model. In addition, EPR measurements of Fe^{2+} are difficult. It was thought that Mossbauer experiments might resolve these questions about coordination in glasses for both Fe^{2+} and Fe^{3+}. Careful and extensive experimentation have resolved this quandry for a few glasses. On the basis of a comparison of isomer shifts of Fe^{3+} in crystalline compounds for which the coordination is known with the isomer shift of this ion in various glasses, Kurkjian[52] suggests that in alkali silicate and alkali iron silicate glasses Fe^{3+} is fourfold coordinated while in three phosphate glasses the coordination is sixfold. The isomer shifts for Fe^{2+} are two to three times larger than those for Fe^{3+} and in various compounds range from 0.053 cm sec^{-1} for four-fold coordinations in a plapar array to 0.110 for a distorted 8-fold coordination.[53] In phosphate glasses the Fe^{2+} isomer shift is larger than in silicate glasses.[50] However, in silicate glasses the distribution of values covers the range from four-fold (tetrahedral) to six-fold. There are apparently no obvious trends in the magnitude of quadrupole splitting as a function of coordination. Hence these effects have not been particularly helpful in assigning coordination numbers.

Uhrich and Barnes[54] have measured the Mossbauer spectra of Tm^{3+} in an alkali silicate glass. They find that the Tm^{3+} appears to preferentially occupy sites of well defined symmetry similar to one of the sites in Tm_2O_3. On the basis of the temperature dependence of the quadrupole splitting, they suggest that a degree of short range order exists. There is also evidence that the bonding of the Tm^{3+} is partially covalent.

Martarese, et al.,[55] in a preliminary report on an EPR spectrum of Fe^{3+} is synthetic quartz suggested that the Fe^{3+} was in Si sites. In a later more detailed analysis[56] they could not exclude the possibility that it was in an interstitial site although their analysis favored the substitutional site. On the other hand Fe^{2+} in natural silicate minerals has not been found in Si sites.[57,58] Whether this substitution is favored for Fe^{3+} is not known. It is evident that the much larger Fe^{3+} ion in a Si site would introduce distortions in site symmetry. Oxidation of olivine which contains ∿5 atomic percent Fe^{2+} causes a precipitation of iron rich phases[59,60] and hence it is difficult to obtain such minerals with Fe^{3+} randomly distributed in the structure.

CONCLUSION

It is useful in this discussion to consider some simple relations. These are the radii[61] of some atoms which are given in Table V. The point to note is that the radii of the charge states, Mg^{2+}, Ca^{2+}, Ti^{3+}, V^{3+}, Cr^{2+}, Mn^{2+}, Fe^{2+}, Co^{2+}, Cu^{2+}, Sn^{4+}, are all ∿1.6 times larger than the radii of Si^{4+}, Ge^{4+}, P^{3+} and P^{5+}. In view of this disparity in radii between Si^{4+}, Ge^{4+}, P^{3+}, P^{5+} and the other ions in Table V, it does not appear reasonable to think of the large diameter ions as substituting for these small diameter ions

Table V. Effective Radii of Atoms* (Å)

Charge State	B	Mg	Ca	Al	Si	Ge	P	Ti	V	Co	Mn	Fe	Co	Cu	Sn
2+		0.66	0.99					0.94	0.88	0.89	0.80	0.74	0.72	0.70	
3+	0.20			0.50			0.44	0.76	0.74	0.63	0.66	0.64	0.63		
4+					0.42	0.53		0.68	0.63						0.71
5+							0.35		0.59		0.60				

$R(O^{2-}) = 1.40$ R(alakli halides^{+1}) ~ 0.7

*Values taken from Handbook of Physics and Chemistry, p. 3507, 44th. Edition.

in their tetrahedral configuration. Considering them as hard spheres, a square configuration of O^{2-} ions has an ∿0.66 Å diameter hole and all of the ions in Table V have radii of this order. In a tetrahedral configuration the diameter of the hole is ∿0.35 Å. Hence, tetrahedral coordinations of the transition group of elements require larger volumes than the same coordination for Si, Ge, P. One approach is to regard the SiO_4, GeO_4, PO_4 tetrahedral molecules as relatively undisturbed in a glass, ions such as Fe^{2+} and Fe^{3+} may then be accommodated in the spaces between these tetrahedra in which spaces coordination may be four, five, or six and perhaps higher depending upon the other compounds added in making a particular glass. Since Zn^{2+} and Mn^{2+} have similar radii the substitution of Mn^{2+} for Zn^{2+} in $ZnCl_2$ does not produce a large distortion. In this case, the Mn^{+2} did occupy the four-fold coordinated Zn site.

It is well to recall the NMR experiments[18] in which there is evidence for the existence, in many glasses, of a local atomic order that is not greatly different from the local order existing in crystalline compounds most similar in composition to the glasses. Deviations from this order are clearly evident for example in the spectrum of Mn^{2+} added as a structure probe.[35,36] These deviations appear as small changes in the parameters of the terms of the electric field potential from one EPR, or NMR or Mossbauer site to the next. This distribution of values is usually assumed to have a gaussian distribution about some particular value.[37] which is determined from the data. These average values have been found to be the same as the values determined from crystalline material[35,18] in some cases.

ACKNOWLEDGEMENT

Dave Griscom very generously supplied copies of two of his manuscripts prior to publication. His permission to use some of the data in these manuscripts is deeply appreciated. John Barrett kindly permitted the use of his lecture notes on the Mossbauer effect.

REFERENCES

1. H. I. Schiff, Quantum Mechanics, pp. 112 & 292, McGraw Hill Book Co., New York, 1949.
2. L. Pauling and E. B. Wilson, Introduction to Quantum Mechanics , p. 321, McGraw-Hill Book Co., New York, 1935.
3. P. Pringsheim, Fluorescence and Phosphorescence , Interscience, New York, 1949.
4. J. W. Orton, Electron Paramagnetic Resonance , London Iliffe Books, Ltd., London; J. A. McMillan, Electron Paramagnetism , Reinbold Book Corp., N.Y., 1968; W. Low and E. L. Offenbacker, Electron Spin Resonance of Magnetic Ions in Complex Oxides. Review of ESR Results in Rutile , Perovskites, Spinel and Garnet Structures, in Solid State Physics 17, 136, Ed. by F. Seitz and D. Turnbull, Academic Press, N.Y., 1965.
5. A. Abragam, The Principles of Nuclear Magnetism , Oxford University Press, London, 1963; G. E. Pake, Nuclear Magnetic Resonance, in Solid State Physics 2,1 (1954); Spin Temperature and Nuclear Relaxation in Solids, in Solid State Physics 15, 409 (1963).
6. H. Frauenfelder, The Mossbauer Effect, W. A. Benjamin Inc., N.Y., 1962; G. K. Wertheim, Mossbauer Effect: Principles and Applications, Academic Press, N.Y. 1964 .
7. G. Burns, J. Appl. Phys. 32, 2048 (1961).
8. E. Sonder and W. A. Sibley, Defect Creation in Polar Crystals, in Solids, Edited by J. H. Crawford, Plenum Publ. Corp., N.Y., to be published.
9. Y. Chen and W. A. Sibley, Phys. Rev. 154, 842 (1967); W. D. Compton and G. W. Arnold, Disc. of Faraday Soc. 31, 130 (1961).
10. W. A. Sibley and Y. Chen, Phys. Rev. 160, 712 (1967).
11. T. Purcell and R. A. Weeks, J. Phys. Chem. Glasses 10 (5), 201 (1969).
12. R. A. Weeks, J. Appl. Phys. 27, 1376 (1956); R. A. Weeks, Phys. Rev. 130, 570 (1963); R. A. Weeks and E. Sonder, The Relation between the Magnetic Susceptibility, Electron Spin Resonance and Optical Absorption of the E_1' center in Fused Silica, in Paramagnetic Resonance Vol. 2, p. 869, Edited by W. Low, Academic Press, N.Y., 1963; R. A. Weeks and E. Lell, J. Appl. Phys. 35, 1932 (1964); J. G. Castle, D. W. Feldman, P. G. Klemens and R. A. Weeks, Phys.

Rev. 130, 577 (1963). In this latter paper an oxygen divacancy model was tentatively proposed for the E_1' center. Subsequent research has shown that a single oxygen vacancy which has trapped an electron has properties in excellent agreement with the available data.

13. J. Anderson and J. Weil, J. Chem. Phys. 31, 427 (1959)
14. R. A. Weeks and T. Purcell, J. Chem. Phys. 43, 483 (1965).
15. R. A. Weeks and P. J. Bray, J. Chem. Phys. 48, 5 (1968).
16. D. L. Griscom, BAPS 15, 314 March 1970 .
17. J. Biscoe and B. E. Warren, J. Amer. Cer. Soc. 21, 287 (1938).
18. P. J. Bray, Magnetic Resonance Studies of Bonding, Structure and Diffusion in Crystalline and Vitreous Solids, in Interaction of Radiation with Solids , p. 25, Edited by A. Bishay, Plenum Press, N.Y., 1967.
19. D. K. Stevens, W. J. Sturm and R. H. Silsbee, J. Appl. Phys. 29, 66 (1958).
20. M. C. Wittels, Phil. Mag. 2, 1445 (1957).
21. R. Comes, M. Lambert and A. Guinier, "Mechanism of the Transformation of Crystalline Quartz into Amorphous Silica by Neutron Irradiation", in Interaction of Radiation with Solids, p. 319, Edited by A. Bishay, Plenum Press, N.Y. 1967 .
22. E. Lell, N.J. Kreidl and J. Raymond Hensler, "Radiation Effects in Quartz, Silica and Glass", in Progress in Ceramic Science 4, 4-9, Edited by J.E. Burke, Pergamon Press, 1966 .
23. R. A. Weeks and C. M. Nelson, J. Amer. Cer. Soc. 43, 399 (1960).
24. R. A. Weeks and M. M. Abraham, BAPS 10, 374 (1965).
25. T. Purcell and R. A. Weeks, unpublished data.
26. R. A. Weeks, Phys. Rev. 130, 570 (1963).
27. R. A. Weeks, unpublished data. At this distance, the hyperfine interaction is of the same order of magnitude as the interaction between the hydrogen nucleus and the magnetic field,i.e., 5 gauss. One consequence of this equivalence is that the doublet structure shown in Fig. 7 is replaced by a quartet of equally spaced (5 gauss separation) lines of almost equal intensity. That this structure is due to the hyperfine interaction of the E' electron with a hydrogen nucleus is confirmed by measurements at a higher frequency for which the interaction between magnetic field and

the hydrogen nucleus is ∿2.5 times larger. At this field the expected doublet structure is observed with a 5 gauss splitting. The outer two lines are detected but with greatly reduced intensity (∿0.1 the intensity of the two central lines) and separated from the two central lines by ∿12 gauss, the magnitude of the interaction between the hydrogen nucleus and the applied magnetic field.

28. R. A. Weeks, "Some Defect States of Pure Four-Fold Coordinated Oxides: Expectations and Realization", in Interactions of Radiation with Solids, p.55, Edited by A. Bishay, Plenum Press, New York, 1967.
29. D. L. Griscom, P. C. Taylor, D. A. Ware, and P. J. Bray, J. Chem. Phys. 48, 5158 (1969).
30. D. L. Griscom, P. C. Taylor, and P. J. Bray, Submitted to J. Chem. Phys.
31. J. O. Edwards, D. L. Griscom, R. B. Jones, K. L. Watters and R. A. Weeks, J. Am. Chem. Soc. 91, 1095 (1969).
32. V. M. Kim and P. J. Bray, private communication, a paper on this topic has been submitted for publication.
33. R. J. Landry, J. J. Fournier and C. G. Young, J. Chem. Phys. 46, 1285 (1967); G. Hochstrasser, Phys. Chem. Glasses 7, 178 (1966); G. Hirayama, J. G. Castle, and M. Kuriyama, Phys. and Chem. Glasses 9, 109 (1968); H. G. Hecht, Phys. Chem. Glasses 9, 179 (1968).
34. T. Purcell and R. A. Weeks, accepted for publication in J. Chem. Phys.
35. A. Chatelain and R. A. Weeks, J. Chem. Phys. 52, 3758 (1970).
36. D. L. Griscom and R. E. Griscom, J. Chem. Phys. 47(8), 2711 (1967).
37. There are many references. A recent one containing detailed computer programs is "Lineshape Program Manual" by P. C. Taylor and P. J. Bray, Department of Physics, Brown University, Providence, Rhode Island. This manual contains programs applicable to both EPR and NMR problems. There is also an excellent introduction to this type of problem in Electron Paramagnetism by Juan A. McMillian, p. 152-162, Reinhold Book Corporation, New York, 1968.
38. M. H. Cohen and F. Reif, Nuclear Quadrupole Effects in Solids, in Solid State Physics 5, 321, Edited by Seitz and Turnbull, Academic Press, 1957.
39. A. M. Stonehan, Rev. of Mod. Phys. 41, 82 (1969).
40. J. F. Baugher, P. C. Taylor, T. Oja, and P. J. Bray, J. Chem. Phys. 50, 4914 (1969).

41. P. J. Bray, "N. M. R. Studies of Glasses and Related Crystalline Solids" in Magnetic Resonance Edited by C. K. Cougan, N. S. Ham, S. N. Stewart, J. R. Pilbrow and G. V. H. Wilson, Plenum Press, New York, 1970.
42. B. E. Warren, J. Amer. Cer. Soc. 24, 256 (1941).
43. H. M. Kriz, S. C. Bishop, and P. J. Bray, J. Chem. Phys. 49, 557 (1968).
44. N. Bloembergen, E. M. Purcell and R. V. Pound, Phys. Rev. 73, 679 (1949).
45. S. G. Bishop and P. J. Bray, J. Chem. Phys. 48, 1709 (1968).
46. C. P. Slichter, Principles of Magnetic Resonance, Harper and Row, N.Y. 1963 .
47. R. A. Weeks, A. Chatelain, D. Kline and J. L. Kolopus, Geochimica et Cosmochin. Acta, Proceedings of the Apollo 11 Lunar Science Conference, Houston, Texas from 5-8, 1970. Supplement 1, volume 3, p. 2467.
48. D. W. Jones, R. S. Mathews, N. Ruddlesden and D. J. Williams, J. Am. Cer. Soc., 51 (11), 664 (1968).
49. D. E. Woessner and B. S. Snowden, J. Chem. Phys. 50 (4), 1516 (1969).
50. C. R. Kurkjian, J. of Non-Cryst. Solids 3, (2), 157 (1970).
51. B. H. Zimmerman, H. Jena, E. Isshinko, H. Kiban and D. Segboth, Phys. Status Solidi 27, 639 (1965); C. Czyjek, J. L. C. Ford, Jr., J. C. Love, F. E. Obenshain and H. H. F. Weggener, Phys. Rev. 174, 331 (1968).
52. C. R. Kurkjian and E. A. Sigety, Phys. Chem. Glasses 9, 73 (1968).
53. M. G. Clark, G. M. Bancroft, and A. J. Stone, J. Chem. Phys. 47, 4250 (1967).
54. D. L. Uhrich and R. G. Barnes, Phys. Chem. Glasses 9, 184 (1968).
55. L. M. Martarese, J. S. Wells, and R. L. Peterson, BAPS 9, 502 (1964).
56. L. M. Martarese, J. S. Wells, and R. L. Peterson, J. Chem. Phys. 50, 2350 (1969).
57. W. A. Deer, R. A. Howie, and J. Zussman, Rock Forming Minerals, Vol. I., John Wiley and Sons, New York, 1962.
58. D. Vinro and S. Hafner, Mineral Soc. Amer. Special Paper 2, 67 (1969).
59. P. E. Champness and P. Gay, Nature 218, 157 (1968).
60. R. A. Weeks, J. L. Kolopus, A. Chatelain, and D. Kline, "Paramagnetic Resonance Spectra of Some Silicate Minerals, Semiannual Technical Progress Report,

December 31, 1968, ORNL CF No. 69-3-5.
61. Handbook of Chemistry and Physics, 44th Edition, p. 3507-8, Chem. Rubber Pub. Co., 1962.
62. Selected from Table 3-1 p. 37-39. The Mossbauer Effect, H. Frauenfelder, W. A. Benjamin, New York, 1962, and from Table 2, p. 173, Mossbauer Spectroscopy in Inorganic Glasses, C. R. Kurkjian, J. Non-Crystalline Solids **3**, 157 (1970).

General Aspects of the Crystallization of Glass

C. G. Bergeron
Dept. of Ceramic Engineering
and Materials Research Laboratory
University of Illinois at
Urbana-Champaign

Crystallization is generally viewed as a two-step process. The first step involves the formation of a stable nucleus and the second the growth of the nucleus to form a crystal. Nucleation may be classified as either homogeneous or heterogeneous. If nucleation occurs within a pure phase, where the probability of such an occurrence is the same throughout the phase, the process is described as homogeneous. Heterogeneous nucleation involves the formation of a nucleus at an interface which serves to catalyze the reaction. Such discontinuities as container walls, bubble surfaces, or particles of a second phase may provide an energetically favorable site for a nucleus to form.

HOMOGENEOUS NUCLEATION

The formation of a crystal nucleus within a liquid phase involves not only a liquid-solid phase transformation, but it also requires that a solid-liquid interface be formed. Consider the formation of a spherical nucleus from a supercooled liquid at constant temperature and pressure (Figure 1). The free energy change which occurs in forming the spherical particle can be written as follows:

$$\Delta G_r = 4\pi r^2 \gamma_{LS} + 4/3\pi r^3 \Delta G_v \qquad (1)$$

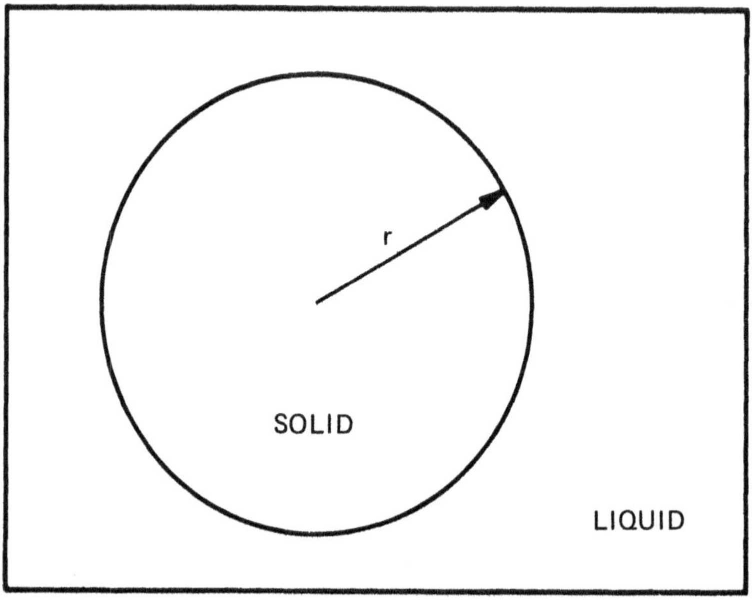

Figure 1. Spherical Nucleus of Radius r Formed from a Supercooled Liquid at Constant Temperature and Pressure.

Where r = radius of particle

γ_{LS} = liquid-solid interfacial energy (assumed to be independent of orientation)

ΔG_V = free energy change per unit volume accompanying the liquid-solid phase transformation.

The first term on the right side of the equation represents the energy required to form the solid-liquid interface and is always positive. The second term represents the energy change associated with the liquid-solid transformation and is always negative below the liquidus temperature. In Figure 2 the two terms are plotted as a function of the radius of the particle. The summation is shown as the solid line in the figure.

For very small particles, where the ratio of the surface area to the volume of the particle is large, the first term predominates and the net free energy change is positive. For larger particles the interfacial energy term is small compared to the volume energy term and the total free energy change is negative. The critical radius, r^*, is determined by setting the derivative of the free

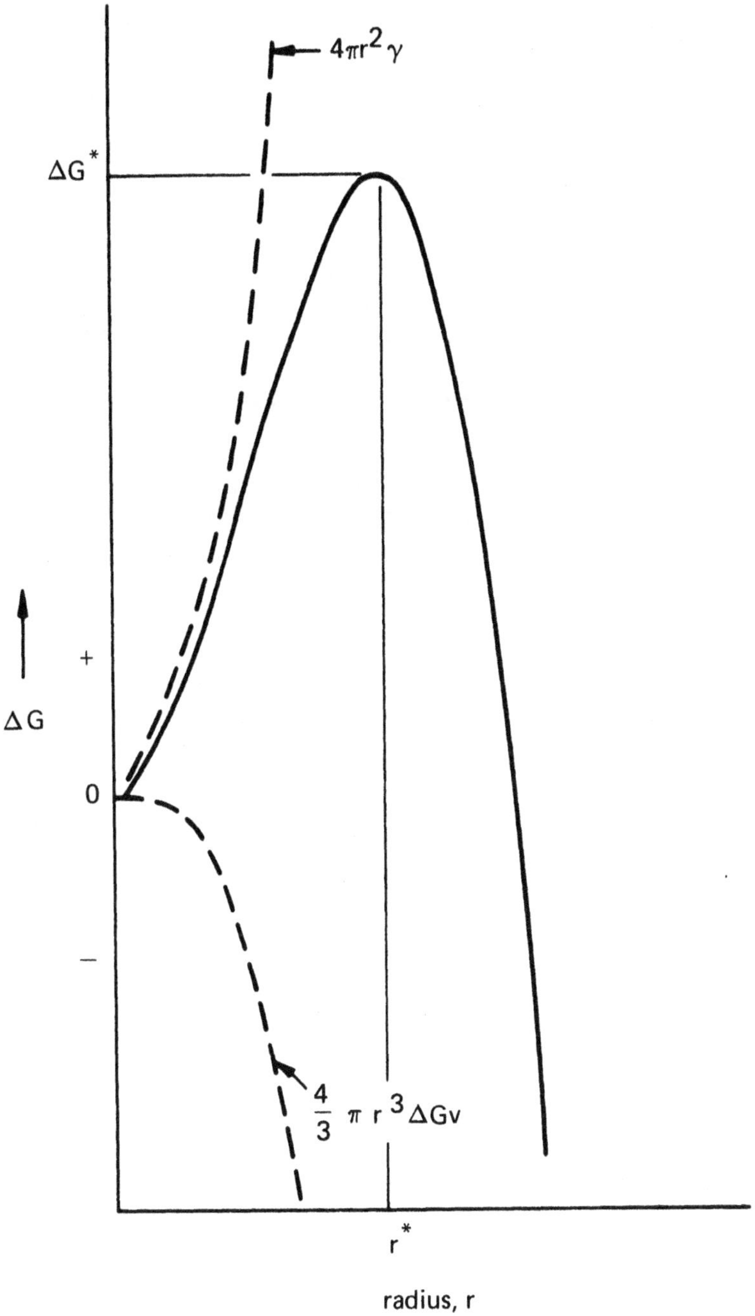

Figure 2. Change in Free Energy of a Spherical Nucleus as a Function of Radius.

energy change with respect to r equal to zero and solving for r:

$$\frac{d(\Delta G_r)}{dr} = 8\pi r \gamma_{LS} + \frac{12}{3}\pi r^2 \Delta G_v = 0$$

$$r^* = \frac{-2\gamma_{LS}}{\Delta G_V} \tag{2}$$

Particles of radius smaller than r* are called embryos and are unstable owing to the decrease in free energy which accompanies their reduction in size. Particles of radius greater than r* are called nuclei and are stable since growth is accompanied by a decrease in free energy.

The free energy change, ΔG_r^*, associated with the formation of the critical-sized embryo is found by substituting the expression for r* into equation (1).

$$\Delta G_r^* = \frac{16\pi \gamma^3{}_{LS}}{3(\Delta G_V)^2} \tag{3}$$

The number of atoms comprising a particle of radius r is given by

$$n_p = \frac{4/3\ \pi r^3}{V} \tag{4}$$

where n_p = number of atoms in particle
V = volume of one atom in the solid phase

When the critical radius is substituted for r in equation (4) the number of atoms in the critical-sized embryo is seen to be

$$n^* = \frac{32\pi \gamma^3}{3V(\Delta G_V)^3} \tag{5}$$

Since the formation of an embryo involves a positive free energy change, it would seem that the probability of such an occurrence would be quite low and that few embryos would be present. However, the entropy of the solid-liquid system can be increased by having some number of atom clusters in equilibrium with the atoms of the liquid. Assuming ideal mixing, this change in entropy is the entropy of mixing and therefore the change in free energy of the system per unit volume due to the formation of such clusters is given by:

$$\Delta G_n = n_r \Delta G_r + kT\left[\frac{n_r}{n_r+n}\ln\frac{n_r}{n_r+n} + \frac{n}{n_r+n}\ln\frac{n}{n_r+n}\right] \quad (6)$$

where n_r = number of clusters of size r per unit volume
n = number of atoms or molecules per unit volume

At equilibrium ΔG_n is a minimum. The number of clusters is expected to be small relative to the total number of atoms in the system. The number of clusters of a given size is:

$$n_r = n \exp\left[-\frac{\Delta Gr}{kT}\right] \quad (7)$$

By the addition of single atoms from the melt, embryos of critical size may grow to form nuclei. The rate of formation of nuclei is given by the product of the number of critical-sized nuclei present per unit volume of liquid and the rate at which atoms are attached to the embryo.

Referring to the schematic representation in Figure 3, an atom in the liquid phase at the interface must acquire an activation energy, ΔG_a, in order to detach itself from the liquid and move across the interface to the crystal.

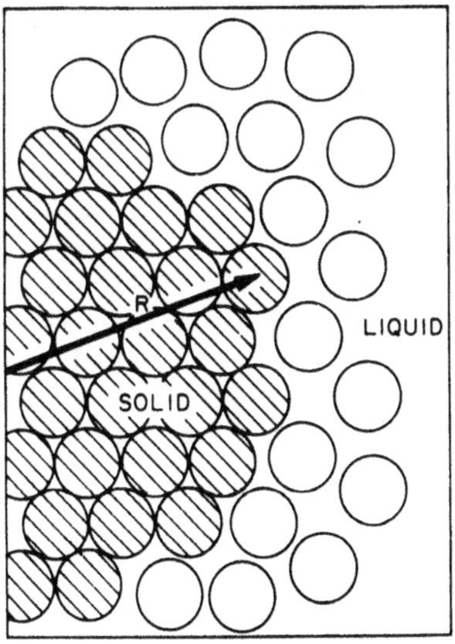

Figure 3. Growth of an Embryo by the Addition of Atoms from the Liquid Phase

The number of times per second that an atom attempts to cross the interface is given by the vibration frequency, ν_0, of the atom. In reaction rate theory, ν_0 is set equal to (kT)/h where h is Planck's constant. The probability that an attempt is successful is given by:

$$P = \exp\,(-\Delta G_a/kT) \qquad (8)$$

Only those atoms which are adjacent to the interface are in a position to attach themselves to the embryo. If this number is n_s, then the number of atoms crossing the interface per second is:

$$\frac{dn_i}{dT} = n_s \nu_0 \exp\,(-\Delta G_a/kT) = n_s \frac{kT}{h} \exp\,(-\Delta G_a/kT) \qquad (9)$$

The above expression assumes that the reverse reaction never occurs, i.e., an atom remains on the embryo once it makes the transition. If the reverse reaction is admitted, equation (9) should be multiplied by a factor of approximately 1/2.

The rate of nucleation is given by the product of equations (7) and (9).

$$I = n \exp\,(-\Delta G_r^*/kT)\, n_s \frac{kT}{h} \exp\,(-\Delta G_a/kT) \qquad (10)$$

It is assumed that the nucleation is sufficiently slow so that the number of embryos attaining critical size each second is equal to the number of nuclei formed per second and thus the number of critical-sized embryos present is always the equilibrium number calculated from equation (7). In an experimental situation, one would expect this number to be somewhat less because the melt must be supercooled appreciably to provide a driving force for nucleation and those critical-sized embryos which were present at the previous temperature will begin growth before the equilibrium number are present at the new temperature.

The temperature dependency of the nucleation rate is shown in Figure 4. The shape of the curve can be understood in a qualitative way by examining the exponential terms in equation (10). ΔG_V is proportional to ΔT, the undercooling, and, therefore, ΔG_r^* varies as $1/(\Delta T)^2$. The term

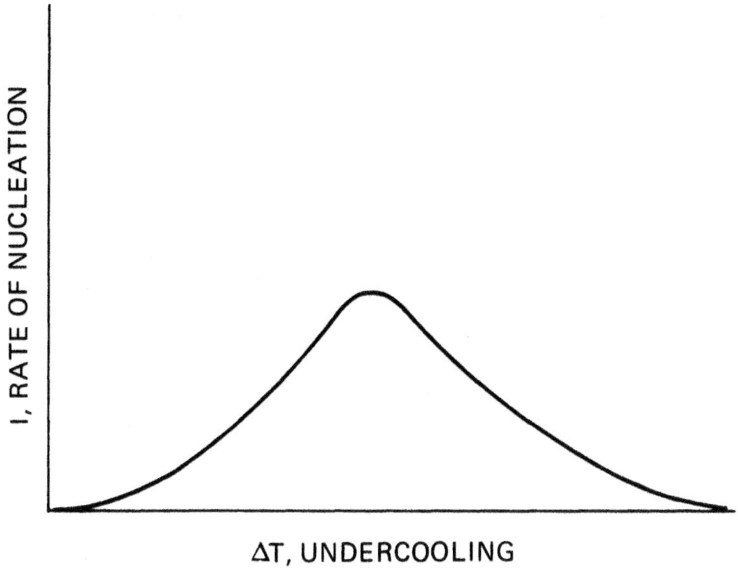

Figure 4. Rate of Homogeneous Nucleation as a Function of Undercooling.

$\exp(-\Delta G_r^*/(kT))$ will be zero at the liquidus temperature and will increase as ΔT increases. At large undercoolings, the term $\exp(-\Delta G_a/kT)$ is dominant and the nucleation rate decreases with further cooling after having passed through a maximum.

HETEROGENEOUS NUCLEATION

Consider the formation of a crystal nucleus on a solid substrate in contact with a supercooled liquid phase (Figure 5). The free energy change associated with the formation of the two new interfaces is:

$$\Delta G_s = \gamma_{LX} A_{LX} + \pi r^2 (\gamma_{XS} - \gamma_{LS}) \qquad (11)$$

where γ_{LX}, γ_{XS}, and γ_{LS} are the interfacial energies of the liquid-crystal, crystal-substrate, and liquid-substrate interfaces, respectively. A_{LX} is the interfacial area of the liquid-crystal interface.

The liquid-substrate interface is reduced by πr^2 when the new surfaces, LX and XS, are formed. If $\gamma_{LS} > \gamma_{XS}$,

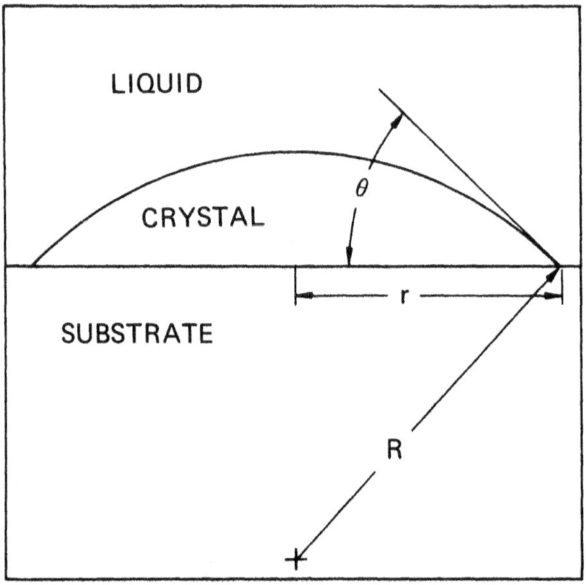

Figure 5. Schematic Representation of a Nucleus Formed on a Solid Substrate within a Liquid Phase.

ΔG_s is less than $\gamma_{LX}A_{LX}$ and the total surface energy required to form the nucleus on a substrate is less than that required for homogeneous nucleation.

The relationship between the contact angle θ and the interfacial energies is:

$$\cos\theta = \frac{\gamma_{LS} - \gamma_{XS}}{\gamma_{LX}} \qquad (12)$$

Substitution of equation (12) in equation (11) yields the following expression:

$$\Delta G_s = \gamma_{LX}A_{LX} - \pi r^2 \gamma_{LX} \cos\theta \qquad (13)$$

If the crystal nucleus is assumed to be in the form of a spherical cap, the volume is:

$$V = \pi r^3 \left(\frac{2 - 3\cos\theta + \cos^3\theta}{3}\right) \qquad (14)$$

The surface area of the nucleus is

$$A_{LX} = 2\pi R^2 (1-\cos\theta) \quad (15)$$

and the radius of the contact area is

$$r = R\sin\theta \quad (16)$$

The free energy change associated with forming the volume, V, of the nucleus is

$$\Delta G = V\Delta G_V \quad (17)$$

As shown in the previous section dealing with homogeneous nucleation,

$$R^* = \frac{-2\gamma_{LX}}{\Delta G_V} \quad (18)$$

and from eq. (16) $r^* = \dfrac{-2\gamma_{LX}\sin\theta}{\Delta G_V}$ (19)

The total free energy change to form the nucleus is given by

$$\Delta G^* = \gamma_{LX}A_{LX} - \pi r^2\gamma_{LX}\cos\theta + V\Delta G \quad (20)$$

Substitution of equations (14), (15), (16), and (18) into eq. (20) gives the following expression:

$$\Delta G^* = 2\gamma_{LX}\pi R^2 - 2\gamma_{LX}\pi R^2\cos\theta - \pi R^2\sin^2\theta\gamma_{LX}\cos\theta -$$

$$2\pi R^2\gamma_{LX}\left(\frac{2-3\cos\theta + \cos^3\theta}{3}\right) \quad (21)$$

which can be reduced to

$$\Delta G^* = \frac{16\pi\gamma^3_{LX}}{3(\Delta G_V)^2}\left[\frac{(2+\cos\theta)(1-\cos\theta)^2}{4}\right] \quad (22)$$

The first term on the right side of the equation is identical to that determined for homogeneous nucleation, eq. (3). When the contact angle θ is zero (complete wetting of the substrate by the crystal in the presence of the liquid phase), $\cos\theta = 1$, the last term of the equation becomes zero and $\Delta G^* = 0$; there is no barrier to nucleation.

When $\theta = 90^o$, $\cos\theta = 0$, and the energy barrier to nucleation is reduced by one-half. If $\theta = 180^o$, no wetting occurs, $\cos\theta = -1$ and equation (22) reduces to eq. (3).

It can be seen that an effective catalyst involves a small contact angle. The nucleus-catalyst interfacial energy should be low. Thus a similar atomic arrangement between nucleus and catalyst with strong attractive forces across the interface would provide favorable conditions for nucleation to occur.

The form of the equation for the rate of heterogeneous nucleation is similar to that of equation (10) for homogeneous nucleation. Differences are as follows: ΔG_r^* in eq. (10) is replaced by the expression given in eq. (22), ΔG_a becomes the activation energy required to disengage an atom or molecule from the liquid phase and move it across the interface to the catalyst, and n becomes the number of possible nucleation sites per unit volume. When these substitutions are made, the rate of heterogeneous nucleation is given by

$$I_{HET} = n \exp\left(-\Delta G_r^*\right) \left[\frac{(2+\cos\theta)\ (1-\cos\theta)^2}{4}\right] n_s \frac{kT}{h} \exp(-\Delta G_a/kT) \qquad (23)$$

It is generally agreed that crystallization in glass-forming systems almost always occurs by heterogeneous nucleation, usually proceeding from an external surface or an interior bubble or other discontinuity. Those systems in which extensive internal nucleation has been observed have been shown to involve liquid-liquid separation. The surface between the discrete phase and the matrix serves as a nucleation site from which crystallization may proceed.

CRYSTAL GROWTH

Once a stable nucleus has been formed, crystal growth may proceed by the addition of atoms or molecules to the nucleus at a rate which is determined by conditions within the melt and at the interface. Figure 6 illustrates schematically the arrangement at the crystal-melt interface for a congruently melting crystal. The derivation of the rate equation follows that of Turnbull[1] for a

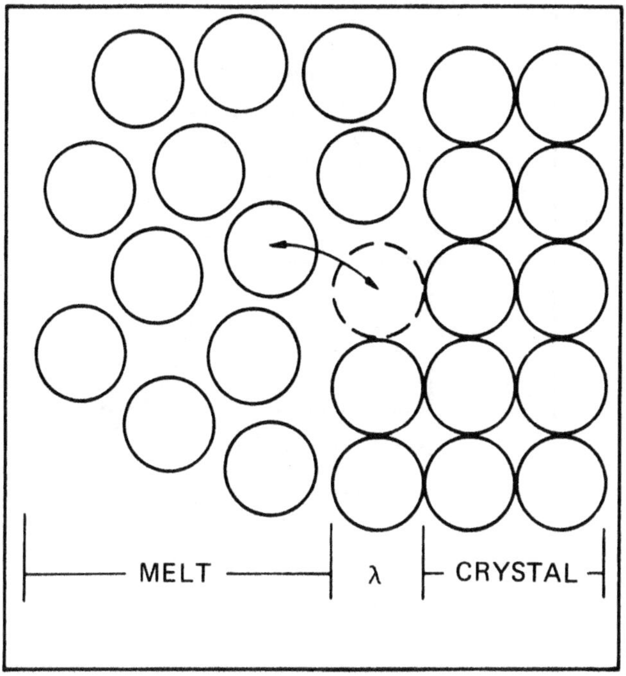

Figure 6. Schematic Representation of Crystal-Melt Interface During Crystal Growth.

diffusion-like growth process.

An atom or molecule must acquire an activation energy ΔG_a (Figure 7) in order to detach itself from the liquid and move across the interface to the crystal. The rate of transfer of atoms from the liquid to the crystal is equal to the number of atoms, s, at the interface times the frequency of attempted jumps times the fraction of atoms which acquire sufficient thermal energy to make the jump. The rate in the forward direction is thus given by

$$\frac{dn_f}{dt} = s\nu_o \exp(-\frac{\Delta G_a}{kT}) \qquad (24)$$

For the reverse direction, crystal to liquid, the rate is given by

$$\frac{dn_r}{dt} = s\nu_o \exp - \left[\frac{(\Delta G_a + V\Delta G_v)}{kT}\right] \qquad (25)$$

and the net rate of transfer from the liquid to the crystal is

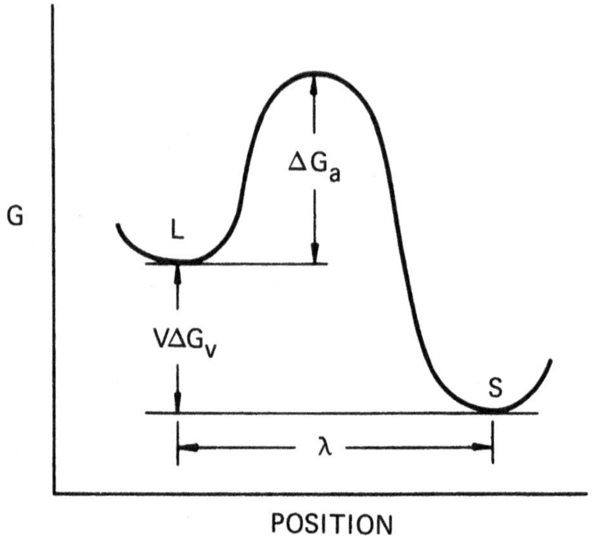

Figure 7. Free Energy Change Associated with the Transfer of an Atom from Liquid to Solid Phase.

$$\frac{dn}{dt} = \left(\frac{dn_f}{dt} - \frac{dn_r}{dt}\right) = s\nu_o \quad \exp\left(-\frac{\Delta G_a}{kT}\right) - \exp\left(-\frac{\Delta G_a + V\Delta G_v}{kT}\right)$$

$$\frac{dn}{dt} = s\nu_o \exp\left(-\frac{\Delta G_a}{kT}\right) \quad 1\text{-}\exp\left(-\frac{V\Delta G_v}{kT}\right) \qquad (26)$$

U, the linear growth of the crystal, is equal to the number of atoms transferred per unit time divided by the interfacial area, s, times the increase in length, λ, due to the addition of atoms.

$$U = \frac{\lambda}{s}(s)\nu_o \exp\left(-\frac{\Delta G_a}{kT}\right) \quad 1\text{-}\exp\left(-\frac{V\Delta G_v}{kT}\right)$$

$$U = \lambda\nu o \exp\left(-\frac{\Delta G_a}{kT}\right) \quad 1\text{-}\exp\left(-\frac{V\Delta G_v}{kT}\right) \qquad (27)$$

When the degree of undercooling is small, $V\Delta G_v$ is a small negative number and exp $(V\Delta G_v)/kT$ can be expanded as a series of the type $e^x = 1 + x + x^2/2! + x^3/3! + \ldots$ in which all but the first terms may be neglected. Hence

$$\exp\left(\frac{V\Delta G_V}{kT}\right) = 1 + \frac{V\Delta G_V}{kT} \tag{28}$$

and
$$1 - \exp\left(-\frac{V\Delta G_V}{kT}\right) = -\frac{V\Delta G_V}{kT}$$

The growth rate at small undercoolings becomes

$$U = \lambda\nu_o \exp\left(-\frac{\Delta G_a}{kT}\right) - \left(\frac{V\Delta G_V}{kT}\right) \tag{29}$$

If the molar free energy change, ΔG, for the transition from liquid to solid is used, the last term in eq. (29) becomes $-\Delta G/RT$.

At constant temperature $\Delta G = \Delta H - T\Delta S$ and at the liquidus temperature T_1, $\Delta G = 0$.

$$\Delta H = T_1 \Delta S$$

$$\Delta G = T_1\Delta S - T\Delta S$$ (assuming ΔH constant over the temperature range being considered)

$$\Delta G = \Delta S(T_1 - T)$$

$$\Delta G = \Delta S \Delta T \tag{30}$$

Substituting eq. (30) in eq. (29)

$$U = \lambda\nu o \exp\left(-\frac{\Delta G_a}{kT}\right)\left(-\frac{\Delta S\Delta T}{RT}\right) \tag{31}$$

The exponential term in eq. (31) changes very little for small undercoolings and consequently the growth rate varies approximately as ΔT.

At large undercoolings, ΔG_v for transition liquid to crystal becomes large and the driving force term in eq. (27),

$1 - \exp\left(-\frac{V\Delta G_V}{kT}\right)$, approaches unity. The growth rate is limited by the rate at which atoms can diffuse across the interface. The diffusion coefficient for transport across the interface may be written as

$$D'' = \lambda^2\nu_o \exp\left(-\frac{\Delta G_a}{kT}\right) \tag{32}$$

and eq. (27) can be simplified to

$$U = \frac{D''}{\lambda} \quad (33)$$

Thus the growth rate for large undercoolings would vary exponentially as (-1/T).

All sites on the crystal surface may not be available for molecular attachment, thus a factor, f, defined as the fraction of sites to which molecules can be attached, is introduced as an additional factor in eq. (31).

$$U = f\lambda\nu_0 \exp\left(-\frac{\Delta G_a}{kT}\right)\left(-\frac{\Delta S \Delta T}{RT}\right) \quad (34)$$

For large undercoolings where the thermodynamic driving force is large and the critical-sized two-dimensional nucleus is therefore quite small, growth can occur almost everywhere on the crystal surface and the factor f would be approximately unity. However, at small undercoolings, f would be expected to exhibit a temperature dependency which would vary with the growth mechanism.

GROWTH RATE DEPENDENT UPON RATE OF TWO-DIMENSIONAL NUCLEATION

If the solid-liquid interface is smooth on an atomic scale, the growth rate of the crystal may depend upon the rate at which two-dimensional nuclei[2] are formed. The factor f would be expected to have a temperature dependency similar to that for the formation of two-dimensional nuclei.

The free energy change accompanying the formation of a two-dimensional nucleus of radius r and one atomic diameter, a_o, in thickness (Figure 8) is

$$\Delta G = 2\pi r a_o \gamma_{SL} - \pi r^2 a_o \Delta G_v \quad (35)$$

Differentiating eq. (35) with respect to r, setting the resultant expression equal to zero and solving for r, the critical radius is found to be

$$r^* = \frac{\gamma_{SL}}{\Delta G_v} \quad (36)$$

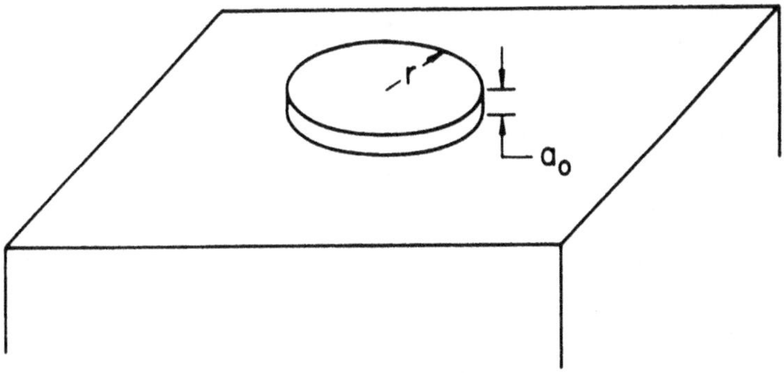

Figure 8. Formation of a Two-Dimensional Nucleus on a Crystalline Surface.

and

$$\Delta G^* = \frac{\pi a_o \gamma^2_{LS}}{\Delta G_v} \qquad (37)$$

where ΔG^* is the free energy change accompanying the formation of the nucleus of radius r^*.

The equilibrium concentration of such nuclei will be

$$n_r = n \exp\left(-\frac{\Delta G^*}{kT}\right) \qquad (38)$$

where n = the number of possible nucleation sites.

The nucleation rate will be

$$I = n \exp\left(-\frac{\Delta G^*}{kT}\right)\left(\frac{2\pi r^*}{a_o}\right)\nu \qquad (39)$$

where $\frac{2\pi r^*}{a_o}$ = the number of edge sites on a nucleus of radius r^*

ν = frequency of transfer of atoms from the liquid to the nucleus

From eq. (30), $\Delta G_v = \Delta S \Delta T$, therefore

$$\Delta G^* = \frac{\pi a_o \gamma^2_{LS}}{\Delta S \Delta T} \qquad (40)$$

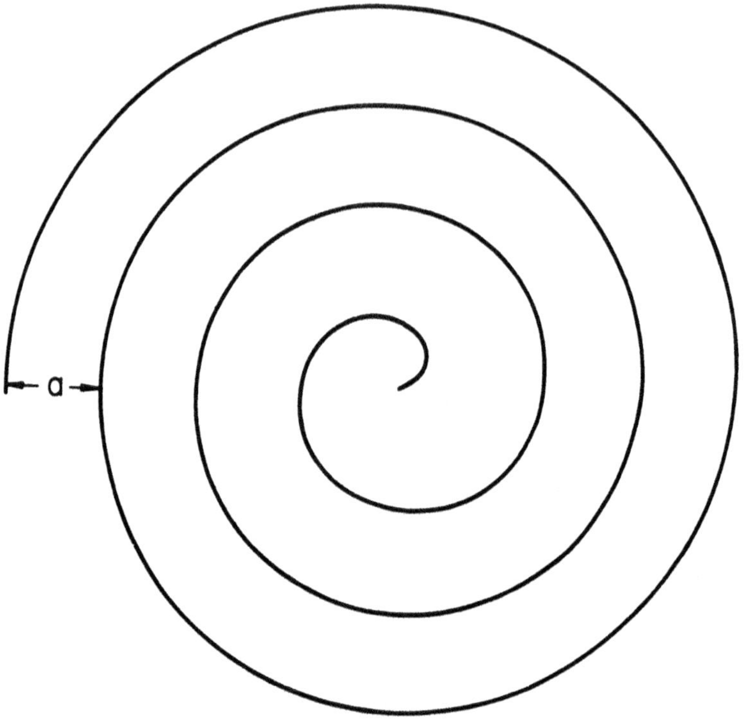

Figure 9. Archimedian Spiral.

Substituting eq. (40) in eq. (39), the nucleation rate is found to be directly proportional to exp $(-1/\Delta T)$, and therefore the factor f varies exponentially as $-1/\Delta T$.

SCREW DISLOCATION MECHANISM FOR CRYSTAL GROWTH

Hilling and Turnbull[3] have shown how the presence of screw dislocations at the surface may provide step sites for molecular attachment. The temperature dependence of the factor f when screw dislocations are present has provided an explanation for the experimentally observed temperature dependency of the growth rates of water, glycerine, salol, tin, and other substances.

Consider the Archimedian spiral shown in Figure 9. The equation for the spiral is

$$r = a\theta \tag{41}$$

where θ = angular coordinate of the spiral
a = a constant equal to the separation of the spirals
r = the radius

If the radius of the central island of the spiral is set equal to the critical radius for a two-dimensional nucleus at undercooling ΔT, $r^* = a/2$ and the equation becomes

$$r = 2\theta r^* \tag{42}$$

From eq. (36), $r^* = \dfrac{\gamma_{SL}}{\Delta G_V}$

and from eq. (17), $\Delta G_V = \dfrac{\Delta G}{V}$. From eq. (30), $\Delta G = \Delta S \Delta T$,

therefore
$$r^* = \frac{\gamma_{SL} V}{\Delta S \Delta T} \tag{43}$$

The spacing between the spirals is inversely proportional to ΔT. An increase in ΔT corresponds to a smaller r^* and an increase in the total length of edge available for molecular attachment. Thus, the total step length is directly proportional to ΔT.

The factor f can be calculated by dividing the total number of atom sites on the spiral step (assumed to be one atom spacing high) by the total number of atom sites on the area beneath the spiral.

$$f = \frac{4\pi^2 r^*/a_o}{8\pi^3 r^{*2}/a_o^2} = \frac{a_o}{2\pi r^2} \tag{44}$$

Substituting eq. (43) for r^* in eq. (44)

$$f = \frac{a_o \Delta S \Delta T}{2\pi\gamma_{SL} V} \tag{45}$$

Substituting eq. (45) for f in eq. (34) yields an expression for the temperature dependency of the growth rate when screw dislocations are present to provide nucleation sites.

$$U = \frac{a_o (\Delta S)^2 \; (\Delta T)^2}{2\pi\gamma_{SL} VRT} \nu_o \exp\left(-\frac{\Delta G_a}{kT}\right) \tag{46}$$

The growth rate is proportional to $(\Delta T)^2$. In many systems the growth rate is found to be approximately proportional to the square of the undercooling.

OBSERVATIONS IN GLASS-FORMING SYSTEMS

The foregoing discussion has considered systems which involve only one kind of atom and simple melt structures such as one would expect to encounter in metals. Inorganic glass-forming systems offer some interesting contrasts in that the melts involve several kinds of atoms and are considered to contain ionized chains and/or rings of covalently-bonded glass formers such as SiO_2 or B_2O_3. The growth rates in metallic systems are generally many orders of magnitude greater than in ceramic systems and frequently exceed one centimeter per second at only a few degrees of undercooling. In ceramic systems, growth rates have been measured over 200-300 degrees of undercooling and are usually, with a few exceptions, less than 10^{-3}/cm/sec. The maximum growth rate usually occurs at about 100 degrees of undercooling.

Our very limited success in arriving at an equation to quantitatively predict the growth rate as a function of temperature for a given crystal has been due to a number of factors which at present are very difficult to evaluate.

1. Interface Temperature

The driving force for crystal growth is the free energy difference between the crystal and melt. This value may be estimated if the undercooling is known. However, a knowledge of the temperature at the crystal-melt interface is required to compute the undercooling. The heat of fusion liberated at the interface causes a temperature rise so that the undercooling at the interface is always less than that of the bulk melt. In a sodium disilicate melt[4,5] where the growth rate is small (max. = 0.87 x 10^{-4}cm/sec.) and the enthalpy of fusion for the crystal is also relatively small (8 kcal/mole), the rise in temperature at the interface has been estimated to be less than a degree.[6]

For the crystal $PbO \cdot 2B_2O_3$ which has a large enthalpy[7] of fusion (31.5 kcal/mole), the rise[8] in temperature at the interface at the maximum growth rate[9] (2 x 10^{-4} cm/sec.) has been estimated to be 22 degrees. The crystals, $BaO \cdot 2B_2O_3$ and $SrO \cdot 2B_2O_3$ have enthalpies[10] of fusion of 27.6 and 33.4

kcal/mole, respectively. The maximum growth rates[11,12] for these two crystals were found to be 43 x 10^{-4} and 160 x10^{-4} cm/sec., respectively (Figure 10). The high growth rates of these crystals coupled with their large enthalpies of fusion would be expected to produce a significant temperature rise at the interface and thus the undercooling at the interface would be considerably less than that of the melt.

2. Melt Structure

In inorganic glass-forming systems, the structures of the melts are not known with certainty. It is frequently assumed that the structure of the melt is similar to that of its glass although the former would be expected to have a greater fluidity owing to the greater thermal energy available for the breaking of bonds and the subsequent greater mobility of the chain-like structures in the melt.

If the entropy of fusion of a crystal is small, the crystal and its melt are thought to have a high degree of structural similarity. In such systems, crystal growth from the supercooled melt would be expected to occur with less difficulty because of the lesser amount of re-arrangement required to incorporate the structural units of the melt into the crystal. For example, the entropies of fusion of $BaO \cdot 2B_2O_3$ and $PbO \cdot 2B_2O_3$ are 17.5 and 30 e.u., respectively. At their respective melting points, the viscosity of $BaO \cdot 2B_2O_3$ is more than three times greater than that of $PbO \cdot 2B_2O_3$. The maximum growth rate of $BaO \cdot 2B_2O_3$ is approximately 35 times that of $PbO \cdot 2B_2O_3$. It has been proposed that the $BaO \cdot 2B_2O_3$ crystal grows by a chain-folding mechanism (similar to that in organic polymers) in which the chains in the melt fold back and forth onto the crystal surface. The $PbO \cdot 2B_2O_3$ melt is thought to contain structural groupings not found in the crystal and hence requires partial dissociation of these groups before they can be incorporated into the crystal.

Another interesting contrast is found between the isomorphous crystals $SrO \cdot 2B_2O_3$ and $PbO \cdot 2B_2O_3$. Their entropies of fusion are 26.3 and 30 e.u., respectively. Their viscosities at the liquidus temperatures are the same, but the maximum growth rate of $SrO \cdot 2B_2O_3$ is 130 times greater than that of $PbO \cdot 2B_2O_3$. On the basis of the differences in glass structures shown by the NMR studies of Bray[13,14] and co-workers, it seems likely that the

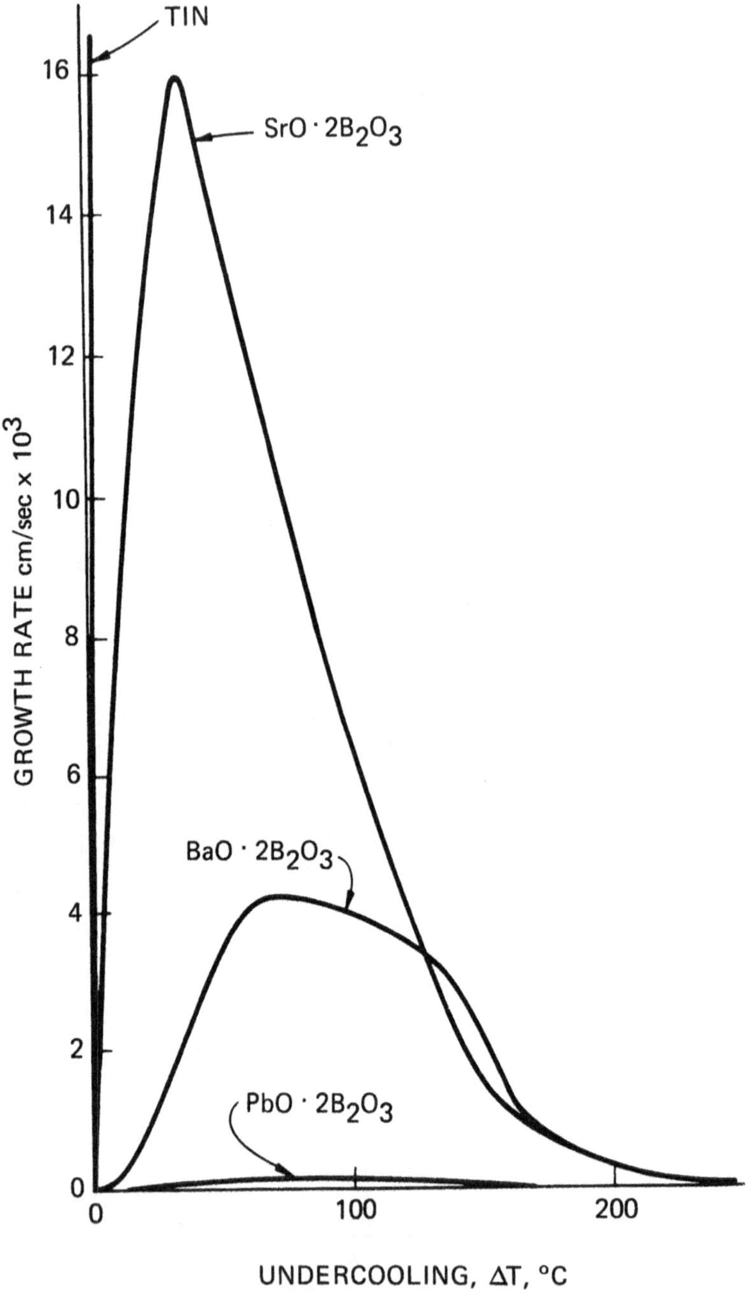

Figure 10. Crystal Growth Rates in Undercooled Melts.

disparity in growth rates between these two crystals may be due to differences in their melt structures. A portion of the lead ions in the lead tetraborate glass are thought to coordinate with four oxygen ions in a pyramidal configuration not found in the crystal, while the strontium tetraborate glass contains structural units which are more readily incorporated into its crystal.

3. Non-Stoichiometry

If the crystal grows in a melt which is different in composition from that of the crystal, the composition of the melt near the interface will be different from that of the bulk melt. This compositional difference might be expected to reduce the growth rate because of the decrease in the availability of those constituents which make up the crystal. The maximum crystal growth rate of sodium disilicate[3] was shown to decrease as the SiO_2 content of the melt exceeded the stoichiometric amount. The effect may be due to the increase in viscosity of the silica-rich melt. However, in the case of a congruently melting compound, the driving force for crystallization is reduced owing to the lower liquidus temperature of the non-stoichiometric compositions. The effective undercooling (T_l-T_i), where T_l=liquidus temperature and T_i = the interface temperature, becomes less at the interface.

The growth rate of $PbO \cdot 2B_2O_3$ has been determined in melts in which the composition was varied[15] from an excess of B_2O_3 to an excess of PbO. The maximum growth rate was observed in those melts with a 1-2% excess of PbO. Although the effective undercooling was reduced, it was shown that the greater fluidity of the PbO-rich melt was sufficient to overcome the smaller driving force and thus the growth rate was higher in the non-stoichiometric melt.

The presence of small amounts of impurities may serve to enhance the growth process by providing irregularities which may serve as nucleation sites on the crystal surface. Impurities in the melt may also increase the fluidity of the melt at the interface and permit more rapid orientation of the crystallizable species. Water in the melt has been observed to increase the fluidity in some silicate glasses. Inhibition of the growth process might be expected if the impurity atom is rejected by the crystal at the interface and accumulates to form a barrier layer.

SUMMARY

Nucleation in supercooled inorganic melts is usually heterogeneous in nature, occurring externally or at surfaces within the melt. Separation of a second liquid phase is frequently the precursor to internal nucleation. Assuming an activated process, the rate of nucleation increases to a maximum value as the undercooling is increased; with further undercooling the rate decreases exponentially as the proportion of atoms with sufficient thermal energy to make the jump across the melt-crystal interface is decreased.

The temperature dependency of the growth rate of crystals in a supercooled melt follows a curve which is similar in shape to that for nucleation. At small undercoolings the growth rate would be expected to vary as ΔT, the undercooling. Departures from this relationship are frequently observed. In some instances the rate of two-dimensional nucleation may be the rate-controlling step, while in others the presence of screw dislocations at the crystal surface may enhance the growth rate by providing sites suitable for molecular attachment.

At large undercoolings, the growth rate is limited by the rate at which atoms can diffuse across the melt-crystal interface. The activation energy for viscous flow is frequently used to approximate the activation energy for diffusion. In some systems, but not all, this assumption provides reasonable agreement between theory and experiment.

The temperature at the crystal-melt interface during crystal growth, the structure of the melt and its dependence upon temperature, and the effects of non-stoichiometry and impurities on crystal growth are some of the key factors which must be studied in greater detail to increase our understanding of the mechanism of crystal growth.

ACKNOWLEDGEMENT

A portion of the experimental work reported in this paper was supported by the Advanced Research Projects Agency under contract ARPA SD-131.

REFERENCES

1. Turnbull, D., "Phase Changes", Solid State Physics, Vol. 3, 225-306, Academic Press, N.Y. (1956).
2. (a) Volmer, M., and Marder, M., Z. Physik Chem. 154, 97 (1931).
   (b) Cahn, J. W., Hilling, W. B., and Sears, G. W., "The Molecular Mechanism of Solidification", ACTA Met. 12, 1421-39, (1964).
3. Hilling, W. B., and Turnbull, D., "Theory of Crystal Growth in Undercooled Pure Liquids", J. Chem. Phys., 24, 914 (1956).
4. Scott, W. D., and Pask, J. A., "Nucleation and Growth of Sodium Disilicate Crystals in Sodium Disilicate Glass", J. Am. Ceram. Soc., 44, 181-7 (1961).
5. Morley, J. G., "Crystal Growth in Glass", Contemp. Phys., 6, 349-64 (1965).
6. Meiling, G. S., and Uhlmann, D. R., "Crystallization Kinetics of Sodium Disilicate", International Conference on Crystal Growth, Boston, 1966, Edited by H. S. Peiser, 1st Ed. Pergamon Press, N.Y., 1967.
7. Ricker, J. M., Bergeron, C. G., and Eagan, R. J., "Enthalpy of Fusion of $PbO \cdot 2B_2O_3$", J. Am. Ceram. Soc., 51,(7) 410 (1968).
8. Laird, J. A., and Bergeron, C. G., "Interface Temperature of a Crystallizing Melt", J. Am. Ceram. Soc., 51, (1) 60 (1968).
9. DeLuca, J. P., Eagan, R. J., and Bergeron, C. G., "Crystallization of $PbO \cdot 2B_2O_3$ from Its Supercooled Melt", J. Am. Ceram. Soc., 52, 322-6 (1969).
10. Nagel, S. R., Eagan, R. J., and Bergeron, C. G., "Enthalpy of Fusion of $SrO \cdot 2B_2O_3$", J. Am. Ceram. Soc., 53, 4, (1970).
11. Laird, J. A., and Bergeron, C. A., "A Chain-Folding Mechanism for the Growth of a $BaO \cdot 2B_2O_3$ Crystal in Its Melt", to be published.
12. Nagel, S. R., "The Crystallization of $SrO \cdot 2B_2O_3$ from Its Melt", M. S. Thesis, University of Illinois, Urbana, February 1970.
13. Bray, P. J., Leventhall, M., and Hooper, H. O., "Nuclear Magnetic Resonance Investigations of the Structure of Lead Borate Glasses", Phys. Chem. Glasses, 4 (2), 47-66 (1963).
14. Bray, P. J., "The 'B' NMR Study in Strontium Borate Glasses", Abstract, Bull. Am. Ceram. Soc., 47, (9) 851-53 (1968).

15. Eagan, R. J., Deluca, J. P., and Bergeron, C. G., "Crystal growth in the System $PbO-B_2O_3$", J. Am. Ceram. Soc., 53, (4) 214-219.

GENERAL REFERENCES

Fine, M. E., "Introduction to Phase Transformations in Condensed Systems", MacMillan Company, N.Y., 1964.

"Crystal Growth", Discussions of the Faraday Society, No. 5, Gurney and Jackson, Ltd., London, 1949.

Chalmers, Bruce, "Principles of Solidification", John Wiley and Sons, New York, 1964.

Doremus, R. H., Roberts, B. W., and Turnbull, D., Editors, "Growth and Perfection of Crystals", John Wiley and Sons, N.Y., 1958.

EDITOR'S NOTE

Many of the topics discussed in this chapter play an integral role in "The Kinetic Theory of Glass Formation". Further information on this view of glass formation may be found in the following references:

1. Rawson, H.; "The Kinetic Theory of Glass Formation", Inorganic Glass Forming Systems, Academic Press, London, New York, Chapter 3, 1967, pp. 31-43.
2. Staveley, L. A. K.; "Homogenous Nucleation of Supercooled Liquids", The Vitreous State, Glass Delegacy of the University of Sheffield, 1955, pp. 85-90.
3. Turnbull, D. & Cohen, M. H.; "Concerning Reconstructive Transformation and Formation of Glass", Journal of Chemical Physics, 29, (5), 1958, pp. 1049-1054.
4. Turnbull, D. & Cohen, M. H.; "Molecular Transport in Liquids & Glasses", Journal of Chemical Physics, 31, (5), 1959, pp. 1164-1169.
5. Turnbull, D. & Cohen, M. H.; "Crystallization Kinetics and Glass Formation", Modern Aspects of the Vitreous State, (J. D. MacKenzie, ed.), Butterworths, London, 1960, pp. 38-62, Vol. 1.
6. Turnbull, D. & Cohen, M. H.; "Composition Requirements For Glass Formation in Metallic & Ionic Systems", Nature, 189, (4759), 1961, pp. 131-132.

Phase Separation of Simple Glasses

Harrie J. Stevens

SUNY College of Ceramics
Alfred University
Alfred, New York

INTRODUCTION

Immiscible systems like oil and water, have been known for many years. It was not until some sixty years ago that this same type of immiscibility was found in glass systems.

In the 1920's Greig tried to melt a glass of approximately 28% CaO and 72% SiO_2 below 1700°C since the cristobalite liquidus is 1723°C and the glass liquidus should slope away from this temperature. Much to his amazement this glass did not form at 1700°C. He then began a detailed study at higher temperatures and with different compositions. Observation of the quenched melts under an optical microscope showed the presence of two immiscible phases. By index of refraction measurements one phase was estimated at 99.4% SiO_2 and the other phase was rich in calcium. Since this discovery many detailed studies of silicate, borate, and other glasses have been launched into the nature of phase separation. This is still a topic of extensive research and debate.

In this chapter the nature and occurrence of immiscible glass systems will be considered and the relationship of free energy diagrams to this type of phase separation will be explored. With the insight gained from discussions in these areas, various theories of phase separation will be examined. During this discussion the CaO-SiO_2 system will be used as the example. Correlations can then be easily made with other systems. Figure 1 shows the phase diagram for the CaO-SiO system. The immiscibility region is illustrated by the dome in the upper left hand corner.

Immiscible systems may occur above or below the liquidus. The latter is often referred to as sub-liquidus or metastable immiscibility. Common silicate melts which exhibit the above liquidus type are SrO, CaO, FeO, ZnO, and MgO. Sub-liquidus immiscibility is often found in silicate melts that have an s-shaped liquidus, such as K_2O, Na_2O, Li_2O and BaO.[2]

In Figure 1 the normal type of phase separation is illustrated.[3,4] If the phase separation was sub-liquidus the dome would be below the 1705°C liquidus.

FREE ENERGY DIAGRAMS

The spontaneous occurrence of immiscibility implies that the free energy of the system has been lowered by the separation into two phases. Yet a normal free energy curve does not suggest anything of this sort. Let's first examine the normal case. The free energy for mixing can be expressed as:

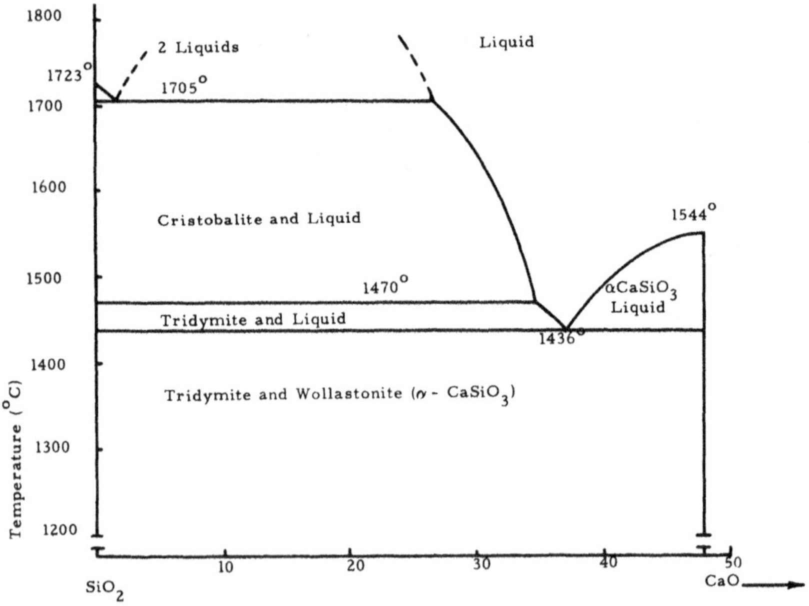

Figure 1. The CaO·SiO_2 system.

$$\Delta G_m = \Delta H_m - T\Delta S_m \quad (1)$$

at a given temperature.[5]

By plotting the free energy as a function of composition at any one temperature, we develop a free energy diagram for a typical system. If ΔH_m has a negative or a very small positive value we get the general type curve shown in Figure 2. This mixing curve is used to represent the free energy of the liquid. Nothing about it implies or suggests phase separation.

If we consider what happens when ΔH_m is very positive, the curve shown in Figure 3 will result. This curve permits the lowering of the free energy by the separation of any liquid between B & C into two liquids of composition B & C. (See Figure 4). By starting with 1 mole of composition A this can be illustrated. If composition A divides into composition B & C the relative amount of each phase can be determined by the lever rule. The free energy of the separated system is then given by summing the free energy of each phase times the mole fraction of that phase. Since the free energy of each separated phase is lower than the free energy of A, the total free energy will be lowered

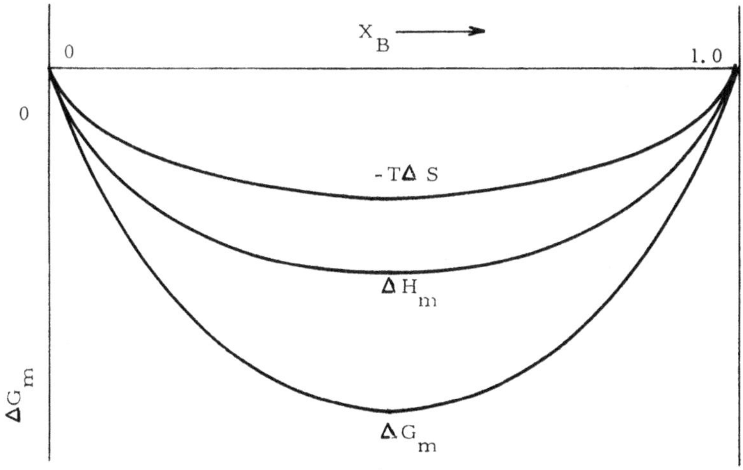

Figure 2. Free energy vs composition for a solution when ΔH_m is negative.[5]

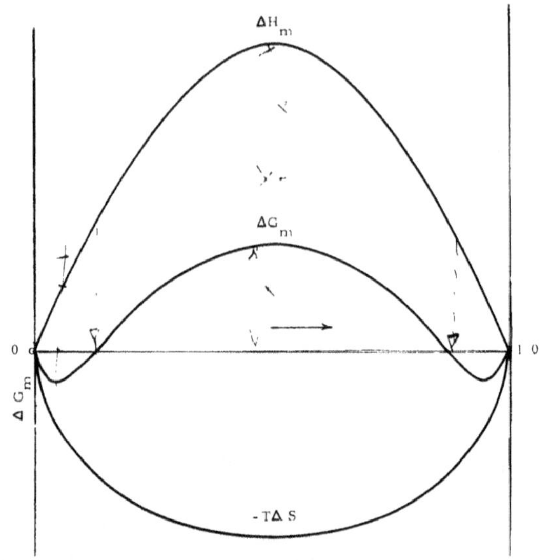

Figure 3. Free energy vs composition for a solution when ΔH_m is positive.

$$\frac{Y_2}{Y_1 + Y_2} + \frac{Y_1}{Y_1 + Y_2} = A$$

$$\frac{Y_2}{Y_1 + Y_2} \times E_c + \frac{Y_1}{Y_1 + Y_2} \times E_b = A \times E_a$$

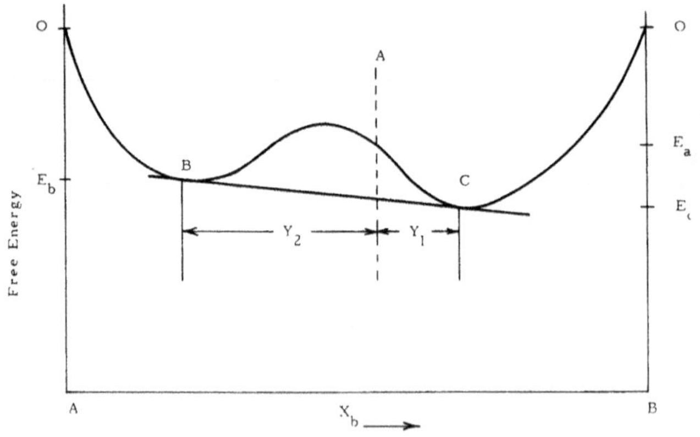

Figure 4. Free energy vs composition for an immiscible system.

if the system separates.

As the temperature increases the $-T\Delta S_m$ term plays a larger role and the magnitude of this deflection decreases. When the temperature is high enough it disappears. Consequently, we can plot the limiting phase compositions as a function of temperature and arrive at a phase diagram with an "immiscibility dome," Figure 5.

This figure also indicates an area within this immiscibility dome known as the spinodal. The spinodal is a dome within the immiscibility dome. This inner dome

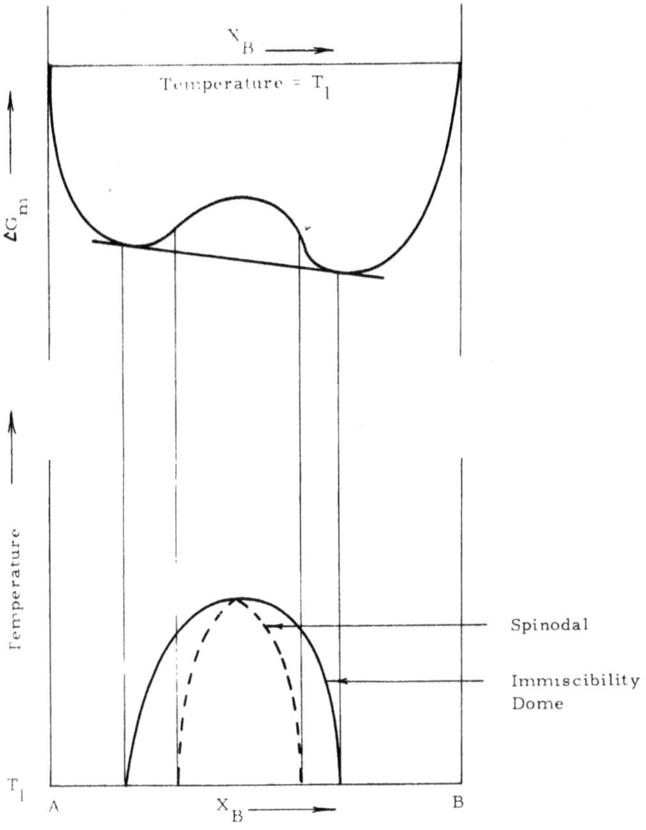

Figure 5. Relationship between the spinoidal and the free energy diagram.

is a result of the points of inflection of the free energy curve. A melt within this dome will spontaneously separate if the mobility of the ions is great enough. If the initial composition of the glass is outside of this area, the phases will not separate spontaneously, but require the formation of nuclei. These nuclei can be formed by composition variation in the sample. The degree of the variation necessary for nuclei formation, and consequently phase separation, depends on the distance of the original composition from the spinodal.

Spontaneous phase separation will occur when the free energy of the system is lowered. When a composition outside of the spinodal begins to separate, the free energy of one phase increases as its composition proceeds toward the spinodal. This is because the slope of the free energy curve in that area is increasing. Once the composition reaches the spinodal, further change will decrease the free energy, since the slope is decreasing. There is, therefore a certain required separation that must occur before the immiscibility can occur spontaneously. This is equivalent to saying a nuclei of some critical size must be created for phase separation outside the spinodal. The size of this initial separation or nuclei will depend on the distance the initial composition is from the spinodal.

Within the spinodal, there is no critical sized nuclei that must be created before phase separation will proceed spontaneously. Any slight variation in composition will result in a lower free energy. Phase separation in this area results in a matrix of intertwining phases with a gradual variation in composition from one phase to the next. This differs from non-spinodal phase separation where there is a sharp boundary between the two phases due to the formation of the critical nuclei.

Although the theory of phase separation and spinodal decomposition has been developed to a very fine degree, recently some debate has arisen among noted scientists concerning the applicability of this theory to the practical case. The step by step process of going from one phase to two phases needs further investigation before the theory can be definitely confirmed. The lack of 100% confirmation does not, however, invalidate the theories, and they are still the best explanation of phase separation to date. An awareness of this debate should be noted, but the fruitfulness of these theories may be still exploited.

A quasi chemical description of the spinodal compositional deviation is given by Swalin.[5] A homogeneous solution of n atoms with average composition X_B will deviate from the average composition by an amount ΔX_B due to thermal fluctuations. We are concerned with the probability of occurrence of this fluctuation X_B. As previously discussed a certain size fluctuation is necessary for phase separation outside the spinodal whereas within the spinodal there is no critical size.

$\overline{G}$ is defined as the average free energy of the n atoms and G is the free energy of any one atom. The probability that any atom will have energy G different from the average free energy per atom in solution, $\overline{G}$, is given by Boltzman statistics.

$$P\Delta G = A \exp\left[-(G-\overline{G})/KT\right] = A \exp[-\Delta G/KT] \tag{2}$$

The probability that n atoms will have the same free energy simultaneously will be

$$P\Delta G^n = A \exp[-n\Delta G/KT] \tag{3}$$

Let us consider for ease in calculation, the situation where ΔX_B is small and, therefore, ΔG is small. "G" may be expanded by a Taylor series about X_B. Thus

$$G = \overline{G} + \frac{1}{2}\left(\frac{d^2G}{dx_B^{\,2}}\right)(X_B-\overline{X}_B)^2 + \ldots \tag{4}$$

If $X_B - \overline{X}_B \ll 1$, higher order terms may be neglected. By substituting this equation in equation (3) the probability of a small fluctuation involving n atoms is

$$P\Delta G^n = A'\exp\left[\frac{-n(X_B-\overline{X}_B)^2(d^2G/d\overline{X}_B^{\,2})\overline{X}_B{}'}{2KT}\right] \tag{5}$$

From this equation it is seen that a high probability of a composition fluctuation is associated with a small value of $(d^2G/dX_B^{\,2})\overline{X}_B$, particularly if it is negative (Figure 6). Negative values of the second derivative of the free energy will give a positive exponent and will give a probability greater than one, which is to say, the separation will occur. If, on the other hand, the second derivative is positive the exponent will be negative and the probability

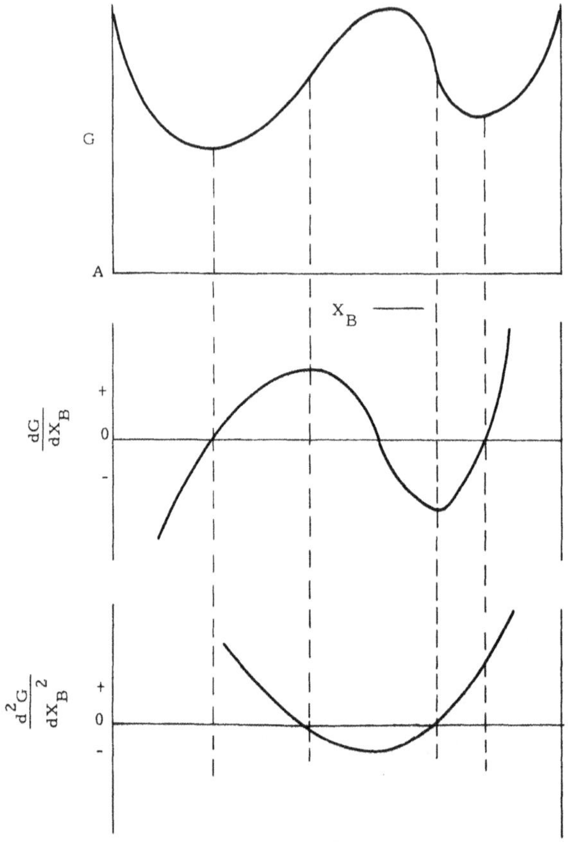

Figure 6. Spinoidal phase separation.[5]

will be less than one. As the composition moves away from the spinodal the probability gets smaller and smaller. Consequently, phase separation is less likely to occur for such compositions. It is important to remember, however, that if a composition with this improbable separation separates to the point that one phase reaches the spinodal, it will then proceed spontaneously.

Figure 7 is an electron micrograph of a spinodal decomposition, while 8 shows the results of a nucleated decomposition.

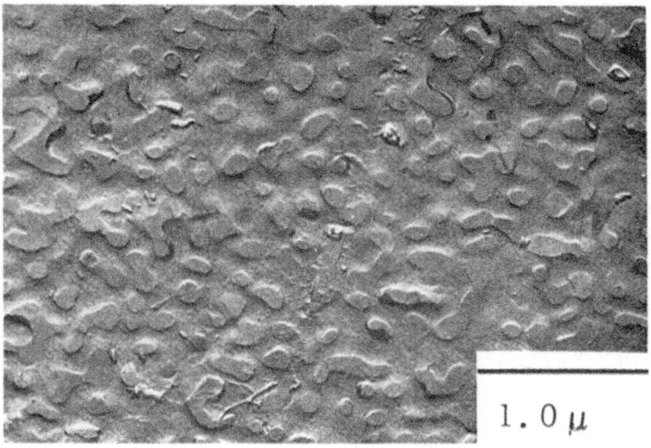

Figure 7. Spinoidal phase separation.[6]

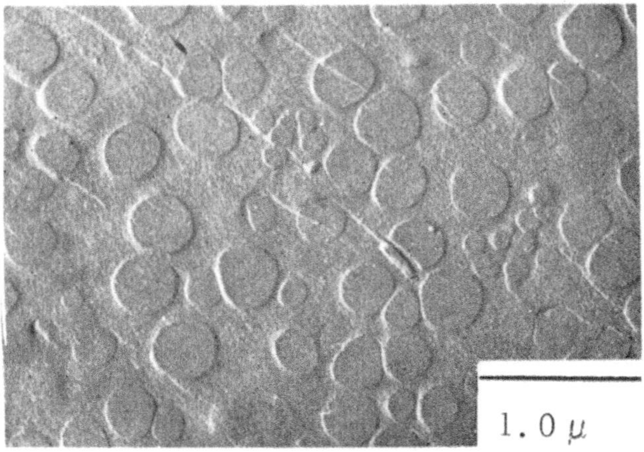

Figure 8. Nucleated phase separation.[6]

RELATIONSHIPS BETWEEN PHASE DIAGRAMS AND FREE ENERGY DIAGRAMS

Up until now only free energy diagrams of the liquid have been considered. The actual occurrence of the liquid phase, regardless of its nature, will depend on the location of the free energy curves of the other phases of the system. For instance, if the free energy curve for the crystalline phase is below the free energy curve for the immiscible systems, then the phase that will actually appear will be the crystalline phase; the phase of the lowest free energy. (See Figure 9).

If the solid phase free energy curve falls below the liquid curve in only a limited range, then the phases present will result from a combination of the two free energy curves. (See Figure 10). As we change the temperature from T_0 to T_1 to T_2, the relative positions of the two free energy curves vary and the phase diagram changes accordingly.

Figure 10 illustrates what happens when the immiscibility dome is just below the liquidus. The liquidus has a tendency to become s-shaped.

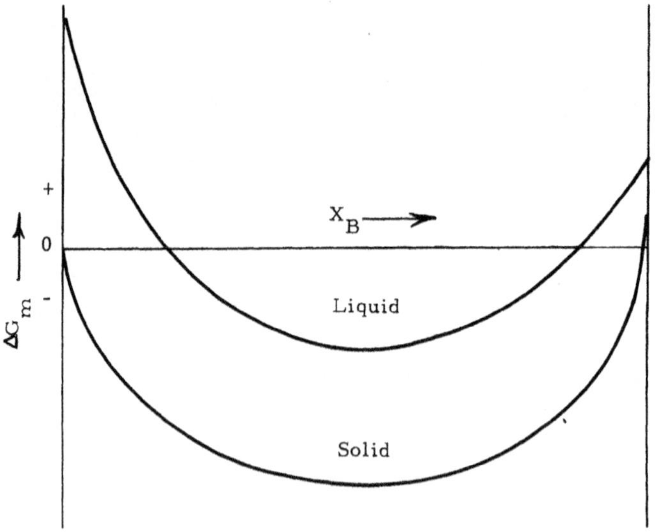

Figure 9. Sub-liquidus crystallization.[5]

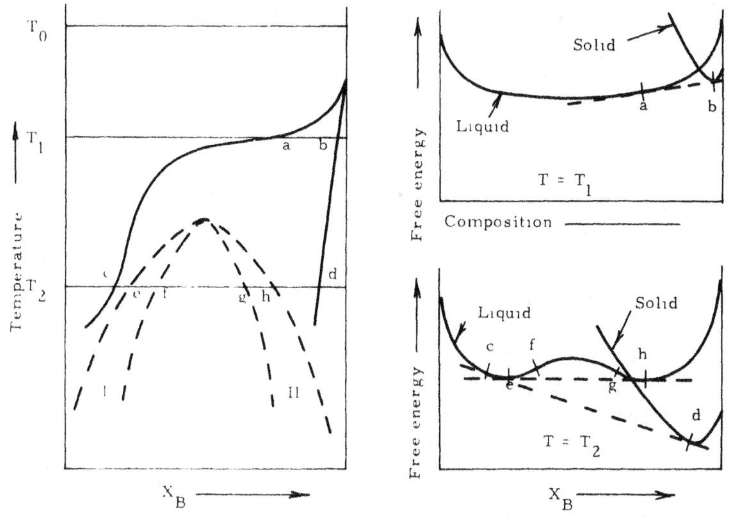

Figure 10. Sub-liquidus phase separation.[6]

As before, common tangent points between the two curves mark compositions of two co-existing phases. When this s-shaped liquidus occurs, metastable or sub-liquidus phase separation may exist. Generally, we must cool the liquid to a glass fast enough so that crystallization doesn't have a chance of occurring. Upon heat treating this glass at a proper temperature, phase separation will initially develop without crystallization. Further heat treatment, however, may result in crystallization of the glass.

This procedure illustrates the importance of chemical kinetics in sub-liquidus phase separation. The thermal history must provide enough time at the proper temperature to allow a slight separation to occur. There will result, therefore, some degree of ordering on a very small scale. If the thermal treatment is too long ordering on a much greater scale may result.

Several workers have criticized the use of the term phase separation in reference to sub-liquidus immiscibility. There are, however, a far greater number of scientists who have accepted this term as an accurate name for the phenomena. The problem arises from the interpretation of the sub-liquidus immiscibility dome in regard to the phase rule. The phase rule states that at equilibrium the number of components (C) in a system plus 2 equals the sum of the number of phases (P) and the number of degrees of freedom (F) in that system or C + 2 = P + F. In the area of the sub-liquidus immiscibility dome, there is also a primary crystalline phase. This suggests that two liquid phases and a crystalline phase could exist in equilibrium in a divariant pressure-temperature region. If this were true a revision of either the phase rule or the present definition of a phase would be needed.[7] If the present definition of a phase seemingly contradicts the phase rule, then the term phase-separation would also be a contradiction and thus erroneous.

The problem is easily resolved by noting that this is not an equilibrium system. The crystalline phase does not exist in equilibrium with the two liquid phases and that is why the term metastable immiscibility is often used instead of sub-liquidus phase separation. By thermodynamic principles equilibrium would be achieved only when the lowest free energy phase is present or when only the crystal phase exists.

It should also be pointed out that all simple glasses might have a free energy curve that indicates immiscibility. The occurrence of this curve might, however, appear at such a low temperature that kinetically it is impossible to separate the two phases. This temperature would be below the transformation range of the glass and the glass would not be able to rearrange its atoms to phase-separate.

THEORIES OF PHASE SEPARATION

Up to now only the very general nature of phase separation has been discussed. Nothing has been said as to why one system might phase separate where another system will not. Coupled with these questions is what happens on an atomistic scale when a system phase separates. There are many theories, some very closely related, that have been supposed to answer these questions. By reviewing them in a cursory manner one may be better able to develop his own ideas as to the nature of this phenomena.

THE WORK OF WARREN AND PINCUS[8]

In 1940 an attempt was made to apply the techniques of crystal chemistry to the explanation of liquid immiscibility in oxide glass systems. The atomic bonding of the glass system was considered to be largely ionic in nature. It was, therefore, felt that the bonding of the glass could be related to the electrostatic bonding energy, E, which equals $Z_1 . Z_2 . e^2/R_{12}$; where Z_1 and Z_2 are the charges of ions 1 and 2, e is the charge of an electron, and R_{12} is the distance between ions 1 and 2. The higher the value of the electrostatic bonding energy, the stronger the bond. The Si-O bond is extremely stable because the charges are four and two and the silicon oxygen distance is 1.62Å. On the other hand, the Na-O bond is relatively weak, with charges of one and two and with an interatomic distance of 2.35Å. Glass formers generally have larger bond energies and the modifiers have lower bond energies. It was postulated that immiscibility results when the bond energies of the second oxide are relatively high and there is, consequently, a competition between the glass former and the high bond energy modifier in the determination of the structure.

If only oxide systems are compared, the bonding energy equation can be reduced to an expression of the nature Z/r; where Z and r equal the charge and the radius of the cations respectively. The remainder of the terms of the bonding energy equation apply to oxygen and consequently remain the same for each oxide considered. Table I shows the ratio of the cation valence to the radius for a number of different ions. Also indicated is the type of liquidus curve these oxides display when melted with SiO_2. The s-shaped curve is suggestive of so-called metastable immiscibility or sub-liquidus immiscibility. The relationship between Z/r and the existence of immiscibility is clear.

They also, with a few simple assumptions, made a theoretical calculation of the limiting composition and compared it to the real case. In a single phase lime-silica melt, there will be oxygens bonded to two silicons (bridging oxygens) and oxygens bonded to one silicon (non-bridging oxygens). Each calcium ion tries to surround itself with several of these non-bridging oxygens. Since there are twice as many non-bridging oxygens as calciums, and in order to have a reasonable number of surrounding oxygens,

Table I. Ratio of Cation Valence To Radius

	R	Z	Z/R	Type of Curve
Cs	1.65Å	1	0.61	
Rb	1.49	1	0.67	Nearly straight
K	1.33	1	0.75	
Na	0.99	1	1.02	
Li	0.78	1	1.28	s-shaped
Ba	1.43	2	1.40	
Sr	1.27	2	1.57	
Ca	1.06	2	1.89	Immiscibility
Mg	0.78	2	2.56	

it is necessary for each non-bridging oxygen to be bonded to more than one calcium. It was assumed that if two calciums are bonded to the same oxygen, then they could not be further apart than the Ca-O-Ca distance (2 x 2.38Å or 4.76Å). The cube of this linear separation ($108Å^3$) was assumed to be the approximate volume of glass that contains one calcium.

If the composition is represented by x percent by weight of CaO and (100-x) percent by weight of SiO_2, the volume of 100 grams of glass will be:

$$V_{100} = (100/\rho)\ (10^{24})\ (Å^3) \qquad (6)$$

where ρ is the density.

The number of Ca atoms in x grams of CaO is given by:

$$N_{Ca} = (x/\text{weight of CaO}\cdot\text{Avagadro's number}) =$$

$$= (x/56.1)(0.602)(10^{24}) \qquad (7)$$

The volume of glass per calcium atom in $Å^3$ is then equal to:

$$V_{Ca} = V_{100}/V_{Ca} = (100/\rho)\left[\frac{(56.1)(10^{24})Å^3}{(0.602)(10^{24})(x)}\right] = 108Å^3. \qquad (8)$$

The density of the glass in this range is approximately 2.6 grams per cc. A solution can then be made for x:

$$x = \frac{(100)(56.1)}{(108)(0.602)(2.6)} = 33\% \qquad (9)$$

This approximate geometrical calculation is in satisfactory agreement with the experimental value of 28 percent. The preceding work led to the theory that the limiting composition is that composition in which the lime content is high enough so that the Ca^{++} ions and the non-bridging oxygens are close enough to each other to have suitable surroundings. A similar treatment can be made for other oxide systems.

THE WORK OF VON A. DIETZEL[9, 10]

Dietzel used an approach similar to Warren's. However, he felt that the Coulombic Equation of Force ($F=Z_1Z_2e^2/R_{12}^2$) would give better correlations to the properties of oxides in glasses. The Coulombic Equation of Force can also be reduced if oxygen is always considered as the second species. He, therefore, used the reduced form, Z/a^2, the ionic field strength, in his consideration of the oxides. In this equation Z equals the charge of the cation and a is the anion to cation distance. This ionic field strength was calculated for a series of oxides and compared with the glass forming properties of these oxides. The results, in his opinion, gave better relationships than the electrostatic bonding energy used by Warren (see Table II). The explanation of immiscibility that he offered was similar to Warren's. Dietzel suggested that there is a tendency for calcium to surround itself with many non-bridging oxygens in a preferred coordination. Silicon also demands a structure which is compatible to its needs. As a result of the different needs of the two cations, two different structures will result until both demands can be satisfied with the same structure.

THE WORK OF LEVIN AND BLOCK[2, 11-13]

A. Analysis & Calculation of Immiscibility

Calculations concerning the limiting composition for immiscibility were presented in a four part series. The structures of the limiting compositions were also postulated for many silicate and borate structures.

An expression was developed for the average separation of the cations (S_p) in a glass by using the treatment of Warren and Pincus[8]. This distance was given as a function of the molecular weight of the modifier, the density of the

Table II. Classification of Cations By Increasing Field Strengths

Element	Valence Z	Ionic Radius CN=6 r	Interatomics Distance CN=6	True Coordination No.	Interatomic Distance for True CN a	Field Strength Z/r^2	Elect. Potential Z/r	Field Strength Considering Anions Z/a^2
K	1	1.33	2.65	8	2.77	0.56	0.75	0.13
Na	1	0.98	2.30	6	2.30	1.04	1.02	0.19
Li	1	0.78	2.10	6	2.10	1.65	1.28	0.23
Ba	2	1.43	2.75	8	2.86	0.98	1.40	0.24
Pb	2	1.32	2.64	8	2.74	1.15	1.51	0.27
Sr	2	1.27	2.59	8	2.69	1.24	1.57	0.28
Ca	2	1.06	2.38	8	2.48	1.78	1.88	0.33
Mn	2	0.91	2.23	6	2.23	2.42	2.2	0.40
Fe	2	0.83	2.15	6	2.15	2.90	2.4	0.43
Zn	2	0.83	2.15	4	2.03	2.90	2.4	0.49
Mg	2	0.78	2.10	6	2.10	3.3	2.6	0.45
				4	1.96			0.53
Zr	4	0.87	2.19	8	2.28	5.3	4.6	0.77
Be	2	0.34	1.66	4	1.53	17.4	5.9	0.86
Fe	3	0.67	1.99	6	1.99	6.7	4.5	0.76
				4	1.88			0.85
Al	3	0.57	1.89	6	1.89	9.2	5.3	0.84
				4	1.77			0.96
Ti	4	0.64	1.96	6	1.96	9.8	6.3	1.04
B	3	0.20	1.52	4	1.50	75	15	1.34
				3	1.36			1.63
Si	4	0.39	1.71	4	1.60	26	10	1.57
P	5	0.34	1.66	4	1.55	44	15	2.1

glass and the limiting composition. They then created two probable cation coordinations that might exist in the glass Figure 11. For each of these structures a separation (S_g) can be calculated and compared to the average separation calculated from the formula. By this method the most likely coordination (A or B) was found. The results are summarized in Table III. Column 5 of this Table is the approximate limiting immiscibility composition as found experimentally. Columns 6 & 7 are calculated using experimental density values whereas column 8 & 9 were obtained using additive density values. For calcium in a silicate glass an "A" type coordination is indicated.

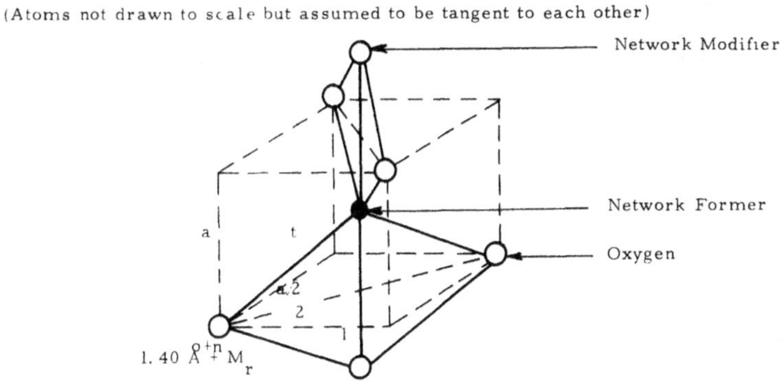

Type A

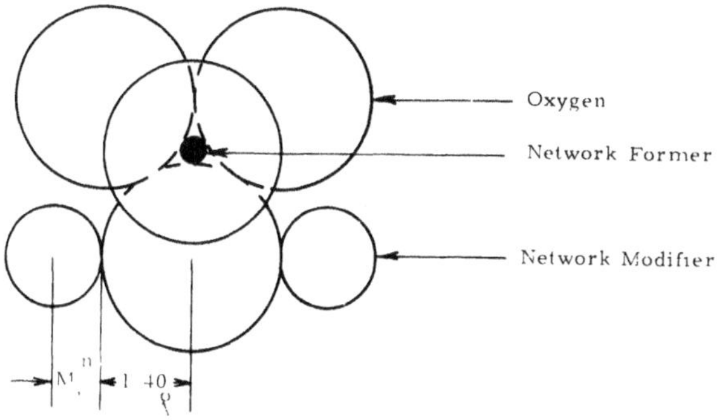

Type B

Figure 11. Possible coordinations of a modifier cation.[2]

Table III. Comparison of Experimental and Geometrically Calculated Values For Separation of Modifier Cations and For Composition of Modifier-Rich Liquids in Binary Silicate Systems.

Modifier	Cation Separation (a.u.) Physical S_p	Geometrical $S_g(A)$	Geometrical $S_g(B)$	Limiting composition (mole % of modifier) Experimental	From experimental and $S_g(A)$	From experimental and $S_g(B)$	From additive and $S_g(A)$	From additive and $S_g(B)$	Coordination Type
(1)	(2)	(3)	(4)	(5)	(6)	(7)	(8)	(9)	(10)
(A) S-shaped Liquidus									
K_2O	7.4	5.4	6.6	~5.5	14.8	7.9	15.0	8.1	B
Na_2O	6.1	4.7	5.7	~9.5	21.1	11.5	21.6	11.9	B
Li_2O	5.1	4.0	4.8	~15.3	29.3	17.4	30.8	18.2	B
BaO	6.5	5.5	6.6	~15.2	30.2	13.9	27.0	15.2	B
(B) Complete Immiscibility									
SrO	5.9	5.0	6.1	19.8	35.0	17.2	31.7	18.1	B
CaO	5.0	4.7	5.8	28.9	34.2	18.9	29.1	16.0	A
FeO	4.4	4.3	5.2	35.8	39.8	21.0	38.0	23.6	A
ZnO	4.6	4.2	5.2	35	45.8	23.4	40.4	24.8	A
MgO	4.3	4.1	5.0	40.1	47.1	27.9	44.1	28.0	A

Another method for the calculation of the limiting composition was also suggested. This method is called the oxygen volume method and it does not require a density approximation. The factors involved are ionic radii, the appropriate coordination type and the average volume of space occupied by an oxygen atom in a close-packed arrangement. A volume commonly used by crystallographers for the space occupied by an oxygen in a close-packed crystal arrangement is 17Å^3. The volume of the space occupied by the modifier can be calculated by multiplying 17Å^3 by the cubed ratio of the radius of the modifier (r) to oxygen (1.40). For calcium, the volume equals 6.0Å^3. Since the network former ions are assumed to be situated in the oxygen interstices, their volume of space is included in the oxygen volume. Therefore, for each type (A and B) it is possible to calculate the number of oxygens per cation. Since $(S_g)^3$ approximately equals the volume of each coordination type and the difference between this and the volume of the modifier will equal the oxygen volume, the number of oxygens will then equal:

$$\text{No. oxygens} = (S_g^{\,3} - 6.195r^3)/17.0 \qquad (10)$$

Using this method, the number of oxygens per calcium was calculated to be 11.4 for "B" type and 6.1 for "A" type. With 11.4 oxygens per calcium, the mole percent of the modifier is 16.1 percent. For 6.1 oxygens, it is 28.2 percent. Thus "A" type coordination again agrees more closely with the experimental limiting composition of 28.9 percent.

On the basis of these calculations, calcium was assumed to be "A" type with a coordination of six oxygens. Other systems were treated similarly. The results for these oxides are shown in Table IV.

B. Coordination Principles Applied to Immiscibility

The first part of the work of Levin and Block(A.) was based essentially on geometrical considerations. In this part, well-established principles of ionic-crystals were incorporated into their discussion of immiscibility.

The quantity, Z/(c.n.), where Z equals the charge of the ion and c.n. is the coordination number, was defined by Pauling as the electrostatic bond strength. In this

Table IV. Calculation Of Composition Of Modifier-Rich Liquids in Binary Silicate Systems by Oxygen-Volume Method

Modifier	Type	S_g (a. u.)	S_g^3 (a. u.^{3})	Vol. of metal ion (a. u.^{3})	Vol. of oxygen (a. u.^{3})	No. of oxygen	Formula	Mole % of modifier Cal.	Exp.
(1)	(2)	(3)	(4)	(5)	(6)	(7)	(8)	(9)	(10)
				(A) S-shaped Liquidus					
K_2O	B	6.6	292.8	14.6	278.5	16.4	$K_2O \cdot 15.9\ SiO_2$	5.9	5.5
Na_2O	B	5.7	190.1	5.3	184.8	10.9	$Na_2O \cdot 10.4\ SiO_2$	8.8	9.5
Li_2O	B	4.8	115.5	1.3	114.2	6.7	$Li_2O \cdot 6.2\ SiO_2$	13.9	15.3
BaO	B	6.6	299.4	15.2	284.2	16.7	$BaO \cdot 7.85\ SiO_2$	11.3	15.2
				(B) Complete Immiscibility					
SrO	B	6.1	236.0	8.9	227.1	13.4	$SrO \cdot 6.2\ SiO_2$	13.9	19.8
CaO	B	5.8	200.2	6.0	194.2	11.4	$CaO \cdot 5.2\ SiO_2$	16.1	
	A	4.7	109.2	6.0	103.2	6.1	$CaO \cdot 2.55\ SiO_2$	28.2	28.9
FeO	A	4.3	79.5	2.6	76.9	4.52	$FeO \cdot 1.76\ SiO_2$	36.2	35.8
ZnO	A	4.2	78.4	2.5	75.9	4.46	$ZnO \cdot 1.73\ SiO_2$	36.6	35
MgO	A	4.1	68.9	1.7	67.2	3.95	$MgO \cdot 1.48\ SiO_2$	40.3	40.1

treatment, the bond strengths were calculated compared with the glass properties of that oxide (Table V). Those with small bond strengths, i.e. 1/12, are modifiers which form completely miscible systems. On the other hand, higher bond strengths, i.e., greater than 1/4, indicated immiscibility. Calcium has a bond strength of 1/3 for six fold coordination and 1/4 for eight fold. It is also indicated, from the previous calculations, that if the bond strength is 1/3 or more, "A" type of coordination results; whereas if the bond strength is less than 1/3, "B" type of coordination results.

C. Ionic Field Strength

The experimental data supported the structural interpretation for the occurrence of immiscibility as representing the competition for oxygens between the network former and the modifier. Both cations would like to govern their own coordination. Thus a measurement of the difference of relative bond strengths would be helpful in comparing different systems. The ionic field strength expression used by Dietzel is again considered.

$$IFS = Z^{+}/(r_1 + 1.40)^2 \qquad (11)$$

Different network formers can be compared if compensation is made for the difference in the field strength of the network formers. If the difference between the ionic field strength of the modifiers and the ionic field strength of the network former is taken, such a compensation is made. It was found by this method that if the difference is greater than 0.8 to 0.9 or less than 0.06 to 0.02, immiscibility will not occur in borate, silicate or germanate systems.

Calcium's ionic field strength is 0.35. The difference between it and the network former field strength, the delta IFS, is 0.724 for germanates, 0.779 for borates and 0.858 for silicates. Observed immiscibility occurs in each of these systems.

THE WORK OF GLASSER, WARSHAW AND ROY[14]

Another attempt to correlate the existing experimental evidence on liquid immiscibility in silicate systems was made by plotting ionic potential, Z/r, versus mole percent

Table V. Electrostatic Bonding Energy (Z/c.n.) of Some Cations with Oxygens, Together with Shape of Liquidus Curves and Type of Modifier Coordination for Binary Borate and Silicate Systems

Ion	Radius (Pauling) (a.u.)	Radius ratio R/1.40	c.n.	Z/c.n.	Shape of liquidus SiO_2	Shape of liquidus B_2O_3	Coordination type with SiO_2 or B_2O_3
(1)	(2)	(3)	(4)	(5)	(6)	(7)	(8)
Cs^+	1.69	>1	12	1/12	Straight line		
Rb^+	1.48	>1	12	1/12	Straight line		
K^+	1.33	0.95	8	1/8	S		B
Na^+	0.95	0.68	6	1/6	S		B
Li^+	0.60	0.43	6	1/6	S		B
Ba^{++}	1.35	0.96	8	1/4	S	2 liquids	B
Pb^{++}	1.21	0.86	8	1/4	?	2 liquids	B
Sr^{++}	1.13	0.81	8	1/4	2 liquids	2 liquids	B
Ca^{++}	0.99	0.71	6, 8	1/3, 1/4	2 liquids	2 liquids	B and A
Cd^{++}	0.97	0.69	6	1/3	2 liquids	2 liquids	A
Fe^{++}	0.75	0.54	6	1/3	2 liquids	2 liquids	A
Zn^{++}	0.74	0.53	6	1/3	2 liquids	2 liquids	A
Co^{++}	0.72	0.51	6	1/3	2 liquids	2 liquids	A
Mg^{++}	0.65	0.46	6	1/3	2 liquids	2 liquids	A
Be^{++}	0.31	0.22	4, 3	1/2, 2/3	1 liquid	1 liquid	
Cr^{3+}	0.64	0.46	6	1/2	2 liquids		
Ti^{4+}	0.68	0.49	6	2/3	2 liquids		
Network formers							
B^+	0.20	0.14	3, 4	1, 3/4			
Si^{4+}	0.41	0.29	4	1			

of the limiting composition (see Figure 12). Lines can be drawn through ions of equal valences. The lines for each valence are approximately parallel and can be used to predict immiscibility in other systems, if the ionic potentials are within the ranges of the graph.

If the value of the ionic potential in a binary silicate system is less than 3.0, there are certain common features in the phase diagrams.

1. The primary cyrstalline phase field underlying the two liquid regions is SiO_2 (cristobalite).
2. One of the liquids has a composition close to pure SiO_2.
3. The binary eutectic between SiO_2 and the highest silica compound stable at the eutectic temperature is located on the silica-poor side of the liquid immiscibility region.

If the ionic potential is greater than 3.0, the phase diagrams of the systems are different. They share the common features that:

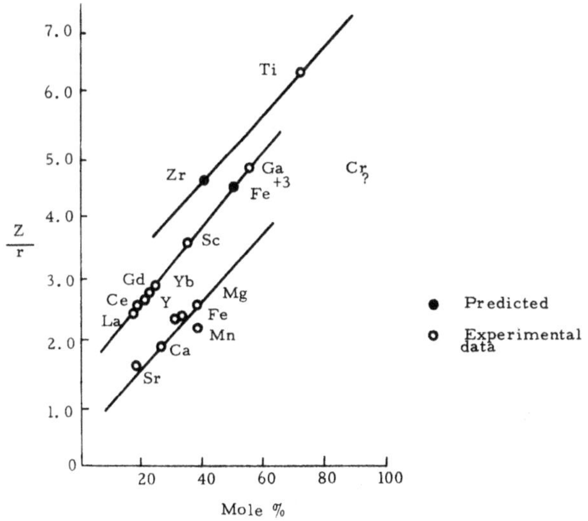

Figure 12. Correlation between ionic potential (Z/r) and extent of liquid immiscibility.[14]

1. The primary crystalline phase will not be silica, but either a silicate compound or the M_xO_y end member itself.
2. The high silica liquid is not necessarily nearly pure SiO_2. Its composition will be variable over relatively wide limits and the composition is fixed by the properties of the second oxide.
3. The binary eutectic between silica and the M_xO_y end number or the silicate compound underlying the two liquid region will be at a higher silica composition than that of the silica rich liquid.

It is also noted that the melting point of the modifier oxide below the immiscibility dome is important. It must be refractory; if it isn't, the silica liquidus must drop very sharply. Such a rapid lowering of the silica liquidus is unlikely, and the immiscibility dome is not likely to exist.

WORK OF J. W. CAHN[6, 15-19]

In a preceding section the probability of achieving a certain size fluctuation in composition within the immiscibility dome was disucussed. This treatment was greatly extended by Cahn. He showed through eloquent mathematics and computer calculations what would happen if this fluctuation in composition was expressed as a sinusoidal function.

The free energy was initially expressed as

$$F = \int [f(c) + k\ (\nabla c)^2]\ dv \qquad (12)$$

for an inhomogeneous solution whose composition everywhere differs only slightly from the average composition (C_o) and with small composition gradients. f(c) is the free-energy density of a homogeneous material of composition c, and is similar to the free energy expression developed by Swalin.[5] $k(\nabla c)^2$ is the additional free energy density if the material is in a gradient composition. The free energy difference between the initially homogeneous solution and the inhomogeneous solution is then given as

$$\Delta F = \int [1/2(\frac{d^2f}{dc^2})\ (c-C_o{}^2 + k(\nabla c)^2]\ dv \qquad (13)$$

In order to have a positive surface tension k must be positive. In this case if

$$\frac{d^2f}{dc^2} > 0 \tag{14}$$

then the solution is stable to all infinitesimal fluctuations. If instead

$$\frac{d^2f}{dc^2} < 0 \tag{15}$$

then the solution is unstable to all fluctuations such that the first term dominates. By choosing a large enough distance scale such a fluctuation can always be found. It was then assumed that the composition fluctuations vary as sinusoidal functions. The composition was varied mathematically by the superposition of sine waves of a fixed wavelength but random in orientation, phase and amplitude. If the composition at any point exceeded C_o a mark was printed out by the computer, and if the composition was less than C_o it was left blank. (See Figure 13) The two dimensional plots created in this manner show

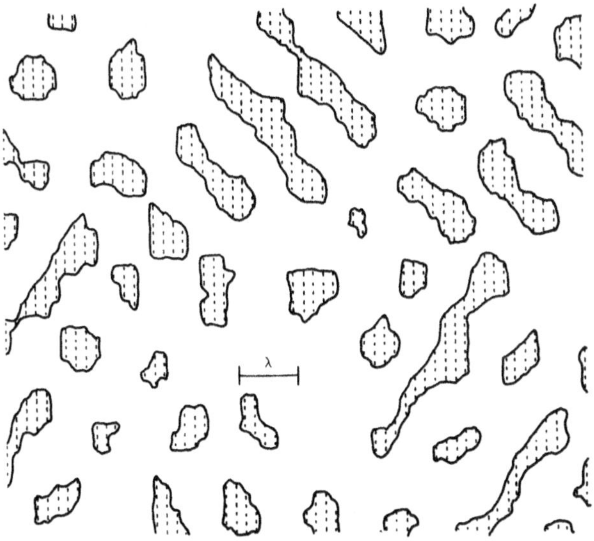

Figure 13. Computer simulated spinoidal phase separation.[6]

remarkable resemblance to actual spinodal phase separation. Cahn thereby showed how spinodal decomposition can occur through a gradual variation in composition, and not the distinct composition separations that exist outside the spinodal.

WORK OF STEVENS & SMYTH [23]

A treatment similar to the one presented by Smyth (this book) was developed for the $CaO \cdot SiO_2$ system. Simplifications were made in the entropy term, and consequently, the temperature dependency of the free energy function was slightly less than in previous developments. With the proper assumption of the energies involved, free energy curves were generated from 0 to 50 mole % CaO. (Figure 14) From these free energy curves, the immiscibility dome shown in Figure 15 was drawn. This dome fall is in the proper composition and temperature range for the $CaO \cdot SiO_2$ system.

The role of the sphere of influence is indicated by comparison with the free energy curves in Figure 16. In this figure the same calculations were made, except the sphere of influence was omitted. All non-bridging oxygens were allowed to bond to calciums. In these curves, no

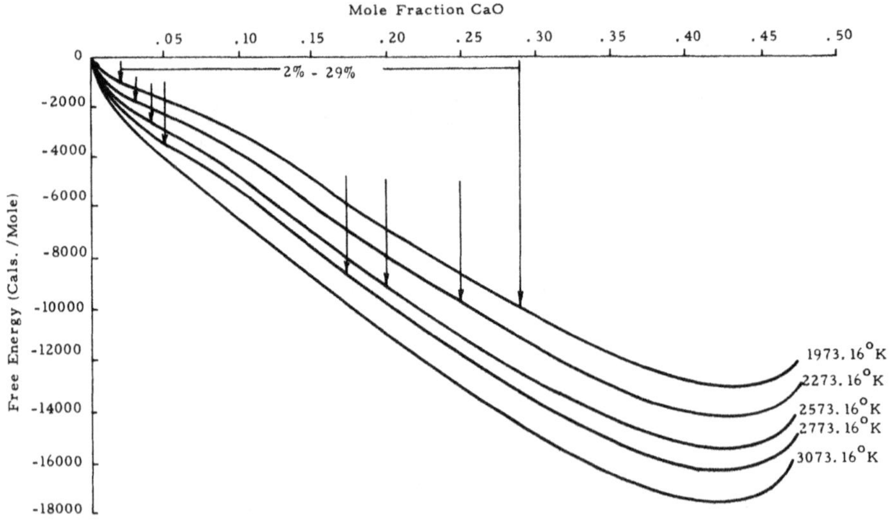

Figure 14. Free energy vs composition in the $xCaO \cdot (1-x)SiO_2$ system - model IV (arrows indicate limiting composition).

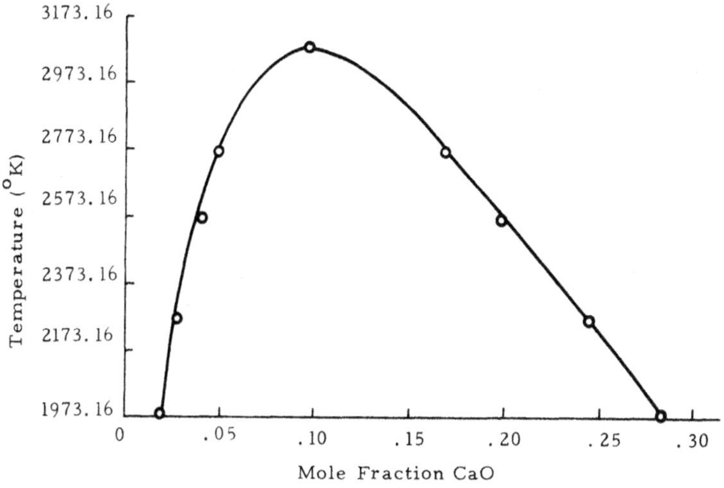

Figure 15. The immiscibility dome as calculated from Figure 14.

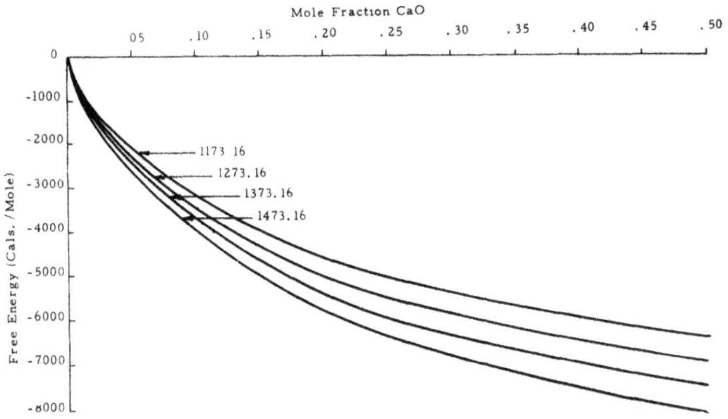

Figure 16. Free energy vs composition for a typical model II calculation.

immiscibility region is indicated. This indicates that the availability of non-bridging oxygens to calcium plays an important role in the creation of the immiscibility dome.

The development of the free energy "bump" which indicates immiscibility can be contributed to the presence of non-bridging oxygens that are not bonded to calciums and consequently exist in a high energy state. Initially the free energy curve decreases due to entropy considerations. As the concentration of non-bridging oxygens and calciums increase a larger percentage fall within the sphere of influence and the free energy curve again turns down.

A mathematical treatment of this study is presented in Appendix I.

WORK OF R. J. CHARLES[20, 21,22]

A departure from the traditional experimental method of investigation was made by R. J. Charles. An attempt was made to calculate the free energy curves for the Li_2O, K_2O, Na_2O-SiO_2 systems, and the $BaO-Li_2O-SiO_2$ systems. The activities were calculated from the liquidus and extrapolated below the liquidus to show metastable immiscibility. These calculations were made on the assumption of sub-regular solution behavior. From these activities, the free energy of mixing curve was calculated, and with the proper adjustments for the standard state, free energy curves displaying immiscibility regions were created. On the basis of these calculations, predictions were made concerning the occurrence of sub-liquidus, metastable immiscibility in these systems.

SUMMARY

By combining the preceding theories conclusions can be drawn about phase separation in glasses. The first obvious conclusion is that phase separation is very complex and complicated. The exact nature of this phenomena still needs a detailed explanation. Those who are deeply involved in the area, will be the first to admit the need for further research.

Beyond this first conclusion other general conclusions

may be drawn.

1. The shape of the free energy curve for the liquid must be such that it has an immiscibility deflection.
2. The position of this free energy curve must be favorable with respect to the free energy curves of the possible phases. Favorable means below, as in the case of simple immiscibility, or adjacent, as in the case of sub-liquidus immiscibility.
3. The relationship of the composition to the spinodal must be considered. Inside the spinodal phase separation will occur by composition variation. Outside the spinodal a nuclei must be first formed.
4. The relative bond strengths of the atoms present must be such that there is a competition for the determination of the structure.
5. The preferred structure around the cations must be such that the structures are incompatible with each other.
6. At some concentration there must exist a geometrical situation where the desired structures cannot be achieved due to the unavailability of certain components. (This is the situation when all the non-bridging oxygens are not close enough to the calciums to be associated with them).
7. Finally the existence or non-existence of phase separation will depend on the thermal history of the sample. Since many of the above requirements are thermodynamic in nature, their fulfillment requires a state approaching an equilibrium. Consequently the time-temperature treatment of the composition must be such that it is allowed to approach this thermodynamic equilibrium.

APPENDIX I.

The starting assumptions are:

1. All Si's have the same 4 fold coordination.
2. The oxygen coordination will be either bridging or non-bridging.
3. All Ca's have 8-fold coordination, some bridging oxygens and some non-bridging.

The 8 fold coordination of calcium is suggested by x-ray work[24] and radius ratio calculations. The coordination of the oxygens will vary to accommodate the Ca's and the Si's. Using these assumptions various structural units can be postulated. (Figure 17). To each unit an energy is assigned.

A free energy expression can be written in terms of the structural units and the appropriate energies.

The internal energy is composed of the sum of the number of each unit times the energy of that unit. The

	Coordination with Calcium	Number Present in Glass	Energy of Each Unit
Bridging Oxygen Groups			
Si —— O —— Si	0	Z(21)	U(21)
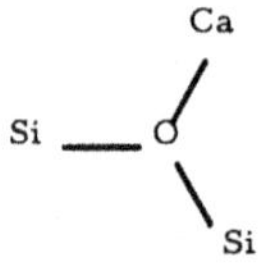	1	Z(11)	U(11)
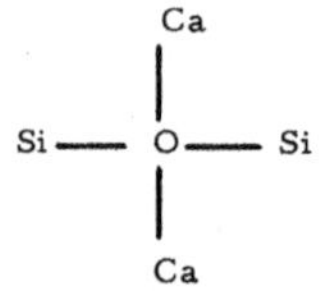	2	Z(12)	U(12)
etc. until	10	Z(20)	U(20)
Non-bridging Oxygen Groups			
Si —— O ——	0	Z(22)	U(22)
Si —— O —— Ca	1	Z(1)	U(1)
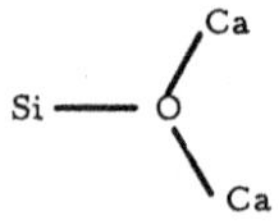	2	Z(2)	U(2)
etc. until	10	Z(10)	U(10)

Figure 17. The structural elements.

entropy portion of the free energy function is found through use of statistical mechanics.

$$S = k \ln\omega \tag{16}$$

k = Boltzmann's constant
ω = Probability of the existance of a given state
S = Entropy contribution of that given state

$$\omega = {}_N C^n = N!/_{n!(N-n)!} = \text{the number of} \tag{17}$$

combinations of N things taken n at a time.

If there are many states of different types

$$\omega = N!/_{n_1!n_2!n_3!} \ \ldots \tag{18}$$

An entropy expression can then be written for the different bridging and non-bridging oxygen states.

$$\omega = [\ (2xA)!/Z(1)A!\ Z(2)A!\cdots Z(10)A!Z(22)A!]$$

$$[(2-3x)A!/Z(11)A!Z(12)A!\cdots Z(20)A!Z(21)A!] \tag{19}$$

where A is Avagadro's number.

2x is the number of non-bridging oxygens in a glass whose composition is expressed as $x\text{CaO}\cdot(1-x)\ \text{SiO}_2$

2-3x is the number of bridging oxygens.

Employing Sterling's Approximation.

$$\ln N! = N\ln N - N \tag{20}$$

$$\ln\omega = A\ [2x\ln 2x - Z(1)\ln Z(1) - Z(2)\ln Z(2) -$$

$$\cdots Z(10)\ln Z(10) - Z(22)\ln Z(22) + (2-3x)\ln(2-3x)$$

$$-Z(11)\ln Z(11) - Z(12)\ln Z(12) - \cdots Z(20)\ln Z(20)$$

$$-Z(21)\ln Z(21)] \tag{21}$$

To convert to a free energy expression on a molar

basis Avagrado's number is multiplied by Boltzmann's constant to yield the gas constant R. $S = R\ln\omega$ is then the entropy expression for a mole of the glass.

The free energy expression can then be written as.

$$\text{Free Energy} = U_1Z_1 + U_2Z_2 + U_3Z_3 + \cdots U_{22}Z_{22} - RT$$

$$[2x\ln 2x - Z(1)\ln Z(1) - Z(2)\ln Z(2) - Z(3)\ln Z(3)$$

$$- \cdots Z(10)\ln Z(10) - Z(22)\ln Z(22) + (2-3x)\ln(2-3x)$$

$$- Z(11)\ln Z(11) - Z(12)\ln Z(12) - \cdots Z(20)\ln Z(20)$$

$$- Z(21)\ln Z(21)] \qquad (22)$$

Before this free energy equation can be solved the concept of the sphere of influence must be introduced. If a calcium is placed in the center of a sphere of radius R, then the number of non-bridging oxygens falling within that sphere can be calculated. When the concentration of CaO is very low it isn't likely that a non-bridging oxygen will fall within the sphere, if the radius is of a reasonable size; such as 3.00Å. This chosen radius is between one and two Ca-O bond lengths. As the amount of CaO increases more and more n.b.o's fall within the sphere. The exact number of non-bridging oxygens can be calculated from.

$$\text{Number of n.b.o's/Ca} + \left(\frac{4\pi R^3}{3}\right)\left(\frac{2xA\rho}{g}\right) \qquad (23)$$

where $\frac{4\pi R^3}{3}$ = volume of the sphere

$$\frac{2xA\rho}{g} = \frac{\text{number of n.b.o}}{\text{unit volume (cc)}}$$

ρ = density $\left(\frac{\text{gms}}{\text{cc}}\right)$ g = molecular weight (gms/mole)

This number is the set equal to the average number of Ca to n.b.o. bonds.

To save the free energy expression certain dependent variables must be omitted prior to minimization by partial differentiation. By transferring the assumptions to mathematical equations these dependent variables can be ferreted out.

$$\text{Composition} = x\text{CaO}\cdot(1-x)\text{SiO}_2 \tag{24}$$

Average Coordination of Calcium by Oxygen

$$= 8x = Z(1)+2Z(2)+3Z(3)+\cdots 10Z(10)+Z(11)+2Z(12)+\cdots 10Z(20) \tag{25}$$

$$\text{Number of Non-bridging Oxygens} = 2x = Z(1)+Z(2)+Z(3)+\cdots Z(10)+Z(22) \tag{26}$$

$$\text{Number of Bridging Oxygens} = 2-3x = Z(11)+Z(12)+Z(13)+\cdots Z(20)+Z(21) \tag{27}$$

Average Number of Non-bridging Oxygens to Calcium Bonds =

$$\text{ANO} = \frac{Z(1)+2Z(2)+3Z(3)+\cdots 10Z(10)}{Z(1)+Z(2)+Z(3)+\cdots Z(10)+Z(22)} = \frac{Z(1)+2Z(2)+3Z(3)+\cdots 10Z(10)}{2x} \tag{28}$$

on rearrangement of these equations the dependent variables are determined.

$$Z(1) = \text{ANO}(2x) - 2Z(2) - 3Z(3) - \cdots 10Z(10) \tag{29}$$

$$Z(11) = 8x - \text{ANO}(2x) - 2Z(12) - 3Z(13) - \cdots 10Z(20) \tag{30}$$

$$Z(21) = 2-3x - 8x + \text{ANO}(2x)+Z(12)+2Z(13)+\cdots 9Z(20) \tag{31}$$

$$Z(22) = 2x - \text{ANO}(2x)+Z(2)+2Z(3)+3Z(4)+\cdots 9Z(10) \tag{32}$$

The expressions for these dependent variables are then substituted into the free energy expression. The partials with respect to each variable are then taken and set equal to zero for minimization of the free energy function.

Non-bridging Oxygen Relations:

$$U(22)-U(1)+U(2)-U(1)-RT\left[\ln\frac{Z(1)}{Z(22)}+\ln\frac{Z(1)}{Z(2)}\right]=0 \quad (33)$$

$$U(22)-U(1)+U(3)-U(2)-RT\left[\ln\frac{Z(1)}{Z(22)}+\ln\frac{Z(2)}{Z(3)}\right]=0 \quad (34)$$

or for n = 2, 10

$$U(22)-U(1)+U(n)-U(n-1)-RT\left[\ln\frac{Z(1)}{Z(22)}+\ln\frac{Z(n-1)}{Z(n)}\right]=0 \quad (35)$$

Bridging Oxygen Relations:

$$U(21)-U(11)+U(12)-U(11)-RT\left[\ln\frac{Z(11)}{Z(21)}+\ln\frac{Z(11)}{Z(12)}\right]=0 \quad (36)$$

$$U(21)-U(11)+U(13)-U(12)-RT\left[\ln\frac{Z(11)}{Z(21)}+\ln\frac{Z(12)}{Z(13)}\right]=0 \quad (37)$$

or for n = 12, 20

$$U(21)-U(11)+U(n)-U(n-1)-RT\left[\ln\frac{Z(11)}{Z(21)}+\ln\frac{Z(n-1)}{Z(n)}\right]=0 \quad (38)$$

These equations can then be reduced to a more workable form by rearrangement and the substitution of two new variables: $C=Z(1)/Z(22)$ and $D=Z(11)/Z(21)$

Then

$$Z(n-1)C\exp-\left[\frac{U(22)-U(1)+U(n)-U(n-1)}{RT}\right]=Z(n) \quad (39)$$

So

$$Z(1)C\exp-\left[\frac{U(22)-U(1)+U(2)-U(1)}{RT}\right]=Z(2) \quad (40)$$

$$Z(2)C \exp - \left[\frac{U(22)-U(1)+U(3)-U(2)}{RT}\right] = Z(3) \qquad (41)$$

$$Z(1)C^2 \exp - \left[\frac{2U(22)-2U(1)+U(3)-U(1)}{RT}\right] = Z(3) \qquad (42)$$

Therefore, for

n=2, 10

$$Z(1)C^{n-1} \exp - \left[\frac{n-1(U(22)-U(1))+U(n)-U(1)}{RT}\right] = Z(n) \qquad (43)$$

and for

n = 12, 20

$$Z(11)D^{n-1} \exp - \left[\frac{n-1(U(21)-U(11))+U(n)-U(11)}{RT}\right] = Z(n) \qquad (44)$$

By using equations (43) & (44), equations (26) & (28) can be rewritten as functions of Z(1), and C and consequently solved.

Number of Non-Bridging Oxygens = $2x = Z(1)+Z(2)+Z(3) + \cdots$

$$Z(10) + Z(22) - f(Z(1),C) \qquad (45)$$

$$\text{Since } C = \frac{Z(1)}{Z(22)} \qquad (46)$$

Average Number of Non-Bridging Oxygen to Calcium Bonds=

$$ANO = \frac{Z(1)+2Z(2)+3Z(3)+\cdots 10Z(10)}{Z(1)+Z(2)+Z(3)+\cdots Z(10)+Z(22)} = f(Z(1),C) \qquad (47)$$

Similarly equations (27) & (25) can be rewritten and solved.

Number of Bridging Oxygens =

$$2-3x = Z(11)+Z(12)+Z(13)+\cdots Z(20)+Z(21) = f(Z(11),D) \qquad (48)$$

$$\text{Since } D = Z(11)/Z(21) \qquad (49)$$

Average Coordination of Calcium by Oxygen =

$$8x = Z(1)+2Z(2)+3Z(3)+\ldots 10Z(10)+Z(11)+2Z(12)+$$

$$\ldots 10Z(20) = f(Z(1), C, Z(11), D) \qquad (50)$$

The solutions from these equations are the fed back into the original free energy equation. The only problem at this point is the determination of the appropriate energy values.

Figure 18 illustrates how the energies are expected to behave. The standard state (pure SiO_2) was set at zero

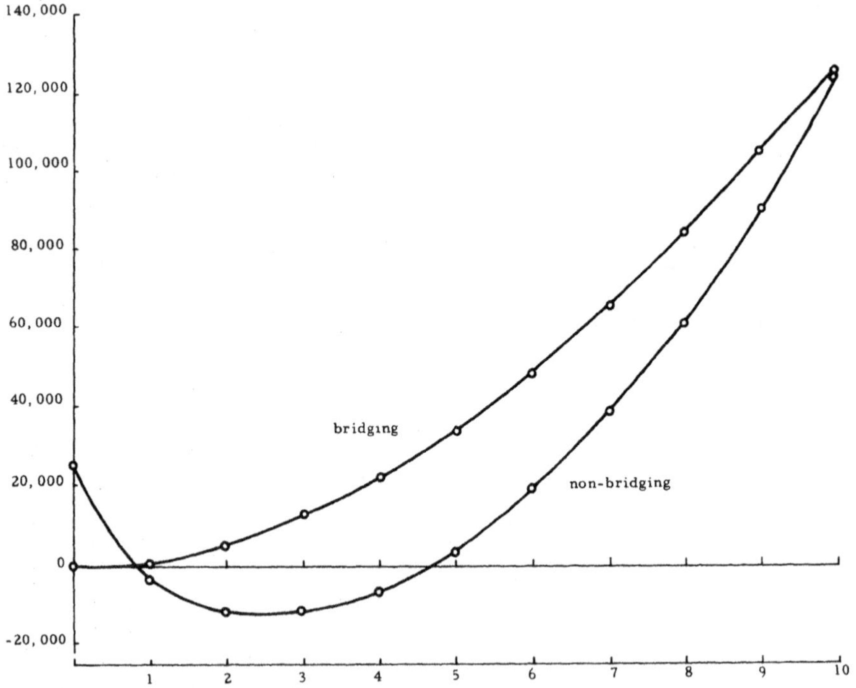

Figure 18. Energy of oxygen coordinations.

energy. This would correspond to bridging oxygens with no calciums or U(21). As bridging oxygens are surrounded with calciums the energy is expected to increase parabolically. The non-bridging oxygens offer a different type of problem. Smyth (this book) indicated that 36 Kcal were needed to form a non-bridging oxygen and a three coordinated silicon. Therefore, if this energy is proportioned between these two units there is approximately 25 Kcal associated with a non-bridging oxygen. Therefore, 25 Kcal was assigned to U(22). As non-bridging oxygens are surrounded with calciums this energy is initially decreased until the calciums begin to crowd each other. This reasoning leads to a curve that parabolically decreases to a coordination of about four and then increases and approaches the bridging oxygen curve at higher coordinations.

These energies were plugged into the free energy equation and the free energy was calculated as a function of composition. This was plotted to give the shape of the free energy curve, Figure 14. This is the immiscibility region the free energy curves offer. (Figure 15)

REFERENCES

1. J. W. Greig, "Immiscibility in Silicate Melts, I-II," American Journal of Science [5th Series], 13 [73], 1927, pp. 1-44; [74], 1927, pp. 133-154.
2. E. M. Levin and S. Block, "Structural Interpretation of Immiscibility in Oxide Systems: I, Analysis and Calculation of Immiscibility," Journal of the American Ceramic Society, 40 [3], 1957, pp. 95-106.
3. G. A. Rankin and F. E. Wright, American Journal of Science, [4th series], Vol. 39, 1915, p. 5.
4. B. Phillips and A. Maun, "Phase Equilibria in the System Calcium Oxide - Iron Oxide - Silica in Air", Journal of the American Ceramic Society, 42, [9], 1962, p.414.
5. R. A. Swalin, "The Free Energy of Binary Systems," Thermodynamics of Solids, John Wiley and Sons, Inc., New York, London, Sydney, Chapter 11, pp. 156-179.
6. R. J. Charles and J. W. Cahn, "Initial Stages of Phase Separation in Glasses," Physics and Chemistry of Glasses, 6, [5], 1965, pp. 181-191.
7. R. Roy, "Metastable Liquid Immiscibility and Subsolidus Nucleation" Journal of the American Ceramic Society 43, [12], 1960, p. 670-671.

8. B. E. Warren and A. G. Pincus, "Atomic Considerations of Immiscibility in Glass Systems," Journal of the American Ceramic Society, 23, [10], 1940, pp. 301-304.
9. Von A. Dietzel, "Structurchemie des Glasses," Die Naturwissenschaften, 29, 1941, pp. 537-545.
10. Von A. Dietzel, "Beziehungen zu Entglasungsvorgangen, Zur Verbindungsbilding und zu den Schmelzpunkten von Silicaten," Ztschr. Electrochem. Bd. 48, Nr. 1, 1942, pp. 9-23.
11. S. Block and E. Levin, "Structural Interpretation of Immiscibility in Oxide Systems: II, Coordination Principles Applied to Immiscibility," Journal of the American Ceramic Society, 40, [4], 1957, pp. 113-118.
12. E. M. Levin and S. Block, "Structural Interpretation of Immiscibility in Oxide Systems: III, Effect of Alkalies and Alumina in Ternary Systems," Journal of the American Ceramic Society, 41, [2], 1958, pp. 49-54.
13. E. M. Levin, "Structural Interpretation of Immiscibility in Oxide System: IV, Occurrence, Extent, and Temperature of the Monotectic," Journal of the American Ceramic Society, 50, [1], 1967, pp. 29-38.
14. F. P. Glasser, I. Warshaw and R. Roy, Physics and Chemistry of Glasses, 1, [2], 1960, pp. 39-45.
15. J. W. Cahn, "Phase Separation by Spinodal Decomposition in Isotropic Systems," Journal of Chemical Physics, 42, [1], 1965, pp. 93-99.
16. J. W. Cahn and J. E. Hillard, "Free Energy of a Non-uniform System: I, Interfacial Free Energy," Journal of Chemical Physics, 28, [2], 1958, pp. 258-267.
17. J. W. Cahn, "Free Energy of a Non-uniform System: II, Thermodynamic Basis," Journal of Chemical Physics, 30, [5], 1959, pp. 1121-1124.
18. J. W. Cahn and J. E. Hillard, "Free Energy of a Non-uniform System: III, Nucleation in a Two-Component Incompressible Fluid," Journal of Chemical Physics, 31, [3], 1959, pp. 688-699.
19. J. W. Cahn, "On Spinodal Decomposition," Acta Metallurgica, Vol. 9, 1961, pp. 795-801.
20. R. J. Charles, "Metastable Liquid Immiscibility in Alkali-Metal-Oxide-Silica Systems," Journal of the American Ceramic Society, 49, [2], 1966, pp. 55-62.
21. R. J. Charles, "Activities in Li_2O, Na_2O and K_2O-SiO_2 Solutions," Journal of the American Ceramic Society, 50, [2], 1967, pp. 631-640.
22. R. J. Charles,"Metastable Immiscibility in the

$BaO-Li_2O-SiO_2$ System," Physics and Chemistry of Glass, 8, [5], 1967, pp. 185-189.
23. H. Stevens, Ph.D. Thesis, Rutgers University (1969).
24. J. Biscoe, "X-ray Study of Soda-Lime-Silica Glass," Journal of the American Ceramic Society, 24, [8], 1941, pp. 262-264.

Concepts of Glass-Ceramics

Daniel R. Stewart

Owens-Illinois
P. O. Box 1035
Toledo, Ohio 43601

INTRODUCTION

It has been pointed out in previous sections[1] that, while cooling from the melting temperature, all liquids traverse a temperature range in which the liquid becomes thermodynamically unstable with respect to one or more crystalline phases. Whether or not crystals will form as a result of this instability is dependent on the rate of cooling and the kinetics of crystallization of the particular crystalline species that are involved.

In addition, a nucleus must be present in order for crystallization to occur. The number and location of the nuclei determine whether the crystallization is controlled or uncontrolled. If only a few nuclei are present, the crystallization process is difficult to control and will generally result in a body with a few large crystals (Figure 1a). Such uncontrolled crystallization is referred to as devitrification.

If, on the other hand, the formation of nuclei is enhanced by the addition of so-called "nucleating agents" or by using special glass formulations which are "self-nucleating," the crystallization process results in a body with many, very small crystals (Figure 1b). Such polycrystalline solids resulting from the controlled crystallization of glasses are called glass-ceramics. Glass-ceramics are also referred to in the technical literature as "vitrocerams,"

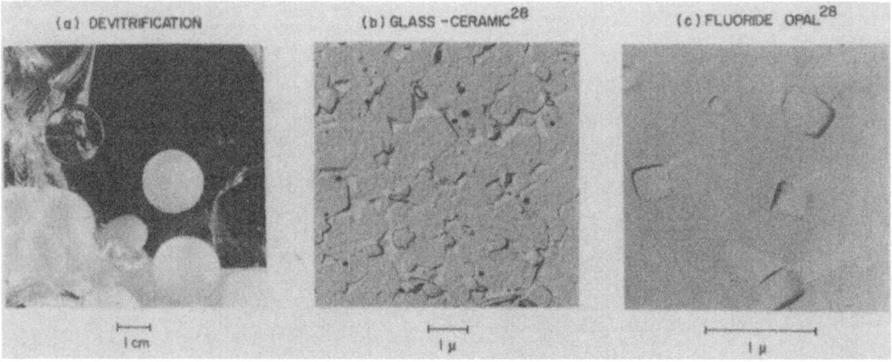

Figure 1. Types of Crystallization.

"devitrocerams,", "sitalls," and "melt-formed ceramics."

The term glass-ceramic is generally restricted to materials which are highly crystalline. There are several other types of glass products, such as fluoride opals and copper ruby glasses, in which minor amounts of a crystalline phase are produced (usually less than 10%) by controlled crystallization[2] (Figure 1c). These are generally not considered to be glass-ceramics and will not be discussed further.

There are other commercial products, more crystalline than the opal glasses, which are generally not considered to be glass-ceramics which utilize some degree of controlled crystallization. For most glasses, if nucleating agents are not purposefully added, crystallization still occurs if they are heated in a certain temperature range. However, in this case, the crystals generally are nucleated at the surface and grow into the body. One can accelerate the nucleation process by grinding the glass into a fine powder before heating. This procedure (along with proper composition formulation) is the basis of a family of products known as devitrifying solder glasses which are used primarily for sealing applications.[3]

HISTORICAL DEVELOPMENT

For centuries the glass maker has been trying to avoid devitrification, since it degraded his product and he did not know how to control it. Morey[4] described devitrification as "the chief factor which limits the composition range of practical glasses, and is an ever-present danger in all glass manufacture and working."

There were a few scientists who evidently thought otherwise. One of the earliest references to an attempt to use the crystallization of glass to advantage dates back to the early 1700's when deReaumur[5] in France attempted to crystallize soda-lime-silicate bottles. By packing the bottles in mixtures of sand and gypsum and heating them for long times, he succeeded in crystallizing them. However, the crystals started to grow from the surface, and the resulting bodies were weak and deformed.

Little further work was done until the early 1900's. At this time Tammann[6] started to investigate crystallization in detail and, through theoretical and experimental investigations, supplied the basis for much of our present understanding of the nucleation and crystallization process.

In the same time period, a new industry arose in Europe based on the controlled crystallization of molten rocks, particularly basalts. Using many of the principles derived by Tammann, rock-like ceramics in a great variety of shapes were produced and utilized in a wide range of applications. Portevin,[7] in particular, gives a very detailed summary of the principles of melting and crystallization of fused basalts with photographs of production items.

In the 1930's, Blau[8] illustrated the importance of controlled crystallization; and in the 1950's and 1960's, there was accelerated interest in controlled crystallization aided by nucleation promoters. Stookey[9] demonstrated that colloidal dispersions of metals could be used as nucleants. The effectiveness of certain refractory oxides as nucleation promoters was discussed by Stookey,[2] Albrecht,[10] and others too numerous to mention. These latter aspects will be discussed in more detail in subsequent sections.

THE GLASS-CERAMIC PROCESS

(A) General Comments

Inherent in the definition of a glass-ceramic is that it must start out as a glass. Glasses for subsequent crystallization are batched and melted much like any other glass. They are generally more difficult to melt and form than a soda-lime glass but can be successfully manufactured with existing technology or reasonable extrapolations thereof. After forming, the glass is subjected to a carefully controlled thermal treatment to bring about the nucleation and crystallization.

The glass-ceramic process offers several distinct advantages over conventional ceramic processes.

1) It is possible to produce a wide range of properties. i.e., there is some degree of tailorability. In principle, any composition that can be cooled to the glassy state can be crystallized under the proper heat treatment conditions if the proper nucleant is found. In practice, present knowledge is insufficient to do this to all compositions. Nevertheless, quite a range of properties can be obtained. Table 1 is a conservative listing of some ranges obtained for several properties.

Table 1. Range of Glass-Ceramic Properties

Thermal Expansion	$\sim$ -20 to +200 x $10^{-7}/°C$
Strength	6 to 60 kg/mm^2
Transparency	Transparent to opaque
Chemical Durability	Water soluble to inert
Electrical	Semiconductor to insulation

2) Glass-forming techniques can be used. This means that bodies can be shaped by pressing, blowing, drawing, casting, or combinations thereof. This has economic implications since many of these processes are amenable to high-speed, automatic production and close control of dimensions.

3) The dimensional changes during the transformation of the glass to the glass-ceramic are small compared to the

shrinkage of most conventional ceramics during drying and firing.

4) Zero porosity is routinely obtained.

5) Uniformity of properties is generally outstanding.[11] This homogeneity can be attributed to the nature of the nucleation process and the degree of homogeneity that can be obtained in the initial molten state.

6) Crystalline phases can be produced which are not obtained using other reaction routes such as solid-state sintering.

(B) Heat Treatment

As has been inferred in previous comments, in the crystallization process, the phase transformation does not occur spontaneously throughout the parent phase. The transformation proceeds from distinct centers called nuclei. Two parts of the crystallization are, therefore, distinguished: nucleation and crystal growth.

Tammann[6] and Blau[8] illustrated that the formation of crystals depended on (a) the number of nuclei formed, (b) the rate of crystal growth, and (c) the viscosity of the glass. If the temperature ranges of nuclei formation and crystal growth overlap, only partial control is possible. This is illustrated schematically in Figure 2a where there is significant overlap of the nucleation and growth rate curves. Between T_2 and T_3, as soon as a nuclei forms, it can grow. This produces comparatively few crystals which exhibit a wide range of crystal sizes.

Control is possible when the overlap of the two curves is minimized such as is shown in Figure 2b. Between T_1 and T_2, only nuclei form. Therefore, by holding the glass at a temperature between T_1 and T_2, it is possible to control the number of nuclei. After the required number of nuclei have formed, the glass is heated in the range T_3 to T_4 where crystal growth occurs. This is the reason for the commonly used two-stage heat treatment such as is shown in Figure 3, schedule a.

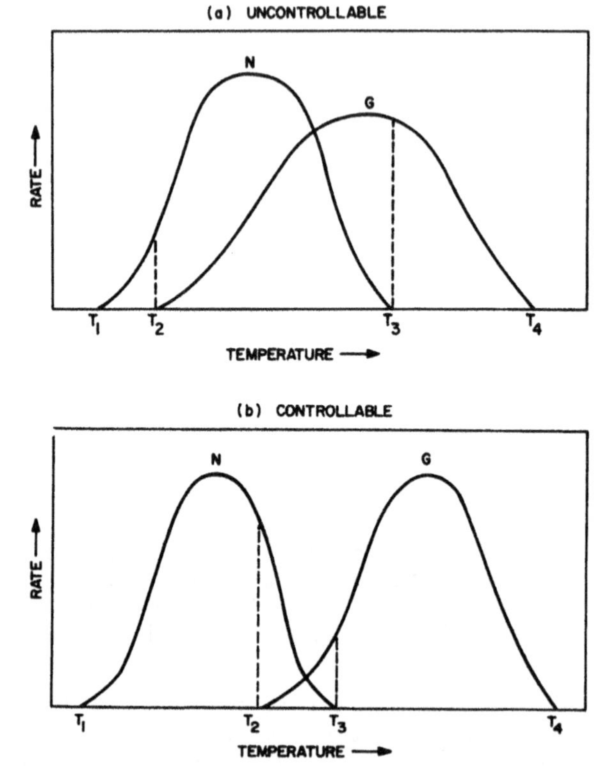

Figure 2. Nucleation and Crystallization Rate Curves.

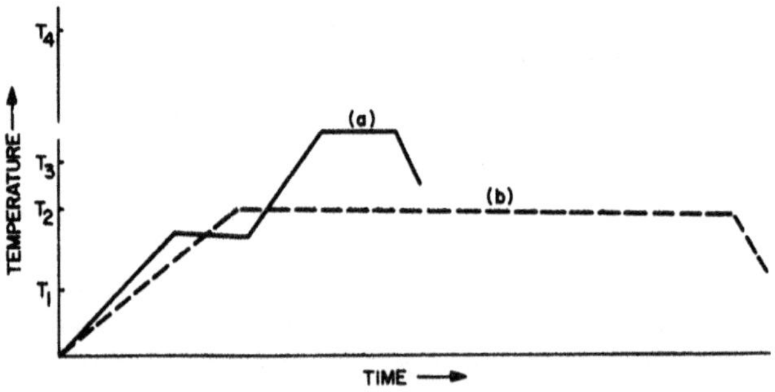

Figure 3. Heat Treatment Cycle for Controlled Crystallization (a) Two Stage (b) Isothermal.

The crystallization is accompanied by the release of energy in the form of heat. The magnitude of the heat released varies from composition to composition; but in the case of the crystallization of a high-quartz solid solution phase, the heat of crystallization was estimated to be 60 to 70 cal/gm.[12] This can result in a significant temperature increase which, in turn, can produce undesirable temperature gradients in the piece being heat treated. One way to minimize this problem is to select a composition which has a tail of the growth rate curve extending into the nucleation region. This overlap can be tolerated if the rate of growth is sufficiently low. A one-stage, i.e., isothermal, heat treatment can then be used in which the nuclei formation is followed by a slow rate of crystal growth (Figure 3, schedule b). This spreads the heat evolution over a longer time base and allows the evolved heat to dissipate without producing undesirable gradients.

1) Nucleation.

The mathematical treatment of homogeneous and heterogeneous nucleation is well developed.[1,13] However, despite considerable research, the role of the nucleating agents in catalyzing the reaction is still not completely resolved. In other words, on an atomic scale, how does a nucleating agent nucleate?

Numerous authors have pointed out that the controlling factor which allows one to make conventional glasses without crystallization is the very slow nucleation rate which implies the presence of a large energy barrier to nucleation. One cannot hope to develop a generalized picture of the nucleating mechanism, since it undoubtedly differs from nucleant to nucleant. However, it is probably safe to say that, by one means or another, the nucleant causes a disproportionation of the binding forces[14] which, in turn, introduces sites of lower thermodynamic stability. The present state of the art does not permit a priori predictions of the most effective nucleant for a given base composition.

The nucleant which is generally a noble metal or refractory oxide is incorporated in the batch and becomes an integral part of the glass structure during the melting. By heating the glass approximately 50° to 100°C above the annealing point for times on the order of one hour or more, nucleating sites form. These nucleating sites are randomly distributed throughout the glass in concentrations of 10^9 and 10^{15} nuclei/cc of glass and serve as sites for subsequent crystal growth.

Stookey[15] has listed some of the characteristics of an effective nucleating agent.

a. The nucleant should be readily soluble at melting and forming temperatures but only sparingly soluble at low temperatures.

b. The nucleant itself should have a low free energy of activation for homogeneous nucleation.

c. The nucleant atoms or ions must diffuse rapidly at low temperatures, compared to the major constituents of the glass.

d. The interfacial energy between the glass and nucleant must be low.

e. The crystal structure and lattice parameters of the nucleant and the crystal phase to be developed should be as nearly the same as possible.

The first three factors govern the formation of the nucleus, the latter two govern the effectiveness of the nucleant as a crystal growth promoter. Item e is not a requirement for all modes of nucleation.[16]

Much of the information on nucleation is derived from experimental observations of what occurs just prior to crystallization. In many cases the experimental tools are insufficient to adequately define the sequence of events. In addition, it is not always possible to identify a cause-effect relationship.

Figure 4 illustrates some of the paths traversed in going from the homogeneous glass to the initial crystallization of the major phase. There is very little evidence for a direct transformation of a homogeneous glass to the major crystalline phase (path 1) without the appearance of

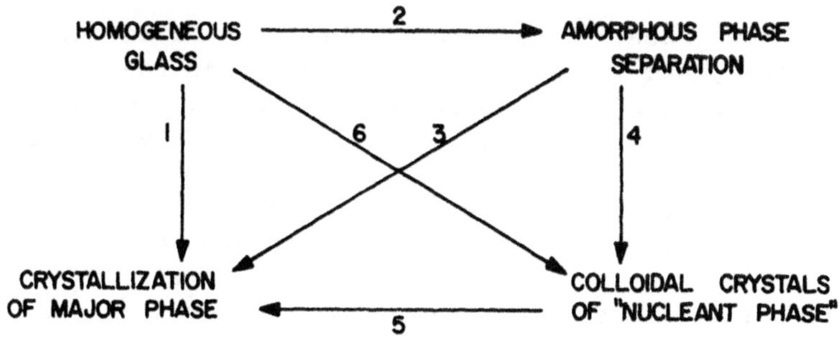

Figure 4. Reaction Paths Between the Initial Glass and the Final Glass-Ceramic.

intermediate processes. However, Neilson[17] using small-angle x-ray scattering techniques to study various ZrO_2-nucleated magnesia aluminosilicate glasses has observed a transition approaching this. His initial glass which had a small degree of localized clustering of Zr atoms transformed directly to a high-quartz solid solution phase under certain conditions, without the intermediate formation of a crystalline Zr-containing phase or further amorphous phase separation.

The path labeled (6) (5) probably represents the best understood sequence for the case in which the colloidal crystals are metallic. Small amounts of metals such as Ag, Au, Pt, Cu, etc.,are incorporated in the glass.[2] By exposure to actinic radiation followed by heat treatment, colloidal particles of the metal form by a diffusion controlled mechanism. On further heat treatment, these colloidal particles act as nuclei for the epitaxial growth of the major phase. For example, by using this procedure on a glass having the composition (in wt %) 81.5 SiO_2, 12.0 Li_2O, 3.5 K_2O, 3.0 Al_2O_3, 0.03 CeO_2, 0.02 Ag, a lithium metasilicate phase forms on the metallic Ag particles.[9] Maurer[18] has shown that the lithium metasilicate does not start to crystallize until the metal crystals grow to a size of about 80 Å.

In the early 1960's, for TiO_2-nucleated systems, it was widely believed that an initial nucleation phase of crystalline TiO_2 or a titanate catalyzed the crystallization of the major phase.[19] Failure to observe such an initial crystalline phase led Roy[20] to postulate that the separation of a second liquid phase enriched in Ti was a crucial nucleating step, i.e., that path (2) (3) in Figure 4 was responsible for the crystallization. Subsequent work has shown that glass-in-glass separation (path (2) (3)) is at least a common event in the nucleation sequence.

Illustrations of the presence of a glass-in-glass separation are numerous. To cite just two, MacDowell and Beall[21] in a study of Al_2O_3-SiO_2 glasses observed amorphous phase separation followed by crystallization without the addition of any nucleating agents. McMillan's successful use of P_2O_5 as a nucleant[22] is attributed to its effectiveness in promoting phase separation.

There have been several proposals to explain why amorphous phase separation leads to favorable conditions for nucleation. It has been suggested that the interfacial energy between two glassy phases is less than that between a glass and a crystalline phase, thereby reducing the barrier to nucleation.[23] It has also been pointed out that the composition of one of the amorphous phases may more closely resemble the crystalline phase than the original homogeneous glass. This could lead to easier nucleation by reducing the activation energy.[20] There are cases in which the crystal phase appears to initiate within one of the amorphous phases and other cases where the crystallization starts at the boundary area between the two amorphous phases. Numerous examples can also be cited where massive phase separation can occur without enhancing nucleation.[24]

Path (2) (4) (5) in Figure 4 is also commonly observed. In this case, the amorphous phase separation is followed by the crystallization of small particles of a metastable crystalline phase which, in turn, is followed by crystallization of the major phase or phases. This has been demonstrated by Doherty et al[25] who studied a lithia aluminosilicate glass containing titania using transmission electron microscopy. They found that, on cooling, phase separation of the glass occurs on a scale of about 50 Å. On reheating, an aluminum titanate crystalline phase forms which appears to be $Al_2Ti_2O_7$ with a particle size of about 50 Å. On further heating at higher temperatures, the crystallization of the major phase occurs. They point out that the formation of the $Al_2Ti_2O_7$ is undoubtedly influenced by the phase separation of the glass, possibly by an enrichment of one of the phases in titanium. A similar sequence was observed by Neilson[26] in a study of a zirconia-containing magnesium aluminum silicate.

Blinov[27] believes that the colloidal crystalline phase is a vital part of the nucleation mechanism in at least the lithium and magnesium aluminosilicate systems since the crystalline compound which is first precipitated is isomorphous with the predominant crystal phase that is subsequently developed in the glass-ceramic. Nakagawa and Izumitani[24] conclude that the thermodynamic barrier for the formation of the major phase is lower at the surface of a crystalline nucleant phase than at a glass-glass interface. However, Neilson[17] has suggested that the colloidal crystals of the "nucleant phase" which occur after the amorphous phase separation and before the crystallization of the major phase are not necessarily an integral part of the nucleation of the major phase.

2) Crystallization.

After nucleation, the material must generally be heated to higher temperatures in order for crystal growth to proceed. The growth step also represents a complex process for a number of reasons. First, there may be a number of crystalline phases appearing simultaneously. Commonly, the composition of the crystals differ from the composition of the glass. This means that the composition at the crystal-glass interface is constantly changing. In addition, portions of the primary crystal phase may start to transform by solid-state reaction to a new structure type.

The complexity can be illustrated by an isothermal transformation plot.[28] Figure 5 is a characteristic isothermal transformation plot for a glass-ceramic material. This plot relates the phases produced to the time and temperature of the heat treatment and is a non-equilibrium diagram.

The lower line represents the release of residual stress in the glass. Region A is the nucleation range. Glass-in-glass phase separation or the appearance of the nucleating phase may occur in this region. The first line above the stress release line represents the appearance of the first crystalline phase, Phase 1. The line separating B and C marks the first appearance of a second crystalline phase,

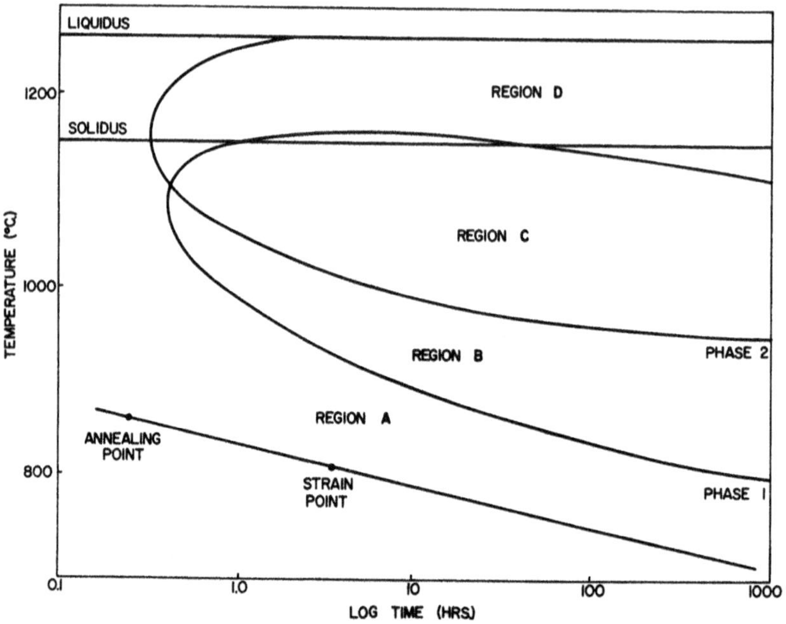

Figure 5. Schematic Isothermal Transformation Plot for a Glass-Ceramic Material.

Phase 2. In Region C, Phase 2 may be formed by crystallization of the remaining glass or by a solid-state transformation of Phase 1 to Phase 2. Phase 2 may be located within the grains of Phase 1; it may segregate at the Phase 1 grain boundaries; or by massive recrystallization, it may completely obliterate the grain boundaries of Phase 1. With a few exceptions,[29] very little attention has been directed to this aspect of the crystallization process. In Region D, Phase 1 has largely disappeared and appreciable amounts of liquid start to form because Region D is above the solidus.

An isothermal transformation plot can be useful for helping establish the required heat treatment conditions. For example, it defines the lowest temperature at which crystallization starts to occur and the highest temperature permissible to avoid the formation of a liquid phase. If Phase 1 has the desired properties and Phase 2 does not (or vice versa), the diagram indicates the temperature-time matrix that will produce the desired phase. Such a plot also contains valuable information on phase stability.

The transformation plot, however, is only a first

approximation for proper heat treatment development since it does not contain any representation of crystallization kinetics. From a practical standpoint, it is important that the rate of crystal growth be "just right." If the crystal growth rate is very slow, there is a danger of deformation of the sample during heat treatment. At temperatures in and above the optimum nucleation range, the viscosity of the glass may be such that the body can deform under its own weight before sufficient crystals form to retard flow. If the crystal growth rate is too fast, the heat evolved during the crystallization cannot be dissipated to the surroundings fast enough to avoid detrimental thermal gradients. This can result in stresses sufficient to break the ware.

FACTORS THAT INFLUENCE THE FINAL PROPERTIES OF A GLASS-CERAMIC

Table 2 summarizes the primary factors which determine the final properties of a glass-ceramic.

Table 2. Factors Affecting the Properties of the Glass-Ceramic

1. The properties of the crystalline phase
2. The grain size
3. The percent crystallinity and the distribution of the glass
4. The intergranular bonding
5. Crystal orientation

It is obvious that the properties depend to a large extent on the properties of the crystalline phase or phases that develop. For example, one would not try to produce a low-expansion body by crystallizing out a phase which has inherently a high-expansion coefficient.

The grain size also plays an important role, particularly in regard to optical and mechanical properties. Tashiro[30] has found that the strength increases with decreasing grain size. With some compositions, it is possible to make transparent, translucent, or opaque glass-ceramics all from the same composition by proper control of the grain size. This is shown in Figure 6.

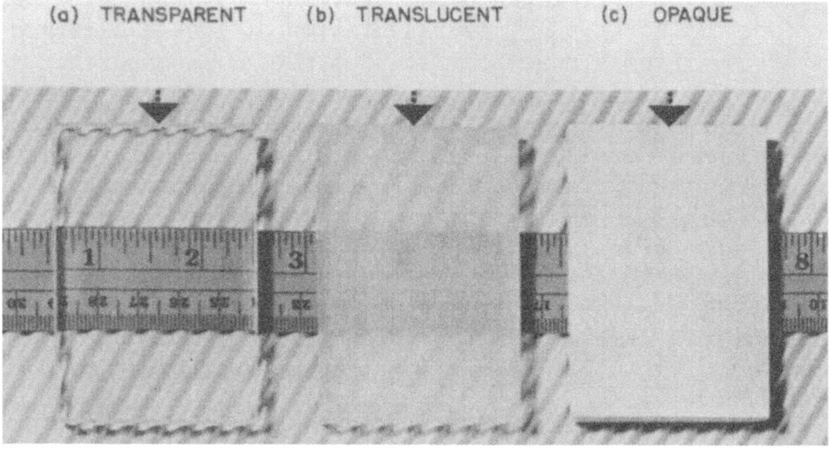

Figure 6. Glass-Ceramic Materials Demonstrating the Range of Transparencies that can be Achieved.

The percent crystallinity affects all the properties. A material that is 10% crystalline will behave much differently than one that is 90% crystalline. This is illustrated in Figure 7 by the change in the coefficient of expansion with increasing crystallinity.[31] There are undoubtedly other factors changing at the same time that the crystallinity is changing, but the percent crystallinity is having the largest effect. In addition to the amount of glass, the distribution of glass must be considered. If the glass occurs as a coating on each grain, its influence is much different than if it is distributed randomly in more isolated pockets where the grains intersect.

The bonding between the phases has an important influence on the transport properties and mechanical properties. However, it is a difficult subject to treat quantitatively. Chemically, the intergranular material is very complex because of depletion of some atomic species due to diffusion of atoms into the crystalline phase as well as enrichment in other species due to exsolution by the developing crystal. It is also a difficult area to study experimentally because of the extremely fine grain size of most glass-ceramics, i.e., the tools to study the intergranular phase are very limited.

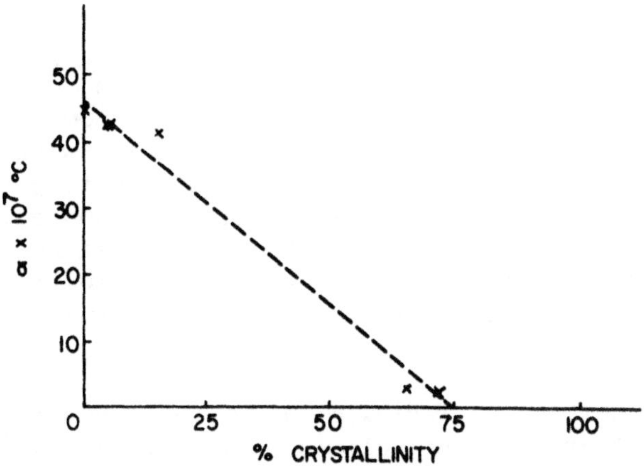

Figure 7. Dependence of the Coefficient of Expansion on Per Cent Crystallinity for a Li_2O-Al_2O_3-SiO_2 Glass-Ceramic.

In most glass-ceramics, crystal orientation (Table 2, Item 5) does not play a role because the crystals are randomly oriented. However, it is possible to get selective orientation of some phases which can result in a degree of directionality of properties.

The items listed in Table 2 which define the microstructure of the material are strongly influenced by the variables listed in Table 3.

Table 3. The Controllable Parameters

1. The base composition
2. The nucleants
3. The heat treatment

Each of these will be discussed in subsequent sections.

(A) Base Composition

Chemically, a commercial glass-ceramic is very complex. There are generally at least seven or eight atomic species, each one added to fill a particular role. In a later section, some of the factors involved in selecting each constituent and its concentration will be discussed. The term base composition refers to those phase-determining constituents other than the nucleants. To illustrate the importance of the base composition, four systems will be discussed.

First, however, a review of a few basic concepts is necessary. A crystalline solid consists of an indefinitely extended array of atomic species arranged in a regular geometric pattern. The basic structural unit of a crystal, i.e., the unit cell, is the smallest atomic grouping that still possesses the characteristic properties of the crystal. The properties of a crystal are determined by the arrangement of the various atomic species in the cell. The unit cell must obey the laws of constant proportions and be electrically neutral, but it can be modified by the incorporation of selected foreign atoms without destroying the basic structural framework. It is thus possible to get a family of compositions whose atomic arrangement is derived from the basic structural unit. The resulting structures are known as derivative solid solutions[32] to emphasize that they are derived from basic structures by the incorporation (solution) of the foreign atoms.

In silicate structures the Si^{4+} site is frequently occupied by an Al^{3+}. The substitution of a 4^+ ion by a 3^+ ion results in a charge imbalance which must be counterbalanced by the addition of other cations. These added cations which are frequently alkali or alkaline earth ions attempt to locate themselves in the interstices of the structure near the charge imbalance to bring about localized charge neutrality. Such structures are known as stuffed derivatives and typify the majority of silicate framework structure types.

1) Lithium Silicate.

Glasses and crystals of the same nominal composition may vary greatly in their properties. In the lithium silicate system, the difference in chemical durability between the glass and some of the crystalline phases is very great. This led Stookey[9] to utilize the lithium silicate system for the manufacture of photosensitive glasses which can be chemically machined. By suitable masking and irradiation of lithium silicate glasses containing photosensitive noble metals as nucleants, it is possible to nucleate and crystallize only preselected areas. The crystallized portion contains lithium metasilicate crystals which are much more soluble than the glass. Due to this difference in solubility, the metasilicate crystals can be etched away leaving the uncrystallized portions essentially intact. By this means, it is possible to form intricate patterns in the glass. It was mainly by exploiting this one property that lithium silicates found their way into the commercial market.

Lithium silicates have also been the base composition used in innumerable scientific investigations delving into their phase separation tendencies, their nucleation and crystallization properties, and their crystallography. The other binary alkali silicates are less inclined to bulk crystallize and as a result have attracted less attention.

2) Li_2O-Al_2O_3-SiO_2

The family of glass-ceramics that has been most widely exploited is based on the lithia aluminosilicate system and modifications thereof. The crystal chemistry, the crystallography, and the kinetics of phase transformation in this system have been extensively studied because of the low-expansion materials that can be produced. Two phases are of particular importance: high-quartz solid solutions and keatite solid solutions. Both of these are derivative solid solutions[32] based on phases of pure silica in which some of the silicon atoms have been replaced by aluminum atoms with lithium atoms occupying interstitial sites to maintain charge neutrality.

In pure silica, the low form of quartz has a very high expansion coefficient. After the structural rearrangement to high quartz, the thermal expansion coefficient is slightly negative. The low expansion high-quartz solid solutions are a result of successful stabilization of the high-quartz structure to room temperature by suppressing the high-low inversion by ionic substitution. The high-quartz structure can also be stabilized by the addition of other ions such as magnesium and zinc. A variety of names have been used to identify high-quartz solid solution phases.

In compositions having the general formula $Li_2O{\cdot}Al_2O_3{\cdot}nSiO_2$ where n represents the number of moles of silica, heating the high-quartz solid solution to higher temperatures causes it to transform to the keatite solid solution structure for n greater than about 3.5.[33] The keatite solid solution which is also commonly referred to as a β-spodumene solid solution is a derivative structure based on the silica polymorph keatite and is also a low-expansion phase.

The sequence of events that generally occurs during heating are glass → high-quartz solid solution → keatite solid solution → liquid. This is shown in Figure 8 which is a schematic isothermal transformation plot[34] such as was shown earlier. The high-quartz and the keatite solid solutions are readily nucleated by titania, zirconia, or mixtures of titania and zirconia. As far as properties go, the outstanding feature is their resistance to thermal shock. Expansion coefficients can be developed which range from negative values through zero to low positive values. The optical properties are also rather interesting in that transparent materials can be made by controlling the crystal size. Interesting colors can also be produced. Based on these favorable properties, a number of commercial products are being sold such as cookware, stove tops, massive mirror substrates, pump parts, etc.

3) $MgO\text{-}Al_2O_3\text{-}SiO_2$

Glass-ceramic compositions in this system have also been studied in considerable detail. The most commonly developed phases are high-quartz solid solution phases[35] and cordierite ($2MgO{\cdot}2Al_2O_3{\cdot}5SiO_2$). Glass-ceramics from this system generally have good dielectric insulating characteristics at high frequency, high strength, and moderate-to-high expansion. They have been nucleated by TiO_2, ZrO_2, and SnO_2; but generally the nucleation conditions are a little more exacting than in the lithia aluminosilicate system.

The main application of the system to date has been in the production of radomes.

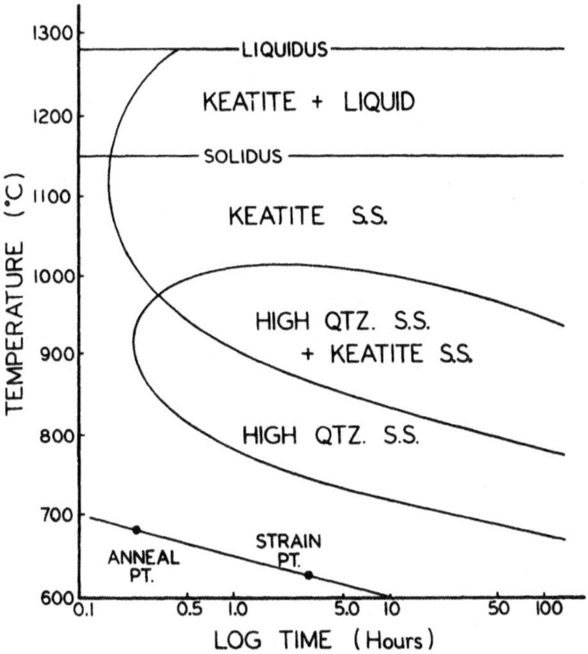

Figure 8. Schematic Isothermal Transformation for a Lithia Aluminosilicate Glass-Ceramic Material.

4) Na_2O-Al_2O_3-SiO_2

The last system to be discussed is the soda aluminosilicate system. Glasses in this system have been catalyzed with titania alone or titania in combination with other divalent oxides.[36] The primary cyrstal phase observed is nepheline which is a stuffed derivative of the silica form, tridymite. The nepheline composition is quite variable since it can take part in a number of solid solution series with ions such as potassium, barium, calcium, etc.

Glass-ceramics from this system can be generally characterized as having moderate chemical durability, expansion coefficients greater than 100 x $10^{-7}/^{o}C$, poor electrical resistivity, and moderate strength.

The main application of this svstem to date has been in a glazed dinnerware.

(B) Nucleants

Nucleants have been discussed in detail in Section

3-B-1. It probably suffices to say that, in general, without the nucleants, there probably would be no glass-ceramic process. From this standpoint, then, the nucleants affect all the properties.

One point that was not discussed earlier is that the nucleants can affect final phase assemblage in a rather dramatic manner. Sack and Scheidler[37] have shown that merely by varying the nucleating agents, titania and zirconia, in the same base glass, it is possible to obtain glass-ceramics with different appearances, different crystalline phase contents, and different crystallite sizes.

(C) Heat Treatment

The importance of the heat treatment has been illustrated by the comments on nucleation and crystallization and illustrated by Figures 5 and 8. As a glass-ceramic is heated, it can undergo a series of structural transformations which, in turn, affect the properties. Developing the optimum combination of temperature and time for the heat treatment can be experimentally tedious. The size of the piece being crystallized must also be considered.

(D) Compositional Design

Arriving at a useful composition is based on the ability to make compromises. The desired physical, thermal, and chemical properties must be adequate; the economics must be right; the initial glass should have suitable formability for automatic forming processes. Optimizing all of these properties is no easy task. To illustrate some of the factors involved, the next few paragraphs will use the work of Smith[38] to illustrate some of the considerations that are important. The first step in developing a glass-ceramic is to decide on the requirements in the way of properties. Table 4 lists some of the goals of Smith's work.

The first step is the selection of a composition field that has the potential of meeting items 1, 2, 3, and 4 in Table 4. This is based largely on existing knowledge of the properties of various crystalline species. In this case, selection boiled down to one of the low expansion phases

Table 4. Goals in the Development of a Low Expansion Glass-Ceramic Material

1. Maximum melting temperature should be ~ 1595°C

2. Liquidus temperature should be below 1350°C; the temperature at which log η = 4 should be about the same or higher

3. The glass should be stable enough to be lampworked

4. Thermal expansion coefficient should be < 25×10^{-7}/°C, preferably < 15×10^{-7}/°C

5. Modulus of rupture should be ≥ 10 kg/mm^2

6. Appearance should be opaque and glossy

in the lithia aluminosilicate system or the magnesium aluminosilicate system. Prior work indicated that the required glass-working properties were not available in the magnesium aluminosilicate system, and previous experience in the lithia aluminosilicate had demonstrated that the region of suitable working properties was outside the range of lithium concentrations which could be provided economically. Since the desired properties were not available in either ternary system, Smith tried to get the best of each system by exploring a mixture of Li_2O and MgO, i.e., the Li_2O-MgO-Al_2O_3-SiO_2 system. By systematically varying the lithia, magnesia, alumina, and silica, the composition ranges which gave the desired forming conditions and the proper thermal expansion were defined experimentally. For example, there was a trade off in the MgO concentration based on two opposing effects: increasing the MgO decreased the viscosity as desired but increased the expansion.

Systematic experiments were then carried out to empirically determine the best nucleant or nucleants. The nucleant(s) must bring about controlled cyrstallization to achieve goals 4, 5, and 6 without detrimentally affecting the already attained goals of 1, 2, and 3 in Table 4. Smith encountered problems with discoloration at high titanium concentrations. There was also objectionable devitrification during lampworking. These problems could be avoided if the titanium level was held well below 3%. This concentration would effectively nucleate the system, but the heat treatment became very critical. Zirconia was also investigated; and it was found that with increasing zirconia, the liquidus temperature increased rapidly. At a level of about 1.5 to 1.7%, the liquidus temperature exceeded the set specifications. At this level,it was not adequate to give good nucleation. Many other nucleants were tried with little or no success. Combinations of titania, zirconia, and, in some cases, SnO_2 were then investigated in an attempt to get the advantages of each nucleant without any of the disadvantages. A combination of nucleants gave favorable results as long as the TiO_2 was below 2.4% and ZrO_2 was below about 1.7%. Smith had then arrived at a six-component composition which had adequate properties.

However, there was still room for improvement. By evaluating other additives, it was found that small additions of P_2O_5 appeared to aid in the forming of glossy ceramic surfaces. It also contributed to making the heat treatment less critical. Although the reason for this was not quite clear, it was decided to incorporate a small amount. Fluorine was added because of its role as a fluxing agent and its role as a promoter of crystallization. The small amount of fluorine ($<0.4\%$) was found to be beneficial to the forming, the heat treatment, and the strength. Too much fluorine caused breakage during heat treatment.

The effect of soda was established since it occured in appreciable quantities in the petalite that was used as the prime raw material. Additional soda was added with the refining agents. It was found that as long as the soda was less than 1.5% and preferably less than 1% it could be tolerated. In concentrations above 1.5%, Na_2O increased the expansion coefficient and made the heat treatment conditions more critical. A number of other constituents were added or substituted to determine if they produced any significant advantage. Some of these were found to be helpful; some were harmful.

Thus, by using a more or less empirical approach, a composition containing 8 or 9 cationic species and 2 or 3 anionic species was arrived at which had the desired properties.

PROPERTIES OF GLASS-CERAMIC MATERIALS

(A) Thermal

1) Thermal Expansion

The thermal expansion coefficient can be varied over wide ranges. It is possible to develop negative-expansion materials, zero-expansion materials, materials with very high expansions, and materials with expansion coefficients matching those of many metals and alloys. The wide range of possible thermal expansion coefficients is the feature of the glass-ceramic process that has been exploited commercially. This is particularly true for the low-expansion glass-ceramics[2,12,39] and for glass-ceramics with expansion coefficients matching those of metals and alloys.[40]

2) Thermal Stability

The maximum use temperature for a glass-ceramic is dependent on the degree of crystallinity and the phases that are present.[34] The intergranular glass is probably the controlling factor. This residual glass at elevated temperatures may further crystallize or it may enhance deformation due to viscous flow occurring at the grain boundaries. In addition, under certain conditions, the primary crystalline phase may undergo transformation to other phases or a change in grain size may occur. Both of these phenomena may produce an observable degradation of the properties.

3) Thermal Conductivity

The thermal conductivity of glass-ceramics is slightly higher than that of conventional glasses and is generally higher than such silicate ceramics as porcelain. However, the conductivity is much lower than that of the pure oxide ceramics such as alumina, magnesia, and beryllia. The experimentally observed variation in thermal conductivity is about a factor of 3.

(B) Optical Properties

The transmission in the visible can vary over wide ranges. Transparent glass-ceramics have been reported in a number of systems,[41] but the majority of glass-ceramics are translucent or opaque. To date, transparency has been mainly exploited for inspection purposes in such items as the low-expansion, transparent glass-ceramics used in reflective optics.[39] The transmission in the near infrared of most glass-ceramics is typical of that of silicate glasses (see Figure 9).

Some compositions have been shown to display electro-optical activity.[42]

(C) Mechanical Properties

Glass-ceramics are brittle materials at room temperature like ordinary glasses and the ceramics. The strength exhibits many of the same features as that of glasses, i.e., there is a statistical distribution of strength for any given set of samples and the strength depends on the testing conditions and the condition of the surface. Watanabe et al.[43] have demonstrated for a given composition that increasing heat treatment brings about a dramatic reduction in the unabraded flexure strength but a significant increase in the abraded flexure strength (Figure 10).

The abraded strengths of most glass-ceramic materials generally fall within the range of 7 to 14 kg/mm^2. This contrasts with strengths of 4 to 7 kg/mm^2 for glass under identical abrasion and test conditions. The glass-ceramics are probably stronger for a number of reasons. They appear to have a higher abrasion resistance than ordinary glasses which means they are less prone to surface damage. The fracture propagation in a glass-ceramic probably requires considerably more energy than that of a glass because of the nature of the fine-grained microstructure. Tashiro[30] has shown for a limited size range that the strength increases with decreasing grain size.

The data on the effects of temperature are limited. It appears, however, that at least in the case of lithia aluminosilicate glass-ceramics, some compositions can be heated to around $1000^{\circ}C$ without appreciable degradation of their strength.

A number of glass-ceramic compositions have been discovered which exhibit exceptionally high flexure strengths. These flexure strengths have often been attributed to the formation of a surface compressive layer during heat treat-

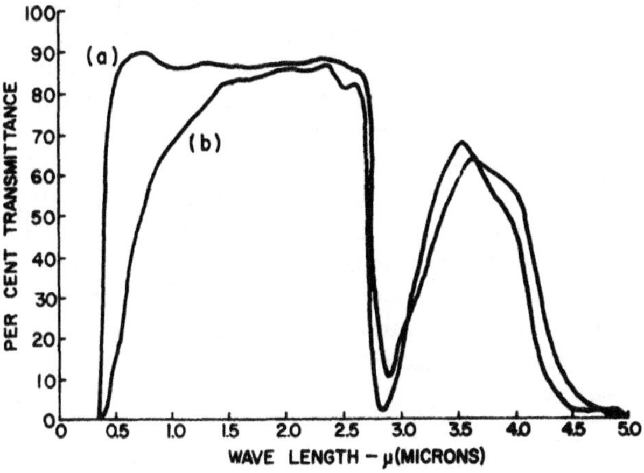

Figure 9. Transmittance Curves of Two Low-Expansion Lithia Aluminosilicate Glass Ceramics (a) Transparent (b) Translucent.

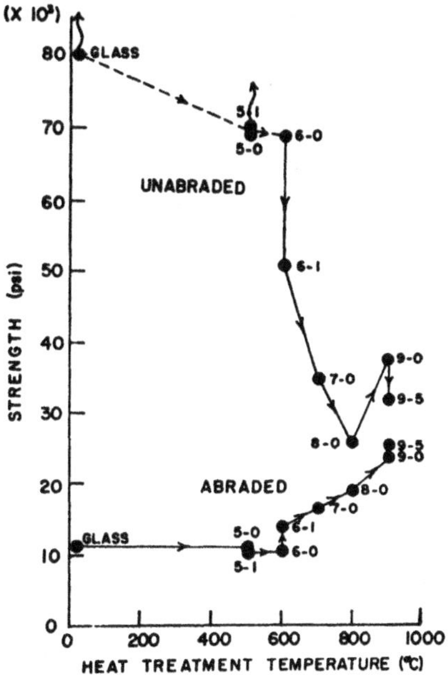

Figure 10. Strength of Abraded and Unabraded Glass-Ceramic Versus Heat Treatment.

ment. Kanbara and Tashiro[44] have shown that in the lithia aluminosilicate system the addition of fluorine to the original batch can increase the flexure strength up to as much as 60 kg/mm^2 simply through controlled heat treatments. Henry and Smith[45] have reported on high strength glass-ceramics in the magnesium aluminosilicate system in which flexure strengths of 70 kg/mm^2 are attainable with 25 to 40 kg/mm^2 being standard. Stookey[46] has reported on similar materials having somewhat lower strength. The high strength appears to be due to a difference in expansion between the surface and bulk due to a different phase assemblage in each of these areas which occurs as part of the heat treatment.

A number of techniques have been reported for increasing the strength of glass-ceramics. Duke et al[47] applied compressive glazes to glass-ceramics in the Na_2O-BaO-Al_2O_3-SiO_2 system nucleated with TiO_2 and reported up to fivefold increases in the moduli of rupture. The compression was generated by the glaze having a coefficient of expansion which was 30 or 49 x $10^{-7}/^oC$ lower than that of the glass-ceramic.

Duke et al[48] ion exchanged similar glass-ceramics by treatment in molten potassium salts to effect a $K^+ \rightarrow Na^+$ exchange. By this exchange, the nepheline was transformed to kalsilite on the surface which resulted in a compressive layer on the surface due to volume differences. Abraded moduli of rupture above 140 kg/mm^2 were reported.

Strengthening due to "crowding" has been observed for keatite solid solution and high-quartz solid solution phases after $K^+ \rightarrow Li^+$ and $Na^+ \rightarrow Li^+$ exchanges.[49] Moduli of rupture up to 70 kg/mm^2 were realized. High-quartz solid solutions in the MgO-Al_2O_3-SiO_2 system were reported to be strengthened by an exchange of 2 Li^+ ions from an external source for each Mg^{2+} ion in the surface.[50] The abraded flexural strengths ranged from 20 to 110 kg/mm^2.

(D) Electrical Properties

Most glass-ceramics are insulators. The conduction mechanism is primarily ionic as opposed to electronic and is much more complicated than that in a glass of the same composition because of the complex microstructure. A diffusing ion may encounter several different structural environments. The resistivity decreases with temperature. Table 5 lists some data for three glass-ceramics.

Table 5. Data for Three Glass-Ceramics

		0-I C-101	0-I C-117	Corning 9606
Volume Resistivity	25°C	10^{14}	7×10^{14}	$\sim 10^{16}$
(ohm-cm)	200°C	---	10^{12}	10^{11}
	250°C	10^{8}	---	10^{10}
Surface Resistivity	25°C	10^{13}	3×10^{16}	---
(ohm/sq	200°C	---	4×10^{11}	---
0% relative humidity)	250°C	3×10^{10}	---	---

The dielectric properties likewise can be varied over wide ranges. The losses are influenced by the mobility of the alkali ions that are present as well as the presence of the modifying ions. The intergranular residual glass contributes to the lossiness. In general, the dielectric loss of the glass-ceramic is less than that of a glass of the same composition since the alkali ions are tied up more tightly in the crystal lattice.

Materials with high dielectric constants can be produced by the proper compositional formulations.[51]

(E) Magnetic Properties

It is possible to controllably crystallize a magnetic phase using a glass-ceramic approach. Unfortunately, most magnetic phases of interest do not contain the common glass-forming elements in significant amounts. Moreover, inherent in the glass-ceramic process is the necessity for glass formation. As a result of these two factors, the magnetic properties of most of the magnetic glass-ceramic materials studied to date are considerably diluted since the compositions are generally formulated by mixing the atoms of the desired phase with glass-forming constituents. After melting and heat treatment, the glass-forming constituents remain as a glassy matrix or a non-magnetic crystalline matrix. Magnetic glass-ceramics have yet to be commercially exploited.

EXPERIMENTAL TOOLS

The characterization and understanding of glass-ceramic materials represent a challenge to the experimentalist. This is due to the involvement of a glassy phase and the extremely fine size of the particles under study. It is inherently difficult to study amorphous materials due to their lack of long-range structure. This is evidenced by rather meager knowledge of the structures of rather simple glasses. The crystalline particles involved are beyond the resolution of the microscope and, in the early stages, beyond the resolution of the electron microscope. However, despite these difficulties the limitations of any one tool can be partially circumvented by utilizing data obtained by a variety of techniques. The next few paragraphs will very succinctly describe some of the more important experimental tools. More detailed information can be found in various texts.

(A) X-Ray Techniques

X-ray diffraction in some form or other probably represents the one universal investigatory tool. Its uses include:

a. The determination of the onset of crystallization
b. The identification of the crystalline species present
c. The estimation of the % crystallinity
d. Structural determination of the crystalline species.

The lower limit of crystallinity that can be detected varies with the x-ray technique that is used and the crystal species that are present. Camera techniques are the most sensitive and can detect crystallinity in excess of approximately 1%.

The identification of the crystal species is a well established procedure in which the x-ray diffraction pattern of the unknown is compared to that of known crystals. The main complications that arise are due to the presence of solid solution phases or the presence of two or more crystalline phases with interfering diffraction peaks.

The quantitative evaluation of % crystallinity is a more complicated task, especially when more than one crystalline phase is present. Generally, the x-ray pattern of the unknown is compared with a series of patterns obtained on known mixtures. The sensitivity is on the order of ±5-10%.

The structural determination of the crystal species is time consuming and somewhat tedious. However, it can supply considerable insight into changes in properties due to changes in the base composition as well as a better understanding of the observed phase transformations.

(B) Electron Microscopy

Electron microscopy is used to characterize the spatial distribution of the phases and their size and concentration. Two techniques are used: replication and direct transmission. In the replication technique, fractured or polished surfaces are etched to develop a topography by differential etching. The surface is then replicated with a thin vapor deposited film such as carbon and shadowed with a material opaque to electrons to increase the contrast. The replica is then examined in the electron microscope.

The replication techniques are more commonly used than transmission microscopy because of the relative ease of sample preparation, but there are some drawbacks and dangers involved. Under normal conditions, the resolution is approximately 100 Å compared to approximately 10 Å for direct transmission. In addition, the microstructure is sensitive to the etchant used, i.e., the observed microstructure can be varied by varying the etchant or its concentration. Extraneous phases can also be precipitated from the etching solution onto the surface before replication, and they can sometimes be seen on the final micrographs. Despite these shortcomings, replication is a valuable technique and with care representative microstructures are obtained.

Transmission electron microscopy utilizes differences in electron density within a sample to produce intensity variation in the electron beam when it is passed through the sample. Important information can be obtained but only extremely thin sections (approximately 1000 Å) can be examined because of the limited penetrating power of the electrons. The preparation of thin sections is an art in itself, but several procedures have been devised, e.g., grinding and polishing, ionic bombardment, chemical etching, etc. Many of them require a steady hand and there is the ever-present danger that the preparation procedure itself may alter the microstructure by such things as heating and chemical attack. It is also possible to obtain electron diffraction patterns of selected areas of the sample. This permits phase identification as well as differentiation of crystalline regions from amorphous regions on a very fine scale.

The electron microprobe has limited usefulness in microstructural studies of glass-ceramics, since the grain size of the crystals are generally smaller than the smallest beam size in the probe. It provides useful information on larger grains which have formed inadvertently or which have been purposely developed by modifying the heat treatment.

(C) Small-Angle X-Ray Scattering

When an x-ray beam passes through a material containing a system of submicroscopic particles, scattering occurs at very low angles in the vicinity of the direct beam. Small-angle scattering occurs when the particles have an electron density different from that of the matrix, and the scattering is determined primarily by the size of the particles. The value of the technique is that it can detect heterogeneities on a scale from less than twenty to thousands of angstroms and can supply information on the size, electron density, and concentration of the particles.

Small-angle x-ray scattering represents a promising tool when used in conjunction with electron microscopy and x-ray diffraction. It is one of the few tools that can look at the early stages of nucleation and crystallization, but there are some difficulties involved. For example, the alignment of the apparatus is critical and somewhat tedious; a number of corrections must be made to the raw data to take account of parasitic scattering in the apparatus; the analysis of of the data is mathematically complex and a number of assumptions are used; occasionally, surface effects can arise from the preparation of the thin samples that are required ($\sim$0.1 mm).

(D) Light Scattering

Light scattering is similar to small-angle x-ray scattering in the sense that it relates to the interaction of electromagnetic energy with materials. In light scattering, the intensity and the spatial distribution of a monochromatic light beam is measured after it has passed through the sample. The incident light is scattered from its path due to its interaction with regions whose refractive index differs from that of the bulk. By characterizing the angular dependence of the scattering and the degree of depolarization of the beam, it is possible to characterize heterogeneities ranging from $\sim 0.05\ \mu$ to $\sim 1\mu$ in size with respect to size, index of refraction, and concentration. The technique is sensitive to the presence of particles much finer than $0.05\ \mu$ (down to molecular sizes) but little size information can be obtained.

Light scattering complements small-angle x-ray scattering. It is experimentally a simpler technique; it supplies information in low-angle scattering regions where x-ray scattering is less reliable; and it is sensitive to phase anisotropy and strain. The major difficulty is the interpretation of the scattering data. This is particularly true at the larger particle sizes where multiple scattering becomes a problem.

(E) Differential Thermal Analysis (DTA)

When a material passes from one state to another, heat is either absorbed or evolved. DTA is an instrumental technique for measuring the magnitude of the heat effect and the temperature at which it occurs. This is done by comparing the temperature of the sample being tested to the temperature of an inert standard as they are heated at the same rate. Any heat effect shows up as a temperature differential which is proportional to the heat evolved or absorbed.

DTA is used to:

1. Characterize the onset and the magnitude of heat effects due to phase transitions

2. Obtain kinetic information on transformation rates

3. Determine the low temperature viscosity of glasses.

Its main utility resides in the ease of data generation, good sensitivity, and its value in helping to develop proper heat treatments. Its major drawback is that it is sometimes difficult to interpret the data, particularly when working in new systems.

SUMMARY

During the last several decades a resurgence has occurred in the field of controlled crystallization of glasses. There has been considerable vigor in the research in this area and many new commercial products have been introduced. The glass-ceramic approach often results in a unique combination of properties and it is possible, within limits, to do some tailoring of the material properties. Glass-ceramic materials have been most widely exploited for their thermal properties, their strength, and their homogeneity.

The theoretical background for the nucleation and crystallization is reasonably well developed. However, the applicability of nucleation theory to glass-ceramic materials could be considerably improved. Further research is also needed in grain boundary effects. Some of the properties that make glass-ceramic materials important, such as their fine grain size, also make more quantitative characterization difficult or impossible. However, continued development and utilization of the available experimental tools should considerably enhance the knowledge and utility of this family of materials.

REFERENCES

1. C. G. Bergeron, "General Aspect of the Crystallization of Glass," paper at this symposium.
2. S. D. Stookey and R. D. Maurer, "Catalyzed Crystallization of Glass-Theory and Practice," Progress in Ceramic Science, Vol. 2, Pergamon Press, N.Y., 1962.
3. H. S. Hartman, "Application of High-Temperature Microscopy to Production Problems," Glass Industry, 51 (2) 60-65 (Feb., 1970).
4. G. W. Morey, The Properties of Glass, 2nd Ed., Reinhold Pub. Corp., N.Y., 1954.
5. M. deReaumur, "The Art of Making a New Type of Porcelain by Very Simple and Easy Methods of Transforming Glass into Porcelain, Part I," Mem. Acad. Sci., 370-388 (1739).
6. G. Tammann, The States of Aggregation, D. Van Nostrand Co., N.Y., 1925.
7. A. Portevin, "Molten Basalt," Mem. Soc. Ing. Civils France, 266-300 (1928).
8. H. H. Blau, "Diffusing Glasses for Illumination," Ind. Eng. Chem., 25, 848-853 (1933).
9. S. D. Stookey, "Chemical Machining of Photosensitive

Glass," Ind. Eng. Chem., 45 (1) 115-118 (1953).

10. F. Albrecht, "New Types of Hard Materials from Glass," Beispiele Angewandter Forschung, Herausgegeben von der Fraunhofer Gesellschaft zur Forderung der Wissenschaften Forschung, Munchen, pp. 19-22, 1955.

11. D. R. Stewart and W. L. Davis, "Evaluation of a Large Glass-Ceramic Mirror Blank," Appl. Opt., in press.

12. D. A. Duke and G. A. Chase, "Glass-Ceramics for High Precision Reflective Optics Applications," Appl. Opt., 7 (5) 813-817 (1969).

13. M. E. Fine, Phase Transformations in Condensed Systems, The MacMillan Co., N.Y., 1964.

14. W. A. Weyl, "Nucleation, Crystallization and Glass Formation," Sprechsaal fur Keramik. Glas. Email., 93 (6) 128-136 (1960).

15. S. D. Stookey, "Catalyzed Crystallization of Glass in Theory and Practice," Ind. Eng. Chem., 51, 805-808 (1954).

16. D. Bahat, "Heterogeneous Nucleation of Alkaline Earth Feldspars in Glasses," J. Mater. Sci., 4, 847-854 (1969).

17. G. F. Neilson, "Phase Separation in Glass and Glass-Ceramic Systems," Faraday Society Discussions, in press.

18. R. D. Maurer, "Effect of Catalyst Size in Heterogeneous Nucleation," J. Chem. Phys., 31 (2) 444-448 (1959).

19. W. Hinz and P. Knuth, "Phase Separation and Nucleus Formation in the Preparation of Vitroceramics," Glastech. Ber., 34, (9) 431-437 (1961).

20. R. Roy, "Metastable Liquid Immiscibility and Subsolidus Nucleation," J. Am. Ceram. Soc., 43 (12) 670-671 (1960).

21. J. F. MacDowell and G. H. Beall, "Immiscibility and Crystallization in Al_2O_3-SiO_2 Glasses," J. Am. Ceram. Soc., 52 (1) 17-25 (1969).

22. P. W. McMillan, Glass-Ceramics, Academic Press, N.Y., 1964.

23. S. D. Stookey, "Catalyzed Crystallization of Glass in Theory and Practice," Glastech. Ber., 32K, pp. V/1-8, V. Int. Glaskongress, Frankfurt, 1959.

24. K. Nakagawa and T. Izumitani, "Relationship between Phase Separation and Crystallization in $Li_2O \cdot 2.5SiO_2$ Glass and A Lithium Silicate Containing a Large Amount of TiO_2," Phys. and Chem. of Glasses, 10 (5) 179-184 (1969).

25. P. E. Doherty, D. W. Lee, and R. S. Davis, "Direct Observation of the Crystallization of Li_2O-Al_2O_3-SiO_2 Glasses Containing TiO_2," J. Am. Ceram. Soc., 50 (2) 77-81 (1967).

26. G. F. Neilson, "Small-Angle X-Ray Scattering Study of Nucleation and Devitrification in a Glass-Ceramic Material,"

paper presented at the 1966 Pittsburgh Diffraction Conference.
27. V. A. Blinov, "The Mechanism of Nucleated Crystallization of Glasses in Lithia-Alumina-Silica and Cordierite Systems," J. Mater. Sci., 4, 461-468 (1969).
28. R. H. Redwine and M. A. Conrad, "Microstructure Developed in Crystallized Glass-Ceramics," Chapter 40, Ceramic Microstructures, John Wiley and Sons, Inc., N.Y., 1968.
29. C. K. Chyung, "Secondary Grain Growth of Li_2O-Al_2O_3-SiO_2-Ti(Glass-Ceramics," J. Am. Ceram. Soc., 52 (5) 242-245 (1969).
30. M. Tashiro and S. Sakka, "Some Physical Properties of Glass-Ceramics and Their Relation to Microstructure," Bull. Inst. Chem. Res., Kyoto Univ., 42 (5) 351-365 (1964)
31. J. E. Rapp, personal communication.
32. M. J. Buerger, "The Stuffed Derivatives of the Silica Structures," Am. Mineral., 39, 600-614 (1954).
33. S. Ray and G. M. Muchow, "High-Quartz Solid Solution Phases from Thermally Crystallized Glasses of Compositions (Li_2O,MgO)·Al_2O_3·nSiO_2," J. Am. Ceram. Soc., 51 (12) 678-682 (1968).
34. D. R. Stewart, "Thermal Stability of Low Expansion Glass-Ceramic Materials," pp. 2-6, Preprint Volume, A.I.Ch.E. 67 National Meeting, Feb. 15-18, 1970.
35. W. S. Schreyer and J. F. Schairer, "Metastable Solid Solutions with Quartz-Type Structures on the Join SiO_2-$MgAl_2O_3$," Z. Krist., 116 (1-2) 60-82 (1961).
36. D. L. Evans, "Crystallization of Carnegieite in a CdO-TiO_2 Nucleated Glass," Physics of Non-Crystalline Solids, 57-66, North-Holland Pub. Co., Amsterdam, 1965.
37. W. Sach and H. Scheidler, "Effects of the Nucleating Agents TiO_2 and ZrO_2 on Crystal Phases Formed during Development of Glass-Ceramics," Glastech. Ber., 39 (3) 126-130 (1966).
38. W. E. Smith, U. S. Patent 3,380,818, April 30, 1968; personal communication.
39. G. A. Simmons, "Development of Low-Expansion Glass-Ceramic Materials," Opt. Spectra, 1 (2) 25-29 (April-May-June, 1967).
40. P. W. McMillan, B. P. Hodgson, and G. Partridge, "Sealing Glass-Ceramics to Metals. Part 1," Glass Technol., 7 121-127 (1966).
Part 2 (with H. R. Heap), Ibid., pp. 128-122.
41. G. H. Beall and D. A. Duke, "Transparent Glass-Ceramics," J. Mater. Sci., 4 340-353 (1969).
42. N. F. Borelli, "Electro-Optic Effect in Transparent Niobate

Glass-Ceramic Systems," J. Appl. Phys., 38 (11) 4243-4247 (1967).

43. M. Watanabe, R. V. Caporali, and R. E. Mould, "The Effect of Heat Treatment on the Strength and Abrasion Resistance of a Glass-Ceramic Material," pp. 23-28, Symposium on Nucleation and Crystallization in Glasses and Melts, Am. Ceram. Soc., Columbus, Ohio, 1962.

44. T. Kanbara and M. Tashiro, "Effects of Addition of Fluorine on the Surface Structure of Glass-Ceramics," J. Ceram. Assoc. Japan, 76 (11) 23-31 (1968).

45. K. M. Henry and W. E. Smith, U. S. Patent 3,117,881, Jan. 14, 1964.

46. S. D. Stookey, U. S. Patent 3,231,399, Jan. 25, 1966.

47. D. A. Duke, J. E. Megles, J. F. MacDowell, and H. R. Bopp, "Strengthening Glass-Ceramics by Application of Compressive Glazes," J. Am. Ceram. Soc., 51 (2) 98-102 (1968).

48. D. A. Duke, J. F. MacDowell, and B. R. Karstetter, "Crystallization and Chemical Strengthening of Nepheline Glass-Ceramics," J. Am. Ceram. Soc., 50, (2) 67-74 (1967).

49. B. R. Karstetter and R. O. Voss, "Chemical Strengthening of Glass-Ceramics in the System Li_2O-Al_2O_3-SiO_2", J. Am. Ceram. Soc., 50, (3) 133-137 (1967).

50. G. H. Beall, B. R. Karstetter, and H. L. Rittler, "Crystallization and Chemical Strengthening of Stuffed β-Quartz Glass-Ceramics, J. Am. Ceram. Soc., 50 (4) 181-190 (1967).

51. M. M. Layton and A. Herczog, "Nucleation and Crystallization of $NaNbO_3$ from Glasses in the Na_2O-Nb_2O_5-SiO_2 System," J. Am. Ceram. Soc., 50 (7) 369-375 (1967).

The Melting of Glass

E. Lowell Swarts
Glass Research Center
PPG Industries, Inc.
P. O. Box 11472
Pittsburgh, Pennsylvania 15238

INTRODUCTION

It is the purpose of this review to survey the principal aspects of the melting of commercially important glasses and to articulate those parameters which should be of concern to the successful technologist. No attempt will be made to provide an exhaustive literature survey. Rather, only selected sources which are useful within the context of the discussion will be referenced; special attention will be given recent findings. Nonetheless, at the outset, it is appropriate to refer the reader to previous summaries or reviews which provide a sound background. Jebsen-Marwedel's volume[1] which concerns itself primarily with glass defects also provides useful discussions of the subject areas to be considered here. The compilation of papers presented at the Symposium on the Melting of Glass held in Brussels in 1958,[2] especially the summary by Dubrul, remains a helpful reference. The Handbook of Glass Manufacture,[3] particularly the chapter by Tooley, is useful. More recently, Cable[4] has provided an authoritative review.

The glass manufacturing process consists of bringing together specific starting materials for reaction under controlled conditions. The raw materials and process conditions are chosen in accordance with the desired properties of the end product and vary widely. Glass to be used in a laser ranging system must be orders of magnitude more homogeneous than bottle glass: however, if the bottle is to compete with other container materials it must be orders of magnitude less precious than the laser rod. Fiber glass furnaces can employ a durable refractory, chromium oxide, without concern for the slight color which results from inevitable refractory solution. On the other hand, this coloration is not acceptable in most architectural glass; and thus at least on this count alone chrome refractories are not found in a sheet glass tank.

In spite of the differences between glasses and melting processes, it is essential rather to extract those dominant features which are held more or less in common. Figure 1 provides a sketch in elevation of a continuous glass-melting tank. The principal reactions between raw materials take place within the melting zone which occupies about the first third of the tank and in which the highest temperature of the tank is reached. A typical top temperature for a soda-lime-silica glass might be about 1475°C. Molten glass moves from the melting zone on into the refining section and as in this plan into a cooling chamber. The refiner serves three principal functions: (1) the concentration of dissolved gas and gaseous bubbles is reduced to an acceptable level; (2) requisite chemical homogeneity is obtained; (3) glass temperature is lowered to that point required by the forming process. The concepts of melting, refining, and homogenization as identifiable processes remain valid for virtually all continuous methods and even for the commercially less important batch melting still carried out in large pots.

It is useful then to consider the melting of glass as the superposition of three major processes, as suggested schematically in Figure 2.[3] From the outset, a series of decompositions and dissolutions occur. During this same time, but extending over a longer time period, the gas-removing or fining reactions also proceed. Finally, a still longer time is required for achieving the desired homogeneity. (Heat conditioning of the molten glass also takes place during refining as noted above and in Figure 2, but will not be considered in detail here as a primary melting process.)

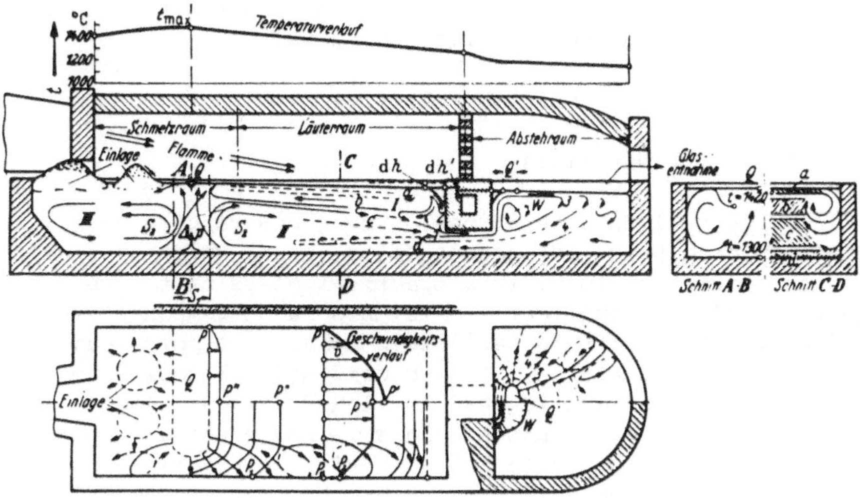

Figure 1. Elevation view of a continuous glass-melting furnace (ref. 1).

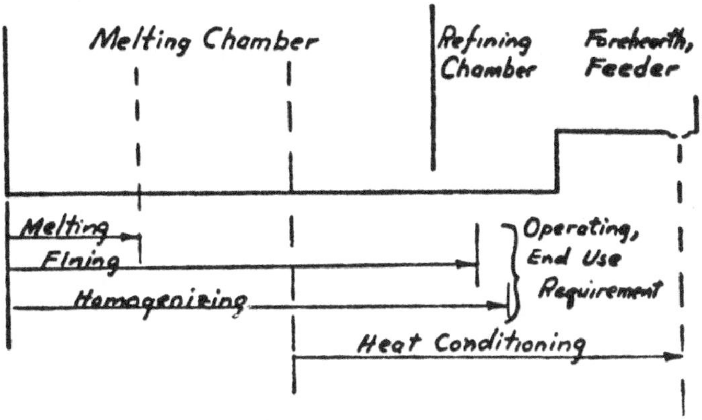

Figure 2. Schematic representation of relations between melting, fining, homogenization and heat conditioning of glass (ref. 3).

While it is convenient to analyze these steps rather independently of each other, they are in fact highly coupled; this close coupling is a primary factor contributing to the difficulty of solving practical problems.

MELTING REACTIONS

(A) Batch Materials

1) General Considerations: The glass batch usually consists of a principal source (e. g.,sand) which by itself is too refractory to yield a practical glass of the required quality economically. To this major component is added fluxing agents to facilitate melting at more readily attainable temperatures. In the conventional soda-lime-silica family, silica is readily fluxed by soda (Figure 3), but since this two-component glass is not especially durable, calcium oxide is commonly added to yield a practical material. Most commercial compositions in this ternary system (Figure 4) correspond approximately to a molar ratio of 1 Na_2O:1 CaO:6 SiO_2 a composition point which long ago was found to represent a good compromise between principal requirements. In addition to these three major components, magnesia, potassia, and alumina may also be added to give additional resistance to chemical attack, to control viscosity within a particular temperature range, or to reduce the tendency toward devitrification. On yet a finer scale, minor additions of transition metals will yield characteristic optical absorption; and oxides of certain polyvalent cations may assist in removal of gas bubbles.

By way of illustrating the diversity of even the most important compositions, a few nominal analyses are given in Table 1.

2) Sources of Components: The choice of a raw material to supply a given component will depend upon factors such as:

a) availability
b) cost (including transportation)
c) impurity contents
d) desired particle size for optimum reaction rates
e) desirability of specific chemical reaction paths, and
f) volatilization characteristics and associated safety or pollution considerations

A few specific illustrations will indicate how these factors

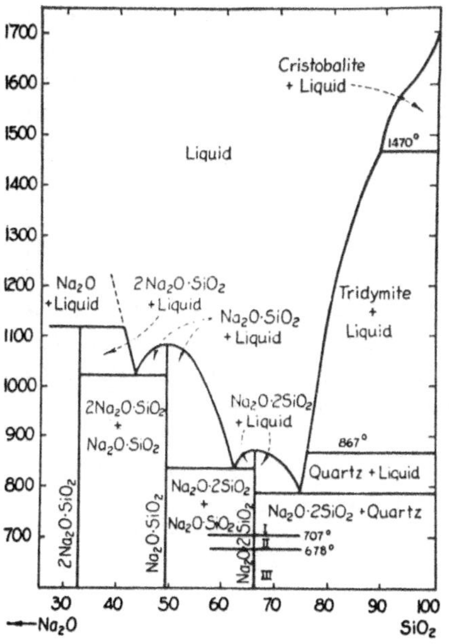

Figure 3. System SiO_2-$2Na_2O \cdot SiO_2$ (ref. 5).

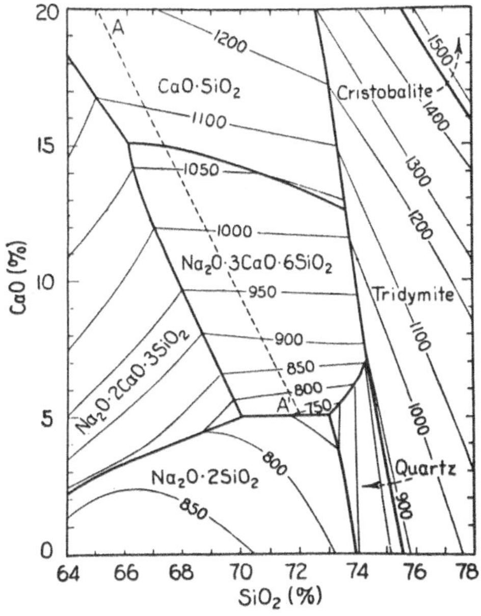

Figure 4. Part of system $Na_2O \cdot CaO \cdot SiO_2$ (ref. 5).

are considered in actual practice. Table 1 shows that from 1/2 to 3/4 by weight of these compositions consist of silica. This component is supplied for the most part from deposits of natural sand or sandstone having widely varying characteristics. At one extreme, some of these deposits contain such high levels of iron or chromium oxides that they are of little use for any melting application. On the other hand, there also exists in the United States commercially important sources of sand which, with special benefication, can be used for fabricating clear fused silica of high quality. The great bulk of clear glass-making sands occupy a middle ground with their available particle size and characteristic impurities dictating their specific applications.

Silica may also be added in the form of minerals which also include other oxides in combination. Thus in the crystallizable and ion-exchange compositions of Table 1 in which alumina and alkali oxides are also required, spodumene,

Table 1. Some Typical Glass Compositions

	Window	Glass-Ceramic	Ion Exchange	Borosilicate	Aluminosilicate	E-Fiber
SiO_2	73.2	70.1	61.2	79.8	61.4	53.8
Na_2O	13.2	0.4	12.2	} 4.2	1.6	0.4
K_2O	--	0.6	4.4		0.2	0.4
Li_2O	--	2.6	--	--	--	--
CaO	8.3	--	--	--	18.2	15.7
MgO	3.5	2.6	3.6	--	--	4.9
Al_2O_3	1.3	17.9	17.5	2.1	17.3	14.5
TiO_2	--	4.4	0.7	--	--	--
ZrO_2	--	--	--	--	--	4.0
B_2O_3	--	--	--	13.6	--	10.3

(several minor constituents are not listed)

petalite, or other feldspar-related minerals have been used. In compositions where the impurity levels of the natural minerals are not tolerable, alumina is added as the calcined or hydrated forms. On the other hand, in some applications such as fiber glass production, relatively impure grades of kaolin are acceptable and economically highly favored.

The alkali oxides have been most commonly added as the carbonates, although here again as indicated above, the feldspar-type minerals can serve as a major source. In borosilicate glasses, borax is a logical source of soda as well as boric oxide. When a specific level of oxidation in the melt is required, as for color control and in refining, sodium nitrate is used. As will be considered in greater detail in a later section, sodium sulfate (salt cake) is widely used to accelerate dissolution of sand and also for refining. A recent trend sees a growing interest in the use of caustic soda (50% aqueous solution of NaOH) as at least a partial replacement of soda ash.[6]

The alkaline earth oxides are usually supplied as carbonates in the form of limestone or dolomitic limestone ($CaCO_3$-$MgCO_3$). For special purposes the "burnt" (oxide alone) or "slaked" (hydrated) products are used.

Boric oxide is usually added as anhydrous borax ($Na_2O \cdot 2\ B_2O_3$) or as boric acid in those compositions such as "E" glass in which alkali contents are minimized. Within recent years, the oxide itself, sold as "anhydrous boric acid," has become commercially important.

A more comprehensive summary of sources for raw materials is the chapter by Scholes included in Reference 3.

(B) Principal Glass-Forming Reactions

A comprehensive analysis of melting reactions can be made only in terms of the specific composition in question. Clearly it is beyond the limits of this review to consider in detail many of even the most important commercial compositions. A compromise will be affected by discussing here the soda-calcia-silica system with enough specificity to pinpoint the important reactions, but with enough generality to allow application of the principles to a wider range of compositions.

As was suggested above in describing the batch materials, the principal glass-forming process in silicates consists of reactions of silica with fluxes. Although careful investigation gives evidence of reaction between components of the batch at temperatures before liquid phases develop, these solid-state reactions are probably not of major importance in conventional compositions. The first significant step in conversion of batch to glass is the appearance of the lowest melting liquid phase. Referring again to Figure 3 and considering only the interaction of soda and silica, this liquid phase corresponds to the eutectic between sodium disilicate and silica which appears at about 800°C. This initially-formed eutectic can react with more soda to precipitate the metasilicate at temperatures below 840°C. However, the most important point is probably the melting temperature of soda ash at 850°C. Hot stage microscopy[7,8] has shown that as the carbonate melts, a rapid and violent reaction follows. In fact, at this elementary stage of the overall reaction, substantial refining occurs as carbon dioxide evolves.

$$SiO_2 + Na_2CO_3 \rightleftharpoons Na_2O \cdot SiO_2 + CO_2 \qquad (1)$$

The coupling of processes is illustrated by the finding[9] that this evolution of carbon dioxide has an important mixing effect, reducing concentration gradients and accelerating elementary reactions. Although most of the CO_2 is indeed removed from the system, the significance of the fundamental equilibrium will be considered further in a later section.

Let us now complicate the binary system by adding lime. An obvious possibility, at least in principle, is the lower melting eutectic between sodium and calcium carbonates at 775°C. In fact, however, one investigator[10] found that when crucible batches were heated at rates corresponding to those encountered within the batch pile of a furnace, limestone resisted wetting by the primary soda-silica phase, and its subsequent decomposition and dissolution proceeded relatively slowly requiring temperatures of up to 1350°C before completion. Partial replacement of limestone by dolomite in order to introduce magnesia had little effect. During this period, carbon dioxide evolution continued strong, not only directly from decomposition of the carbonates, but also from the melts as continuous dissolution of the sand decreases carbon dioxide solubility. However, other experiments have made clear that the time of the limestone decomposition in the overall sequence of reactions and its rate are highly dependent upon particle size and the rate of heating of the batch.

There remains a final stage of melting, the dissolution of the remaining silica residues. In contrast to the initially rapid solution of silica by reaction with soda and dissolution into relatively fluid melts, the dissolution of sand residues into a much more viscous silica-rich melt proceeds many times more slowly. This stage is the limiting step in the overall process and defines the "batch-free time." The rate of disappearance of these sand grains is limited in principle by the diffusion kinetics of silica in the melt, but in practice becomes extremely sensitive to those factors which upset silica concentration gradients.[11,12] Further it should be noted that in alkali-containing melts, quartz inverts to cristobalite relatively rapidly at melting temperatures and may become the dissolving species depending upon its original particle size.[13] The subsequent rate of dissolution of these residues depends of course upon such factors as the initial particle size of the sand and the degree of agglomeration. As these last remnants dissolve, bubbles consisting primarily of carbon dioxide are nucleated at the glass-residue interface, again reflecting the decrease in solubility of carbon dioxide with increasing silica content.

Clearly there exist other important glass-forming systems in which the basic chemistry of batch reactions differs markedly from those in a soda-lime-silica system. An interesting and useful family of new glasses are the ion-exchange or crystallizable compositions which correspond to specific alkali aluminosilicate ratios. If the starting materials are natural alumino silicates, minerals which normally melt to very viscous liquids, the most important melting step is the fluxing by small amounts of free alkali present in the batches. Here the situation is roughly analogous to the fluxing of silica except that the available alkali is generally much less and the resulting melts are more viscous than in a soda-lime-silica system at the usual operating temperatures. The key point here is that the combination, especially in an approximately equi-molar ratio, of alkali and alumina, leads to vastly different properties of the melt.

Going in quite a different direction, phosphate glasses are certainly far less important commercially than silicates; nonetheless, one family of compositions with iron oxide additions has proven useful as infrared filters. In preparing these glasses, one may be primarily concerned with the reaction of the likely starting material, aluminum phosphate, with minor amounts of lime, silica, and boric oxide.

In brief summary, the technologist, in approaching the melting of new compositions, must be able to draw upon fundamental data from the relevant phase diagrams, be able to evaluate possible reaction mechanisms from thermodynamic data, and make use of empirical results such as batch-free times as measured in laboratory experiments.

(C) Other Considerations

1) Particle Size of Batch Materials: Rates of melting and the specific reaction paths depend significantly upon range and distribution of particle sizes in the batch materials. In general, finer particle size and more intimately mixed batch will permit more rapid reaction, at least initially.[8,14] This readily anticipated effect is illustrated in Figure 5. However, as one goes to fine particle size, another factor comes into play. During the initial stages of reaction, carbon dioxide escapes through the voids and channels of the batch. Eventually, liquid phases form, shutting off these channels. As particle size decreases, voids become smaller; there are more points of contact between batch materials; and the first liquid phase closes off open channels earlier. When carbon dioxide can no longer be freely evolved to the atmosphere it suppresses the decomposition reaction of Equation 1. This effect has been verified experimentally in crucible melts with controlled partial pressures of carbon dioxide.[15]

By choosing alkali carbonates with particle size smaller than that of the sand, melting may be accelerated. The finer carbonate probably distributes itself more evenly on the surface of the silica grains, thus permittting relatively early removal of more carbon dioxide before the liquid seals off channels in the batch. However, when sand of too fine a particle size is used, it may agglomerate, leading to slower initial reaction within the batch followed by flotation to the surface and much slower dissolution thereafter.

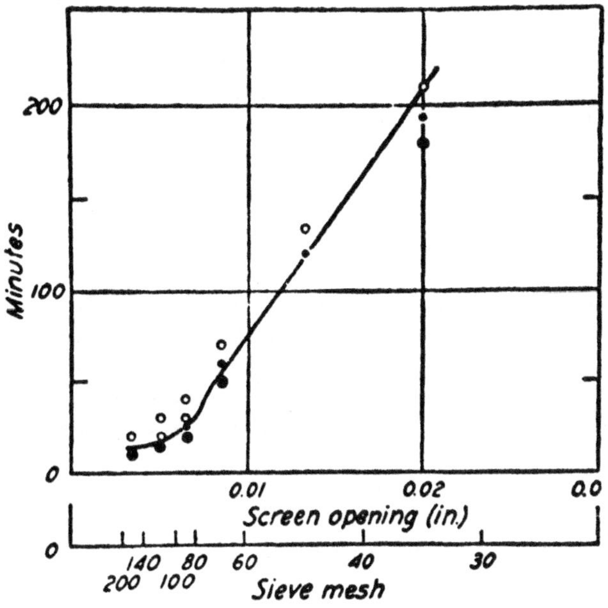

Figure 5. Effect of grain size on melting time (ref. 14).

Particle size of limestone and dolomite should be larger than that of sand or soda ash.[10] Finely divided limestone dissolves more readily, but probably contributes to the cementing together of sand grains, thus leading to their slower final dissolution as mentioned above.

In substantially different compositions, other considerations may prevail. For example, E-composition fiber glass, containing virtually no alkali, about 10% B_2O_3, and 15% Al_2O_3, is more difficult to melt than a soda-lime-silica composition. Here one needs higher melting temperatures, and in order to facilitate melting and homogenizing in particular, fine particle sizes are called for. Similarly, in glass-ceramic compositions melted from major proportions of natural minerals, fine particle sizes are required to accelerate dissolution in viscous melts. Their intrinsically higher viscosities in the face of economically attainable melting temperatures and other complications such as volatilization of constituents at high temperatures lead to special problems in melting these compositions.

Mechanical problems may be encountered with certain

choices of particle size in the batch. For reasons such as are outlined above, the different components of a batch usually do not have the same particle size distribution. In spite of thorough mixing initially, segregation may occur during the course of handling en route from mixer to tank. Segregation has been shown to be one source of longer batch-free and seed-free times and decreased homogeneity.[8]

When very fine particle sizes are used, dusting may occur within the melter, placing fluxes on refractory surfaces above the melt or within the checkers of a regenerative furnace. Segregation and dusting are most simply controlled by addition of 3-5% water to the batch.

A variety of laboratory procedures for comparing reactivity as a function of particle size are to be found in the literature (e.g. Reference 16).

2) Methods of Accelerating Melting Reactions:

a) Addition of water: Going beyond the function described above, water added to some batches also accelerates melting. Since the effect is most noticeable in batches containing water-soluble materials, it probably amounts to partial solution and more uniform deposition of the fluxing agents upon the sand grains.

b) Addition of sulfates: A commonly-used cure-all in large-volume production of soda-lime-silica glass is sodium sulfate (salt cake). In actual practice, of the order of 0.4% by weight of finished glass is added initially as partial replacement of soda ash. Observations and analyses by several investigators[10,17,18,19,20] yield a reasonably coherent picture of the accelerating mechanisms. First of all, primary eutectic phases of sulfate and carbonates form at significantly lower temperatures than in sulfate-free melts. That sulfate which is initially dissolved by the carbonate is subsequently separated out again as the reaction with silica yields a composition approaching the disilicate. The net result is a redispersion of the fluid sulfate phase throughout the batch. At this point the liquid sulfate contacts both solid batch particles and the liquid silicate phase, acting as a surface active agent and greatly enhancing dissolution reactions.

As the temperature of the batch increases, some sulfate decomposes by direct reaction with silica.

$$Na_2SO_4 + SiO_2 \rightleftharpoons Na_2O \cdot SiO_2 + SO_3 \qquad (2)$$

The gas phase, sulfur trioxide or its dissociation products, adds to the effect of evolved carbon dioxide in stirring the batch-glass mixture. Perhaps the most important solubilizing effect of sulfate accompanies its straightforward thermal dissociation at the higher melting temperatures, above 1300°C, yielding more alkali for direct reaction with silica and more gases for vigorous mixing. Figure 6 compares the batch-free times for laboratory melts with sulfate additions.

When salt cake is used at levels which significantly accelerate melting, there inevitably remains substantial undecomposed sulfate which may separate as a second phase, "salt water gall," on the surface of the melt. It has long been known empirically that by adding reducing agents along with salt cake to the batch, one may retain most of the original melting advantages, but with decomposition of the excess sulfate and, in turn, without the troublesome gall layer. Partial reduction of the sulfate starts at low temperatures;

$$Na_2SO_4 + 2\ C \rightleftharpoons Na_2S + 2\ CO_2 \qquad (3)$$

but since the carbon addition usually amounts to about 5% of the sulfate, there results a mixture of both sulfide and sulfate. Sulfide readily dissolves in sulfate with a eutectic melting point at 740°C. At higher temperatures, reaction between sulfide and sulfate becomes appreciable;

$$3\ Na_2SO_4 + Na_2S \rightleftharpoons 4\ Na_2O + 4\ SO_2 \qquad (4)$$

and the resulting alkali produced in intimate dispersion assists in final silica dissolution. Another route to the same final position employs sulfides directly or as included in blast furnace slag.[21]

In alkali-free melts, aluminum sulfate has been used; and reports from Russian workers in particular seem to favor the use of ammonium sulfate. Still another approach has employed the bubbling of sulfur dioxide through the melt.[22]

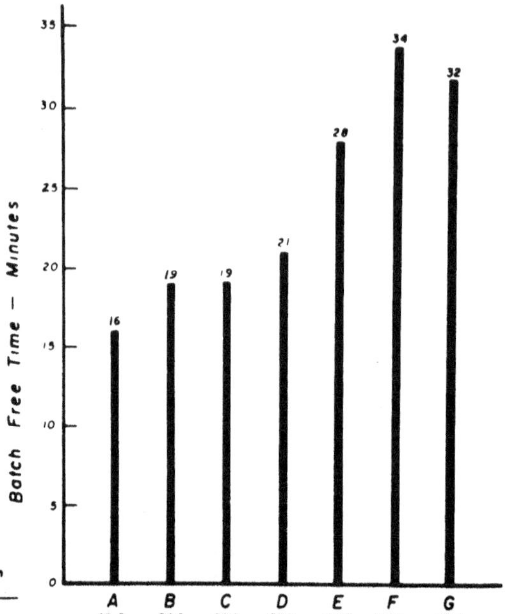

Batch Composition WT %	A	B	C	D	E	F	G
SiO_2	58 6	58 8	58 9	58 9	59 0	58 6	58 9
$CaCO_3$	18 9	18 5	19 0	18 5	18 3	18 9	19 0
Na_2CO_3	22 0	22 1	21.4	22 1	22 1	22 0	22 1
Na_2SO_4			0 6				
$CaSO_4$		0 6					
$BaSO_4$					0 6		
$PbSO_4$						0 6	
$SrSO_4$				0 6			
$(NH_4)_2SO_4$	0 6						
$SO_4^{=}$	0 44	0 43	0 41	0 32	0 25	0 19	

Figure 6. Effect of sulfate compounds on melting rate of standard soda-lime-silica batch (ref. 20).

In actual practice one must strike a balance between sulfate and reductant in order to avoid either a scum of undissolved silica or a "wet" gall layer. Penalties also accompany the advantages of sulfates as melting accelerators. The sulfate dissolved in the molten glass corresponds to a supersaturated state with respect to sulfur trioxide at later stages of the melting process in which the effective partial pressure of SO_3 may be much less than that in the batch pile during the initial reactions. Thus a potential for release of SO_3 or its dissociation product, SO_2, develops which may lead to generation of bubbles in otherwise well-refined glass (reboil). Aside from this problem, sulfur is inevitably driven off in the flue gases to serve as an atmospheric pollutant or as an unwanted flux (condensed

Na_2SO_4) in the refractory structures. Thus a sound case can be made for attempting to achieve comparable melting acceleration by other approaches.

c) Other minor additives: Under specific circumstances, small replacements of soda by lithia or potassia may accelerate melting by initial formation of lower-melting eutectic phases. Small amounts of B_2O_3 or fluorides are also claimed to be helpful, probably by producing lower-melting phases initially or by decreasing viscosity over the entire temperature range of the melting process.[23]

d) Replacement of soda ash by caustic soda: The relatively recent introduction of caustic soda (50% aqueous NaOH) as a practical batch material attracts increased attention as a melting accelerator. With this addition, flux may be distributed more uniformly in contact with the sand grains. A second factor may be the absence of carbon dioxide with its retardation of the carbonate decomposition as described above. Moreover, as will be discussed in a later section, molten glass is generally more tolerant of dissolved water vapor than of carbon dioxide. On the debit side, mixing caustic soda into a batch requires special equipment.

e) Use of soda-silica frit: Going a step beyond the use of caustic soda, several authors have suggested that a portion of both the soda ash and sand of a batch be replaced by a binary glass, corresponding perhaps to the disilicate composition.[24,25] This material would be prepared in a specialized facility designed for economic high volume production without unnecessary concern for quality in the usual sense. One should then expect a batch incorporating this raw material to give faster overall melting.

At the present time this process is not accepted practice. Its ultimate practicality depends upon the trade-off in costs and savings of optimized processes.

f) Gels as batch materials: There exists a totally different route to the production of homogeneous glass, one which thus far seems to have been exploited only in the laboratory. The starting point (for a silicate glass) is a suspension of colloidal silica in an aqueous solution of the alkali and alkaline earth salts.[26,27] This mixture is homogenized, dried, and then melted at a reduced temperature compared with that required by the conventional path. Advantages of this approach are (1) an ultrahomogenous product and (2) easier and more economical melting. On the other hand, the high cost of starting materials and difficulties in handling them have precluded serious consideration of this approach for practical applications.

g) Pre-compaction of batch materials: The literature records much discussion of various techniques of pelletizing or briquetting well-mixed batch materials. In principle, one expects this procedure to decrease segregation and increase the initial rate of reaction through increased contact between materials. Additionally, heat transfer through a densified batch should be greater, providing additional acceleration of melting reactions. This approach has not been widely used heretofore largely because of the disadvantages of additional materials handling. However, new patents in this subject area continue to appear;[28] and production applications are being reported.[29]

h) Addition of cullet to the batch: Even the most successful glass maker must usually dispose of some scrap production. In practice, this scrap cullet will be remelted as part of the batch, commonly up to one-third or even one-half by weight. The admixture of this much cullet into the batch usually influences the course of the melting process significantly.

If melting conditions are otherwise maintained the same, the batch pile will melt down more rapidly since less energy need go into batch reactions. While this in itself is desirable, possible side effects must also be kept in mind. Because the insulating effect of the batch blanket is reduced as its cullet content increases, temperatures near the backwall of the tank may increase, changing flow patterns and possibly leading to greater-than-normal refractory erosion.

Although cullet is discussed here with special reference to accelerating melting, it has been found, at least on the laboratory scale, that finely crushed cullet may actually slow down the melting process.[30] On the basis of hot stage microscopic observations, it has been proposed that alkali carbonates in the batch preferentially react with the finely crushed cullet rather than with sand grains. Thus, more silica is left unreacted initially to dissolve in the final melt and extend the batch-free time. This effect must become insignificant when cullet size is much larger (orders of magnitude) than that of the soda ash, a relation which usually characterizes actual practice.

i) Submerged heating: At the present time, natural gas is the generally preferred energy source in the United States. Conventional practice employs gas-air flames which burn above the surface of the melt. The glass receives energy from the flame directly and indirectly through radiation from the inner surface of the crown of the tank. Most tanks are made rather shallow relative to other dimensions, partially to utilize as efficiently as possible this downward radiation flux. In the melter, the introduction of batch provides an insulating blanket which obviously interferes with downward transfer of heat into the tank. In principle, a corrective strategy is the introduction of energy from sides and bottom below the level of the batch. The literature records all manner of patents and proposals; one of these which has been reduced to at least pilot scale practice employs submerged gas burners positioned in the floor of the tank.[31] Since combustion products are directed upward through the molten glass, thermal efficiency increases, not to speak of the mixing action and obvious enhancement of melting rate. On the other hand, it remains to be determined whether the obvious practical problems associated with installation and maintenance and possible deleterious effects upon fining give this scheme an overall advantage.

An important practical method of introducing energy into the melt from below is by electrical heating through the molten glass from electrodes in sidewalls or bottom of the furnace. Although electricity as an energy source accounts for only a small proportion of the total production in the United States, an increasing number of installations employ some electrical heating of the melt, frequently to heat glass under the batch pile and accelerate melting. It is not within the scope of this review to compare further the merits of electrical versus gas melting; this comparison is the subject of current debate.[32,33]

3) Oxidation-Reduction Equilibria: It is essential that one recognizes the significance of the redox equilibria

which can affect quality of the product and of those factors which in turn influence these equilibria. The oxidation state of the melt can most simply be described in terms of the formal equilibrium

$$1/2\ O_2 + 2\ e^- \rightleftharpoons O^{2-} \qquad (5)$$

where the molecular species designates oxygen dissolved within the glass or the equilibrium partial pressure above the melt which would correspond to this degree of saturation. The species on the right simply designates reduced oxygen generally without structural implications.

The equilibrium partial pressure of oxygen as defined above is a conventional means of describing the redox state of the melt. However, it is essential to understand that only rarely, under carefully controlled conditions designed to insure attainment of chemical equilibrium, does this parameter actually correspond to the real atmosphere above the melt. It is equally important to recall that the glassy state itself is a non-equilibrium state and that during melting, a redox equilibrium is seldom closely approached.

a) Principal determinants: Although one does not readily attain equilibrium under practical conditions, the oxidation state is subject to control. The most obvious parameter is the makeup of the batch itself. Most simply, one may add coal, or going in the other direction, a nitrate. Supplementing this kind of direct batch addition, one may also utilize a more sophisticated approach involving the dependence of certain redox equilibria upon temperature. The equilibrium

$$2\ O^{2-} + 4/n\ M^{(x+n)+} \rightleftharpoons 4/n\ M^{(x)+} + O_2 \qquad (6)$$

between higher and lower states of several cations in a soda-silica glass has been analyzed carefully.[34] The equilibrium constant k is defined by

$$k = \left[\frac{M^{(x)+}}{M^{(x+n)+}}\right]^{4/n} P_{O_2} \qquad (7)$$

where the equilibrium partial pressure of O_2 above the melt, P_{O_2}, replaces the oxygen activity of Equation 6. The variation of k with temperature for several cations is

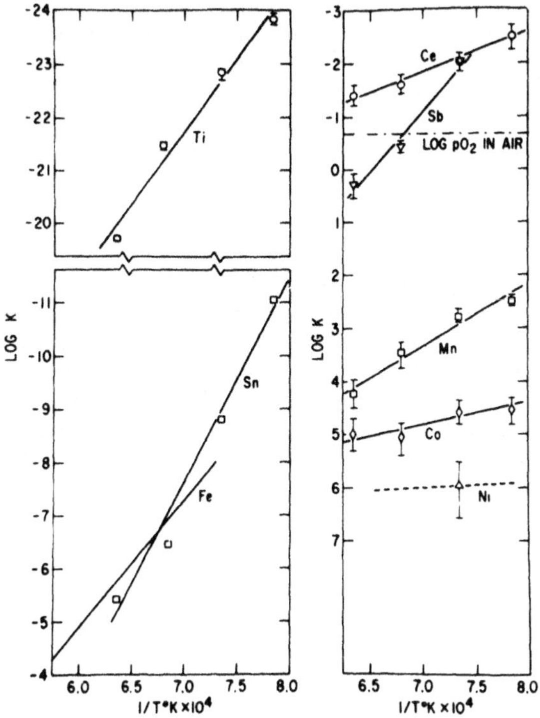

Figure 7. Temperature dependence of equilibrium constants in $Na_2O \cdot 2SiO_2$ glass (ref. 34).

given in Figure 7. The data tell us, for example, that when antimony oxide is dissolved in the base glass, the +5 state is preferred below about 1200°C while the +3 state is more stable at higher temperatures. Thus if a melt containing antimony is heated to 1450°C and allowed to equilibrate with an air atmosphere above it, the antimony will become predominantly reduced. Upon cooling the melt below 1200°C, the preference of antimony for the +5 state provides a potential for reduction within the glass. As will be considered in greater detail in a later section, equilibria of this kind involving antimony and arsenic can be used to remove bubbles.

In particular applications of this approach, a pair of

redox couples such as that of iron and tin may be used. Figure 7 shows that plots of log k for these elements have markedly different slopes. Thus, while the more highly oxidized states of both elements are favored generally, the Sn^{2+}/Sn^{4+} ratio decreases with decreasing temperature much more steeply than does the Fe^{2+}/Fe^{3+} ratio.

It follows that after a silicate glass containing both elements is equilibrated at high temperature and then cooled without allowing equilibration with the external atmosphere, Sn^{4+} forms at the expense of Fe^{3+}, leaving a higher concentration of Fe^{2+} than would otherwise result without the presence of tin. Since the tin species are colorless, this technique has been used to help obtain the high infrared absorption of Fe^{2+}.

Similarly, the large difference in slope of log k plots for antimony and cerium show that cooling of a previously equilibrated melt containing these elements may yield Ce^{3+} in company with Sb^{5+}.

The oxidation state within the melt will be influenced by the atmosphere above the glass, and in particular by the gas-air ratio. All too often the effect may be deleterious. Thus with colors which are highly dependent upon redox state, surface defects may result from the influence of atmosphere in only the surface layers of the melt. For example the flames from most burners are characterized by reducing and oxidizing zones irrespective of gross stoichiometry; thus a single burner positioned so that too much of its flame impinges upon the glass may be damaging.

In contrast to the ease with which oxidation-reduction changes may be generated in surface layers due to inadequate atmospheric control, the atmosphere frequently cannot be depended upon by itself to establish a uniform condition or control a redox equilibrium within the melt which is not already specified primarily by batch composition and temperature of melting. The effects of atmosphere can be utilized within the melt generally only to the extent that diffusion of molecular species (e.g. oxygen) occurs across the glass surface and these surface layers are subsequently mixed throughout the body of the glass by convection. The volume-to-surface ratio, diffusion constants, and flow rates preclude large scale changes occurring rapidly throughout the volume of a tank. In fact, the internal redox state of glasses such as an amber container composition must be maintained many orders of magnitude more reducing than that of the covering atmosphere during much of the melting process.

Glass composition will markedly influence redox equilibria; compositions rich in cations with high field strength (e.g. Si^{4+} or B^{3+}) generally referred to as "acidic" in a very specific sense, have a stronger reducing effect than do "basic" compositions. Thus, since arsenic is less easily reduced than antimony, arsenic is usually preferred over antimony as a fining agent in borosilicate melts while antimony is usually the better choice in soda-lime-silica glasses unless other factors override this initial consideration.

b) Practical problems: In the course of illustrating effects of specific factors in the previous section, practical consequences upon melting were mentioned. It seems worthwhile to mention a few additional problems which may be influenced by the determinants mentioned above.

In melting architectural glass, either a "clear" composition with iron oxide the principal colorant, or tinted compositions containing iron, cobalt, selenium and nickel, reproducibility of color from one day to the next or between different tanks in different years requires close specification of the batch materials, the flame settings, and the temperature schedules.

The so-called carbon-amber color of bottles does not involve carbon directly at all, but rather stems from a complex interaction of ferric with sulfide ions. Here again one must consider the interaction of two redox couples. When the melt is too highly oxidized ($>10^{-8}$ atm O_2) too little sulfide is present; and, conversely, at too high a level of reduction ($<10^{-10}$ atm O_2) too few ferric ions are available.[35] With appropriate amounts of iron, sulfur and reducing agent in the batch, however, and careful maintenance of furnace atmosphere, the internal redox state of the glass is relatively easily confined to the interval specified above, yielding the stable intense amber color. Although this color has long been known to the glass-maker, a fundamental understanding of its origin[36,37] and careful analysis of the influence of batch and melting parameters[38] have come only recently.

In the production of containers particularly, slight coloration due to impurities may be masked by the deliberate addition of small amounts of other colorants. The glass is said to be "decolorized," although in practice the principal effect is to compliment the original color and thus obtain a more or less neutral tint which is less objectionable. Here again, one must closely control equilibria between polyvalent species of such elements as iron, cobalt, cerium, arsenic, and selenium.

Finally, it is worthwhile mentioning a potential melting problem which has become a source of expensive embarrassment even to experienced glass makers. When certain cations from which even very small amounts of the elemental forms may be reduced are included in glass melts and platinum metal is also present, there exists a potential problem. With these melts in platinum crucibles or in contact with the platinum lining of an optical tank, one must deliberately maintain a degree of oxidaiton. For example, elemental forms of arsenic and antimony may result from only mild reducing conditions and readily alloy with the platinum, forming low melting eutectics and catastropic holes.

Summarizing briefly then, glass melting reactions depend not only upon the most obvious factors such as the fundamental chemistry of the batch materials, but also upon other parameters which may exercise more subtle influence. Some of these secondary factors have been discussed; however, there remain others, such as the extent of glass-refractory interactions, heat transfer through molten glass and fluid flow within a tank which, although not within the scope of this review, may also significantly influence melting reactions.

REFINING

For those who may not be familiar with the glass-makers' jargon, it is appropriate at this point to define terms which will be used throughout the following discussion. "Refining" or "fining" shall refer to the removal of bubbles from molten glass. (In some of the literature, these terms refer more or less collectively to all processes occurring between melting and forming processes.) "Seeds" and "blisters" are varieties of bubbles; by one accepted definition, "seeds" describe bubbles less than 0.5 mm in maximum dimension, and "blisters" are larger inclusions. (Some workers reserve this last term for bubbles nucleated from supersaturated melts under specific conditions.)

Again certain literature sources are especially valuable in gaining an overview of this aspect of glass-melting. The comprehensive report of the German Society of Glass Technology presented before the Symposium on the Fining of Glass held in Paris, France[39] is an outstanding contribution; the Proceedings of this Symposium[40] also include many other papers which remain pertinent. Of more recent reviews, references 41 and 42 are especially useful.

(A) Origin of Seeds and Blisters

1) Gases Included in Batch Materials: An obvious first source of bubbles is the air or furnace gas entrapped within the interstices of the granular batch materials. In most glass-melting, the volume thus enclosed is small relative to the volume of gas derived for example from decomposition of the batch. However, in those compositions which may include relatively large amounts of components which do not yield decomposition gases (e.g.,the alkali-aluminosilicates), gases mechanically included with the batch may become more important. In particular, nitrogen may pose a problem because of its relative inertness, at least under oxidizing conditions, as will be discussed in a later section. In principle, the effect of nitrogen may be minimized by flushing the batch just prior to melting with other gases such as carbon dioxide which will be more reactive with molten glass. Melting schemes in which the batch is first reacted under reduced pressure have also been proposed, but these suggestions have not been translated into generally accepted practical technology.

2) Decomposition of Batch Materials: In most glasses, decomposition of the batch, particularly carbonates, is the obvious principal source of those gases which interact with the molten glass. In window glass, the evolved carbon dioxide amounts to about 200 times the volume of the glass. By way of illustrating what must be accomplished during refining, in glass of high quality such as is currently required in automotive windshields, the frequency and volume of bubbles may be of the order of 10^{-9} times the volume of glass. (This corresponds, for example, to one 0.2 mm diameter bubble in a square meter of 3 mm-thick glass.)

Water vapor from boric acid, caustic soda, or hydrated alumina, oxygen and nitrogen from nitrates, and sulfur dioxide from sulfate may also be present. Some of these constituents are added for the explicit purpose of supplying gases which interact physically or chemically with the melt to assist refining; their roles will be considered shortly.

3) Reaction of Glass With Refractories: As one goes in the direction of larger melting facilities with higher volume-to-glass contact ratios, fewer problems with bubbles arising from refractory-glass reactions might be expected. However, in the face of requirements of higher quality, even this source continues then to require attention.

A straightforward empirical approach has proven useful in comparing the seed-generating behavior of materials to be used in contacting molten glass.[43,44] Representative results from one test procedure[45] are shown in Figure 8 and Table 2. From these data, we see that in contact with a conventional sheet glass, dense alumina refractories are logical candidates for those portions of the tank such as feeder canals where one must minimize introduction of bubbles into already well-refined glass.

Bubbles appear at refractory surfaces by means of various mechanisms. Closed porosity in the refractory will yield seeds as these voids are opened during reaction with molten glass. This source is most readily characterized by the presence of nitrogen and argon in the usual atmospheric ratio.

Specific impurities in a refractory may react with molten glass as erosion of the block proceeds. In certain grades of fused-cast oxides, small amounts of reduced impurities may generate carbon dioxide. Carbon dioxide will also come from silicon carbide grains inadvertently left on refractory surfaces after shaping.

Chemical or electrochemical interactions between refractories and molten glass may generate bubbles. Extensive studies of these effects have been reported in several reviews.[46,47,48,49] It is well-known that refractory oxides in contact with glass generate electrical potentials as the consequence of differences in composition of refractories, differences in composition of glass, or differences in

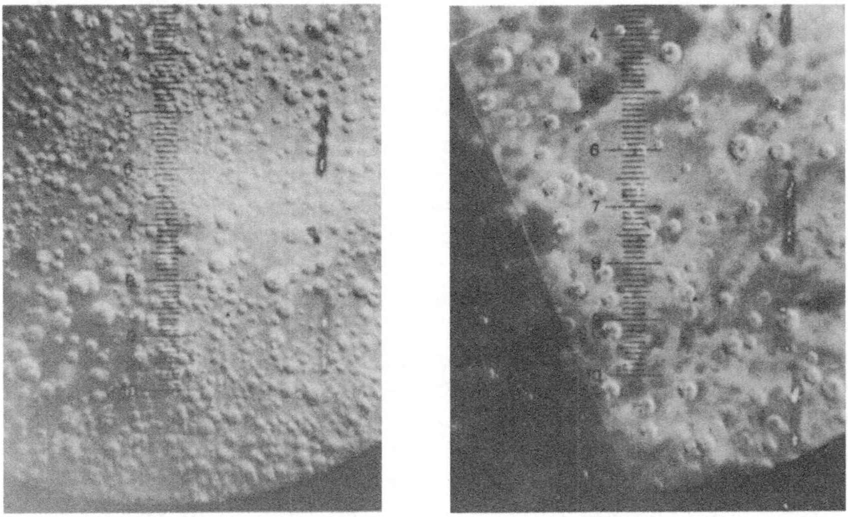

Figure 8. Two examples of seed generation on refractory surfaces (ref. 45).

Table 2. Representative Seed Counts on Refractory Surfaces (I, maximum reported value; II, minimum reported value).Ref. (44).

	Seeds/cm^2
A) Fine-grain matrix, coarse grog particles	
I	3080
II	1462
B) Clay refractories	
I	1235
II	627
C) Fused cast refractories	
I	627
II	95

temperature within the system. However, one significant conclusion, expecially enunciated in reference 49, is overlooked by those who describe oxygen bubbles on refractory surfaces generally as electrochemical in origin. If two refractory oxide electrodes, each with its own polarization potential in contact with the glass, are short-circuited directly or by steelwork of the tank, oxygen bubbles form at the effective anode surface by an electronic mechanism only if that anode material is electronically conducting. Thus one easily accounts for oxygen bubbles on the surface of tin oxide as being electrochemically generated, but common clays and even pure alumina which are orders of magnitude less conductive electronically may also generate oxygen seed in contact with an alkali-containing glass. In the case of alumina, as alkali from the glass diffuses into the refractory, oxygen seeding diminishes. A pretreatment of refractory oxide surfaces with alkali has also been proposed as a method of control.[50] In these systems, electronic conduction within the refractory probably cannot account for oxygen seeding; other chemical mechanisms are likely responsible.

Where platinum metals contact alkali-containing glass, electrochemical generation satisfactorily accounts for seed evolution.[51] Thus thermocouple sheaths or refractory liners may generate oxygen due to temperature differences, as in different parts of a tank or by effective concentration differences, as between the top surface and lower strata of an alkali-containing glass (Figure 9, reference 51). It is accepted practice to isolate electrically, thermocouples or other conductive hardware which may contact the glass at different points. It also follows, consistent with the accepted mechanism, that an oxygen-free atmosphere in the system will minimize seeding.[52]

4) Nucleation and Growth of Bubbles From Supersaturated Solution: Many gases are significantly soluble in molten glass and must be considered essentially as minor glass-making components which may impart characteristic properties to the melt. However, again it must be emphasized that during the glass-melting process, gas solubilities seldom correspond to equilibrium conditions, thus giving rise to possibilities for supersaturation and the potential for bubble formation.

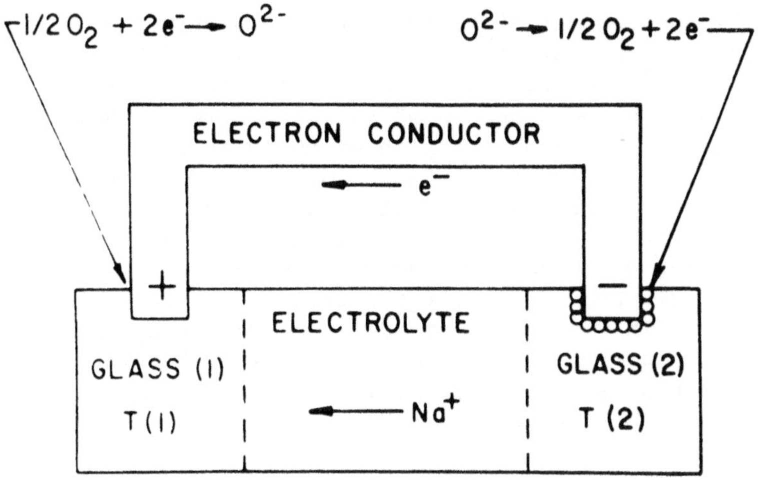

Figure 9. Theoretical model for galvanic oxygen reboil. A galvanic cell is present when glass (1) ≠ (2) or T (1) ≠ (2) (ref. 51).

a) Most gases exhibit a temperature coefficient of solubility; as temperature of the melt changes, equilibrium solubility also changes and in turn the degree of possible supersaturation.

b) If the partial pressure of a gas above the melt decreases with respect to time or position, again the equilibrium solubility changes; and there may occur supersaturation.

c) In general, gas solubilities vary with glass composition; thus, if molten glass relatively rich in a gaseous component changes gross composition, as by refractory solution at a tank wall, supersaturation may again result.

d) The solubility of a gas may depend upon the oxidation-reduction equilibrium. To the extent that redox conditions change during a melting process, as in the burning out of a coal addition to the sheet glass batch, there once again occurs the possibility of supersaturation.

The development of supersaturation within the melt is only a minimum condition for the formation of bubbles. An accompanying requirement of perhaps greater practical importance is the presence of nucleation mechanisms. Homogeneous nucleation has been shown to be an unlikely mechanism of nucleation. On the other hand, heterogeneous nucleation may proceed from at least two different sources. Pre-existing small bubbles such as might remain after dissolution of batch components may serve as sinks and subsequently grow. Clearly a refractory surface also suffices. In fact, one procedure for comparing the dissolved gas contents of laboratory melts calls for the introduction of small blocks of refractory into the melts and observing rates of bubble evolution.[53]

The intermingling of streams of molten glass at different temperatures or of different compositions has also been cited as causing bubbles. Usually it is not made clear (if in fact it is understood) that while this phenomenon may set up a condition of supersaturation, no evidence exists to associate bubble nucleation with this kind of mixing in the absence of an interface with a second phase.

Before proceeding further, the foregoing may be summarized by noting that bubbles will result from

a) air or combustion products entrapped in the batch,
b) decomposition of batch materials,
c) interactions between molten glass and refractory surfaces, and
d) supersaturation of the melt.

B) Reaction Mechanisms

As batch reactions proceed, gases become trapped, forming bubbles within the melt. These bubbles subsequently disappear by floating to the surface or by dissolving in the melt, and in practice various techniques are used to enhance both modes. Thus, glass viscosity may be reduced to speed the rise of bubbles, or volatile batch ingredients are deliberately introduced to sweep out small bubbles by coalescence. Bubbles may dissolve by specific chemical reactions and ultimately by the effect of surface tension.

In practical glass-making, these mechanisms are inevitably

used in combination. Cable[54] found that a complex curve of the form shown in Figure 10 describes the rate of loss of bubbles as a function of bubble diameter. This function may be rationalized as a summation of factors as shown in Figure 11.

1) Flotation of Bubbles: Bubbles which do not interact with one another or with the wall of the container and are small enough to behave as rigid spheres rise according to the classical Stokes' relation

$$\nu = 2/9\ gr^2\ \rho/\mu \qquad (8)$$

where ν = velocity of rise
g = gravitational acceleration
r = bubble radius
ρ = glass density
μ = glass viscosity

(A requirement with respect to the Reynolds' number, describing flow as the bubble rises, must also be met. This will occur under most practical conditions of interest, as for example with bubbles one millimeter or less in diameter and glass viscosities greater than 100 poise.)

From equation (8) it may be seen immediately that (1) glass temperature strongly affects rate of rise by its effect upon viscosity and (2) the bubble size is especially important in light of its second power involvement. Curve A of Figure 11 illustrates the latter effect.

The glass batch is initially heated to the highest temperature of the overall cycle in order to complete batch reactions as quickly as possible, as was discussed in a foregoing section and also to accomplish as much refining as possible, primarily through flotation of the bubbles. At this point of highest temperature, glass viscosity is lowest; and gas bubbles will be large generally. Anticipating some of the discussion which follows, we should note that at this point of highest temperature, solubilities of many gases will be small and diffusion rates will be high (relative to lower temperatures.) Thus bubbles will become enlarged easily from the supersaturated species, facilitating removal by flotation. In outline then, these considerations make clear why the major portion of glass refining should be completed before the lower temperature portion of the melt-

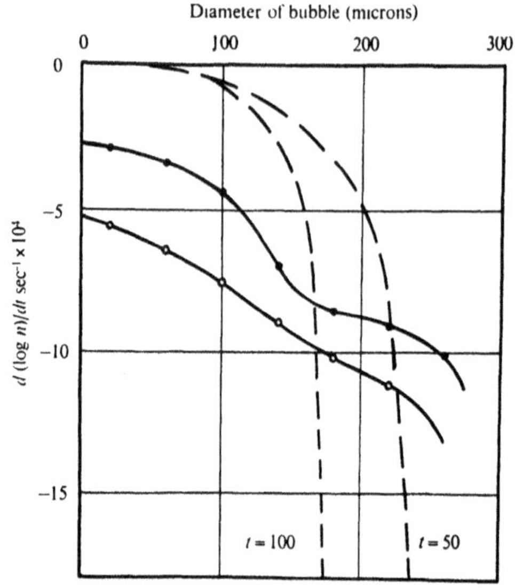

Figure 10. Solid line -- experimentally observed rates for fastest and slowest refining glasses. Dashed line -- theoretically predicted rates of fining from Stokes' rise alone (times of 50 and 100 minutes) (ref. 54).

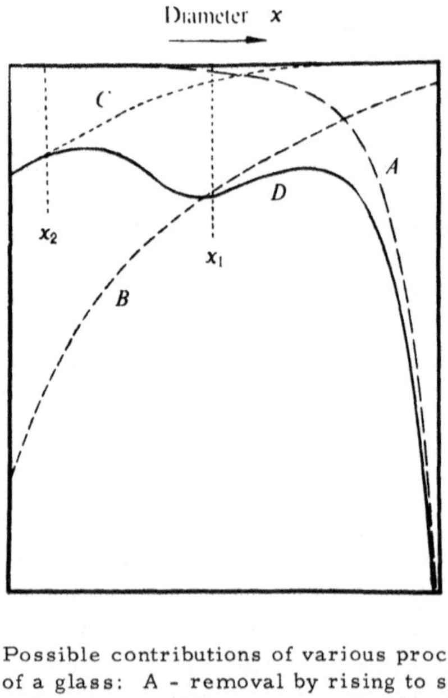

Figure 11. Possible contributions of various processes to the refining of a glass: A - removal by rising to surface of the melt; B - dissolving, diffusion controlled by oxygen; C - dissolving, diffusion controlled by another gas; D. relation expected to be shown by results of experiments (ref. 54).

ing schedule is reached.

In the initial stages of melting, bubbles doubtless coalesce, but this process probably becomes unimportant as the distance between bubbles increases. On the other hand, coalescence may be accelerated as bubble size and hence rate of rise are increased by specific techniques. In early pot-melting technology, pieces of green wood or arsenic oxide were thrust deep under the surface of the melt. Copious generation of steam, CO_2 and arsenic oxide vapor not only stirred the melt, but also combined with and swept out fine seed. Up-dating this approach somewhat, water-cooled bubbling tubes may be dipped into the glass or are mounted permanently in the bottom of a tank.

Another approach to accelerating removal of bubbles consists of adding specific ingredients to the batch which subsequently volatilize and sweep out seed. Halides in particular are useful, as exemplified by the common use of sodium chloride in borosilicate melts.

A recent patent[55] describes a technique which depends upon the relatively large diffusion coefficient of helium. This gas is diffused into the glass melt by bubbling or simply by exposure of the molten glass to a covering atmosphere. Pre-existing fine seed will obviously contain no helium and thus will serve as sinks for helium which has diffused into the glass. These fine seed will grow from helium influx (at least until their helium partial pressure approaches unity) and will rise more rapidly to the surface. A remarkable increase in the rate of refining is claimed, as shown in Figure 12.

Diffusion coefficients for helium and neon have been measured by observing the rates at which small bubbles of these gases dissolve in glass initially free of these species. Figure 13[56] shows results for helium. From these kind of data, the diffusion constants of Figure 14[57] were calculated. By comparison, note that in going from helium with diameter 1.9 Å to neon with 2.2 Å diameter, the diffusion coefficients decrease by an order of magnitude. In going to argon, with diameter 3.0 Å, it seems certain that the diffusion coefficient must be less by at least two more orders of magnitude. The solubilities of the rare gases are similarly determined by their atomic dimensions relative to the sizes

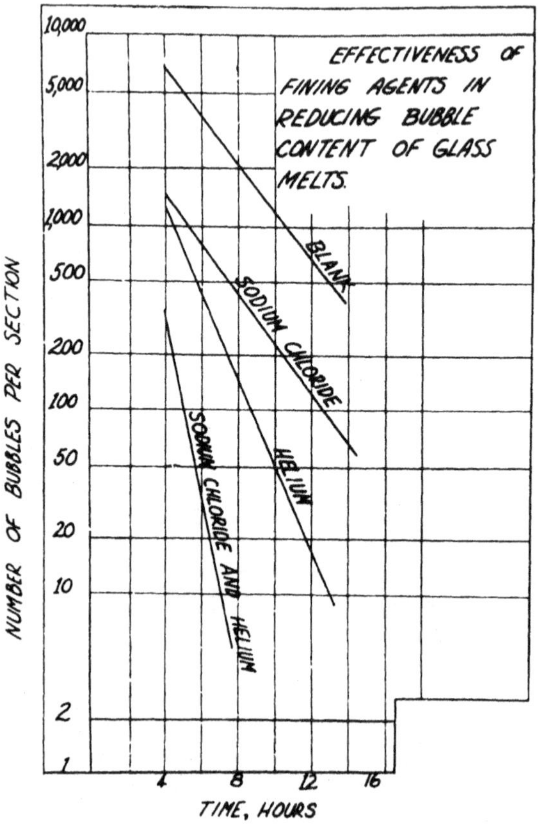

Figure 12. Effectiveness of helium as a refining agent (ref. 55).

of voids in the silicate network (Figure 15, reference 58); it is not surprising that these solubilities are low relative to other gases since chemical interactions are minimal.

2) Dissolution of Bubbles: The rate of disappearance of small seed usually occurs more rapidly than can be rationalized simply by the rise of bubbles to the surface, as was seen in Figure 10. Dissolution must also be considered, especially in light of the many laboratory observations of this process.

a) The effect of surface tension: Irrespective of chemical interactions, all bubbles tend to dissolve when

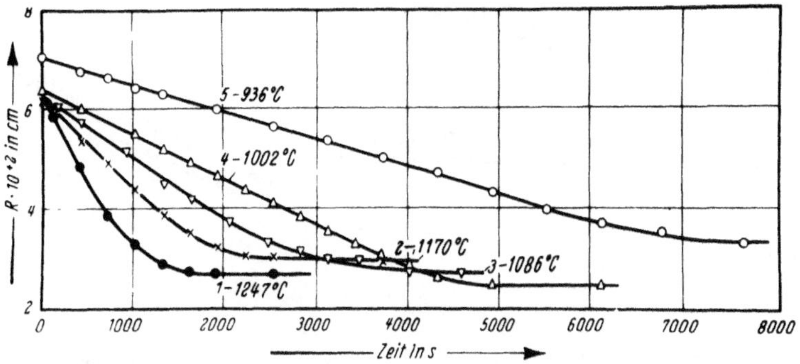

Figure 13. Dissolution of helium bubbles at different temperatures (ref. 56).

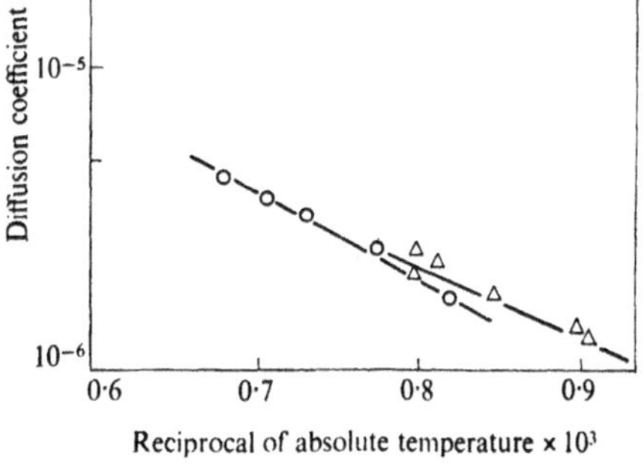

Figure 14. Diffusion of Ne(O, Δ) and He(●) in a glass melt (ref. 57).

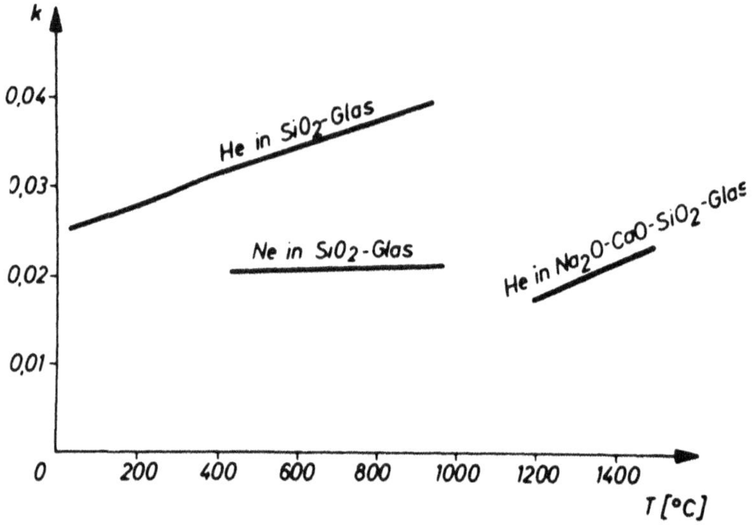

Figure 15. Solubility of He and Ne in glass (k = Ostwald's solubility coefficient) (ref. 58).

they become small enough for surface tension to increase internal pressure significantly. This pressure, p, depends upon surface tension, σ, and bubble radius, r, by the relation

$$p = 2\sigma/r \qquad (9)$$

Figure 16[54] shows a plot of this relation for a surface tension of 300 dyne cm^{-1}, a figure reasonably typical of silicate glasses at conventional melting temperatures. Clearly, as bubble diameter becomes smaller than 0.1 mm, dissolution occurs increasingly rapidly due to this effect alone.

b) Specific chemical solubility and diffusion characteristics:

1. Carbon dioxide: The importance of CO_2 has already been suggested from discussion of the batch decomposition reactions. The solubility of CO_2 is a sensitive function of composition,[59,60] increasing in a soda-silica glass with alkali content, as shown in Figure 17.[58]

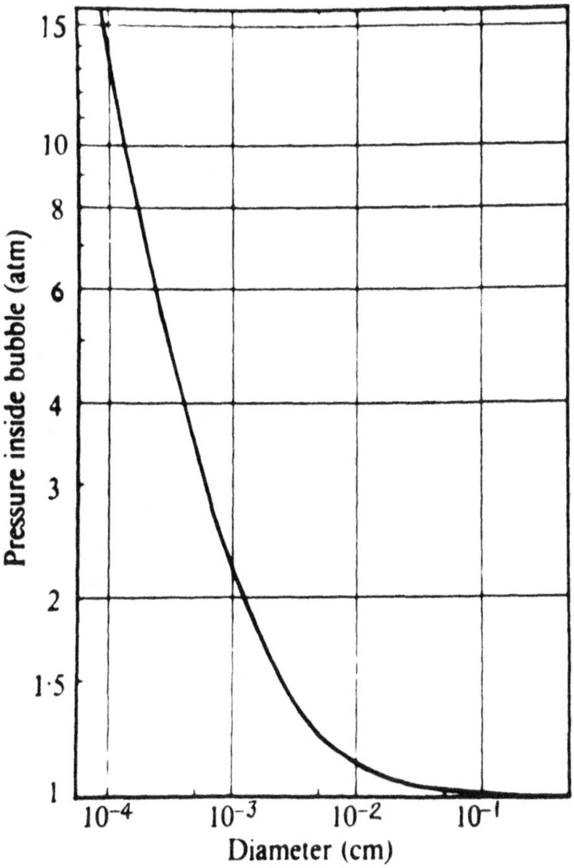

Figure 16. Relation between diameter of a bubble and its equilibrium internal pressure assuming a surface tension of 300 dyne cm^{-1} (ref. 54).

Chemical solubility probably occurs by formation of carbonate ions as in the equilibrium

$$CO_2 + O^{2-} \rightleftharpoons CO_3{}^{2-} \qquad (10)$$

Figure 17 also shows that CO_2 solubility increases significantly as the temperature decreases below about 1200°C. Thus, as glass passes through the highest temperature region of the melting schedule, dissolved CO_2 is minimized; subsequently, cooling of the melt during refining will in principle develop increased capacity for CO_2 and accelerate dissolution of bubbles. The importance of this process depends also upon the diffusion characteristics of CO_2. In

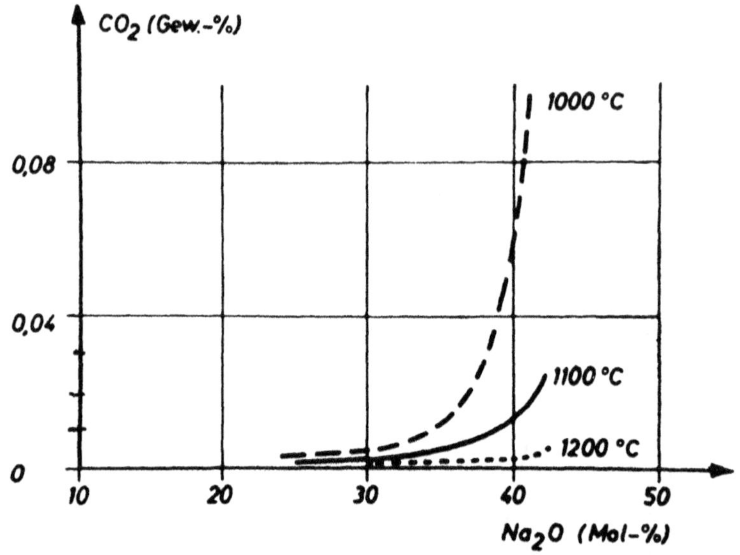

Figure 17. Solubility of CO_2 in $Na_2O \cdot SiO_2$ melts (ref. 58, 59).

comparing the behavior of the rare gases in an earlier section, the large sensitivity of the diffusion coefficient to atomic size was noted. Thus, since CO_2 has a molecular size larger than that of argon, one might expect such a low diffusion constant that bubble dissolution would be quite unimportant. However, whereas the diffusion coefficient for Ar is probably less than 10^{-7} cm^2/sec in the range 1100-1300°C, values ranging from 10^{-4} to 10^{-6} cm^2/sec have been measured for CO_2.[61] This relatively large diffusivity for CO_2 is probably related to its chemical interaction, as opposed to only the physical processes involved with the rare gases, but cannot be specifically accounted for.

Another chemical interaction which apparently enhances solution of CO_2 in molten glass has been reported recently.[62,63] Minor additions of arsenic oxide not only affect oxygen solubility, as will be described below, but also increase the rate of solution of CO_2. As yet, it is not clear what specific chemistry is involved. In any event, experience indicates that carbon dioxide from the batch reactions seldom turns up as a bubble constituent.

2) Water vapor: Water solubility increases with increasing alkali content and with increasing temperature. In general terms, water solution occurs principally through the formation of OH^- groups within the glass structure. These groups are readily identified by means of characteristic optical absorption in the near infrared region, and in fact these absorption bands are commonly used as a measure of dissolved water content. Although liquid water is sometimes reported as being present in bubble defects, positive identification of water is usually lacking. Recent measurements[64] of the rate of solution of water vapor bubbles in molten glass as shown in Figure 18 indicate that water vapor as such is probably never a problem by itself. Any water which may be present in defect seed or blisters probably results from secondary or incidental processes. In the temperature range 1100-1300°C, the diffusion coefficient for water is about two orders of magnitude smaller than for CO_2. Thus the rapid solution of water vapor bubbles must be attributed to the relatively high solubility of water in molten glass.

3. Oxygen: The solubility of oxygen within an oxidic glass melt depends upon possible departures from stoichiometry of the component oxides and upon the presence of certain polyvalent cations. The latter effect is of far greater practical significance. A classic example of this effect is obtained by the use of arsenic oxide in the glass batch; melts with this addition generally become seed-free significantly faster than when arsenic is not included. The key to this effect is illustrated in Figure 19.[65] These measurements show conclusively that when arsenic is present in the glass, oxygen bubbles shrink more rapidly. The mechanism which is believed to account for this behavior depends in turn upon the thermodynamic characteristics of the dissolved arsenic.[66,67] At temperatures in the lower end of the melting range, As^{5+} is the preferred state. However, as the temperature is raised, a significant portion, if not essentially all of the As^{5+}, will be thermally reduced to As^{3+}, giving off oxygen which replaces some of the existing gases of the batch or sweeping out fine seed, leaving primarily oxygen-rich bubbles within the melt. Upon subsequently cooling the melt during the refining period, As^{5+} will again be favored, creating within the glass an oxygen sink which in turn accelerates solution of oxygen from bubbles into the glass. The pertinent equilibria are

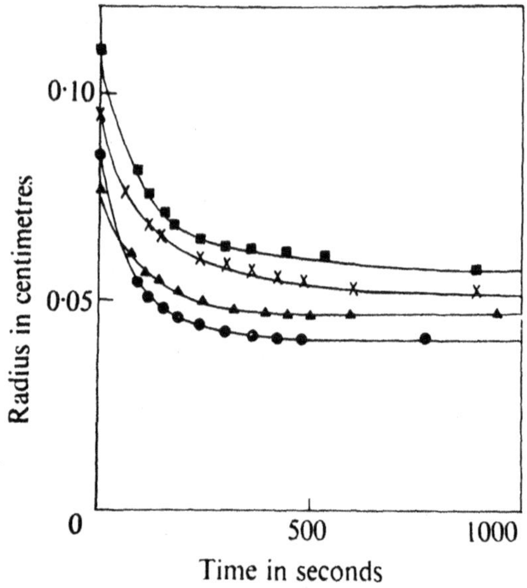

Figure 18. Dissolution of water vapor bubbles (ref. 64). X, ▲ = 1080° C, O, ■ 1130° C.

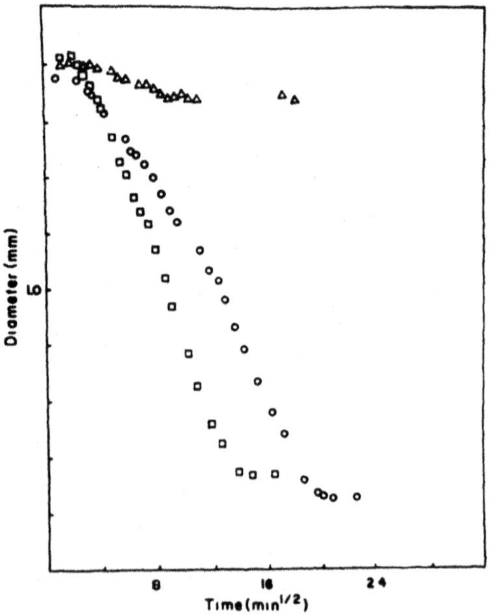

Figure 19. Shrinkage of oxygen bubbles at 1165° C in soda-lime-silica glass (ref. 65). 12G - no fining agent. 13G - aresenic included. 14G - aresenic and nitrate included.

$$O_2 \text{ (bubble)} \rightleftharpoons O_2 \text{ (dissolved)} \qquad (11)$$

$$O_2 \text{ (dissolved)} + As_2O_3 \rightleftharpoons As_2O_5 \qquad (12)$$

In practice, antimony behaves generally in a similar manner, but the effectiveness of both additives depends upon the base composition relative to the required temperature range of melting. The more silica-rich the glass, the lower the temperature at which the reduced valence forms become favored.

In spite of evidence which supports this explanation, puzzles remain. It is not entirely clear why the simultaneous addition of nitrates yields effects which are so concentration-dependent. Nor does there seem to exist any firm understanding of why sulfate additions enhance the effect of arsenic as shown in Figure 20.[68]

4. Sulfur: Sulfate additions have long been known to aid materially in refining. One mechanism probably consists of simply enlarging small bubbles at the highest melting temperature by the sulfate decomposition products, SO_2 and O_2, thus accelerating removal by flotation. On the other hand, dissolution also likely occurs. If there remain bubbles of SO_2 and O_2 in the melt as it cools during refining, increased solubility at the lower temperature will lead to their dissolution. Bubble shrinkage experiments as shown in Figure 21[68] illustrate the point. Here, a bubble of SO_2 and O_2, originally with diameter of 1.7 mm shrank to a diameter of 1 mm within the first minute of the experiment, whereas bubbles of SO_2 or O_2 alone shrank much more slowly. After the initial rapid shrinkage of the $SO_2 + O_2$ bubble, the subsequent much slower dissolution suggests that in this experiment the SO_2/O_2 ratio was not exactly stoichiometric.

Sulfur dissolves in silicate melts as S^{2-} or as SO_4^{2-}, depending upon oxidation-reduction conditions during melting. In either case, the total dissolved sulfur content increases with increasing alkali concentration. It is believed that sulfur substitutes for oxygen of the melt by the following reactions:

$$1/2\ S_2 + O^{2-} \rightleftharpoons S^{2-} + 1/2\ O_2 \qquad (13)$$

$$1/2\ S_2 + 3/2\ O_2 + O^{2-} \rightleftharpoons SO_4^{2-} \qquad (14)$$

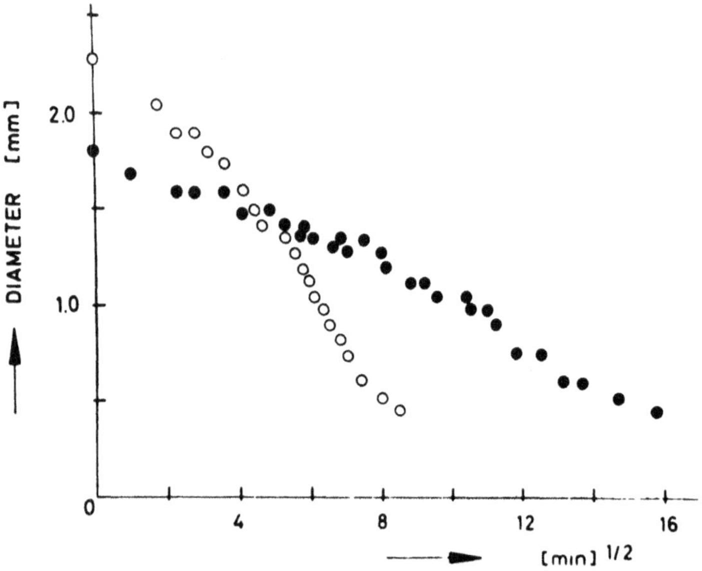

Figure 20. Shrinkage of oxygen bubbles at 1170° C in soda-lime-silica glass (ref. 68). ● - 1% $NaNO_3$ + 1% As_2O_3. O - 1% $NaNO_3$ + 1% As_2O_3 + 1% Na_2SO_4.

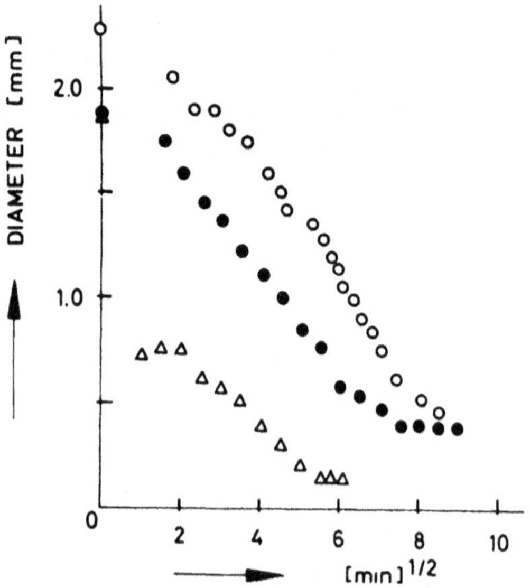

Figure 21. Shrinkage of bubbles at 1170° C in soda-lime-silica glass (ref. 68). O - O_2. ● - SO_2. Δ - SO_2 + O_2.

As has been implied above, at glass melting temperatures, the equilibrium

$$SO_2 + 1/2\ O_2 \rightleftharpoons SO_3 \qquad (15)$$

is displaced to the left. Thus many reports in the literature of SO_2 or SO_3 solubility are meaningless, if not misleading, in the absence of a specified partial pressure of oxygen.

Although SO_2 by itself has been observed to dissolve in molten glass, albeit much more slowly than SO_3, sulfite is probably not the dissolution product. Rather, it appears that SO_2 is oxidized to SO_3 by impurity oxygen or components in the melt, thus yielding dissolved sulfate.

5. Nitrogen: Until relatively recent years, nitrogen was considered an inert unreactive gas in contact with molten glass. Its presence in seeds was usually taken as evidence of air included in the batch or within voids of the refractories. However, it has been shown[69,70] that nitrogen can interact chemically with molten glass. While nitrogen solubility is extraordinarily small under mild oxidizing conditions, this solubility increases substantially when melting conditions become more reducing, as seen in Figure 22.[71]

Evidence for small but finite solubility of nitrogen comes also from another quarter. In bubble shrinkage experiments starting with pure oxygen and pure helium, an impurity gas appeared which most likely diffused "backwards" from molten glass into the bubbles. After the oxygen and helium had diffused out of the bubbles, this unknown gas, now present at a pressure of approximately one atmosphere, diffused back into the glass only very slowly, if even detectably; this unknown gas was most likely nitrogen. In fact, when a deliberate effort was made in one experiment[72] to measure the diffusion constant of nitrogen by this approach, measurable shrinkage was not detected and the diffusion coefficient was estimated to be less than 0.1% of that of helium. The practical consequences are significant. A pre-existing bubble initially free of nitrogen serves as a sink for the collection of nitrogen which may have dissolved in the glass during that time when the glass was at the highest temperature or may have seen reducing conditions. While the partial pressure of nitrogen within the bubble

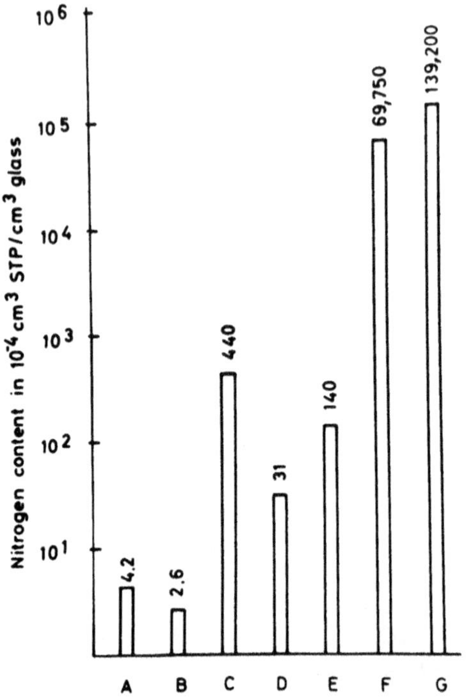

Figure 22. Nitrogen content of soda-lime-silica glass under various experimental conditions (ref. 71). A - bubbling N_2 through melt 16 hours at 1400°C, glass melted in gas furnace, oxidizing flame. B - glass melted in gas furnace, oxidizing flame, 2 hours at 1400°C. C - powdered glass passed through Ar-N_2 plasma. D - glass melted in gas furnace, reducing flame, 2 hours at 1400°C. E - bubbling H_2/N_2 (80/20) through melt 5 hours. F - 3 wt % carbon added to batch, nitrogen bubbled through melt 2 hours at 1400°C. G - ammonia bubbled through melt, 5 hours at 1400°C.

will initially be low, dissolution of the major components (e.g., CO_2 or O_2) will eventually leave nitrogen as the principal constituent. Dissolution of this nitrogen again will now proceed only very slowly. In understanding the refining of glass, it becomes essential to recognize the importance of this counter-diffusion process.

Clearly, if a glass becomes more oxidized during the refining period, a potential for nitrogen seeding may develop. The best corrective is probably the maintenance of as consistent a reducing condition as possible and the minimizing of the number of bubbles which originally survive the high temperature melting zone.

3) Reboil: Throughout the foregoing discussion, it has been pointed out that while the glass-maker's objective is the removal of bubbles, the melt may become supersaturated with respect to some gaseous components. Under these conditions then, there exists the capacity for "reboil", or formation of bubbles in well-refined glass.

Reboil is most commonly associated with the presence of sulfates within a glass. However, most of the observations which are reported in the literature on this subject do not give a clear understanding of the phenomena which were observed; and, in fact, many such reports are seemingly inconsistent. Thus, while sulfates are generally considered to be more soluble in glass at lower temperatures, formation of bubbles in laboratory melts of sulfate-containing glasses during cooling[73] has also been reported. More commonly, sulfate-associated reboil is reported to occur following inadvertent reheating of the glass or exposure to more reducing conditions (e.g.,a malfunctioning burner) after initial equilibration in a more oxidizing atmosphere. Bubbles thus formed usually have very low internal pressures, accompanied by a sodium sulfate deposit on the interior surfaces of the bubble. It is generally accepted that SO_3 which came out of solution as a principal cause of the bubble subsequently reacted with the alkali of the glass bubble's inner surface upon cooling, condensing out as the sulfate deposit (a "smoky blister"). This explanation of sulfate-associated reboil still leaves unaccounted for the mechanism by which such bubbles are nucleated within bulk glass. One likely possibility is the movement of supersaturated glass past refractory surfaces during convective flow.

Another source of reboil blisters is the remelting of devitrified glass such as may accumulate in the cooler regions of a window glass tank.[74] This kind of bubble frequently contains CO_2. These bubbles may result from the lower solubility of gases in devitrified glass or they may result from the remelting of cristobalite crystals, thus locally enriching the melt with respect to silica and creating regions which are in turn supersaturated in CO_2.

Because the solubility of nitrogen and water vapor decrease with decreasing temperature, one might suppose that during the cooling which accompanies refining, supersaturation in these gases might initiate reboil. If this were to occur, doubtless other gases would also enter the bubbles, thus masking the primary cause. In any event, there does not seem to exist unambiguous evidence in support of nitrogen and water vapor as primary factors.

HOMOGENIZING

Having converted batch materials to the molten state and removed the gaseous bubbles, one has gone, in a sense, only two-thirds of the way in melting glass for the forming procedure which must follow. The third fundamental process which is necessarily coupled to melting and refining is homogenization, the reduction of fluid heterogeneity to whatever level is acceptable for the product's end use. In this subject area, references 4 and 75 provide a comprehensive review.

(A) Significance of Homogeneity

Vitreous heterogeneity is distinguished by differences in chemical composition which in turn cause variations in such parameters as color, density, and refractive index. The terms "striae" and "cords" commonly describe essentially two-dimensional layers and one-dimensional veins, respectively, of a composition present within a matrix of slightly different composition. The term "scale" may be used to describe thickness of the heterogeneity, and "intensity" measures the difference in properties of striae and surrounding matrix. These terms are illustrated in Figure 23.[75] The larger the dimensions of the pattern, the larger the scale; as blacks and whites go toward greys, the less the intensity.

In containers, cords and striae are of concern because the stresses which result from uneven coefficients of expansion may lead to degraded strength and even spontaneous rupture. Glass used in vision applications must generally be much more homogeneous than container glass because of the optical distortions which may arise from smaller internal variations. At still a different level, a much higher degree of homogeneity is required in laser rods due to the relatively long light path and because lasing performance depends critically upon loss of energy by internal scattering.

B) Detection of Inhomogeneity

Vitreous inhomogeneity may be detected visually,

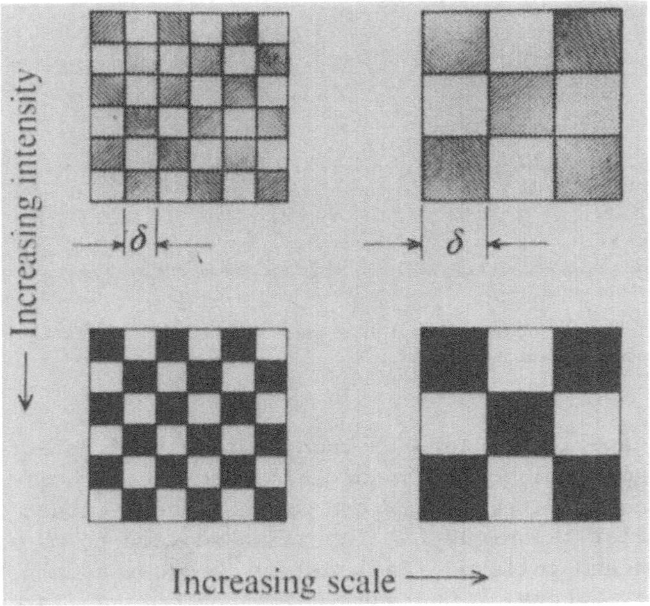

Figure 23. Schematic representation of scale and intensity of segregation in a binary mechanical mixture (ref. 75).

supplemented by photographic recording. One particularly useful procedure employs examination by a schlieren technique of cross-sections of the final product -- for example a ring section cut from a bottle or a strip taken from a plate glass ribbon (Figure 24). Stress patterns from the coarser inhomogeneous structure may even be seen with polarized light.

Off-color streaks in tinted architectural glass are usually seen visually but may correlate with variations in refractive index or density (Figure 25). Optical distortion in flat glass caused by striae may be seen by viewing the glass against a "zebra board," a well-illuminated background of parallel black and white stripes.

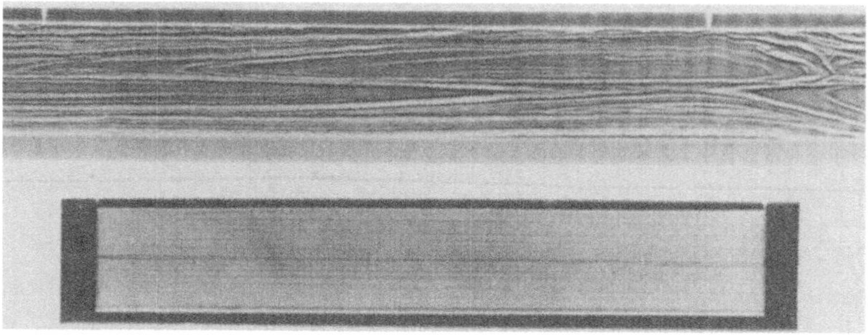

Figure 24. Striagrams of plate glass having different levels of internal inhomogeneity.

An especially useful method, not only of detecting striae and cords, but also of inferring their chemical deviations from the matrix and in turn their cause, is the Loeffler technique.[76] The glass section to be examined is ground and polished, followed by chemical etching with suitable reagents. Upon examination by optical interferometry, the inhomogeneities can be shown to have dissolved either more or less rapidly than the surrounding glass. By comparison with the known etching rates of glass richer in silica, alumina, or other specific components, it is frequently possible to deduce a likely origin of the defect. In Figure 26, both Schlieren and Loeffler techniques illustrate differing degrees of inhomogeneity within two samples of flat glass. Figure 27 illustrates the application

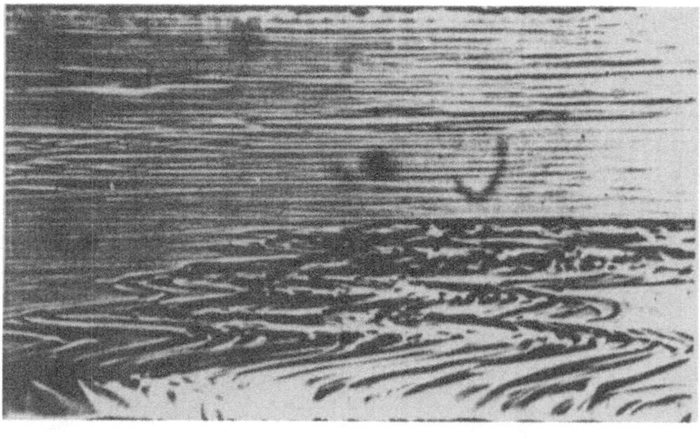

Figure 25. Striagram of sample containing off-color streaks.

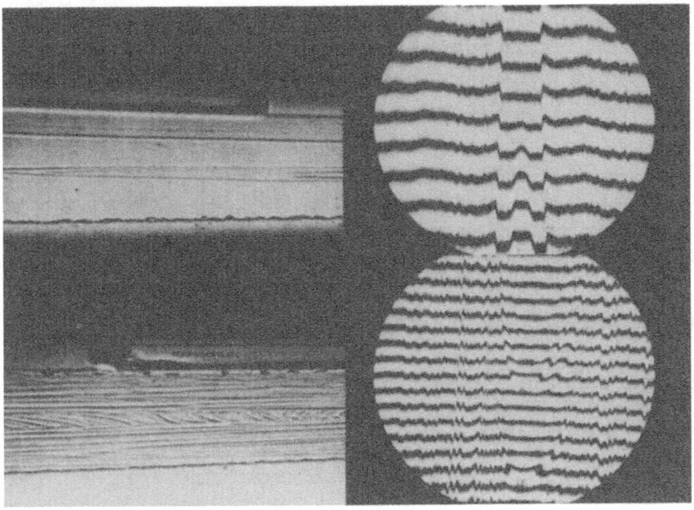

Figure 26. Different levels of inhomogeneity in plate glass as seen by striagram and by interferometry.

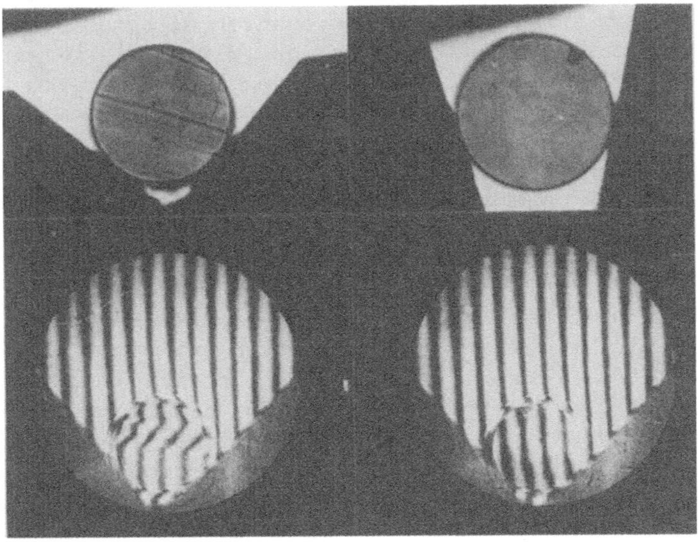

Figure 27. Different levels of inhomogeneity in laser rods seen by striagram and interferometry.

of the same techniques to laser rods. Other techniques depend upon determining the spread in values of a given property such as refractive index[77,78] in powdered samples of the glass which include the defects.

Reference 1 must again be mentioned here as a most comprehensive and useful source on the general subject of defects, including striae and cords, their detection and origins.

(C) Sources of Inhomogeneity

Inhomogeneity is obviously minimized if one starts with well-mixed batch and melts under conditions such that the original uniformity is not perturbed. In this case, the degree of heterogeneity will relate only to the grain size of the batch. The use of colloidal mixtures or gels for obtaining homogeneous glass illustrates this point. On the other hand, as has been discussed in foregoing sections, practical glass-making usually requires raw materials with particular particle size distributions orders of magnitude larger than in colloidal mixtures. Thus, initially poor mixing of the batch, segregation of batch, even as it is being fed into the furnace, and ultimately the inherent microheterogeneity of even a perfectly mixed batch will limit the degree of homogeneity which can be attained in finite times.

A generally more troublesome source of striae and cords are reactions at the glass interface with refractories and the covering atmosphere. During the melting of virtually any practical glass, some components preferentially volatilize if equilibrium partial pressures are not maintained. This latter requirement can almost never be met, and thus compositional changes at the glass surface usually occur. Particularly troublesome is the loss of alkali oxides, PbO, and B_2O_3 by this route. Redox reactions may occur at the surface, as when a container glass colored by the iron-sulfur chromophore is exposed to oxidizing conditions or when window glass is streaked by unintentional generation of the same color.

At the boundary between glass and refractories, and in particular at those junctions of glass, refractory, and atmosphere, refractory dissolution will enrich the glass with silica, alumina, and zirconia.

(D) Homogenization Mechanisms

The glass-maker must recognize that vitreous heterogeneity will inevitably occur in most conventional processes and then proceed to use appropriate means to reduce it to an acceptable level. Thus it is useful first to recognize basic principles and second to know of those procedures which in general have been found most effective.

1) Basic Principles: In liquids even as viscous as molten glass, diffusion will ultimately reduce sharpness of boundaries between striae and the surrounding matrix. Material transport will eventually even out heterogeneity. In terms of Figure 23, the blacks and whites will tend toward the same uniform shade of gray; the intensity will decrease. But the shorter the diffusion paths (the smaller the scale) the more rapidly will occur these changes in intensity. Moreover, the relationship is not simply linear; reducing the thickness of layers by a given factor has a proportionately much larger effect upon reducing the time required for achieving a given level of homogeneity. Figure 28 (Reference 1) illustrates qualitatively the interrelationship.

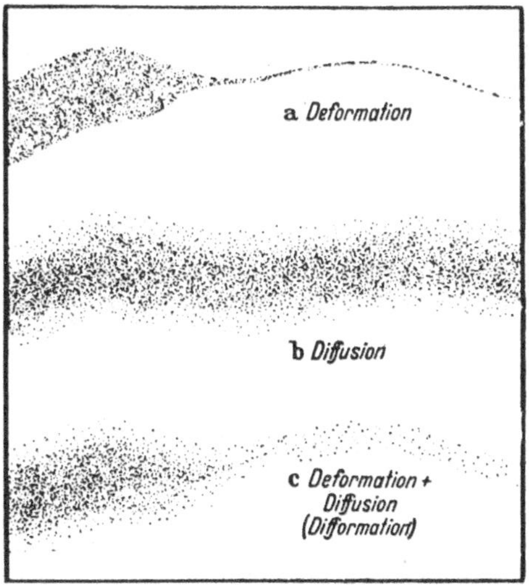

Figure 28. Disappearance of inhomogeneity through deformation and diffusion processes (ref. 1).

This last relation leads one immediately to conclude that the most efficient homogenization calls for extensive attenuation of striae before depending upon mass transport by diffusion. In principle, attenuation may result from simple shear, characterized by a velocity gradient normal to the direction of flow, and stretching, resulting from a positive velocity gradient in the direction of flow.

2) Practical Approaches: In tank melting, the glass-maker depends first of all upon establishing complex patterns by means of specific temperature distributions and furnace geometry. For example, firing of the melter is controlled in such a way that glass temperature at the surface rises to a maximum at a point one-third to one-half the length of the melter down tank from the back wall (where batch enters). This point denotes the "spring zone," where molten glass flows from bottom layers of the tank to the surface and outward on the surface toward the colder regions. By maintaining both back and front ends of the melter at a lower temperature than that of the spring zone, two large convective flow patterns may be established. As glass proceeds through these recirculating flow patterns, attenuation of striae will occur. Figure 1 (Reference 1) illustrates this behavior.

Specific measures may also be taken to separate out unavoidable heterogeneities at some point in the process. In continuous tanks, the so-called skim-kilns, structural appendages on the sides of a tank, serve to draw off surface glass which may be inhomogeneious. Large refractory floats located in the glass at restrictions in the furnace impede surface flow down tank toward the forming process and in turn divert this glass into skim kilns or into the return flow pattern. In operations where surface volatilization may pose a particular problem, as with a borosilicate tank, relatively more siliceous glass may be removed from the surface by drains through the side of the furnace.

An interesting solution to the specific problem of producing ultra-homogeneous glass for laser rods utilizes a bottom-draw furnace of special design.[79] After the glass within the central zone of this refractory-lined furnace is well-homogenized, it is withdrawn from the bottom drain in such a way that the inhomogeneous layer of glass from that region closest to the refractory is retained on the outside of the casting.

Mechanical mixing of glass such as is used for opthalmic and optical components is carried out by means of stirrers fabricated of platinum and more or less conventional refractories. Mixing can also be obtained by bubbling gas, as from orifices in the bottom of the tank or by long lances, but this procedure may also generate other problems such as enhanced volatilization and unacceptable concentrations of gaseous seed.

SUMMARY

The melting of important commercial glasses may be considered essentially as the reaction of a principal refractory component, usually silica or an aluminosilicate, with less refractory oxide fluxes, usually added as carbonates. The course of the decomposition and dissolution reactions can be significantly influenced by the choice of temperature-time schedules, physical properties of the raw materials and minor additives.

A second process, occurring coincident with melting, is the accumulation of gases derived principally from decomposition of the batch into bubbles trapped within the molten glass. Bubbles will be removed by flotation and dissolution, depending upon the time-temperature schedule and the chemical effects of specific additives.

The third principal process is reduction of compositional inhomogeneity by attenuation and diffusion mechanisms.

Finally it should be noted that this survey has dealt with only that process and those compositions which are today commercially most significant. It is increasingly important to be aware, for example, of the use of chalcogenide glasses in new kinds of electronic devices and of the fabrication of glass films by unconventional methods such as sputtering and chemical deposition. While these products will not compete in tonnage with bottles or windshields, nonetheless their potential importance clearly points toward new needs in the science and technology of glass.

ACKNOWLEDGEMENTS

The author is pleased to acknowledge the assistance of many colleagues in the Glass Research Laboratories. In particular, he thanks Mr. William Englert for his help and Dr. H. W. V. Willems, Consultant, for a critical reading of the manuscript.

The author also thanks the following Societies and publishers for permission to reproduce figures as indicated below:

The American Ceramic Society -- Figures 3, 4, 5, 7, 9, 19, 20, 21, 22.

Deutsche Glastechnische Gesellschaft -- Figure 13.

Glass Industry -- Figures 6, 15, 17.

Ogden Publishing Company -- Figure 2.

The Society of Glass Technology -- Figures 10, 11, 14, 16, 18, 23.

Springer-Verlag Berlin/Heidelberg/New York -- Figures 1, 28.

REFERENCES

1. H. Jebsen-Marwedel, Glastechnische Fabrikationsfehler, Springer-Verlag, Berlin (1959).
2. Symposium on the melting of glass, Union Scientific Continental du Verre, Charleroi (1958).
3. F. V. Tooley, Handbook of glass manufacture, Ogden Publishing Company, New York (1961).
4. M. Cable, Eighth International Congress on Glass, 164-178, Society of Glass Technology, Sheffield (1969).
5. E. M. Levin, H. F. McMurdie and F. P. Hall, Phase diagrams for ceramists, The American Ceramic Society, Columbus (1956).
6. J. Moss, Glass Ind. 50, 536-539 (1969).
7. A. Vidal, Glastech. Ber. 36, 305-323 (1963).
8. W. H. Manring and W. C. Bauer, Glass Ind. 45, 354-357, 413-416, 449-451 (1964).
9. W. Buss, Glastech, Ber.32, 89-95 (1959).
10. W. Buss, Ibid 35, 167-176 (1962).
11. K. G. Kreider and A. R. Cooper, Glass Technol. 8, 71-73 (1967).
12. J. Hlavac and H. Nademlynska, Ibid 10, 54-58 (1969).
13. E. F. Riebling, Am. Ceram. Soc. Bull. 48, 766-769 (1969).

14. J. C. Potts, G. Brookover and O. G. Burch, J. Am. Ceram. Soc. 27, 225-231 (1944).
15. C. Kröger, Glastech. Ber. 25, 307-315 (1952).
16. A. Pons and F. Pons, Method of determining the melting rate of soda-silica batches and some applications, Paper No. 56, International Congress on Glass, Brussels (1965).
17. R. Brückner, Glastech. Ber. 34, 289-299 (1961).
18. R. Brückner, Ibid 34, 518-528 (1961).
19. R. Brückner, Ibid 35, 93-105 (1962).
20. A. R. Conroy, W. H. Manring and W. C. Bauer, Glass Ind. 47, 84-89, 110, 133-139 (1966).
21. W. H. Manring, D. D. Billings, A. R. Conroy and W. C. Bauer, Ibid 48, 374-380 (1967).
22. J. P. Poole (Brockway Glass Co.), U.S. 3,375,095 (1968).
23. C. Kröger and F. Marwan, Glastech. Ber. 29, 275-289 (1956).
24. V. Stoicheva, Tsv. Vodenicharova and E. Karaslojanova, Stroitelni Mat. Silikat Prom. 7, 31-34 (1966).
25. L. G. Mel'nichenko, V. P. Sokol'skii and T. S. Ostapenko, Steklo Keram 8, 8-9 (1951).
26. F. de Körösy, J. Am. Ceram. Soc. 20, 162-163 (1941).
27. R. Roy, Ibid 52, 344 (1969).
28. C. J. Brown, Jr. (Libby-Owens-Ford Glass Co.), U.S. 3,325,298 (1967).
29. J. Yamamato and E. Komatsu, Glass Ind. 49, 491-493 (1968).
30. W. H. Manring and A. R. Conroy, Ibid 49, 199-203, 269-270 (1968).
31. J. R. Carroll, J. C. Maginn and G. Y. Anders, The submerged-combustion technique for high-temperature melting, presented at the Eleventh National Heat Transfer Conference, AICHE-ASME, Minneapolis (1969).
32. J. F. G. Hicks, Glass Ind. 47, 140-141, 160 (1966).
33. L. Penberthy, Ibid 47, 319-322 (1966).
34. W. D. Johnston, J. Am. Ceram. Soc. 48, 184-190 (1965).
35. D. Brown and R. W. Douglas, Glass Technol. 6, 190-196 (1965).
36. K. Karlsson, Glastek. Tidskr. 24, 13-19 (1969).
37. R. W. Douglas and M. S. Zaman, Phys. Chem. Glasses 10, 125-132 (1969).
38. R. J. Ryder, J. Can. Ceram. Soc. 38, 95-101 (1969).
39. G. E. Rindone, Glass Ind. 38, 489-493, 516, 526, 528, 561-565, 576, 577 (1957).
40. Symposium on the refining of glass, Union Scientific

Continental du Verre, Charleroi (1956).
41. M. Cable, J. Am. Ceram. Soc. 49, 436-441 (1966).
42. H. Scholze, Eighth International Congress on Glass, 69-83, Society of Glass Technology, Sheffield (1969).
43. I. E. J. Sainsbury, Glass Technol. 8, 138-147 (1967).
44. A. G. Bossard and E. R. Begley, Symposium on defects in glass, 69-81, International Commission on Glass and the Ceramic Association of Japan, Tokyo-Kyoto (1966).
45. D. A. Krohn, Ceramic Ind. 90, 44-46, 65 (1968).
46. J. Chmielenski, Szklo i Ceramika 14, 322-329 (1963).
47. H. Meyer and H. Poehnitzsch, Glastech. Ber. 38, 393-397 (1965).
48. E. Plumat, Reference 40, 349-398.
49. E. Plumat, F. Toussaint and M. Boffee, J. Am. Ceram. Soc. 49, 551-558 (1966).
50. A. G. Bossard (Corning Glass Works), U.S. 2,919,209 (1959).
51. J. H. Cowan, W. M. Buehl and J. R. Hutchins III, J. Am. Ceram. Soc. 49, 559-562 (1966).
52. W. M. Buehl and J. H. Cowan, Jr. (Corning Glass Works), U. S. 3,387,961 (1968).
53. T. Inoue and Y. Ikeda, Study on the fine seeds in flat glass, Symposium on defects in glass, 273-284, International Commission on Glass and the Ceramic Association of Japan, Tokyo-Kyoto (1966).
54. M. Cable, Glass Technol. 2, 60-70 (1961).
55. Corning Glass Works, Br. 1,159,647 (1969).
56. G. H. Frischat and H. J. Oel, Glastech. Ber. 38, 156-166 (1965).
57. G. H. Frischat and H. J. Oel, Phys. Chem. Glasses 8, 92-95 (1967).
58. H. Scholze, Glass Ind. 47, 546-551, 622-628, 670-675 (1966).
59. M. L. Pearce, J. Am. Ceram. Soc. 47, 342-347 (1964).
60. C. Kröger and D. Lummerzheim, Glastech. Ber. 38, 229-232 (1965).
61. C. Kröger and N. Goldmann, Ibid 35, 459-466 (1962).
62. M. Cable, A. R. Clarke and M. A. Haroon, Glass Technol. 9, 101-104 (1968).
63. M. Cable, A. R. Clarke and M. A. Haroon, Ibid 10, 15-21 (1969).
64. L. Nemec, Ibid 10, 176-181 (1969).
65. C. H. Green and H. A. Lee, Jr., J. Am. Ceram. Soc. 48, 528-533 (1965).
66. R. H. Doremus, Ibid 43, 655-661 (1960).

67. M. Cable, Glass Technol. 2, 151-158 (1961).
68. C. H. Greene and D. R. Platts, J. Am. Ceram. Soc. 52, 106-109 (1969).
69. H. O. Mulfinger and H. Franz, Glastech. Ber. 38, 235-242 (1965).
70. T. H. Elmer and M. E. Nordberg, Nitrided glasses, Paper No. 30, Seventh International Congress on Glass, Brussels (1965).
71. H. O. Mulfinger, J. Am. Ceram. Soc. 49, 462-467(1966).
72. G. H. Frischat and H. J. Oel, Glastech. Ber. 40, 311 (1967).
73. M. Cable, C. G. Rasul and J. Savage, Glass Technol. 9, 25-31 (1968).
74. M. Boffe, G. Pecreaux and E. Plumat, Symposium on nucleation and crystallization in glasses and melts, 47-54, American Ceramic Society, Columbus (1962).
75. A. R. Cooper, Jr., Glass Technol. 7, 2-11 (1966).
76. J. Löffler, Glastech. Ber. 30, 457-463 (1957).
77. V. I. Shelyubskii, Symposium on defects in glass, 116-138, International Commission on Glass and the Ceramic Association of Japan, Tokyo-Kyoto (1966).
78. S. M. Budd and M. Blanchard, Glass Technol. 7, 12-14 (1966).
79. E. W. Deeg and C. G. Silverberg, Glass Ind. 51, 164-167 (1970).

Experimental Glass Melting Techniques

G. E. Blair

Eastman Kodak Company
Kodak Apparatus Division Research Laboratory
Rochester, New York 14650

INTRODUCTION

The composition range in which glass formation can take place in most systems, depends on the cooling rate from a temperature above the liquidus to the glass transformation temperature. Therefore, any study of a particular system must refer to a uniform cooling rate. In addition, for practical purposes,it is necessary to define the smallest usable volume. This varies from fractions of a gram (in reflective beads) to many kilograms (for massive optics, etc.) Turnbull[1] states that any body, however formed, which exhibits diffraction behaviour consistent either with the continuous random model or with a microcrystalline model in which the average crystallite size appears to be no greater than about five molecular diameters, will be considered to be a glass. There are many methods to form glasses but glass formation referred to here will be confined to cooling from an inorganic melt of the same composition as the resultant glass.

GLASS-FORMING MATERIALS

Most of the elements have been incorporated into glasses in experimental studies. Rawson[2] has classified glass formers into 6 general types:

1. <u>Single elements</u> - examples: sulfur & selenium
2. <u>Single oxides</u> - examples: SiO_2, GeO_2 B_2O_3 P_2O_5,

and As_2O_3 (Sb_2O_3 can be added when cooled rapidly).
3. Two oxide systems - (Conditional glass formers) examples: TeO_2, SeO_2, MoO_3, WO_3, Bi_2O_3, Al_2O_3, Ga_2O_3, and V_2O_5 will form glasses when combined with another oxide; not necessarily from the single oxide glass formers listed above.
4. "Hydrogen-bond" systems - example: $KHSO_4$
5. Halide systems - examples: BeF_2 and $Zncl_2$.
6. Chalcogenide systems - examples: As-S, As-Se, P-Se, Ge-Se.

In addition to the primary glass forming components as listed above, most glasses are modified by the addition of other elements to obtain desired physical or chemical properties. Wide use is made of the alkalis (Li, Na, K, etc.) for fluxing the melts of the single oxide and two oxide systems and of the alkaline earths (Ca, Mg, Sr, etc.) to impart chemical durability to the systems that have been fluxed with alkalis. Pb, Ba, Ti, La, Ta, Ag, Cu, the transition elements and rare-earths are also used frequently to obtain some specific property.

Except for the elemental or chalcogenide glasses, glass raw materials are usually added as oxides, hydroxides, carbonates, nitrates, sulfates, halides and in some cases as metal-organic complexes (e.g., acetates, to produce a reducing condition in the melt). Many of these are available with a choice of the valence state of the metal ion. Roy[3] has suggested the use of gels as starting materials. This points out an important consideration. Gels have extremely small particle size and intimate mixing is assured by this method. Potts et al.,[4] have shown that in conventional powdered materials, small particle size (0.1-0.25 mm) coupled with the same size for all materials, yields the shortest time to a batch-free condition in melting. In some cases, the purity of the raw materials is of extreme importance. Generally, glass used for opthalmic and optical purposes is required to be free of absorption bands in the visible region of the spectrum. In some of the higher index optical glasses, 10-20 ppm of Fe^{3+} ion will produce a yellow color (blue absorption). Iron is present in all natural silica materials usually to a much greater extent than 10-20 ppm. Therefore, it becomes necessary in most cases to either purify the natural materials or produce synthetic materials. In the case of silica, free iron can be removed by magnetic separators and chemically combined iron by repeated treatment with HCL acid. The usual procedure is to analyze the selected material for impurities by emission spectrography or some similar method. If the material does not meet the purity requirements, either a new source is found or the material is purified by removal of the objectionable constituent. In the case of raw materials for optical glasses, both methods are frequently used. The transition metal ions are the contaminants usually producing absorption bands.

GLASS PREPARATION

After selection of raw materials, the batch is weighed-out based on either the molecular ratios or weight ratios of the oxide constituents. In either case it is necessary to allow for decomposition, (e.g., $Na_2CO_3 \rightarrow Na_2O$ (glass constituent) + $CO_2\uparrow$). After weighing, the batch should be blended. Thorough mixing will help produce the shortest melting time at the lowest temperature with the minimum reaction with the crucible. A satisfactory blender is shown in Figure 1. A typical composition study to map out a glass formation field for a new optical glass would be carried out in a 25 cubic

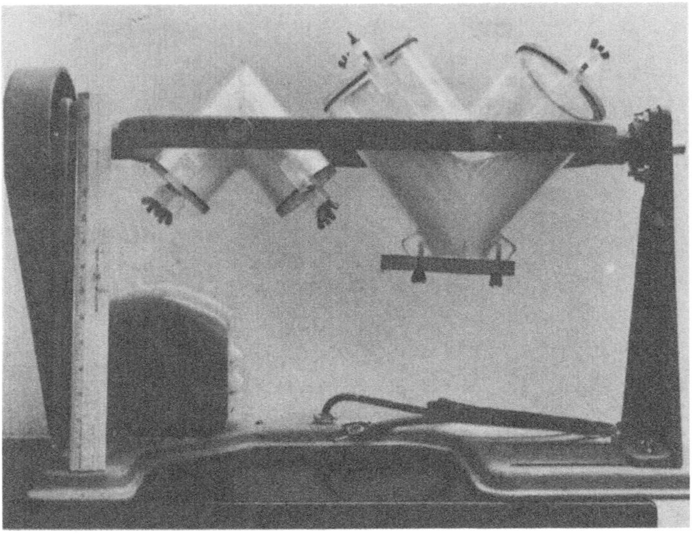

Figure 1 Blender Used for Homogenizing Batch.

centimeter (cc) crucible. Pure platinum crucibles are used wherever possible to minimize contamination. The proper amount of batch is charged into the crucible to yield 25 grams of glass. The melting is typically carried out in a small electrically heated, silicon carbide resistance furnace as shown in Figure 2. Melting temperatures range from 1200-1500°C for the conventional silicate and borate compositions. Tooley et al.,[5] sum up the reactions taking place in a glass melt as -

1. Evaporation of free water
2. Formation and loss of gaseous constituents (e.g., CO_2).
3. Formation of liquid phases from individual batch constituents and formation of eutectics
4. Mutual solution of liquid phases yielding molten glass free of crystalline phases
5. Volatilization of certain compounds from melt (e.g., Na_2O, PbO, SiF_4)
6. Solution of gas phases (e.g., CO_2, SO_2)

Figure 2. Laboratory Furnace, Electrically Heated - SiC resistance elements, used for small-scale exploratory studies of new glasses.

They also point out that there are 4 separate phases involved-

1. Melting
2. Fining
3. Homogenizing
4. Conditioning

For the initial mapping of a glass formation field as represented by Figure 3, only the melting phase is carried out and the melt is cast onto a heated iron plate. As stated earlier, glass formation in most systems is very dependent on the cooling rate and hence on the temperature of the casting table and on the thickness and total volume of the cast. If small volumes only are required, glass formation might be determined by a method shown in Figure 4. Beads of approximately 0.5 gram form at the orifice and drop free onto a rotating graphite table. This method gives reproducible results. After this initial small scale melting is carried out, and it is desired to develop a usable glass from this study, all 4 phases, as explained by Tooley et al.,[5] need to be examined carefully. Typically, a 250 crucible (see Figure 5) is used. The 4 phases will be studied separately.

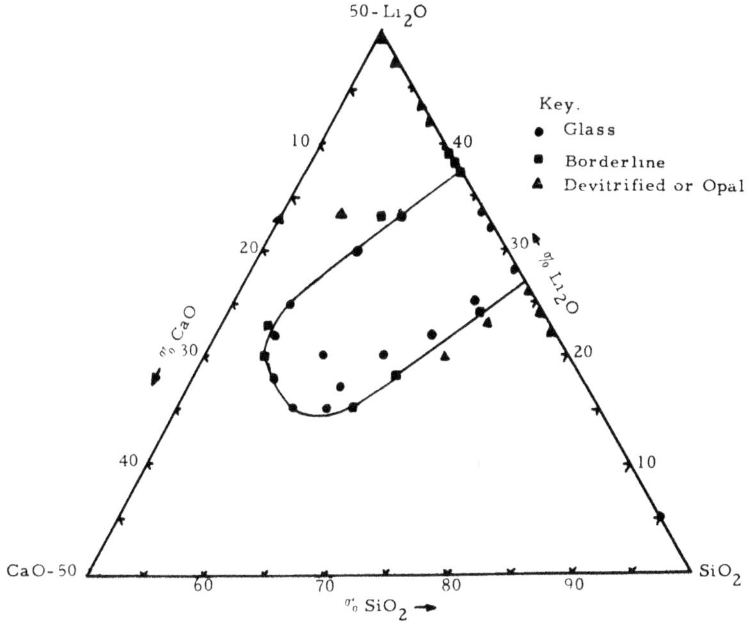

Figure 3. Representation of a 3-Component Formation Field Study

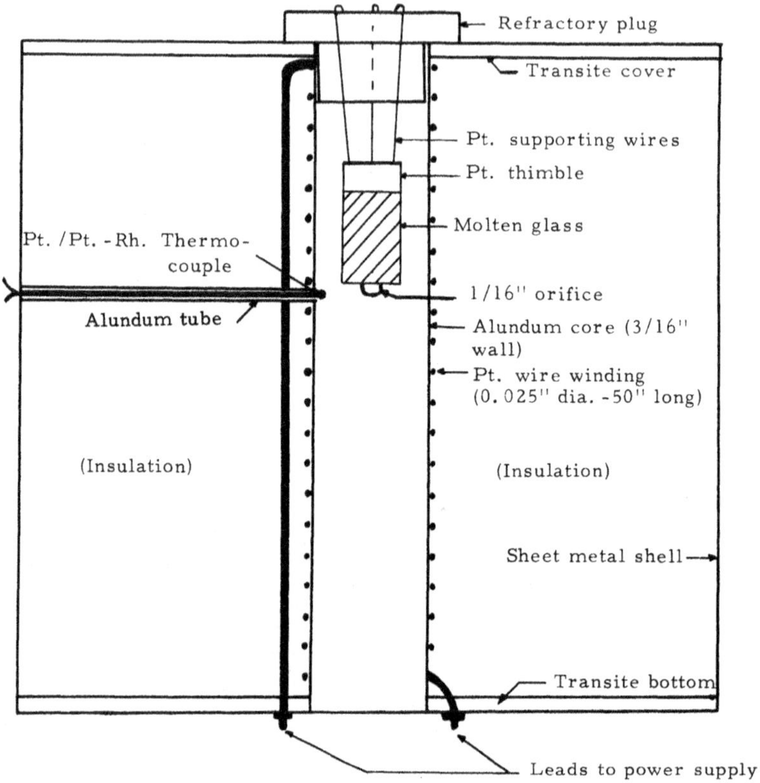

Figure 4. Cross-Section of Bead Furnace used to Produce Glass Beads of Uniform Size for Glass Formation Determination.

Figure 5. Platinum Crucible of 250 cc Capacity with Supporting Refractory Crucible used to Produce Homogeneous Stirred Melts of Glass.

1. Melting:

The first stage of the melting phase is the incorporation of all the batch into the melt. Assuming that it is desired to fill the crucible with molten glass, this requires several additions to reduce the lower density batch materials to the higher density liquid melts. Since a majority of the normal glass raw materials undergo decomposition with a large amount of gas evolution, the temperature during this initial "filling" stage is just high enough to produce a liquid melt without the decomposition and batch reactions becoming too violent. This temperature varies with composition but would be in the range of 1200-1400°C for silicate compositions. After the batch has been added, the temperature is raised (1300-1500°C) to accelerate the reactions and lower the viscosity. The melting phase is considered to be complete when the melt is batch free.

The factors influencing rate of melting are-

a) Temperature
b) Batch formulation
c) Particle size of batch and uniformity of size
d) Size of cullet (if used) and amount
e) Batch homogeneity

This phase is usually completed in 1-2 hours for 250 cc melts. The melting can be carried out in a furnace as shown in Figure 6. This particular furnace allows for mechanical stirring, is of the elevator-hearth type, and is heated with SiC elements. It also allows vacuum melting, or one atmosphere over-pressure of an oxidizing, neutral, or reducing atmosphere. Other special furnaces that are available use induction heating or metallic elements (e.g. molybdenum, tungsten, or tantalum) or a graphite element. The metallic element and graphite element furnaces require a neutral atmosphere around the elements to prevent oxidation. The working zone can have either a vacuum, reducing, neutral, or oxidizing atmosphere by separating the zones with an impervious material such as recrystallized alumina. The induction furnace requires a receptor material, e.g. platinum, or some other electrically conductive material. These furnaces typically will reach maximum temperatures in a few minutes but are usually limited to small volumes. The use of an impervious tube

Figure 6. Laboratory Furnace, Electrically Heated - SiC resistance elements, elevator hearth type, controlled atmosphere (vacuum or one atmosphere over pressure of neutral, reducing, or oxidizing gases), provided with stirring mechanism, used for producing 250 cc - 1500 cc size melts of high quality glass.

allows the production of "pure" atmospheres in the working zones of these furnaces.

2. Fining:

This phase of the glass preparation process is carried out after the melting of the batch is complete and frees the melt of bubbles. The fining process involves both physical and chemical effects. The controlling physical parameters are viscosity, surface tension, and density differences between the bubbles and melt. The chemical effects are controlled by the addition of fining agents. Some of the most widely used are sulfates, As_2O_3, Sb_2O_3, and fluorides. Phosphorus, selenium, selenates, and tellurates have also been used. Chlorides bromides, and iodides have been used in reduced glasses. The fining agents are normally limited to 0.1% additions. They react with the melt in various ways. In some cases they volatilize within the melt producing more bubbles of a lower density and creating a sweeping action. In other cases they chemically combine with the gases within the bubbles. Some of the fining agents can also effect

The viscosity and surface tension of the melt.

3. Homogenizing:

Although this phase is separated from fining, this separation is somewhat artificial in that very often the two processes are going on simultaneously. During the fining process the action of bubble release and the chemical action tend to mix the glass and remove chemical differences within the melt. The glass is said to be perfectly homogeneous when no variation in properties can be determined (e.g., density or refractive index). This, or course, means the absence of striae, cords, etc, and implies chemical homogeneity on a fine scale. Factors influencing the production of homogeneous glass are -

a) Temperature
b) Time
c) Batch formulation
d) Degree of uniformity of batch mixing
e) Segregating tendency on melting
f) Nature and degree of convection and mechanical stirring.
g) Physicochemical relations in the batch-glass-refractory system.

The influence of the melting crucibles should not be overlooked. Any crucible solution, usually of a slow leaching nature, will change the composition and will need to be "stirred-in". Also the drag from the crucible wall will affect the movement imparted by the stirrer. For good homogeneity, it is usually necessary to mechanically stir the melt over a period of time from one hour to several hours depending on the size of melt, viscosity, and various other factors. For a 250 cc melt, stirrers, as shown in Figure 7 have proven to be satisfactory. The design of the stirrer is dictated by the viscosity and size of the melt and by the stirring speed. In the development of a new optical glass, the optical constants and quality are dependent on melt size. More reproducible results and higher quality are usually obtained in larger melts. It is common practice to scale up to a larger size crucible (e.g., 1500 cc). This is shown in Figure 8 with a representative stirrer. Stirring speeds from a few rpm to a few hundred are used. Generally the more fluid melts require higher speeds and more blades or greater pitch to

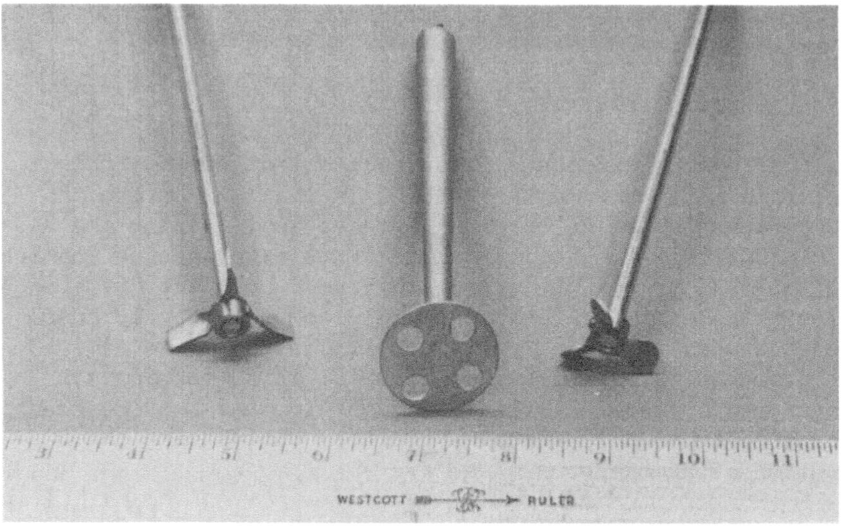

Figure 7. Representative Stirrer Designs That Have Been used with 250 cc Size Glass Melts.

Figure 8. Platinum Crucible of 1500 cc Capacity With Supporing Refractory Crucible and a Representative Stirrer used in Scale-up Studies from 250 cc Size.

the blades of the stirrer. For high viscosity melts, a simple disc stirrer revolving at 20-50 rpm has proven quite satisfactory. Deeg and Silverberg (6) rotated the crucible and stirrer as well as providing for vertical motion of the stirrer in the production of homogeneous glass for laser application. Potts et al.,[4] made 10 gram quantities of glass that were fairly homogeneous by rotating the crucible at one rpm while inclined 12° from the vertical Stirling and Miselbach[7] poured the glass in a fine stream (1-5 mm) from one crucible to another in a random manner a total of 24 times, resting the glass 10-15 minutes between pours, to homogenize small-scale melts. Reference is again made to Roy's study of gels.[3] This consists of chemical mixing followed by gelation which when melted gave a very homogeneous glass without stirring. Because of the extremely small particle size and the intimate mixing provided by this method, glasses have been prepared without ever taking the mixture above the liquidus temperature.

4. Conditioning:

This phase normally consists of lowering the temperature with the glass at rest. This allows the removal of any bubbles stirred in and subsequently increases the viscosity for either casting or forming operations. Typically, the temperature may be reduced to 1100-1200°C over a period of one-half to one hour. This temperature will be determined by the liquidus temperature, or if the melt is held below the liquidus, on the rate of formation of crystalline phases. Having completed the glass preparation, the melt is normally cast into heated molds in experimental studies. These molds are frequently made from thick cross-sections of cast iron. The volume and thickness of the cast is dictated by the end use and by the volume that can be cooled without crystal formation.

ALTERNATE METHODS OF GLASS PREPARATION

The above method of glass preparation is the most common, but other methods are used for special purposes. For example, Lell[8] has produced doped, fused-silica glass by a novel method. In this case, Li^{+} and Al^{3+} were added in solution form to a colloidal SiO_2 soaked in deionized water. This material was dried at 180°C. After drying the resultant powder was sintered at 1500°C for 3 hours in a platinum crucible. This material was then ground in an agate mortar and fused by injection into an oxyhydrogen flame (Verneuil Method). The resultant glass grows as a boule on a block of pure sintered SiO_2 to avoid contamination. In preparing bromide and iodide glasses, Pearson et al.,[9] heat the elemental raw materials (As-S-Br and As-Te-I) in evacuated sealed fused-quartz vials in a horizontal furnace that allows for revolving the vials to mix the melt. The raw materials are weighed out in an inert atmosphere and special handling techniques are described for the bromide glasses. The sealed reaction tube is placed in a steel-pipe bomb during melting at 500-600°C for 2 hrs.

ANNEALING

After any of the above methods of glass formation, it is necessary to anneal the glass. This is to remove stresses that have resulted from temperature gradients during initial cooling and to achieve index homogeneity in optical glass and highest density (lowest fictive temperature) for long-term stability. Hagy[10] describes a method of rate annealing that achieves the above in the shortest time.

SUMMARY

Many other methods of inorganic glass preparation have been proposed. However, most of these are limited to extremely thin sections (e.g., splat cooling, evaporative film, sputtering, etc.) The methods described here have been limited to those producing practical quantities of glass.

REFERENCES

1. D. Turnbull, "Under What Conditions can a Glass be Formed?" Comtemp Physics, 10, No. 5, 473-488 (1969).
2. H. Rawson, "Inorganic Glass Forming Systems", Academic Press (1967).
3. R. Roy, "Gel Route to Homogeneous Glass Preparation", J. Am. Ceram. Soc., 52, No. 6, 344, June (1969).
4. J. C. Potts, G. Brookover, and O. G. Burch, "Melting Rate of Soda-Lime Glasses as Influenced by Grain Sizes of Raw Materials and Additions of Cullet", J. Am. Ceram. Soc., 27, No. 8, 225-231, (1944).
5. F. V. Tooley, M. A. Knight, A. K. Lyle, V. C. Swicker

and A. Ottoson, Handbook of Glass Manufacture - Vol. I, Chapt. IX, Ogden Publishing Co. (1953).

6. E. W. Deeg and C. G. Silverberg, "Performance of New All-Ceramic Glass Melter for Small-scale Production of Optical Glasses", The Glass Industry, 164-167, April (1970).
7. J. F. Stirling and R. G. Miselbach, "A Technique for Improving the Homogeneity of Laboratory Melts", Glass Technology, 9, No. 6, Dec. (1968).
8. E. Lell, "Synthesized Impurity Centers in Fused Silica" J. Am. Ceram. Soc., 43, No. 8, 422-426, Aug. (1960).
9. A. D. Pearson, W. R. Northover, J. F. Dewald, and W. F. Peck, Jr., "Chemical, Physical, and Electrical Properties of Some Unusual Inorganic Glasses", Advances in Glass Technology, Plenum Press (1962).
10. H. E. Hagy, "Fine Annealing of Optical Glass for Low Residual Stress and Refractive Index Homogeneity", Applied Optics, 7, No. 5, May (1968).

Rheological Behavior of Glass

H. E. Hagy
Research and Development Laborabory
Technical Staffs Division
Corning Glass Works
Corning, New York 14830

INTRODUCTION

The viscosity of glass receives much attention in both academic and manufacturing circles of endeavor. It is a property that is very sensitive to composition differences and hence helps the theoretical scientist understand glass structure and, at the same time, gives the manufacturing team a property suitable for quality control. However, quality control is only one aspect of importance that viscosity has in glass manufacture and applications. It is used in melting, forming, annealing, tempering, ion-exchange, and sealing.

The new scientist in a glass field is quickly exposed to the jargon of glass viscosity. What is the softening point, the annealing point, the strain point, the setting point, etc? This jargon can get somewhat confusing, and misconceptions as to the meaning of these terms can easily come about. However, it is fair to say that even some experienced glass scientists have incorrect ideas about the meaning of these points.

The purpose of this paper is to discuss the viscosity of glass more from a practical rather than a theoretical aspect. Emphasis will be placed on the nature of the viscosity - temperature curve, transformation time dependence, reference points, measured procedures, and the practical uses of viscosity data.

VISCOSITY - A DEFINITION

To establish a definition of viscosity consider the diagram in Figure 1. Visualize two rigid plates, separated by a finite displacement, x, with this space separating them filled with a fluid. If the lower plate is fixed and a tangential force, F_t, is applied to the upper plate, the upper plate will move at constant velocity, v. If this fluid follows the simple basic laws of Newtonian flow, the velocity will decrease linearly across the space between the plates and be zero at the lower interface. The viscosity is then defined as:

$$\eta = \frac{F_t/A}{\frac{dv}{dx}} \qquad (1)$$

where: η = viscosity
F_t/A = shear stress
$\frac{dv}{dx}$ = the velocity gradient

The cgs unit of viscosity is the poise which represents one gram per centimeter second (dimensionally $ML^{-1}T^{-1}$). The reciprocal of viscosity is fluidity.

Not only is this a basic way of defining viscosity but it is also a practical scheme for measuring viscosity. Mr. William G. Steiner of Corning Glass Works actually measured the viscosity of devitrifying frit glasses in this fashion. He used two standard microscope slides, presumably of a soda-lime composition, for the rigid plates. A "button" of frit glass approximately 0.025 cm. thick and 1.2 cm. in diameter was placed between these slides. One plate was fixed and the other plate was coupled to a variable speed drive external to the furnace in which this glass "sandwich" was housed. A load cell, with a 23 kilogram capacity was used to measure the force generated at the speed selected. Since the temperature range of interest for the frit glasses was up to 450°C., the microscope slides were sufficiently high in viscosity so that viscous flow in them was negligible. Very convincing data was obtained in this manner and it agreed well with data generated by the parallel plate viscometer, an instrument that is discussed later in the paper. The technique is not adaptable for viscosity below 10^4 poises, however, since it would be difficult to keep

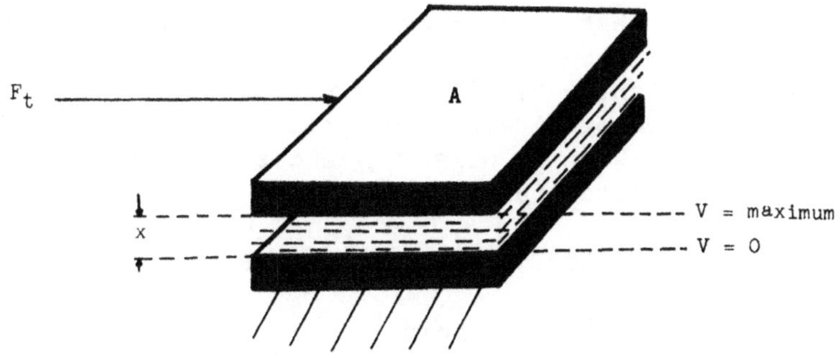

$$\eta = \frac{Ft/A}{dv/dx}$$

Figure 1. Illustrative definition of viscosity.

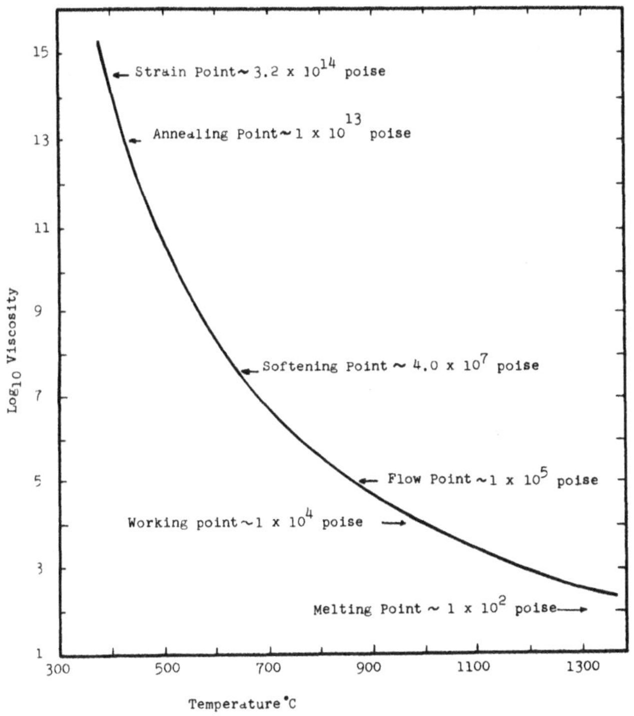

Figure 2. Typical viscosity curve for glass showing common reference points.

the area of the test glass constant due to lateral flow.

Glass apparently behaves as a perfect Newtonian body. That is, the rate of strain is directly proportional to the shear stress with no offset (yield stress) at the origin of the plotted function.

VISCOSITY TEMPERATURE CURVE

a) General Reference Points

Figure 2 shows a typical viscosity-temperature curve for glass. This glass happens to be an alkali-barium-silicate composition. Unlike crystalline materials which show an abrupt change in viscosity at the melting temperature, glass has a smoothly changing viscosity function which increases dramatically as the temperature is lowered.

Throughout the years several reference points have been assigned to glasses on this function. These reference temperatures are helpful in describing in a brief manner the behavior and type of glass and distinguishing between glasses. On the other hand, significance is sometimes assigned to these points which is not justified and it is hoped that such missuse can be corrected with this discussion.

Starting at the lower viscosity region and working upwards, these reference points are:

1. The Melting Point - 100 poises (exactly)
2. The Working Point - 10,000 poises (exactly)
3. The Flow Point - 100,000 poises (exactly)
4. The Softening Point - no exact viscosity can be ascribed since this depends on the density and tension of the glass. Range: 3×10^7 to 1.5×10^8 poises. The value for a soda-lime glass is about 4.2×10^7 poises.
5. The Annealing Point - approximately 10^{13} poises.
6. The Strain Point - approximately 3.2×10^{14} poises (by extrapolation).

One can see that at the higher viscosity levels the established reference points lack exact viscosity definition. The reasons for this will be discussed in detail later.

A graph showing the viscosity curves for several Corning compositions is shown in Figure 3. Note the wide differing composition. The range is actually wider than this graph indicates. Softening points as low as 375°C and as high as 1200°C have been reported for commercially available glasses.

In the parlance of glass science, a high softening point glass is known as a hard glass. A low softening point glass is known as soft glass. One with a steep viscosity-temperature slope is known as a short glass whereas the converse is known as a long glass. Combinations of these terms are also used. For instance, Code 7940 fused silica would be identified as a hard, long glass. These terms are also used to compare two similar glasses, such as: longer or shorter, harder or softer.

b) Mathematical Form

This paper is not the appropriate place for a detailed discussion of the mathematical form of the viscosity-temperature function, but it is worthwhile to take a brief

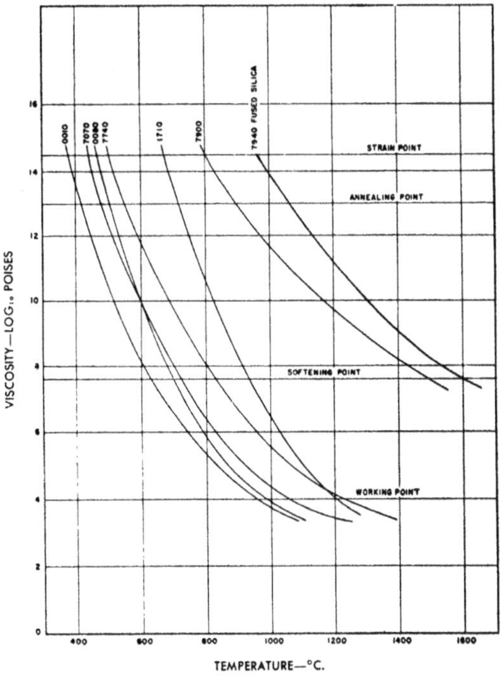

Figure 3. Viscosity curves for several Corning glasses.

look at two equations that have been investigated. One of the earliest equations [1] used is as follows:

$$\log \eta = A + \frac{B}{T} \qquad (2)$$

where: η = viscosity
T = temperature
A and B = constants, B being a measure of the activation energy for viscous flow

This equation holds only for a limited temperature range and it would take many such equations to describe the complete measurable viscosity curve for a glass. A much improved formula was proposed by Fulcher.[2] This equation has the form:

$$\log \eta = A' + \frac{B'}{T - T_o} \qquad (3)$$

Here T_o is added as a third adjustable constant.

This equation fits remarkably well over a wide range of viscosities. It is used to curve fit the round robin test data[3] by the National Bureau of Standards in establishing certified viscosity values for calibrating reference glasses. Figure 4 shows the departures from the fitted Fulcher equation for Standard Reference Material No. 711, a lead-silica glass. Data from eight laboratories is represented, covering a viscosity range from 10^2 to 10^{13} poises. Note that, with the exception of one laboratory, the data follows the equation within 10% of viscosity or 5^oC in temperature. The data for laboratory G obviously represents a consistent error and another set of constants in the Fulcher equation would fit their set of data with good precision.

The Fulcher expression is hyperbolic in nature. A' and T_o are the asymtopes for viscosity and temperature, respectively. Usually T_o is well above room temperature, the value for reference glass No. 711 being 152^oC. This, of course, implies that the viscosity at and below this temperature is infinite. This is not the case. Pearson[4] measured the viscosity of soda-lime-silica glass at room temperature as being 2×10^{20} poises. This points out a

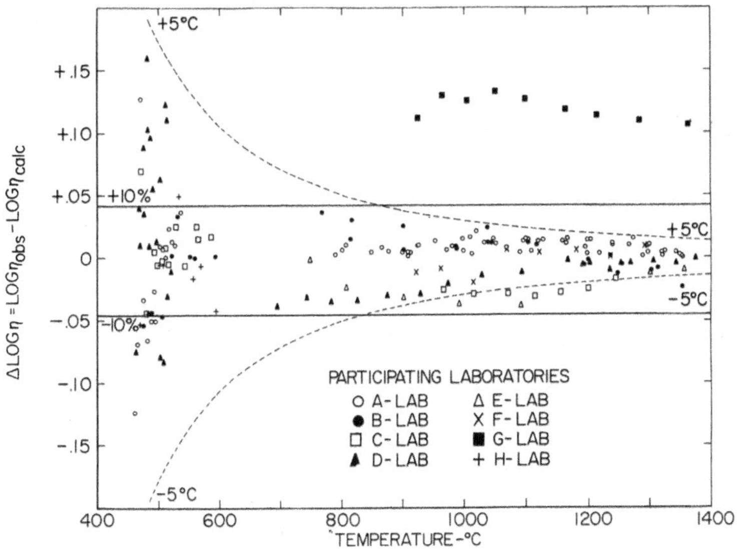

Figure 4. Departures of viscosity data fitted to the Fulcher equation.

limitation of the Fulcher expression.

TIME DEPENDENCE - TRANSFORMATION RANGE

As we shall see later, one of the complicating factors in determining the annealing and strain points relating to exact viscosities lies in the fact that a time dependence comes into play in this region. It is, therefore, necessary to examine phenomenologically the behavior of glass in what is known as the transformation range.

At sufficiently high temperatures, glass experiences a volume change with temperature that is due to two mechanisms. First, as in most solids, a volume increase is brought about owing to increasing amplitudes of vibrations of atoms with increasing temperature. This mechanism responds instantaneously to change in temperature, over all temperature ranges. A second mechanism, operable only at high temperatures, is associated with an atomic redistribution that changes continuously with temperature. This atomic redisbribution contributes the major portion of the total volume change.

Actually, the volume changes can be expressed in terms of expansion coefficients: $\alpha_T = \alpha_V + \alpha_S$, where α_T is the total expansion coefficient, α_V is the vibrational coefficient, and α_S is the atomic distribution or structural coefficient. As stated before, α_V responds instantaneously in all temperature regions whereas α_S becomes significantly time dependent below the annealing point of glass. To gain an idea of how these two mechanisms contribute to the contraction curve for glass, refer to Figure 5, an idealized density-temperature plot for glass. When both expansion mechanisms are fully operable, glass follows the density-temperature line which we call the equilibrium line.

Because α_S is a time dependent mechanism, instanaeous cooling permits only α_V to be operative, and hence the glass will follow a cooling curve shown by the dashed line. If cooled infinitesimally slowly, both mechanisms play complete roles, and the equilibrium curve is followed. Now, since the relaxation time for α_S becomes strongly temperature dependent (it increases by approximately a factor of ten for every 20°C drop in temperature), it becomes obvious that the contraction curve and the room temperature density are cooling rate dependent. Naturally, slower cooling rates produce higher densities as shown in Figure 5. Many physical properties are affected. A higher density will be accompanied by higher numerical values for Young's modulus, electrical resistivity, index of refraction, and viscosity.

The transformation range indicated on Figure 5 can be defined in terms of the measurable activity of the structural coefficient α_S. Below the lower temperature limit, in this range, α_S acts so slowly that its contribution to volume change cannot be experimentally detected; conversely, above the upper limit, it acts so fast that experimentally it cannot be distinguished from α_V. The width of the transformation range in temperature depends on the chemical makeup of the glass and the precision of the experimental technique used to evaluate volume change.

Now that we have introduced the idea of the transformation range and appreciate that a time-dependent mechanism exists in glass, it becomes fairly obvious that if we were to interrupt cooling and hold at a temperature near the lower part of the transformation range a density

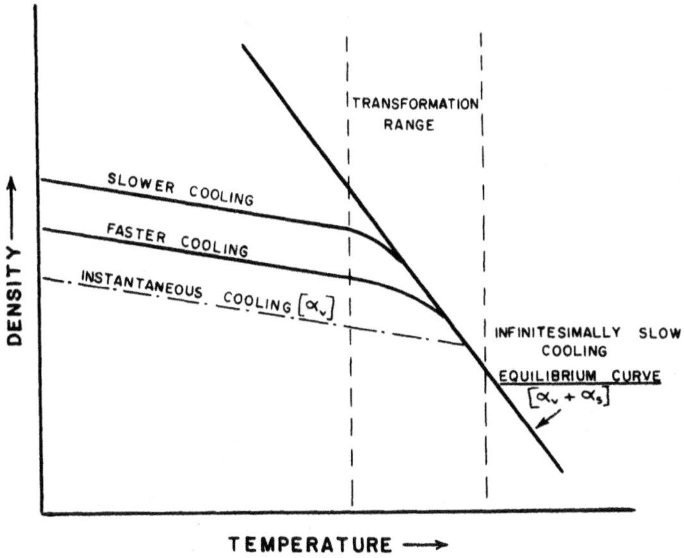

Figure 5. Idealized density-temperature plot for glass.

increase will occur with time. Or, the same thing will happen if we reheat a glass so cooled and, on reaching this temperature, establish constant temperature conditions. Since the structure is becoming progressively more compact, an increase in viscosity with time is expected. This is exactly what Lillie[5] observed in 1933. Figure 6 shows the change in viscosity of a soda-lime-silica glass fiber at 498°C after being reheated from a freshly drawn state.

EXPERIMENTAL METHODS

a) General

It would certainly be outside the scope of this paper to give a complete comprehensive survey of the methods of measuring viscosity. Furthermore, it would hardly hold the interest of the reader. What is planned is a discussion of the more conventional and familiar methods with some input on the newer methods that have proven successful at Corning. Special emphasis will be devoted to the long-standing softening and annealing point methods.

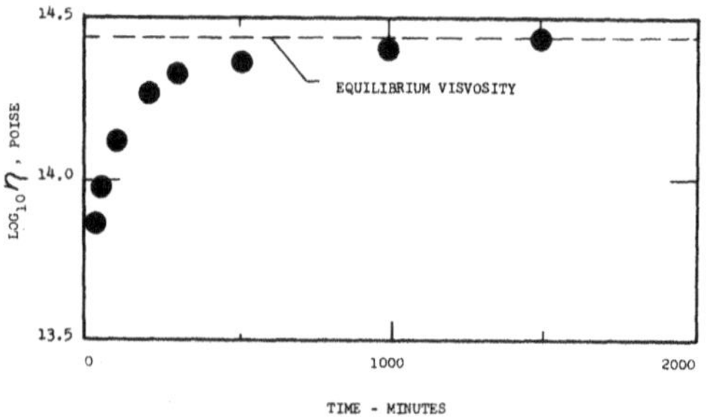

Figure 6. Time dependent viscosity of a soda-lime glass at 498°C.

b) Low Viscosity Range - 10 to 10^7 Poises

The Margules viscometer is probably the most widely used* instrument for determining viscosities below the softening point level. Pioneered by H. R. Lillie[6] in the early 1930's, this technique offers high precision and accuracy of measurement over a wide viscosity range.

The method, also known as the concentric cylinder technique, is shown schematically on Figure 7. In general, the molten glass is contained in a platinum - 10% rhodium crucible. A spindle of special construction and of the same alloy is inserted into the molten glass at the center of the crucible. Details of the crucible, spindle, and immersion depth are shown on Figure 8. A furnace is provided for temperature control and provides uniform heating throughout the volume of test glass. In the scheme at Corning, the crucible is rotated at constant angular velocity during measurement which produces a torque on the spindle. It is also experimentally feasible to rotate the spindle with the crucible fixed. A method for measuring the torque must be provided and our approach to this is to use calibrated wire suspensions and an optical lever. A

---

* All eight laboratories that participated in the round robin study referenced in Figure 4 used a Margules type viscometer.

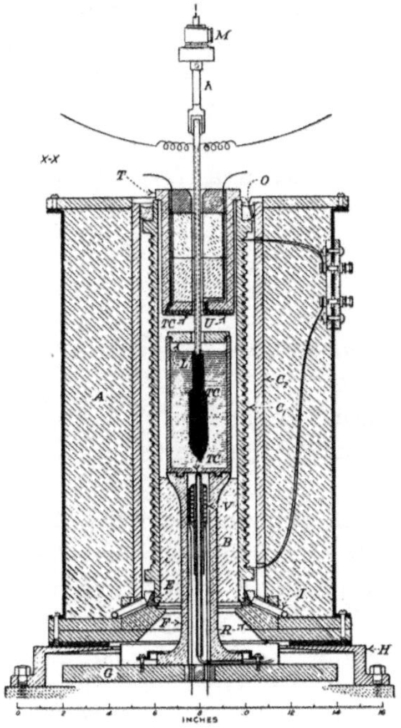

Figure 7. Margules viscometer.

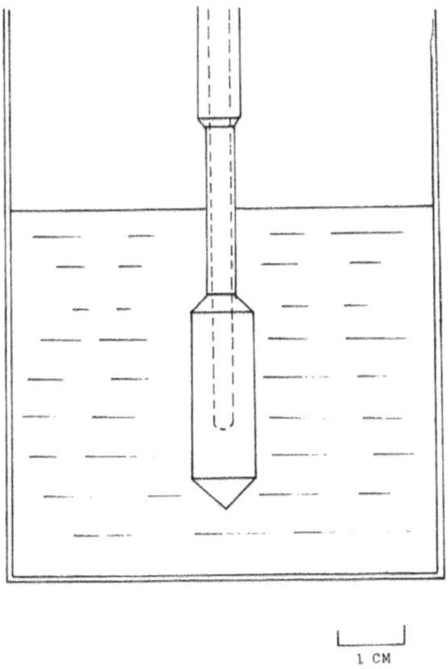

Figure 8. Crucible and spindle details in the Margules viscometer.

thermocouple assembly fits in a hole at the center of the spindle with the junction near the end for temperature measurement. Approximately 200 cm^3 of glass is used for measurement.

The formula for calculating viscosity is:

$$\eta = \frac{KT}{\Omega} \tag{4}$$

where: η = viscosity
T = torque
Ω = angular velocity
K = apparatus constant

This constant rotation method works well up to about 10^4 poises above which level the torques get excessive. For measurements at higher viscosities the aperiodic return method is used. To accomplish this, the crucible is rotated until an angle representing the maximum range of the optical lever is reached. Rotation is then stopped and the wire suspension is allowed to drive the spindle back to the rest point (zero torque). The elapsed time between two selected angles (measured from the rest point) is measured during the return. The viscosity is then calculated from the following equation:

$$\eta = \frac{K'\Delta t}{\ln\left(\frac{\theta_1}{\theta_2}\right)} \tag{5}$$

where: Δt = elapsed time
θ_1 and θ_2 = the selected angles, θ_1 being the larger
K' = apparatus constant.

Thus with these two techniques the complete curve from about 10 poises to 10^7 poises can be determined with an overall precision and accuracy of a few per cent. From this curve the melting point and working point are read as defined in Section IIIA. Normally these points are not used for quality control. This curve is usually run only for periodic checks or when some manufacturing difficulty is experienced.

c) Intermediate Viscosity Range - 10^4 to 10^9 Poises

1. The Softening Point

The most widely used primary control viscosity test is the softening point. This test measures viscosity deviations very precisely for a given composition but it does not indicate a viscosity level that is constant for all glasses. It is a standard method of test adopted by ASTM as Designation C338 [7].

Refer to Figure 9, which shows the details of the softening point apparatus. This is a test which uses a flame drawn fiber (T) as a test specimen. To quote from the ASTM document: "This method of test is intended for determining the softening point of a glass by determining the temperature at which a uniform fiber of the glass, 0.55 to 0.75 mm. in diameter and 23.5 cm long, elongates under its own weight at a rate of 1 mm. per minute when the upper 10 cm. of its length is heated in a specified furnace at the rate of approximately 5C per minute." The elongation is measured with a suitable optical extensometer.

Fiber uniformity requirements and furnace details will not be discussed here. Those interested may consult the ASTM procedure.

At first glance this test would appear to yield a common viscosity for glasses. However, let us examine the tractive force that makes the fiber elongate. This consists of two parts: (a) a downward force due to the weight of the fiber (b) an upward force due to surface tension. Mathematically this can be written:

$$F_t = \frac{\pi d^2 L \rho g}{4} - \frac{\pi \sigma d}{2} \qquad (6)$$

where: F_t = tractive force, dynes
d = nominal fiber diameter, 0.065 cm.
L = length of fiber from center of furnace to lower end, 16.2 cm.
g = acceleration of gravity, 980 cm/sec^2
σ = surface tension, dynes/cm.
ρ = density, gm/cm^3

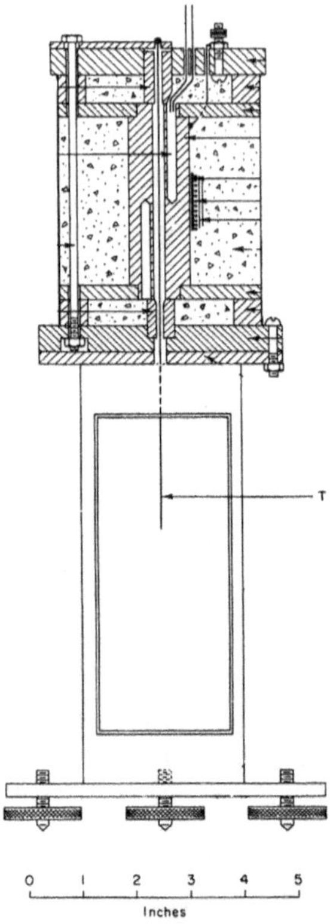

Figure 9. Softening point apparatus.

The viscosity, η, in poises, for the fiber test is given by this equation:

$$\eta = \frac{4 \ F_t \, l}{3\pi \ d^2 \, e} \tag{7}$$

where: l = effective length of fiber, 6.9 cm.
e = defined elongation rate, 0.00167 cm./sec (1 mm/minute).

The effective length of the fiber is determined by a procedure described in the annealing and strain point test.[8] Combining equations (1) and (2), substituting values, and solving for viscosity, we get:

$$\eta = 2.18 \times 10^7 \left[\rho - \frac{\sigma}{516} \right] \qquad (8)$$

This equation now points out the softening point viscosity dependence on density and surface tension of the test glass. Numerical examples for two extremes and a nominal soda-lime-silica glass are given in the table below:

Glass	ρ	σ	η	$\log_{10} \eta$
Vitreous silica	2.2	400	3.1×10^7	7.49
Soda-lime	2.5	300	4.2×10^7	7.62
High lead	6.2	200	1.3×10^8	8.11

The softening point test is generally capable of a precision of ± 2°C. Attempts to push the test to fractions of a degree require extreme fiber dimensional uniformity. Also, equation (3) shows that one percent changes in density and surface tension will make a 1.6% viscosity level shift or 0.3°C. shift exclusive of any real viscosity-compositional change.

2. The Parallel Plate Viscometer

A test which brackets the softening point and permits direct viscosity measurements in this intermediate range is the parallel plate viscometer. This method has now been in use at Corning for several years and has proven quite useful. The test was recently described by Fontana[9].

The test procedure is very simple but experimentally difficult to obtain high accuracy. Basically a specimen of the glass in the form of a solid right cylinder is viscously deformed between two parallel rigid plates in uniaxial compression. A drawing of the apparatus configuration is shown on Figure 10. The load applying plates are of inconel, one of which is supported on a ceramic tube, the other coupled to a ceramic push rod for applying the load. The basic measurement is the determination of the change of the height of the specimen with temperature and time. This is accomplished with a linear variable differential transformer (LVDT). Vitreous silica transfer rods are used for coupling to the LVDT, thus eliminating thermal expansion corrections.

The equation applicable for this method of test is:

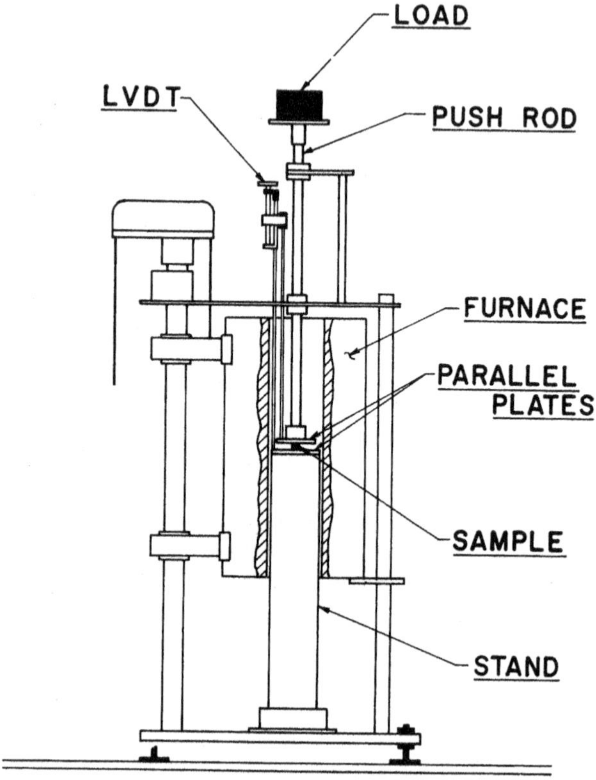

Figure 10. Parallel plate viscometer.

$$\eta = \frac{2\pi \ Mg \ h^5}{3 \ Ve \ (2\pi \ h^3 + V)} \tag{9}$$

where: M = applied load, gms.
h = height of the specimen at the particular time of interest, cm.
V = volume of the specimen, cm^3.
e = rate change of specimen height at time of interest, cm/sec.

This equation points to one of the major difficulties with the method. The high exponential dependence on thickness makes measurement accuracy very critical. Another, not so obvious, problem is that the technique is very sensitive to small temperature inequalities within the specimen. Even so, gratifying performance has been obtained, as shown in Figure 11 where parallel plate data is shown to be in excellent agreement with the standard curve for two NBS reference glasses.

Preferable specimen sizes are small, ranging from 2 to 4 mm. in thickness and from 3 to 6 mm. in diameter. This is an advantage particularly for devitrifying frit glasses. These glasses can easily be pelletized and the dynamic behavior determined on heating. An example is shown in Figure 12 where two lots of the same type of frit, unintentionally slightly different in composition, are shown to exhibit measurably different viscous behavior. The viscosity curves rise sharply as crystalline development accelerates.

d) High Viscosity Range - 10^8 to 10^{15} Poises-
Annealing and Strain Point Tests

1. Historical Background

The history of the annealing and strain points is an interesting one. In 1920 Littleton and Roberts[10] introduced a method of viscously deforming a test rod of glass under tension from which they correlated rates of twist with stress release times. The terms "annealing point" and "strain points" were thus introduced and served as primary definitions for future work.

The torsion test suffered from serious disadvantages. The test rods were time consuming to prepare and frequently broke during test. In addition, the testing time was prohibitive.

As an alternative scheme which overcame these difficulties, Lillie[11] in 1931 introduced a fiber test, which is basically the same test used today. Using fibers from glasses adopted as standards from the torsion tests, annealing and strain points were measured by attempting to arrive at the same viscosity that the torsion tests implied. Slightly different values were then assigned and were chalked off as due to apparatus differences.

Up to this time all torsion and fiber tests were conducted on cooling at a nominal rate of 4^oC per minute.

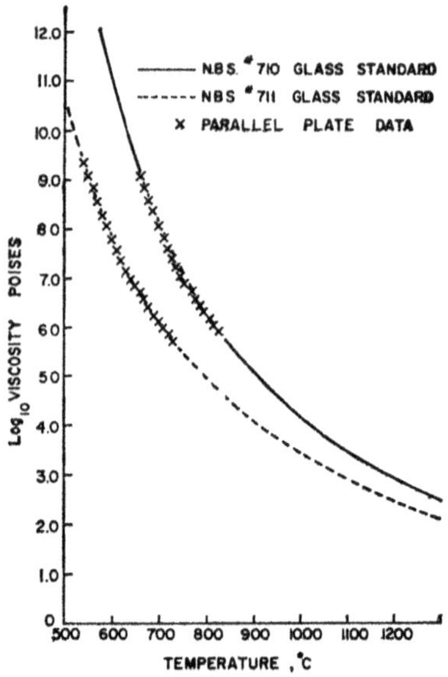

Figure 11. Agreement of parallel plate viscometer data with standard curves for reference glasses 710 and 711.

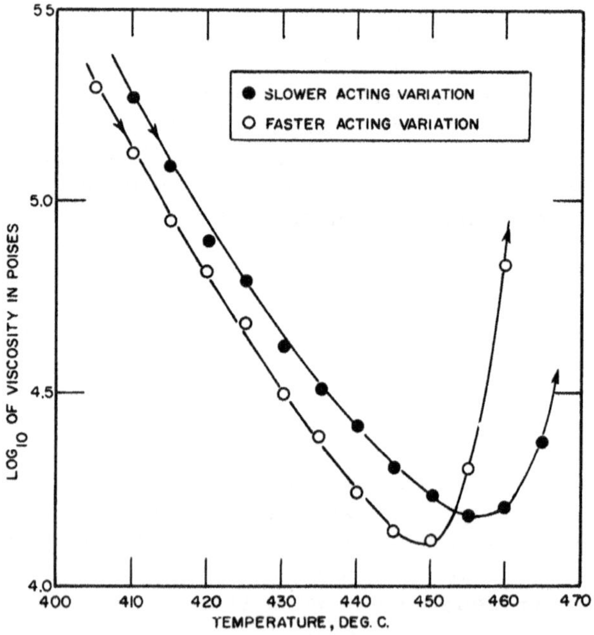

Figure 12. Parallel plate viscometer data for frit glasses.

But to pinpoint the viscosity at the newly assigned values, special tests were carried out at constant temperature. This, as we know now and have discussed in Section 4, will not lead to the same viscosity value due to the time-dependent nature in this temperature range. This was not understood at that time. However, the constant temperature values of $10^{13.4}$ poises and $10^{14.5}$ poises were adopted for the annealing and strain point viscosities, respectively, and an arbitrary 5°C correction was applied to the standard values to compensate for the fact that the elongation rates were not the same at the same temperature under the two thermal conditions.

In 1954 when transformation range phenomena were better understood, Lillie[11] corrected the problem and with the old standard glasses redefined the annealing and strain point viscosities as $10^{13.0}$ and $10^{14.5}$ poises, respectively. To "round and level off" slight adjustments were necessary to the established values and he was criticized for this.

Things were still not in a happy state with this test. The real problem was that laboratories could not agree on values using an assigned viscosity as a basis. During all this, ASTM Committee C-14 was revising the test procedure according to the most recent input and generally accepted truths. Then, in 1962, this committee faced up to two facts: (a) viscosity as such is not required for practical annealing (b) what is required is good interlaboratory agreement. By 1964 a new ASTM method of test was established which eliminated reference to viscosity and recommends direct calibration of test apparatus with standard reference glasses (seven are now available), certified by round-robin testing, and sold by the Bureau of Standards.[12] This is about where things stand today with this test.

2. The Fiber Elongation Test

The basic fiber elongation apparatus is shown on Figure 13 as recommended in ASTM Designation C336-69[13]. A uniform flame drawn fiber, 20 in. long, between 0.55 and 0.75 mm in diameter, is suspended through a specified furnace. A kilogram load is attached to the end of the fiber, the vertical motion (viscous elongation) of which is observed with a suitable extensometer during the test.

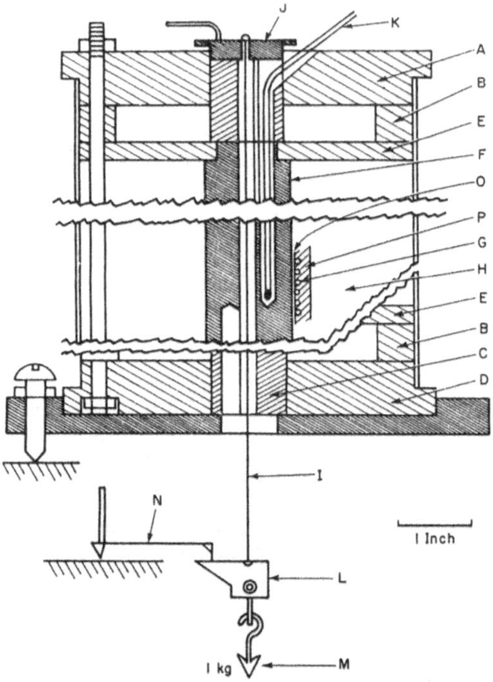

A, B, C, D—Made of asbestos-cement (Transite or equivalent).
E—Webbed asbestos-cement (Transite or equivalent) disk.
F—Copper core 1$\frac{3}{8}$ in OD by 12 in. total length.
G—91 turns No. 22 Nichrome V wire (or equivalent).
H—Diatomaceous earth.
I—Sample fiber.
J—Stainless steel support disk
K—Thermocouple.
L—Lever platform.
M—Load.
N—Optical lever.

Figure 13. Annealing and strain point apparatus.

The procedure calls for heating to about 25°C above the estimated annealing point and then establishing a 4 ± 1°C per minute cooling rate. During cooling, viscous elongation rates are experimentally determined as a function of temperature. Observations are continued to a temperature 10°C below the annealing point. A semi-log plot is constructed as shown on Figure 14. The <u>annealing point</u> is that temperature where the elongation rate is:

$$\frac{2.5 \times 10^{-6}\, l}{d^2} \quad \text{(implying a viscosity of } 10^{13} \text{ poises)} \qquad (10)$$

where: l = effective length of fiber, cm.
d = fiber diameter, cm.

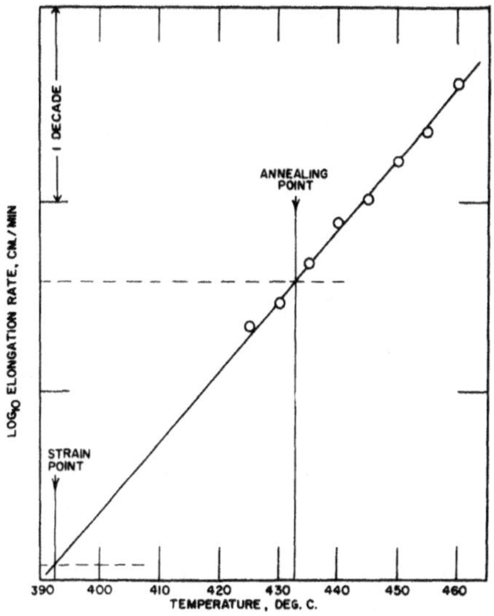

Figure 14. Typical elongation rate vs. temperature plot for annealing and strain point determination.

Or, the elongation rate can be established by observation of a standard reference glass fiber of the same diameter under the same conditions.

The strain point is determined by extrapolating the straight line generated by the semi-log plot to an elongation rate 0.0316 times that established at the annealing point.

The fiber test has several problems. First, the fiber experiences a rather severe temperature gradient making the "effective length" determination a necessity. Secondly, thermal expansion corrections must be applied for both the furnace and the specimen in the usual set-up. Delayed elastic effects have not been evaluated. Nevertheless, the precision of the test is quite good. Annealing points can be measured with a precision of about $\pm 3^{o}C$ and strain point to $\pm 5^{o}C$.

3. The Beam Bending Test

Corning's entry into glass-ceramics prompted the need to look for an alternate method of test for annealing and strain points. Quite simply, by their very nature, these compositions could not be flame worked due to devitrification problems. However, they can be cast and test specimens can be formed by sawing and mill grinding. Beam bending looked promising. Jones[14] had done some work with this method but failed to establish good agreement with the fiber technique.

Work published in 1962[15] established good agreement with fiber data for equilibrium viscosities as Figure 15 demonstrates. The line on this graph represents a Fulcher fit to Bureau of Standards fiber data.

The test has several advantages: (a) the specimen is in a uniform temperature zone (b) expansion corrections are negligible (c) measurements on very hard glasses are easily handled (d) flame drawn specimens can be measured as well (e) other compositions that might be changed by flame working could be measured in the "as formed" state (such as water content changes).

Because of these advantages, ASTM adopted beam bending as a tentative standard (Designation C598) for measuring annealing and strain points in 1967.

Before going further into annealing and strain point measurements, the basic apparatus should be described. A cut-away drawing is shown on Figure 16. A beam of uniform cross section is supported on an alumina muffle, a load is applied at the mid-span with a flame-bent sapphire hook, and an LVDT is used to measure mid-point deflection. Viscosity is calculated from the equation below:

$$\eta = \frac{g\,S^3}{2.4\,I_c v}\left[M + \frac{\rho\,AS}{1.6}\right] \qquad (11)$$

where: S = support span, cm.
I_c = cross-section moment if inertia of test beam, cm^4
v = midpoint deflection rate, cm/min.
A = cross-section area of beam, cm^2
ρ = density of glass, gm/cm^3.
M = applied load, gm.

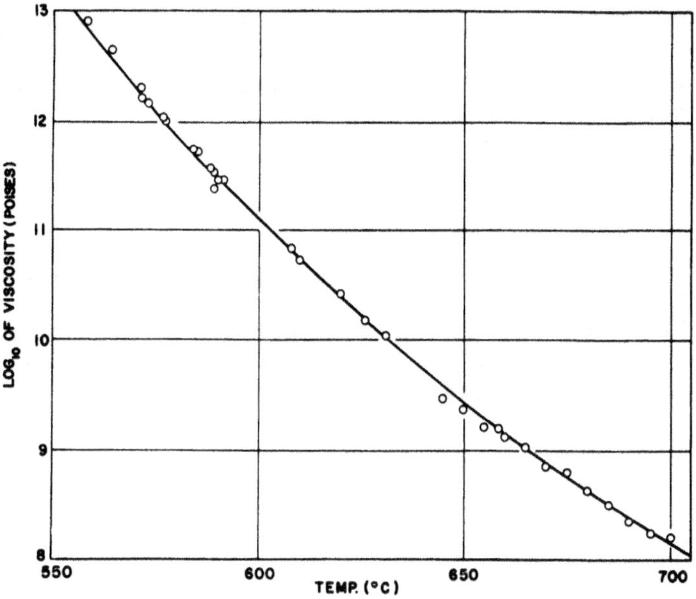

Figure 15. Agreement of beam-bending viscosity data with the curve established by fiber elongation.

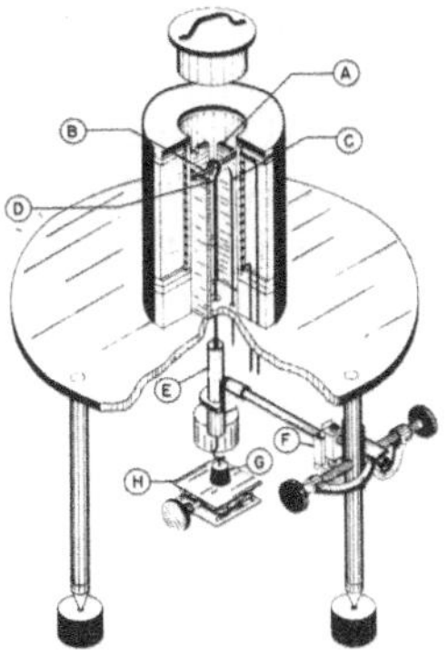

Figure 16. Cutaway drawing of beam-bending apparatus showing (A) alumina support stand, (B) specimen beam, (C) thermocouple, (D) flame-bent sapphire loading rod, (E) linear variable differential transformer, (F) clamping arrangement for transformer, (G) weight, (H) small laboratory jack.

Actually any regular geometrical cross-section can be used for the test beam. Cross-sections successfully employed are: round, elliptical, triangular, square, rectangular, and tubes. For square cross sections, a 2.5 x 2.5 mm size is appropriate for the range 10^{12} to 10^{13} poises and a 7 x 7 mm size is suitable for the 10^8 to 10^9 poise range.

To measure annealing and strain points the procedure is completely analagous to the fiber method. Midpoint deflection data is determined on cooling from 25°C above the annealing point and the annealing point is the temperature where the midpoint deflection rate is:

$$\frac{2.67 \times 10^{-11} S^3 M}{I_c}$$

or that established by direct calibration.

The strain point is established again by extrapolation to a rate 0.0316 times that at the annealing point.

4. More About Annealing and Strain Points

Figure 17 represents a plot of viscosity data taken by beam bending at the cooling rate of 4°C per minute prescribed for the annealing point test. The solid line is the established equilibrium curve for this glass, standard reference No. 710. The dashed line is the usual range of data used for the standard determination. There are two facts that need to be pointed out from this plot:

1a) The data from the annealing point test represents non-equilibrium conditions. This is consistent with the density behavior shown in Figure 5.

2b) The annealing point for this glass is 546°C. The beam bending test indicates a viscosity of 1.5×10^{13} poises at this temperature. The fiber test indicates 1.0×10^{13} poises.

The first point indicates that annealing and strain point data should not be combined with equilibrium data for curve fitting where high precision is required. The second point again points out the difficulty of establishing a viscosity at the annealing point, although the value is

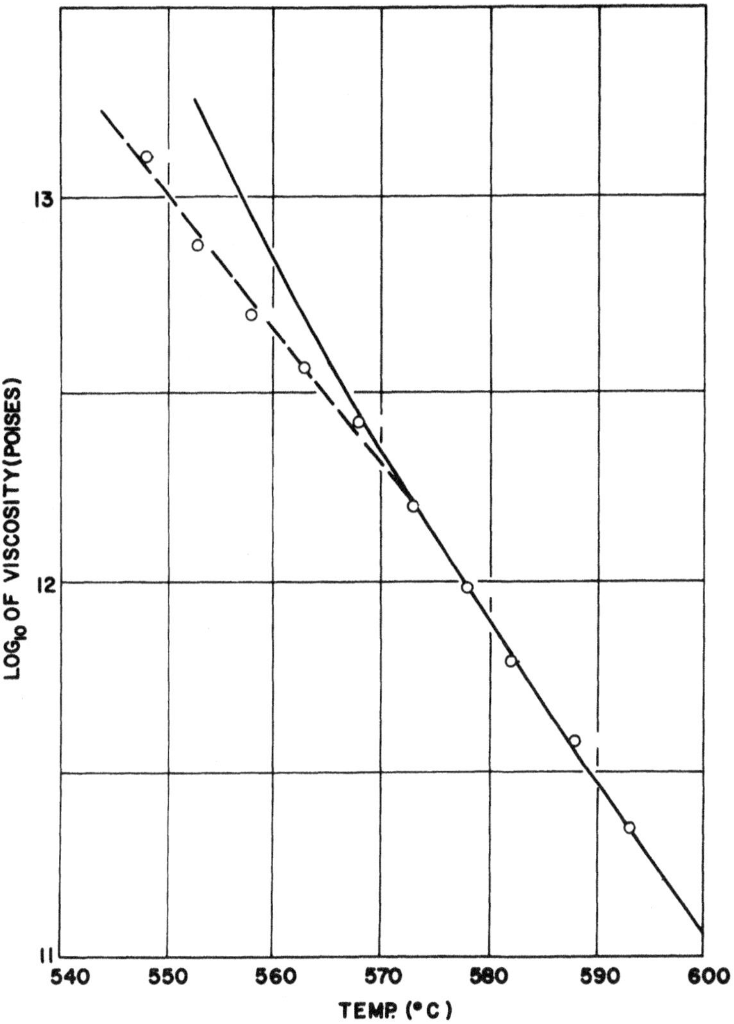

Figure 17. Beam-bending viscosity data taken at 4°C per minute cooling rate. The solid line represents equilibrium viscosity.

somewhere between 1.0 and 1.5 x 10^{13} poises.

The strain point really represents an experimentally undefined viscosity level. The standard extrapolation represents neither an equilibrium condition nor the point through which the data would pass if observations could be continued to lower temperatures (the elongation rates get extremely small). Figure 18 depicts this effect for standard reference glass 710.

The annealing point is that temperature near which slow cooling is started during annealing and the strain point is that temperature below which this rate may be increased.

ASTM defines the annealing point as that temperature at which stress release is substantially complete in a matter of minutes. The strain point definition reads the same but takes a matter of hours.

The strain point is not the lower limit for stress release! Stress release occurs at temperatures well below the strain point. Figure 19 should convince one of this.

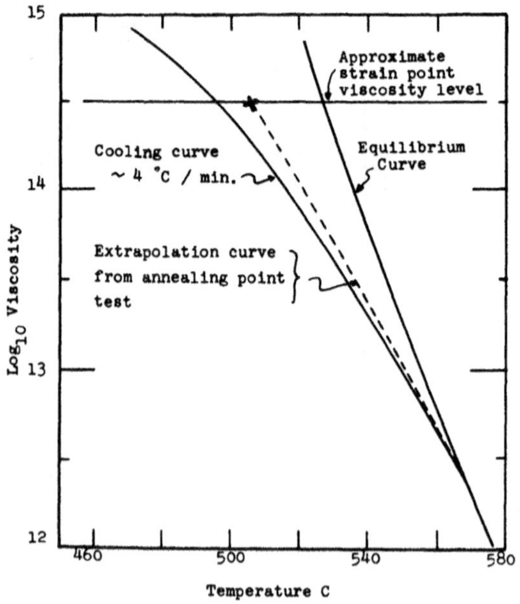

Figure 18. The strain point of reference glass 710 compared to both equilibrium conditions and 4°C per minute cooling.

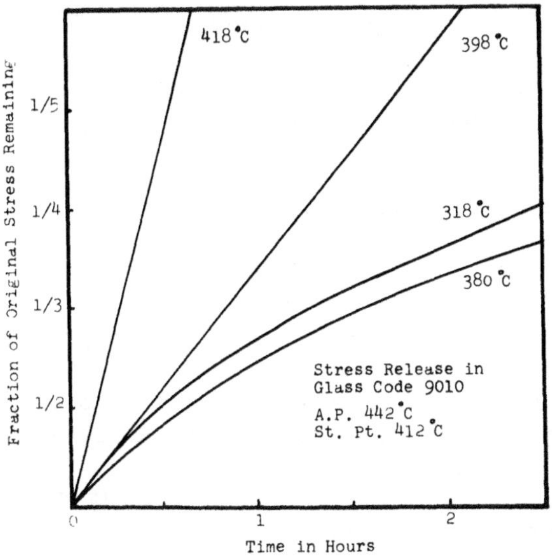

Figure 19. Stress release curves for Corning code 9010 glass.

This data was obtained photoelastically by observing pre-stressed glass specimens in a furnace at the indicated temperatures.

The setting point in sealing is tied to the strain point, generally considered 5°C above it. This is the effective temperature where a glass sealed to another material transforms from a viscous to an elastic body on cooling.

SUMMARY

The rheological behavior of glass has been discussed in a basic manner with the hope of establishing a working knowledge for the uninitiated. The reference points have been described in detail to correct some of the misconceptions that are commonly heard. Experimental apparatus generally used for measuring viscosity and some new approaches have been described to provide a general familiarization.

BIBLIOGRAPHY

An excellent and very complete bibliography on the subject has just been published:

"Viscosity-Temperature Relations in Glass", published by the International Commission on Glass, 1970.

REFERENCES

1. Physical Properties of Glass, J. E. Stanworth, Oxford, London, 1950.
2. Fulcher, G. S., "Analysis of Recent Measurements of Viscosity of Glass", J. Am. Ceram. Soc., 8, 339, 789 (1925).
3. NBS Misc. Pub. 260-11, U. S. Dept. of Commerce, National Bureau of Standards.
4. Pearson, S., "Creep and Recovery of a Mineral Glass at Normal Temperatures", J. Soc. Glass Tech., Vol. 36, 105-114 (1952).
5. Lillie, H. R., "Viscosity-Time-Temperature Relations in Glass at Annealing Temperatures", J. Am. Ceram. Soc., 16, 619 (1933).
6. Lillie, H. R., "Viscosity Measurements in Molten Glass", Journal of Hreology, Vol. 3, 121-126, (1932).
7. ASTM Designation C338-57, Standard Method of Test for Softening Point of Glass, Part 13, p. 294, 1969.
8. ASTM Designation C336-69, 6.2.1. Method A, Part 13, p. 292 (1969).
9. Fontana, E. H., "A Versatile Parallel Plate Viscometer for Glass Viscosity Measurements to 1000°C", to be published in the Am. Ceram. Soc. Bulletin.
10. Littleton, J. T. and Roberts, E. H., "Method for Determining Annealing Temperatures of Glass", J. Opt. Soc. Amer. 4, 224-229 (1920).
11. Lillie, H. R., "Re-Evaluation of Glass Viscosities at Annealing and Strain Points", J. Am. Ceram. Soc., 37, 111-117, (1954).
12. NBS Misc. Pub. 260.
13. ASTM Designation C336-69, Standard Method of Test for Annealing Point and Strain Point of Glass, Part 13, p. 288, (1969).
14. Jones, G. O., "Determination of the Elastic and Viscous Properties of Glass at Temperatures Below the Annealing Range", J. Soc. Glass Technol. 28, 432-462, (1944).

15. Hagy, H. E., "Experimental Evaluation of Beam-Bending Method of Determining Glass Viscosities in the Range 10^8 to 10^{15} Poises", J. Am. Ceram. Soc. 46, 95-97 (1963).

Dispersion, Stress-Optical Effects In Glass, Optical Glasses

Joseph L. Rood
Physics Department
University of Lethbridge
Lethbridge, Alberta, Canada

DISPERSION

INTRODUCTION

The index of refraction of an optical material such as glass (or any other material, optical or not) varies with wavelength: this phenomenon is called dispersion. Sometimes the term dispersion is specifically applied to the derivative $dn/d\lambda$, although this usage is not always followed. A typical qualitative curve of n, index of refraction, versus λ, wavelength, is given in Figure 1 for a large portion of the electromagnetic spectrum for a material largely transparent in the visible region.

The main region of interest in this study is indicated in Figure 1, consisting of the near ultraviolet, visible, and near infrared portions of the electromagnetic spectrum. Less frequently for "optical" materials, and especially for glasses, the vacuum ultraviolet, far infrared, and microwave regions may be of interest. While this study is concerned mainly with optical glasses, much of what is given also applies to other optical materials such as glasses other than optical glasses, optical crystals, and optical plastics.[1]

Optical glasses differ from other types of glasses in having a high degree of index homogeneity and very little

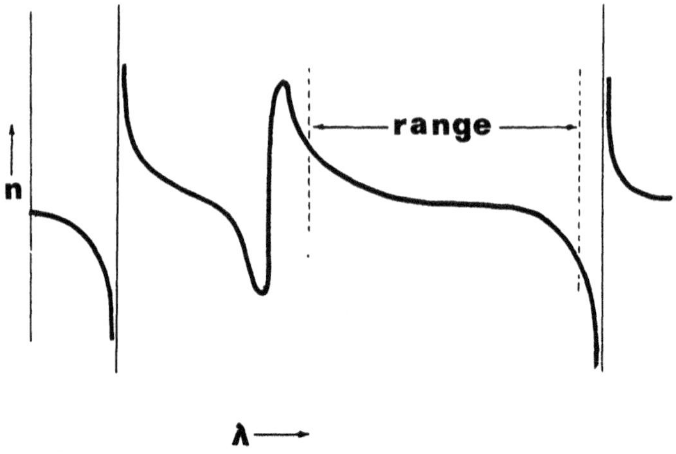

Figure 1. Dispersion curve for an optical material. The range of most interest is from the near ultraviolet to the infrared, 0.25 μ to 4 μ. (μ = micron = 10^{-6} meter.)

strain, requiring precision annealing. They usually also have very low absorption in the spectral regions of interest. A property of optical glasses of great interest is their dispersion. This is also the case of ophthalamic glasses, which may or may not be of optical glass quality.

The relation between dispersion and structure is a complicated one not readily described by a simple theory. It is a subject of considerable practical importance, however, e.g., in the correction of chromatic aberrations of optical instruments. Some basic theory will be given later in Section IE.

OPTICAL GLASSES IN THE VISIBLE REGION

The index of refraction of an optical material such as a glass is normally measured in the visible at a number of wavelengths. Seven standard wavelengths in the visible[2] and their sources are given in Table 1, as well as the C, D, and F wavelengths. The seven standard wavelengths are known and stable to two or three in parts in 10^6. However, it has been customary to specify indices for the C and F

Table 1. Standard Wavelengths

Wavelength, μ	Source
0.435835	Hg
0.479992	Cd
0.4861 (F)	H
0.508582	Cd
0.546074	Hg
0.587561	He
0.5893 (D)	Na
0.643847	Cd
0.6563 (C)	H
0.667815	He

spectral lines of hydrogen and the average of the D lines of sodium (actually a doublet with components at approximately 0.5890 and 0.5896μ). These three lines are not as sharp as the seven others of Table 1, and are therefore inferior for precision work. Nevertheless, it is convenient to measure the C, D, and F indices or else to calculate them (see Section IE ahead), as well as the seven standard wavelengths of Table 1, and possibly others.

In characterizing glasses, the index for the D doublet is important, as are differences of indices, or partial dispersions, and ratios of partial dispersions. The ν value or Abbe number, is

$$\nu = \frac{n_D - 1}{n_F - n_C} \tag{1}$$

which is useful in giving a rough estimate of the dispersion. ν values for optical glasses vary from about 20 to about 80.

A glass is identified by the D index and the ν value. An arbitrary division has been made into 19 different types according to the range of n_D and ν.[1] Thus DBC 623569 is a dense barium crown glass whose nominal n_D is 1.623 exactly, and whose nominal ν is 56.9, also exactly. Likewise the nominal for EDF 689309 (extra-dense flint) is $n_D = 1.689$ and $\nu = 30.9$. Indices of refraction can be measured meaningfully to five or six decimal places, and ν values to two decimals. The nominals merely identify the type of glass, and represent production targets.

The 19 different glass types referred to in the preceding paragraph can be generally categorized as crowns or flints, depending mainly on the ν value. Thus for $n_D < 1.60$, glasses with $\nu \geq 55$ are called crowns, and those with $\nu < 55$ are flints. For $n_D \geq 1.60$, the dividing ν value is 50: crowns are glasses with $\nu \geq 50$, and flints have $\nu < 50$. Crowns are typically alkali silicate glasses, often with B_2O_3 and/or CaO added, and flints are lead alkali silicates but many other compositions are possible. Morey[3] gives complete compositions for a large number of optical and other glasses. For flint glasses the index increases with increasing PbO content. Most optical glasses have SiO_2 as the glass former, but P_2O_5 and B_2O_3 glasses are also used. Phosphate and borate glasses have different partial dispersion ratios (see Section IC) than do silicate glasses.[4]

Figure 2 shows dispersion curves for three optical glasses: EDF689309, DBC623569, and BSC517645 (borosilicate crown); also fused silica, plus two optical crystals, crystalline quartz (O Ray), and fluorite (CaF_2).[5] These curves are merely given as semi-quantitative indications of dispersion. For the fifth or sixth place index accuracy sometimes required, analytical rather than graphical means of analysis are necessary (See Section IC).

MEASUREMENTS

It is only recently, since the advent of extensive use of computers in lens designs, that optical systems have been sufficiently good to require sixth place accuracy for indices of refraction. The problem is made worse because the designs depend largely on differences of indices, and in particular on partial dispersion ratios, which are quotients of differences of indices. Consider P_{FD}, involving the difference of the F and D indices:

$$P_{FD} = \frac{n_F - n_D}{n_F - n_C} \tag{2}$$

a partial dispersion ratio important in determining whether or not optical elements can be useful in correcting for chromatic aberrations.[4] It is desirable to distinguish differences from one optical material to another as small as ± 0.0005 in P_{FD}, which corresponds to uncertainties in the indices themselves of about ± 0.000002. Partial dispersion ratios involving indices at wavelengths other than C, D, and F are also used in evaluating optical materials.

Indices are measured by refractometry with a precision spectrometer, using the method of minimum deviation, at a number of different wavelengths. As might be expected, rather elaborate precautions are necessary to reduce random and systematic errors if accuracies in the fifth place are required, let alone the sixth.[6]

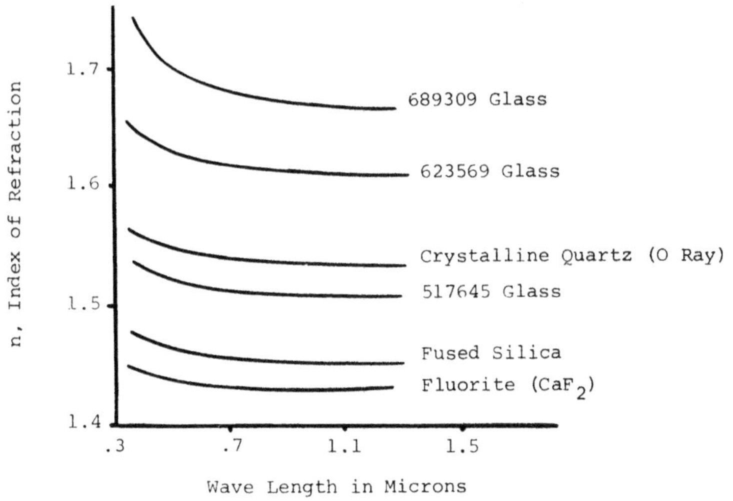

Figure 2. Index of refraction as a function of wavelength of six optical materials.

One possible source of error that should be corrected for is changing atmospheric pressure. This is because the indices are measured relative to air and the index of air changes with pressure. The changes in index on this account may be in the fifth decimal place. Table 2 illustrates the point, for a light barium flint glass. Indices are corrected arbitrarily to 29.92" air pressure as standard. In the hypothetical case shown in Table 2, n_D and n_F are actual measurements made on a piece of glass at a low pressure, and n_C at a high pressure on another day. These values are then corrected. Consider n_C, assumed to be measured at a pressure considerably above 29.92" (actually, 30.71"). The correction for n_C is positive. It is assumed the true but unknown index of the glass is unchanged. The air pressure goes up, so the index of air (assumed proportional to the density) goes up, and the measured glass index goes down, requiring a correction upwards for n_C. Similarly, n_F and n_D have been corrected downwards. Small changes in air temperature are likewise corrected for. The difference in P_{FD} in Table 2 represents an error that would make the work meaningless if uncorrected, since it is hoped to measure differences in P_{FD} as small as 0.0005.

Table 2. Effect of Atmospheric Pressure Changes

	Before Correction	After Correction
n_F:	1.57384	1.573820
n_D:	1.566459	1.566446
n_C:	1.563401	1.563412
ν:	54.29	54.42
P_{FD}:	0.7069	0.7085

It should be emphasized that temperature changes are far more important in changing indices (and introducing possible errors) than atmospheric pressure changes. Temperature coefficients of indices of refraction can be quite substantial.[7] For example, they are around 1×10^{-5} (deg C)$^{-1}$ for fused silica and dense flint glasses, but only around 1×10^{-6} (deg C)$^{-1}$ for crowns. Even more important and perhaps confusing is that for a given temperature the index of refraction depends markedly on the thermal history, with possible changes in the fourth decimal place. Thus, if one is evaluating precise indices for a quantity of optical glass, it is necessary to measure the indices on the pieces to be used, or on a sample from the same lot of glass ("monitored sample") having the same thermal history. The dispersion is not nearly as dependent on the thermal history as are the indices.

OPTICAL GLASSES IN WAVELENGTH REGIONS OTHER THAN THE VISIBLE

Indices of refraction in the near ultraviolet and in the near infrared are often required. The main difference between these spectral regions and the visible as far as measurements are concerned is that the detector in the spectrometer has to be something other than the eye, e.g., a photoconductor.

In general, the variation of dispersion from glass to glass in the near ultraviolet is greater than it is in the visible, and in the near infrared it is less. This regularity of dispersion in the infrared is most noticeable from about 1.0μ to 2.5μ.[8] In this wavelength region the derivative $dn/d\lambda$, while negative as it is in the ultraviolet and visible, is small and almost constant.

Optical materials are sometimes used to modify electromagnetic waves in regions other than the ultraviolet, visible, and infrared, in which case the designation "optical" is usually not used. An example is the microwave region; at these frequencies physical properties of interest that correspond roughly to index of refraction and absorption coefficient, respectively, at higher frequencies are the dielectric constant and the loss tangent.[9]

THEORETICAL AND EMPIRICAL RELATIONS BETWEEN INDEX OF REFRACTION AND WAVELENGTH

Classical dispersion theory,[10-13] and indeed quantum mechanical dispersion theory, has had some success in semi-quantitatively agreeing with dispersion data for crystals, but less success in the case of glasses. The reason is that dispersion is closely related to absorption and the change of absorption with wavelength, and glasses usually have more diffuse absorption bands than do crystals. Thus a simple relation between structure and dispersion is not available. A typical dispersion curve is given in Figure 1. In each of the regions of maximum absorption, "anomalous dispersion" occurs, that is, the index increases with increasing wavelength. Maxima of absorption occur for optical materials in the ultraviolet and in the far infrared, well outside of the visible and near infrared regions.

A dielectric material reacts with and affects electromagnetic radiation (described by Maxwell's Equations) because it contains charged carriers that can be displaced.[14] These carriers are electrons. Both classically and quantum mechanically the electrons behave as far as problems involving light are concerned as though they are harmonic oscillators, but with several natural frequencies, each frequency with its own damping factor. The oscillators are driven by the electric field.[15] The number of polarizable electrons in the material is important, as are their frequencies of oscillation and their damping factors. The relation between these, the angular frequency of the incident radiation ω, and the complex index of refraction is:[11,13]

$$(n - ik)^2 - 1 = \frac{e^2}{m\varepsilon_0} \sum_{j=1}^{\ell} \frac{Nf_j}{\omega_j^2 - \omega^2 + ig_j\omega} \qquad (3)$$

for ℓ absorbing frequencies, where n is the index of refraction, k the measure of absorption, e the electron charge, m the electron mass, ε_0 the permittivity of free space, N the number of atoms per unit volume, f_j the oscillator strength, ω_j the frequency of jth type oscillator, g_j the damping factor, and i is $\sqrt{-1}$. Separating into real and imaginary parts gives:

$$n^2 - k^2 - 1 = A \tag{4}$$

and

$$2nk = B \tag{5}$$

where $$A = \frac{e^2}{m\varepsilon_o} \sum_{j=1}^{\ell} \frac{Nf_j(\omega_j^2 - \omega 2)}{(\omega_j^2 - \omega^2)^2 + \omega^2 g_j^2} \tag{6}$$

and $$B = \frac{e^2}{m\varepsilon_o} \sum_{j=1}^{\ell} \frac{Nf_j \omega g_j}{(\omega_j^2 - \omega^2)^2 + \omega^2 g} \tag{7}$$

Equation (4) and (5) can be solved for n, yielding

$$n^2 = \frac{1}{2}\left[1 + A \pm \sqrt{(1 + A)^2 + B^2}\right] \tag{8}$$

Consider the change in refractive index with change in absorption, for the case where k is very small. Then equation (4) becomes

$$n^2 = 1 + A \tag{9}$$

Differentiating (9) yields

$$2ndn = dA \tag{10}$$

Thus, a small change in n can be calculated from a known small change in A by the formula

$$\Delta n = \frac{\Delta A}{2n} \tag{11}$$

The small change in A could be an additional term in the summation of equation (6) due to the addition of an absorption band of known parameters.

In these equations k is related to α, the absorption coefficient, by

$$\alpha = \frac{4\pi k}{\lambda} \tag{12}$$

where α is defined from

$$I = I_o e^{-\alpha x} \tag{13}$$

if one assumes an exponential decrease of intensity. In equation (13) the original intensity of light is I_o, and this decreases to the intensity I after the light has traversed a thickness x of an absorbing material.

It is hard to relate equation (3) to structure in a particular case. It can be roughly related, as indicated in the following example. A Li -Aℓ - Mg silicate base glass was made, with an index of refraction at 4861A of 1.5495. To this base glass was added a small amount of cerium,[16] which introduces a fairly sharp absorption with a maximum at 3100A. The change in index was calculated roughly from equation (11) to be +0.0074, to give a new index of 1.5569. The new measured index was 1.5591. This is not a good agreement to say the least, but it is of the right order of magnitude, and the rough calculation could no doubt be improved upon somewhat.

It was mentioned in Section IC that index accuracy of two in the sixth place is desired, and can probably be attained in measurements if great care is taken. This is three orders of magnitude better than the accuracy of the calculation in the preceding paragraph, and illustrates the difficulty of calculating index changes from absorption changes. Glass technologists do, however, have strictly empirical data relating index changes to compositional changes, good to the third or fourth decimal place of index.

If k is very small, and the g_j are also very small, then equation (3) becomes Sellmeier's dispersion equation:

$$n^2 - 1 = \frac{e^2}{m\varepsilon_o} \sum_{j=1}^{\ell} \frac{Nf_j}{\omega_j^2 - \omega^2} \qquad (14)$$

Equation (14) is often written in the somewhat different form (15), which is also called Sellmeier's Equation:

$$n^2 - 1 = \sum_{j=1}^{\ell} \frac{A_j \lambda^2}{\lambda^2 - \lambda_j^2} \qquad (15)$$

Equation (15), although related to the theoretical dispersion equations (3) and (14), is often considered a semi-empirical equation, used for "smoothing" and interpolating indices as a function of wavelength, at the U. S. Bureau of Standards[17] and elsewhere. Indices are measured at a large number of wavelengths in the ultraviolet, visible, and infrared, and the A_j and λ_j calculated by least squares from the data.[18]

One can use three or more terms in the sum.

The differences between the measured and calculated indices for four terms on the right of equation (15), with eight constants, average two in the fifth place for the entire wavelength range, but only about seven in the sixth in the visible. This is still more than the hoped for two in the sixth. Furthermore, it has been found[19] that a completely empirical equation such as (16) works as well for many glasses, using eight or nine constants:

$$\log n - A + B \log \lambda + C(\log \lambda)^2 + \ldots + H(\log \lambda)^7 \qquad (16)$$

The measured indices are probably preferable to any calculated ones if the measured indices have been very carefully determined, unless interpolated values are mandatory. There is no real evidence that "smoothed" indices are more accurate than measured indices of less than optimum perfection, although the smoothing process will indicate an occasional index that may be greatly in error.

STRESS - OPTICAL EFFECTS

INTRODUCTION

In the absence of strain glass is isotropic, at least on a macroscopic level. When strain is present, the glass becomes doubly refracting, or birefringent. Strain can be induced by an applied stress, often used during measurements of the stress-optical coefficient; the usual cause of strain in glass, however, is a thermal history that includes temperature gradients within the glass. This may result in residual stress and strain even when the gradient is no longer present. The strain may be removed at least in large part by annealing.

In Section IIB following, some theory[20,21] of stress-optical effects is given that is somewhat simplified, although important in practice. For more general considerations see Morey.[3]

There are two causes on a microscopic level of birefringence from strain.[3] The first is a lattice effect, since the lattice distances become greater between planes normal to a tension, but are unaltered between planes at right angles to these. This produces a negative value of B, where B is the stress-optical coefficient to be defined in Section IIB. The second cause, which produces a positive value of B, is the distortion of the outer electrons of the oxygen atoms. The latter predominates for most glasses, except for very dense flints, for which the lattice effect predominates.

THEORY OF STRESS - OPTICAL EFFECTS

The basic idea is that stress-optical effects are elastic, with strain being proportional to stress. Thus, the phenomenon is called photoelasticity. The ratio of strain to stress is called the stress-optical coefficient, B, or "Brewster's constant". B is a constant, or very nearly so within the elastic limit for a given glass, but varies considerably from glass to glass. From the discussion at the end of the previous section it may be seen that it is possible to make a very dense flint glass with B = O, or nearly so, so that even large stresses would result in almost no strain.

When linearly polarized light is incident on a strained glass, which is therefore birefringent, the velocities of light in different directions, and therefore, the indices of refraction, will be different. Suppose the glass is subjected to a net stress σ, dynes/cm^2, in the y direction. Assume σ to be a compressive stress, as indicated in Figure 3. A linearly ("plane") polarized light beam enters in the x direction, with vibration components of the electric vector in the y and z directions, since light is a transverse wave. The indices n_y and n_z will be different, the difference $n_z - n_y$ being called the "retardation". On emerging from the glass, one component will be behind the other, resulting in elliptically polarized light. The relative retardation is $(n_z - n_y)/n$, where n is the index of the unstressed glass. The relative retardation is of course dimensionless, but it often is given the units of mμ/cm (see Section IIC), in which case it is indicated by r. In any event, $(n_z - n_y)/n$ measures the strain, and σ the stress. The ratio strain/stress is a constant, B, the stress-optical coefficient, usually given in brewsters. The brewster represents a fractional difference in retardation of one part in 10^{13} per dyne per cm^2:

$$\frac{n_z - n_y}{n} = B \frac{\sigma}{10^{13}} \qquad (17)$$

where $(n_z - n_y)/n$ is dimensionless ("cm/cm"), σ is in dynes/cm^2, and B is in brewsters. Stress, like pressure, is sometimes measured in bars, where one bar is defined as 10^6 dynes/cm^2.

If $(n_z - n_y)/n$ is measured in mμ/cm, σ in dynes/cm^2, and B in brewsters, equation (17) becomes

$$\frac{n_z - n_y}{n} = B \frac{\sigma}{10^{6}} \qquad (18)$$

MEASUREMENTS

Measurements of strain and stress in glass are of considerable practical and theoretical interest.[22,23] It is usually the strain that is measured, by polarimetry, and the stress calculated. This requires a knowledge of Brewster's constant (the stress-optical coefficient) for the glass under consideration.

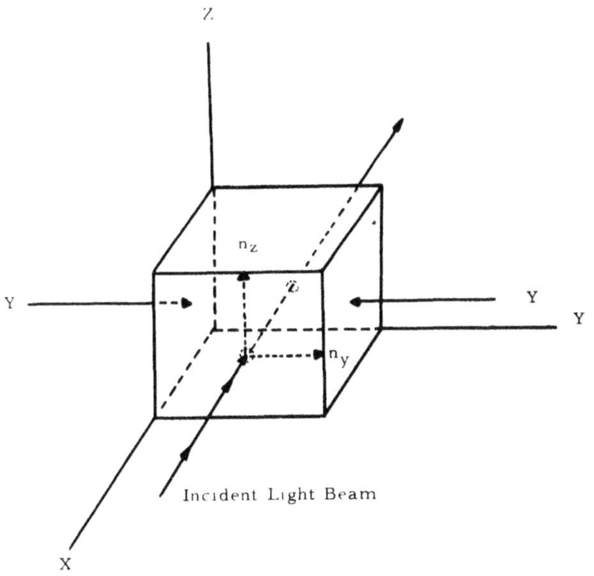

Figure 3. Glass block under compressive stress.

The index of refraction (relative to air) of a glass is the ratio of the velocity in air to that in the glass. Since n is greater than one the velocity in the glass is less than in air. When linearly polarized light enters, the two polarized components n_z and n_y have different velocities, and $(n_z - n_y)/n$ is the relative retardation. If the thickness of the glass is d, the difference in optical path of the two components is $(n_z - n_y)d$. This is compared with (divided by) the optical path nd in the unstressed glass to get $(n_z - n_y)/n$. The difference $n_z - n_y$ corresponds to the difference $n_e - n_o$ for a uniaxial (doubly refracting) crystal, where n_e is the index for the extraordinary ray, and n_o that for ordinary ray.

$n_z - n_y$ is measured, and σ calculated, assuming B known, by sending linearly polarized light, vibrating at 45 degrees to the principal directions of the stressed glass so the two components will be equal in intensity, through the glass, as indicated in Figure 4. A nicol prism is customarily used as polarizer, and also another one as analyzer. Originally the nicols are crossed, so no light gets through. Monochromatic light should be used, e.g., the Hg 546.1 mμ wavelength. The glass sample is between the crossed nicols. This produces elliptically polarized light, so some gets through the second nicol. In addition, a quarter-wave plate is placed between the nicols, and the analyzing nicol turned to again get extinction. The phase difference, which is $2\pi/\lambda$ times the path difference $(n_z - n_y)d$, where λ is the wavelength of the light used, can then be measured, and from this the stress, σ calculated. The phase difference is twice the amount of rotation needed to produce extinction.[24]

If B is not known, a known stress σ can be applied, $n_z - n_y$ measured as above, and B calculated. Some typical values of B for some optical glasses are given in Table 3.

An extra-dense flint glass having approximately 75 percent lead oxide by weight, and n_D of approximately 1.87, would have a value of B very near 0.

Table 3. Brewster's Constant for Various Glasses

Glass type	Glass no.	B, brewsters
Fused silica	458676	3.5
Crown flint	523551	2.8
Crown	524595	2.6
Light barium crown	574535	2.4
Dense flint	616380	2.9
Dense barium crown	617551	2.1
Extra-dense flint	691307	2.1
Lanthanum flint	850324	1.5
Extra-dense flint	915213	-1.4

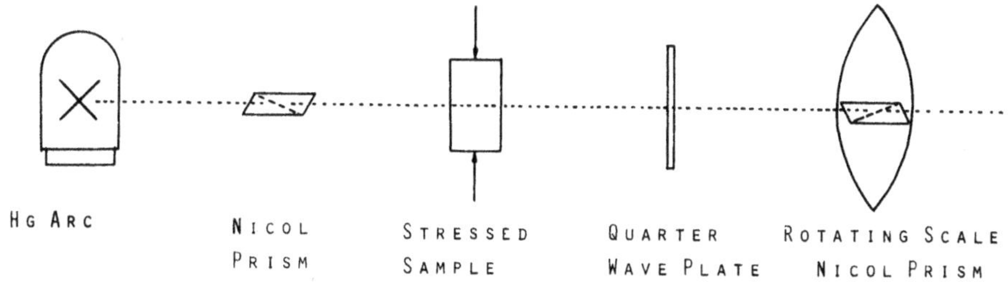

Figure 4. Apparatus for measuring retardation in stressed glass.

SUMMARY

Dispersion of light in optical materials, especially optical glasses, has been considered theoretically and experimentally. The equations of classical dispersion theory have been given; these are most readily evaluated for materials in regions of minimum absorption. In these regions they can be used as semi-empirical equations to interpolate indices to fifth place accuracy. If measurements are made to sixth place accuracy, however, it is preferable to use measured rather than calculated indices.

When glass is stressed it becomes doubly refracting, with a constant for a given glass, called Brewster's constant, relating stress and strain. This has been discussed quantitatively for a simple case. Also, the manner in which the amount of stress in glass can be measured experimentally has been given, using a method originally developed by Friedel.

REFERENCES

1. Kreidl, N. J. and Rood, J. L., "Optical Materials", Chapter 5 in Applied Optics and Optical Engineering, Vol. 1, Academic Press, New York (1965).
2. Habell, K. J. and Jackson, A., "Spectral Lines for Refractometry", Optica Acta, 10, 218, July 1963.
3. Morey, G. W., The Properties of Glass, 2nd Edition, Reinhold Publishing Corp., New York (1954).
4. Brewster, Hensler, Rood, and Weidel, "Partial Dispersion Ratios of Some New Borate and Phosphate Glasses", Applied Optics, 5, 1891, December 1966.
5. Ballard, McCarthy, and Wolfe, Optical Materials for Infrared Instrumentation, Willow Run Laboratories, Ann Arbor, Michigan.
6. Hafner, H. C. and Rood, J. L., "Some Precautions Necessary for Precision Measurements of Index of Refraction", Materials Research Bulletin, 2, 303, March 1967.
7. Molby, F. A., "Index of Refraction and Coefficients of Expansion of Optical Glasses at Low Temperatures", Journal of the Optical Society of America", 39, 600, July 1949.
8. Brewster, Kunz, and Rood, "Dispersion of Some Optical Glasses in the Visible and Infrared", Journal of the Optical Society of America, 48, 534, August 1958.

9. Von Hippel, A. R., Dielectrics and Waves, M.I.T. Press, Cambridge (1966).
10. Born, M. and Wolf, E., Principles of Optics, 3rd Edition, Pergamon Press, Ltd., Oxford (1965).
11. Ditchburn, R. W., Light, 2nd Edition, Blackie and Son, Ltd., Glasgow (1966).
12. Garbuny, M., Optical Physics, Academic Press, Inc., New York (1967).
13. Moss, T. S., Optical Properties of Semiconductors, Academic Press, Inc., New York (1959).
14. Kingery, W. D., Introduction to Ceramics, John Wiley and Sons, Inc., New York (1960).
15. Feynman, Leighton, and Sands, Lectures on Physics, Vols. I and II, Addison-Wesley Publishing Co., Reading, Mass. (1966).
16. Hensler, J. R., Masso, J. D., and Rood, J. L., Bausch and Lomb, Inc., Rochester, New York (unpublished report, 1964).
17. Malitson, I. H. and Dodge, M. J., "Refraction and Dispersion of Lead Fluoride", Journal of the Optical Society of America, 59, 500, April 1969.
18. Sutton, L. E. and Stavroudis, O. N., "Fitting Refractive Index by Least Squares", Journal of the Optical Society of America, 51, 901, August 1961.
19. Rood, J. L. and Wilton, R. D., "Dispersion Equations for Optical Materials", Proceedings of the Montana Academy of Sciences, 30, 89, 1970.
20. Shand, E. B., Glass Engineering Handbook, 2nd Edition, McGraw-Hill Book Co., New York (1958).
21. Lillie, H. R., "Glass", Chapter in Handbook of Physics, E. U. Condon and H. Odishaw, editors; McGraw-Hill Book Co., New York (1958).
22. Waxler, R. M. and Napolitano, A., "Relative Stress - Optical Coefficients of Some National Bureau of Standards Optical Glasses", Journal of Research of the National Bureau of Standards, 59, 121, August 1957.
23. Bateson, Hunt, Dalby and Sinha, "Stress Measurements in Tempered Glass Plates by Scattered Light Method with a Laser Source", Bulletin of the American Ceramic Society, 45, 193, February 1966.
24. Gorenson, R. W. and Adams, L. H., "A Method for the Precise Measurement of Optical Path - Difference, Especially in Stressed Glass", J. Franklin, Inst., 216, October 1933.

The Development of Colors in Glass

Foster L. Harding
Brockway Glass Company, Inc.
Brockway, Pa. 15824

INTRODUCTION

In simplest terms, colors occur in glass because the spectrum of light which is incident upon a piece of glass is altered as it interacts with the glass. Several things may occur when light interacts with glass. Portions of the incident beam will be reflected, absorbed, and scattered. The remainder, if any, will be transmitted through the glass.

For a discussion of color, reflection is of secondary importance. In most commercially important colored glasses, less than ten percent of the incident light is lost due to reflection.

If the glass absorbs or scatters light in one part of the visible spectrum more strongly than another, then it follows that the glass will transmit light more readily in the remaining regions of the spectrum. When this happens, colors are the result. If the absorption or scattering does not occur selectively in certain wavelength regions, the glass may range from perfectly clear through varying shades of gray to perfectly black, depending on the fraction of light transmitted. Thus, if the transmission is the same at all wavelengths, no colors will exist. White, gray, and black are not colors, but merely various degrees of the presence or absence of light.

Kingery[1] has pointed out that a discussion of color should not be undertaken without the realization that color is very much a subjective sensation. Although almost any group of people could agree on which of two pieces of paper was red and which was blue, the actual sensation experienced by the "color detectors", the eyes of the observers, would undoubtedly be different for each of them. Since the observation of color is strictly an individual experience, there is no way to quantitatively compare what any two observers see when they look at a colored object. Fortunately, the functions of the eye are sufficiently similar from one observer to another, so that most of these differences in individual "detectors" are not bothersome. However, it is important to realize that they do exist.

Figure 1 shows how the sensitivity of the eye depends on the wavelength of light in the visible spectrum for a typical observer. The maximum sensitivity occurs in the green region of the spectrum, while the sensitivity decreases to zero as the wavelength approaches either the infrared or the ultraviolet regions of the spectrum. The fact that this distribution of sensitivities is different for different people is one of the reasons for the subjective nature of color perception.

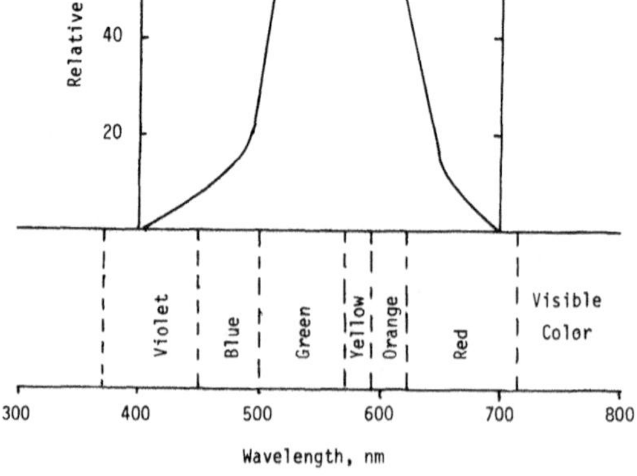

Figure 1. Spectral Sensitivity Curve for the Human Eye.

A second factor in color perception, which should not be overlooked, is the nature of the incident light. Artificial light from an incandescent bulb is quite different from that produced by a fluorescent lamp and both are different from natural sunlight. Light reflected and diffused through the leaves of a tree will contain a much higher portion of green than normal sunlight, while a hazy sunset will contain an unusually large amount of red light. So, a piece of colored glass often will assume many shades of color depending on the conditions under which it is observed.

The physical and chemical processes, which cause the selective absorption or scattering of light, are of primary importance. However, secondary effects, such as light source and individuality of the observer are also important

These remarks lead to the need for distinguishing between the terms transmission and color. The transmission is a physically measurable property which depends on the wavelength at which it is measured. If an arbitrary value of one hundred percent is assigned to the incident intensity at all wavelengths in the visible spectrum, then the transmission spectrum of a glass would be the percent of the incident light which passed through it as a function of wavelength.

Because sample thickness effects transmission values, transmission spectra are usually corrected to the same thickness, often 2 mm. In contrast to this relatively precise definition, the color of a glass specimen is the sensation experienced by the eye and mind of an observer. It is a completely individual experience and cannot be measured or compared in a quantitative manner. The presence of the color sensation is largely caused by the preferential transmission or scattering. However, a portion of the total sensation is also due to individual perception and incident lighting conditions. Since transmission curves are readily measurable, and there is a strong correlation between transmission and color perception, the physical and chemical reasons for selective absorption or scattering of visible light will be reviewed in some detail in the following sections.

For present purposes, the term "color center" will refer to any grouping of ions, atoms, or electrons in the glass which results in absorption of light in some part of the visible spectrum. The visible spectrum comprises wavelengths between about 400 nm and 700 nm, as pointed out by Figure 1. For purposes of discussion, colors will be categorized under two headings; those due to absorption of light and those due to scattering. Most colored glasses owe their color to the fact that the absorption of incident light takes place only in certain regions of the visible spectrum, although several very interesting glasses owe their color to selective scattering of light.

ABSORPTION COLORS

There is one unifying phenomenon which is common to all color centers of the absorption type; namely, that in order for light absorption to occur in the visible region, electron transfer processes must be occurring within the color center. Whenever an electron in any color center is excited from a lower to a higher energy level, the difference in its energy must be supplied by absorption of a photon of light. The frequency of light absorbed is determined by the equation

$$E_1 - E_2 = \Delta E = h\nu \qquad (1)$$

where E_1 and E_2 are the higher and lower electron energy levels, h is Planck's constant (equal to 6.6252×10^{-27} erg-sec), and ν is the frequency of the absorbed radiation in sec $^{-1}$. Wavelength, λ, is given in cm by the reciprocal of frequency, times the velocity of light in cm-sec $^{-1}$.

$$\lambda = (\frac{1}{\nu}) \cdot c = \frac{c}{\nu} \qquad (2)$$

These two equations point out that only a limited range of ΔE values will result in absorption in the visible spectrum. Using the visible spectrum limits of 400 to 700 nm in equations 1 and 2 results in ΔE values between 1.77 and 3.10 electron volts as the approximate energy range. Ions which have noble gas electron configurations, i.e., completely filled electron shells in the ion core, do not result in colors. In this case, the energy required to excite an electron across the quantum jump to the next higher level is sufficiently great, so that its absorption occurs in the ultraviolet region of the spectrum.

In the building up of atoms according to the aufbau principle, the electron levels become closer together in energy as they increase in distance from the nucleus. Thus, the 3d level is quite close in energy to the 4s level. In fact, the 4s level is slightly more stable than the 3d level, so that after the 3p level is filled, the next electron level to be filled is not the 3d level, but rather the 4s level. This results in a situation where electrons are being placed in the 3d shell while it is actually part of an ion core. This series of elements (Sc, Ti, V, Cr, Mn, Fe, Co, Ni, Cu, Zn) is known as the first transition series. It contains many of the coloring ions important in glass technology.

Similar series result upon filling the 4d levels after the 5s orbital and the 5d levels after the 6s orbital. These two series of elements are known as the second and third transition series. They contain only a few ions of importance in colored glass technology.

An additional series of transition-type elements results from filling of the 4f levels after reaching the element lanthanum. This series of elements contains several important coloring ions and is known as the lanthanide or rare-earth series.

A similar series (the actinide series) begins with the filling of the 5f shell after the element actinium. Uranium is the only element in this series which has been of much interest as a colorant.

Thus, of the five transition-type series found on the periodic chart, two stand out as containing most of the coloring ions important to glass technology; the first transition series and the rare-earth series.

There is an important distinction between these two series in that the 3d orbitals of the first transition series project well out into the periphery of these ions, so that electrons occupying them are strongly influenced by the surroundings of the ion. Thus, many of the properties of transition metal ions, including color, are quite sensitive to the number and arrangement of the d-electrons present.

In contrast to this, the 4f orbitals of the rare-earth

elements are rather deeply buried in the ions, so that electrons which occupy them are relatively isolated from their surroundings by the overlying shell of 4d electrons. This accounts for the chemical similarity of many rare-earth elements and also implies relative insensitivity of rare-earth colors to their ionic neighbors.

The importance of the rare-earth and transition elements as coloring ions arises from the fact that under certain conditions, some members of these series experience electron transfer within their partially filled inner shells which results in absorption of light in the visible spectrum.

LIGAND FIELD THEORY

One method of treating the interaction between a given transition metal (or rare-earth metal) cation and its neighbors is known as ligand field theory. A useful treatment of ligand field theory and absorption of light involving transition metal ions has been presented by Bates.[2] Several chapters in Cotton and Wilkinson[3] also cover pertinent material.

As pointed out, ions in both the transition metal series and the rare-earth series are characterized by incompletely filled inner electron orbitals. Ions in the transition metal series contain from one to nine of the ten possible electrons in the 3d shell, while those in the rare-earth series contain from one to thirteen of the fourteen possible electrons in the 4f shell.

If any one of the ions in the first transition series was found in free space, all of the orbitals within its 3d shell would be degenerate (i.e., with no near neighbors, the 3d orbitals all have the same energy). In this case, an electron shifting from one 3 d orbital to another would involve no change in energy, so no light absorption would result. In a liquid or solid environment, the proximity of ligands (neighboring anions, neutral atoms, or molecules) alters this situation. A discussion of ligand field theory as applied to ions of the first transition series follows. A completely analagous development is possible for the other transition series as well as for the rare-earth and actinide ions; however, the larger number of orbitals (seven as opposed to five) involved makes visualization much more difficult in the latter two series.

Ligand field theory, as applied to the first transition series, is based on the assumption that the energy levels of the partially occupied 3d shell are split due to the disturbance arising from the field of the surrounding ligands. In a silicate glass environment, the only two types of coordination for cations which are significant are octahedral (six ligands at the corners of an octahedron with the cation in the center) and tetrahedral (four ligands at the corners of a tetrahedron with the cation in the center). The reasons for 3d ortibal splitting can be visualized by considering the spatial location of the various 3d orbitals in relation to the ligand sites surrounding the cation. Figure 2 shows the spatial arrangement of the five different 3d orbitals about the nucleus of a transition metal ion. The orbitals d_{xy}, d_{xz}, and d_{yz} all have regions of high probability for the presence of electrons at 45 degrees to the x and y, x and z, and y and z axes respectively. The $d_{x^2-y^2}$ orbital has an electron distribution probability which is high along the x and y axes, while the d_{z^2} orbital has a high probability mainly along the z axis, although there is a ring of high probability in the x-y plane as well.

Bates[2] has considered what would happen if a transition metal ion, with 3d orbitals as illustrated in Figure 2, found itself in octahedral or tetrahedral coordination with a group of ligands. The salient points of his treatment are as follows:

Figure 3 illustrates the spatial relation between the $d_{x^2-y^2}$ and d_{xy} orbitals and ligands in octahedral coordination. Since the ligands always present a negative charge towards the central ion, an electron would have a higher energy (i.e., be less stable) in an orbital oriented in the direction of a ligand. Thus, if an electron had a choice of occupying either the $d_{x^2-y^2}$ or d_{xy} orbital, it would preferentially occupy the d_{xy} orbital since its lobes are located in directions midway between ligands. It is obvious by symmetry that the d_{xz} and d_{yz} orbitals will be analagous to and degenerate with the d_{xy} orbital. Calculations show that the d_{z^2} orbital is degenerate with the $d_{x^2-y^2}$ orbital, so these latter two are equally destabilized by the ligand field with respect to the former three. The effect of an octahedral field is to remove the degeneracy of the five 3d orbitals, so that they are separated in energy into a lower group of three (d_{xy}, d_{xz}, d_{yz}) and an upper group of two ($d_{x^2-y^2}$, d_{z^2}).

In Figure 4, the same two orbitals are pictured in tetrahedral coordination. In this case the situation is the reverse of that for octahedral symmetry. The lobes of the $d_{x^2-y^2}$ orbital now lie in the directions between

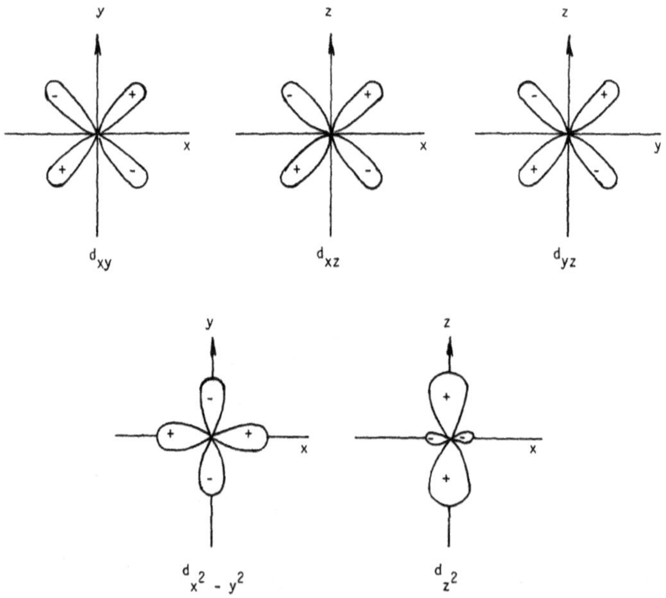

Figure 2. The d-orbitals.

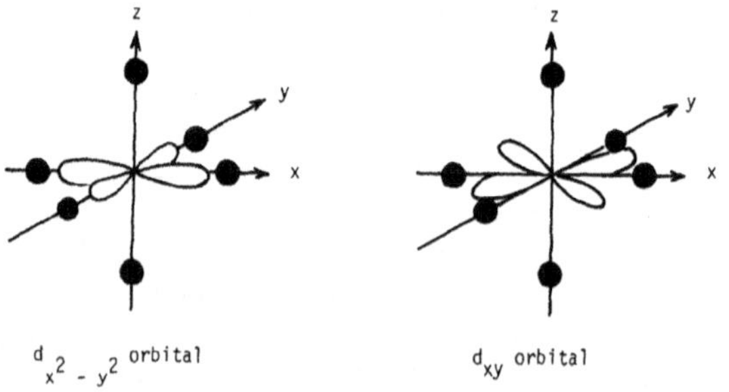

Figure 3. The $d_{x^2-y^2}$ and d_{xy} Orbitals in an Octahedral Ligand Field.

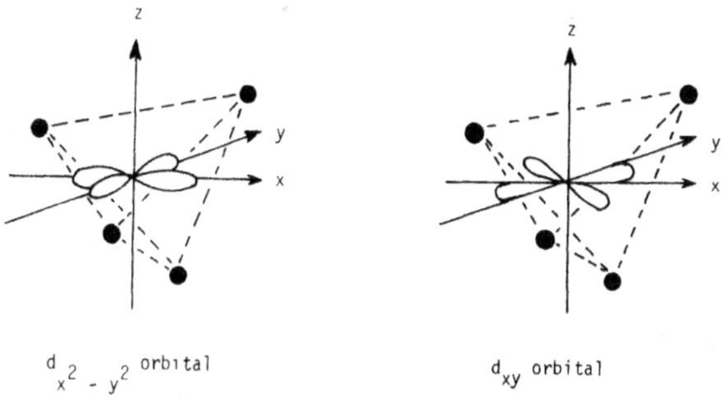

Figure 4. The $d_{x^2-y^2}$ and d_{xy} Orbitals in a Tetrahedral Ligand Field.

ligands, while the lobes of the d_{xy} orbital lie closer to, though not directly in line with, the ligands. Thus, the d_{xy} orbital is destabilized with respect to the $d_{x^2-y^2}$ orbital. The same degeneracy relations hold in tetrahedral as were found in octahedral coordination. Thus the effect of a tetrahedral field is also to split the five 3d orbitals into two groups, (d_{xy}, d_{xz}, d_{yz}) and ($d_{x^2-y^2}$, d_{z^2}), but in this case, the order of the levels is inverted with respect to the octahedral case.

Figure 5 shows the 3d energy levels for the free ion, octahedral, and tetrahedral coordination. In this figure, Δ represents the difference in energy of an electron in the upper and lower energy levels of the split 3d orbitals. Thus, if Δ were substituted into equations (1) and (2), the wavelength of light absorbed by electron transfer in a particular ligand environment would be defined. With transition metal ions, it often turns out that Δ is of the proper magnitude to result in an absorption band in the visible spectrum.

As pointed out by Figure 5, the value of Δ in a tetrahedrally coordinated cation is less than in an octahedral case. It can be shown by geometric calculations that, for the same cation and ligands at the same bond distance, the following equation holds:

$$\Delta\ (\text{tetrahedral}) = -\frac{4}{9}\ \Delta\ (\text{octahedral}) \qquad (3)$$

The negative sign symbolizes the reverse order of the split orbitals in the two cases. This means that the same cations

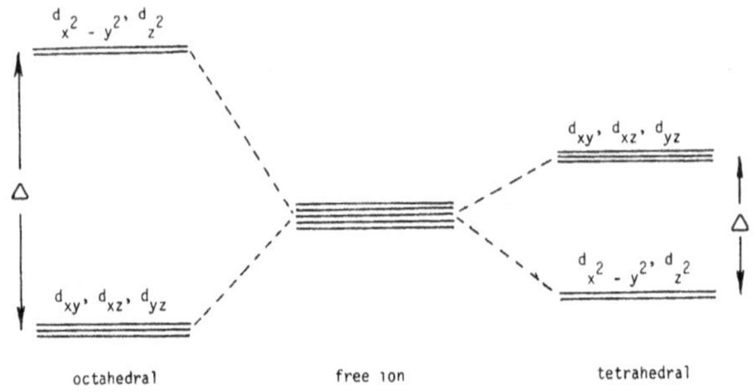

Figure 5. Energy Levels of the d-orbitals in Octahedral and Tetrahedral Fields.

and ligands could give two different colors depending on the coordination number.

In order to discuss absorption in more depth, it is necessary to understand the significance of the molar extinction coefficient, ε , as defined by the Beer-Lambert equation:

$$\frac{I}{I_o} = e^{-\varepsilon c x} \qquad (4)$$

where I is the intensity of transmitted light, I_o is the incident light intensity, c is the concentration of color centers in the glass in gram-moles per liter, and x is the thickness of the specimen in centimeters. The extinction coefficient is essentially a measure of the absorbing strength of an individual color center. The higher the absorbing power of a given color center, the higher will be its extinction coefficient. Thus, the extinction coefficient is directly related to, and an indirect measure of, the ease of electron transfer between the different energy levels involved.

The absorption spectra of transition-metal color centers generally fall into two groups depending on whether the absorption is weak ($\varepsilon \sim 0.01$ to 200) or strong ($\varepsilon \gtrsim 10^3$). Weak bands are due to transitions between the split 3d levels of the central cation. In a static situation (no molecular vibrations), transitions between d levels, where both ground and excited states have the same number of d electrons, are forbidden by a quantum mechanical selection rule known as Laporte's rule.[2] However, in the presence of molecular vibrations, this rule is relaxed and the d level transitions can occur often enough to account for the observed absorption intensities. It is this type of absorption band which can be successfully explained by the ligand field approach.

Strong absorption bands arise from electron transitions between the central cation and the ligands. The intensity of such bands is usually two or three orders of magnitude greater than true ligand field bands because electron transitions between the central ion and its ligands are not forbidden by Laporte's rule. This means that the probability of such an electron transition taking place is two or three orders of magnitude greater in the latter case. Strong absorption bands must be treated by molecular-orbital theory, since the basic assumptions of ligand field theory do not apply in the case of strong cation-ligand interactions. That is, in the case of strong interactions, pertubation theory can no longer be applied. Unfortunately, molecular orbital theory is so complex that meaningful application has been possible only in the very simplest of molecules such as hydrogen. Thus the usefulness of existing theories for strong cation-ligand interactions is extremely restricted.

Fortunately, many transition-metal absorption centers are due to true ligand-field (weak) bands. Equally fortunate, it is possible to attain a good understanding of the basic concepts of color formation without using detailed quantitative calculations based on either ligand field theory or molecular orbital theory. For those who wish to pursue a more detailed study of ligand field theory, the aforementioned reference by Bates[2] is a good starting point.

TRANSITION-METAL ABSORPTION SPECTRA

Before discussing some transition-metal color centers which are of interest in glass technology, a general discussion of absorption spectra in glasses will be presented.

The lack of long-range order in a glassy structure results in many variations in the sites occupied by color centers. Although the ligand symmetry may be nearly identical in all cases, there are many slight differences in the way individual ligands are bonded to their central cation. This is caused by two main factors; (1) the wide variety of bond angles and geometrical irregularities inherent in glass structure and (2) the several possible cations which may be bonded to the ligand to complete its charge balance. Since the ligands are bonded to two cations, the central ion of the color center and another cation outside of the color center, the nature of the transition metal-ligand bond is influenced by the nature of the bond between the ligand and its "other" cation.

The result of this variation in individual ligand-central ion bonds, leads to many slightly different values of d orbital splitting energy, Δ. This, in turn, results in a broadening of the absorption band beyond that which would occur in a similar, but crystalline, environment. As mentioned previously, this effect is much more pronounced in the first transition series than in the rare-earth series, due to the closer proximity of the 3d orbitals to the ligands as contrasted to the 4f orbitals in the rare-earth series.

In addition, there are several other factors which contribute to the complexity of studying color centers in glass. Since ligand field theory is based on central cation-ligand bonding being largely ionic in character, the coordination number of ligands about a given central ion is determined to a great extent by geometric considerations. In glass, most ions of the transition metal series are either in tetrahedral or octahedral coordination with oxygen anions. Equation (3) has pointed out that the value of Δ, and hence the absorption, will be quite different if a given ion is compared in octahedral and tetrahedral coordination. Since the cation/anion radius ratio would be fixed for a given cation-anion combination, it would be quite unlikely for a given cation to enjoy both octahedral and tetrahedral coordination unless the radius ratio should fall near the limiting value separating the two coordination states. In this case, it could result that octahedral and tetrahedral coordination were not greatly different in energy so that both might exist simultaneously.

In some cases, addition of certain cations to the glass may alter the structural relationships between species so that the coordination of a given color center cation is changed. For example, Weyl[4] has stated that about 10 percent TiO_2 in a silicate glass will cause Fe^{3+} in octahedral coordination (almost colorless) to switch over to tetrahedral coordination (deep brown). This interpretation may be in doubt, since recent work by Kurkjian and Sigety[5] has indicated that Fe^{3+} is predominantly in tetrahedral coordination in silicate glasses, even when no TiO_2 is present. Although the reasons for the color change when adding TiO_2 to iron-containing glasses are in doubt, Weyl's interpretation does serve to point out that coordination number is not as simple a matter as merely packing together spherical ions based on geometry alone.

A classic example of an ion producing markedly different colors depending on coordinaiton is that of Co^{2+}. In a silicate glass, Co^{2+} enjoys tetrahedral coordination with oxygen anions. The resulting blue color is one of the best known and easily recognized of all colored glasses. Yet, according to Weyl[4], in a metaphosphate glass, where Co^{2+} is forced into octahedral coordination with oxygen anions, a pink color is produced. Thus, the same cation-ligand combination produces two very different colors depending on coordination number. Figure 6 shows the absorption spectra for the two cases mentioned.

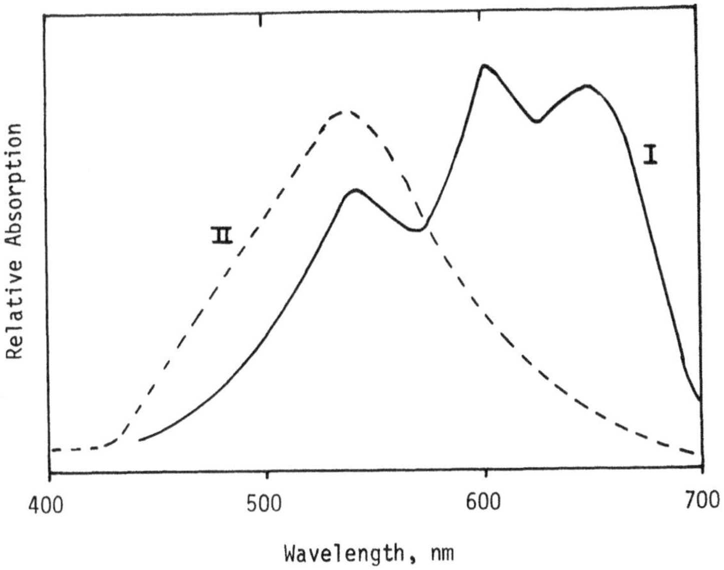

Figure 6. Light Absorption of Co^{+2} in Silicate (I) and Metaphosphate (II) Glasses.

The charge on the central cation is of utmost importance in determining the absorption spectrum caused by a color center. If a given cation suddenly finds its charge changed from +2 to +3, its ionic radius will become smaller. The resulting change in radius ratio will often cause a change in coordination number. If this occurs, the absorption spectra in the two cases will be totally different. If the coordination should remain unchanged, the increased charge will still cause some significant changes. The central ion-ligand attraction will be greater, thus increasing the ligand proximity to the split 3d orbitals. This will result in a stronger ligand field and increase the value of Δ. This, in turn, will shift the absorption to higher energy (shorter wavelength).

In many cases, more than one valence will be present in the glass for a given transition metal. For example, most commercial glasses contain iron impurity ions in both the +2 (octahedral coordination) and +3 (tetrahedral coordination) states. The relative amounts of these species are largely determined by the oxygen partial pressure in the melt system. Simply stated, greater availability of oxygen in the melt system will result in higher contents of the oxidized species. The reverse is also true. The general equation shown was found by Johnston[6] to always be applicable to these redox changes.

$$\frac{4}{n} M^{x+} + O_2 \rightleftarrows \frac{4}{n} M^{(x+n)+} + 2\, O^{2-} \qquad (5)$$

In this expression, n is the number of electrons involved in the valence change of the metal, M, whose original valence was x+.

Equation 5 indicates that if the oxygen availability in the melt could be controlled to the proper value, the amounts of oxidized and reduced species could be controlled at any desired level. In point of fact, control of oxygen availability in the melt is one of the most important methods used to control commercial glass colors.

There are, in fact, several factors which can influence the concentration of a color center in glass. The most obvious factor is the total concentration of a transition metal in all its forms. A particular absorption band will be caused by one particular ionic species, so the relative amounts of the two (or more) species is extremely important.

The effective partial pressure of oxygen in the melt largely determines these relative amounts, so it is an important consideration. The oxygen availability is a function of batch composition and temperature, so both of these variables must be considered. Finally, it must be considered that the ratio of reduced and oxidized species present in the batch materials will quite likely be different from the equilibrium ratio at the temperature of glass melting. The adjustment of the original ratio to the ratio demanded by the melt redox conditions (oxygen partial pressure) is a time-dependent process. Thus, time of melting is important. Since, in some cases, time is not available in the glass melting process for the ratio to attain equilibrium with the effective oxygen pressure of the melt, the oxidation state predominant in the raw materials may be present at a level higher than it would be if equilibrium had been attained. When this situation exists, it should be possible to help attain the desired color by adding the coloring ion in the oxidation state desired for the final color center.

The number of variables mentioned so far indicates the need for maintaining a stable operation in manufacture of colored glasses. It is desireable to hold as many as possible of these variables constant. For instance, batch composition, furnace temperature, and melting time should all be held as constant as possible. In a modern, continuous operation, this may not be strictly enforceable; however, changes in operating parameters should be made carefully. In many cases, minor fluctuations in colored glass transmission are corrected by varying an ingredient which will regulate the oxygen availability in the melt. Usually, variations in carbon, arsenic oxide, nitre, saltcake, or another redox ingredient are used for minor color adjustments.

Since the 3d orbital splitting is dependent upon the ligand field strength, changes in the ligand field can cause shifts in the absorption wavelength. Thus, substitution of one anion for another in the coordination sphere of a given transition-metal ion can cause significant changes in its light absorption. In addition to changing the strength of the ligand field, a partial substitution of one ligand species for another will disrupt the symmetry of the ligand field. This has the effect of relaxing Laporte's rule restricting electron transitions within the 3d shell.

In some cases, electron transfer occurs between the central cation and the substituted anion. Recent developments have shown that the absorption center responsible for amber color is of this type.[7,8] A typical transmission curve is shown in Figure 7. Because of its strong ultraviolet absorption, amber color finds wide use in light protective applications. Beer, pharmaceuticals, and chemicals are commonly packaged in amber containers for this reason.

Because amber is the most commonly produced of all colored glasses, it will be discussed in some detail. Harding and Ryder[7] have recently confirmed that the amber color center model of Douglas and Zaman[8] can be applied to typical production glasses, as well as the two-component alkali silicate glasses studied by the latter authors. In an independent study, Karlsson[9] has also established that amber absorption is caused by a tetrahedrally coordinated ferric ion with a sulfide ion substituted for one of the oxygen anions in its coordination sphere. The extinction coefficient of this amber color center has been estimated[8] at 9000 liter/cm-mole, so it is apparent that electron transfer is occurring between the central cation and one or more of its ligands. Since ferric iron, tetrahedrally coordinated with oxygen alone, does not give amber color, it is concluded that electron transfer occurs between levels in the ferric ion and the sulfide ion.

Accepting this model of the amber color center, some interesting conclusions can be drawn regarding production and control of amber glasses. One very important factor to be noted is that the color center involves both an oxidized species (ferric iron) and a reduced species (sulfide sulfur). Thus, there is only a limited range of oxygen partial pressure where both species can exist simultaneously in the glass. According to Brown and Douglas,[10] this range of P_{O_2} is approximately from 10^{-10} to 10^{-8} atmospheres. This means that when the P_{O_2} is less than 10^{-10} atmospheres, no ferric iron can exist in the glass; while a P_{O_2} greater than 10^{-8} atmospheres oxidizes all the sulfide sulfur. Thus, exceeding either limiting value removes one of the essential components of the amber color center.

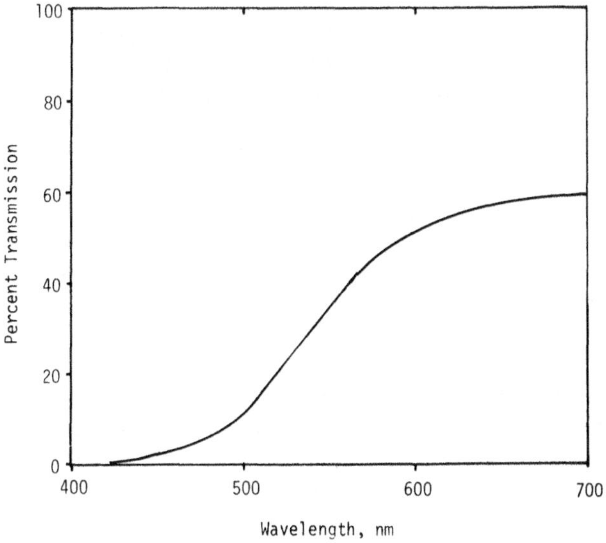

Figure 7. Typical Amber Transmission Curve.

In spite of the above restrictions, amber color is controlled in most instances by varying the effective oxygen pressure of the melt, often by adjusting the amount of carbon in the batch. The following will explain how this control works.

A purely statistical approach would predict that the probability of amber color center formation should be proportional to the product of the ferric iron and sulfide sulfur activities in the glass. Since the concentrations are small, it may safely be assumed that the probability is proportional to the product of their concentrations. Harding and Ryder[7] have shown this to be a good approximation in commercial glasses.

The preceding infers that amber color would remain constant with changing oxygen partial pressure if the ferrous/ferric and sulfide/sulfate redox couples responded equally, but oppositely, to a given P_{O_2} change. If this was the case, a given increase in P_{O_2} would cause an increase in ferric iron which would be exactly matched by a corresponding decrease in sulfide sulfur. Thus their product, to which amber absorption is proportional, would change only slightly. The fact that amber color can be controlled by carbon adjustments indicates that this situation does not exist. In fact, Figures 8 and 9 show that a given change in P_{O_2} causes a larger change in the sulfide sulfur content than it does in the ferric iron content.[7] Thus, increased P_{O_2} makes the glass lighter while decreased P_{O_2} makes it darker. This statement holds over most of the P_{O_2} range where amber color exists. Near either end of the range the glass gets lighter due to almost total absence of one of the essential species.

Figure 10 is a sketch illustrating the effect of carbon (P_{O_2}) variations on the content of ferric iron and sulfide sulfur in the melt. It points out that there should be a specific P_{O_2} value where the product of Fe^{3+} and S^{2-} is maximized for a given batch formula.

Unfortunately, there are other considerations of amber glass melting and forming which impose restrictions on the P_{O_2} control range. Ferrous iron (Fe^{2+}) has a very strong absorption band in the near infrared. Thus, its presence is responsible for poor heat transfer into the glass during melting, and out of the glass during refining, conditioning, and forming. From this point of view it might seem advantageous to lower the total iron level of the glass. The proper transmission could be maintained by making the melt more reducing to increase the sulfide level.

However, Budd, et al[11] and Manring[12] have shown that the tendency for glass to re-evolve gases in the forehearth area of a glass tank after once being well fined, increases in direct proportion to the sulfide sulfur content. This phenomenon, known as reboil, imposes an effective ceiling on the amount of sulfide which can be retained in the glass without causing a stability problem. Since high iron levels impose a heat transfer problem and high sulfide levels impose a stability problem, these two considerations very effectively restrict the P_{O_2} range which may be used to control amber color.

A final factor of interest in absorption character is the temperature of the glass when its color is being observed. A classic example of a glass which changes color upon being cooled or heated is cadmium sulfide yellow. At room temperature, these glasses are a bright yellow. However, Weyl[4] has reported and the present author has

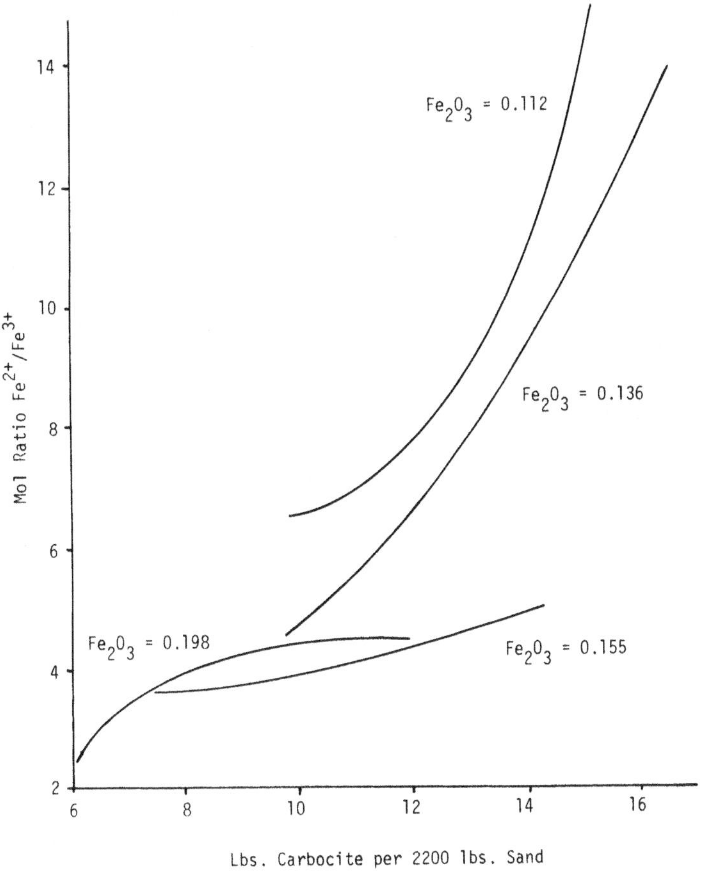

Figure 8. Dependence of Iron Redox Ratio on Carbon in Batch.

observed that such a glass becomes progressively more red until it reaches a bright red-orange at a few hundred degrees above room temperature. Weyl also reported that this same glass would be a very pale yellow at liquid nitrogen temperatures.

These effects can be explained as follows. The higher temperature increases the thermal vibrations of the Cd-S bonds making them, on the average, longer than at lower temperatures. This lowers the ligand field strength and reduces the 3d splitting by a proportionate amount, which, in turn, causes the absorption to be shifted to longer wavelengths. The color changes progressively from yellow to red.

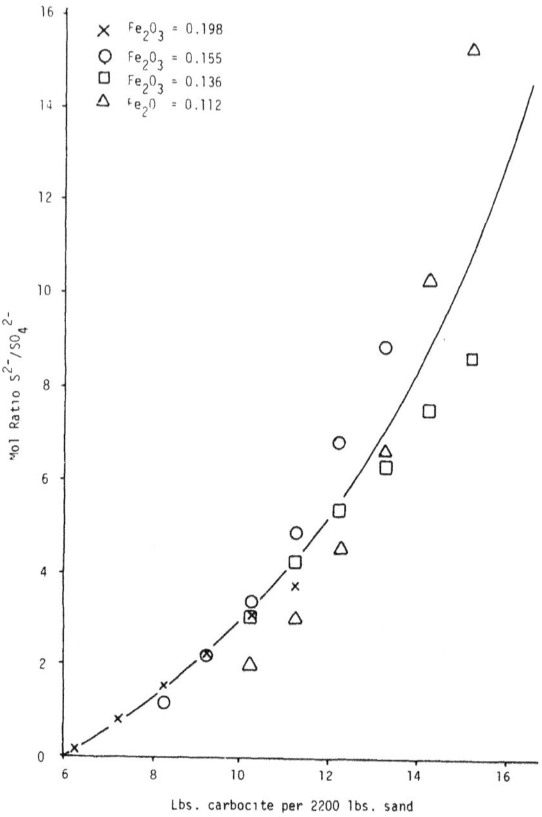

Figure 9. Dependence of Sulfur Redox Ratio on Carbon in Batch.

The effect of cooling below room temperature would be to shorten the cation-ligand bond distance and increase the ligand field strength. Thus, in this case, the absorption is gradually shifted out of the visible spectrum into the near ultraviolet. The effect is reversible with temperature.

If the above logic is accepted, it becomes easy to understand the reasons for the red color in so-called selenium ruby glasses. The color center in such a ruby glass contains cadmium sulfide units, with selenide ions substituted for some of the sulfide. The geometrical arrangement is not well understood. Selenide ions are very similar to sulfide ions in chemical character, as they are

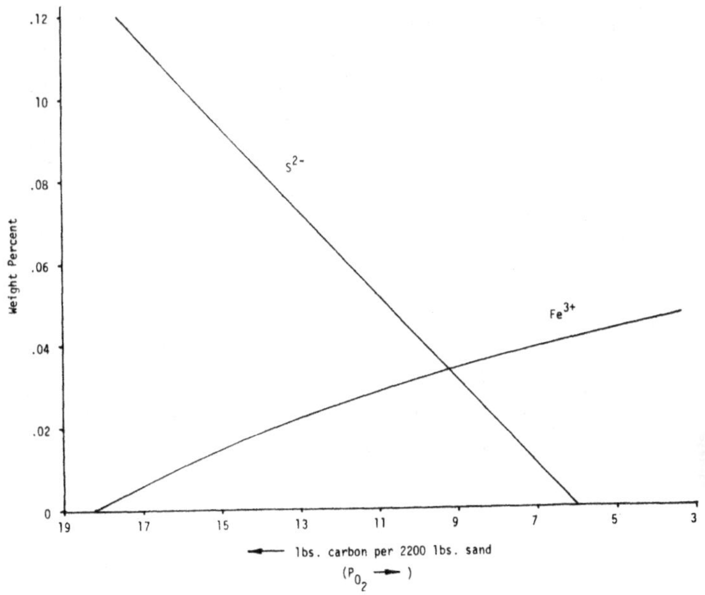

Figure 10. Dependence of Fe^{3+} and S^{2-} Contents on Melt Oxygen Pressure.

in the same group on the periodic chart. However, the selenide ion is larger than the sulfide ion. Thus, when a sulfide is replaced by a selenide ion, the central cation-ligand bond is lengthened and the ligand field weakened. This in turn, decreases the value of Δ in the d orbital splitting of the cadmium ion and shifts the absorption to a longer wavelength. The color is changed from cadmium sulfide yellow to selenium red. This explanation points out that substitution of a selenide for a sulfide in the color center has an effect which is very similar to heating the glass. A similar analogy applies in several other situations. A discussion of the development of selenium red colors has been published by Pavlish and Austin.[13]

The preceding pages have given some indication of the complexities involved in the understanding of color centers in glass. Keeping these in mind, some of the more important and interesting absorption spectra in the transition metal series will be discussed.

Strictly speaking, the first transition series begins with the element scandium. However, since it exists only in the +3 oxidation state,[3] it has no 3d electrons and is thus of no interest as a coloring ion.

Similarly, the Ti^{4+} ion has no 3d electrons and by itself does not cause development of color. Under certain conditions the Ti^{3+} may exist in glasses. Its single 3d electron does undergo excitation within the split 3d levels causing an absorption at about 575 nm. The Ti^{3+} species is not stable in silicate glasses, so the resulting violet color is found only in acidic phosphate and borosilicate glasses melted under slightly reducing conditions. A typical absorption spectrum is shown in Figure 11. This color is presently unimportant in commercial glass manufacturing.

According to Weyl[4], glasses can assume several different colors due to the presence of various color centers containing vanadium. Vanadium can exist in glasses as V^{3+}, V^{5+}, and VO^{2+} ions. In addition, Weyl has referred to the existance of polyvanadates which would be aggregates of vanadium ions in network-forming sites. It must be remembered that V_2O_5 itself is a glass-forming oxide.[14] V^{3+} ions in glass result in two absorption bands at about 425 and 625 nm. The presence of V^{5+} ions results in absorption in the near ultraviolet end of the visible spectrum. Typical spectra for vanadium-containing glasses melted under reducing and oxidizing conditions are shown in

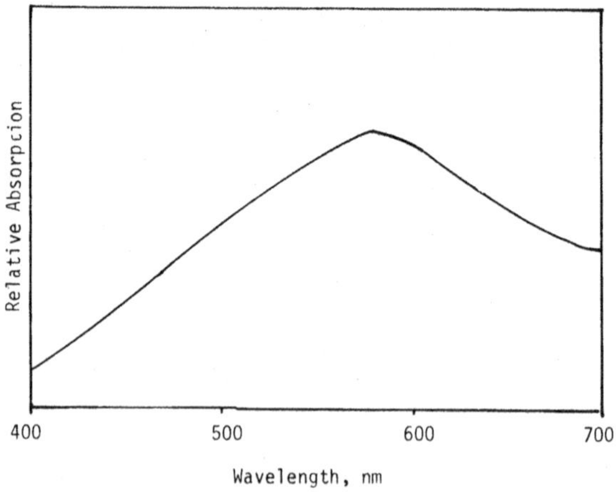

Figure 11. The Spectrum of Ti^{3+} in Glass.

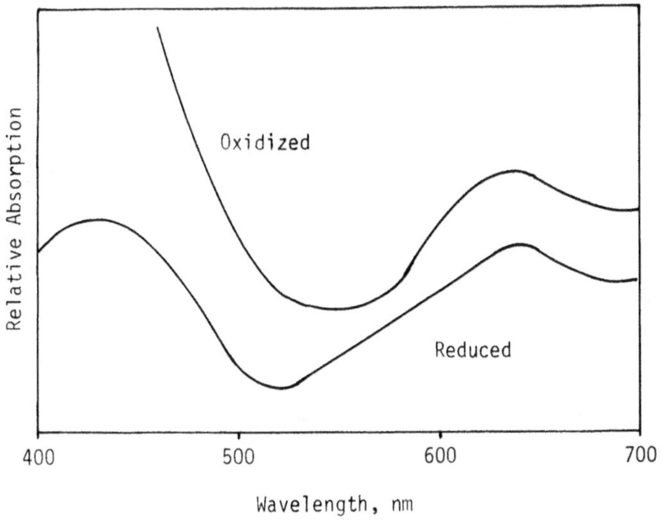

Figure 12. Absorption Spectra for Vanadium in Oxidized and Reduced Glasses.

Figure 12. The reduced glass, containing mainly V^{3+}, resulted in a green color, while the more oxidized glass, containing more V^{5+}, resulted in a brown color. In applying the model presented earlier, it is logical that V^{5+} should attract its ligands more strongly and thus undergo a more severe splitting of its 3d orbitals. This accounts for the shift of absorption toward higher energies (shorter wavelengths) in the case of V^{5+}. Because it is an expensive material, vanadium colors have not achieved practical importance.

Chromium exists in two valence forms in silicate glasses, +3 and +6. As would be expected, the ratio between the two oxidation states depends on the oxygen partial pressure of the melt. In addition, the basicity of the melt (which effects its oxygen partial pressure) can cause shifts in the redox equilibrium, so glass composition is also a factor. A discussion of the Cr^{3+} - Cr^{6+} equilibrium has been presented by Nath and Douglas.[15]

The characteristic emerald green color due to trivalent chromium is widely used commercially for beverage containers. Because of the ultraviolet light protection required by some soft drink beverages, glass for this purpose is melted under more oxidizing conditions which give a higher level of hexavalent chromium. The Cr^{6+} ions result in absorption in the short wavelength region of the visible spectrum and on into the ultraviolet, thus providing light protection. The color of the glass is often somewhat more yellowish than a normal emerald green. Figure 13 shows typical absorption spectra for normal emerald green and ultraviolet absorbing green. It should be pointed out that ultraviolet absorbing green does not protect beer adequately since it does not absorb sufficiently in the near ultraviolet.

It has been established over the years that, in spite of its large number of oxidation states in solution chemistry, manganese enjoys only two valencies in glass;[4] +2 and +3. Paul and Lahiri[16] have discussed the effect of composition on the ratio between the two oxidation states. Weyl[4] has reported that Mn^{2+} in glass results in a weak yellowish brown color while Mn^{3+} gives a strong, characteristic violet color. Bates[2] has stated that Mn^{2+} exists in glasses in both tetrahedral and octahedral coordination. Up to the present, manganese has not found significant application in commercial glass manufacturing. MnO_2 can be

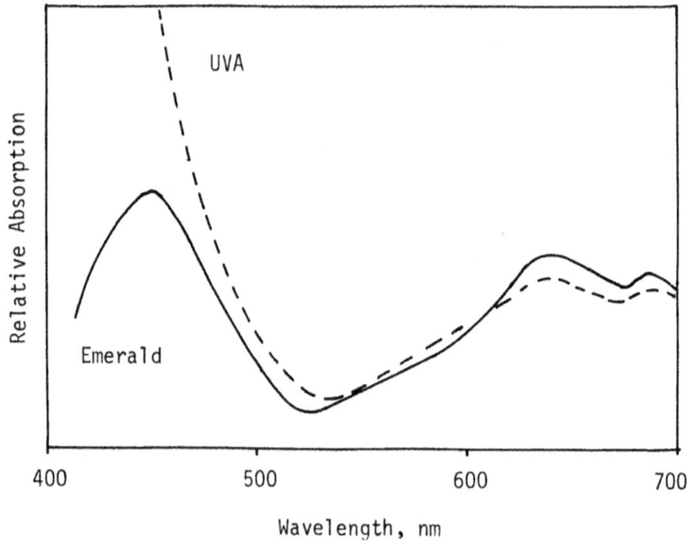

Figure 13. Absorption Spectra for Emerald and Ultraviolet Absorbing Chromium Green Glasses.

used for decolorizing glasses,[4] and this will be discussed briefly in a later section. Figure 14 shows absorption spectra for both oxidation states of manganese.

In terms of importance and influence on glass technology, the most important colorant in glasses is iron. It is so widespread in nature, that raw materials almost always contain it as a contaminant. Thus, it is not practical to eliminate the slight blue-green tint due to iron contamination by eliminating iron from the system. Iron exists in glass in two oxidation states, Fe^{2+} and Fe^{3+}. The Fe^{3+} is a much weaker colorant than the Fe^{2+}, so much of the problem can be eliminated by melting under oxidizing conditions. However, even when using low iron raw materials and controlled melting conditions, sufficient color exists so that decolorizing of the glass by other means is common commercial practice.

The basis of decolorizing is to add very small quantities of colorants in addition to the iron contaminant. These decolorizing ingredients are chosen so that they absorb in regions of the visible spectrum where the iron species transmit. This balances the transmission so that it is even, throughout the visible spectrum. The key is to keep all of these ingredients at a low enough level so that the glass remains clear in appearance. Too high an iron level will require such a high concentration of decolorizers that the whole visible transmission curve is significantly decreased. A grayish glass will result. Figure 15 shows absorption curves for both Fe^{3+} and Fe^{2+}. The need for spectral compensation to give a colorless glass is apparent.

The most widely used system to decolorize glasses consists of an oxidizing agent to maintain as much of the iron as possible in the Fe^{3+} state. Saltcake, or nitre and arsenious oxide are commonly used. To compensate for the residual yellow-green tint, very low levels of cobalt oxide and selenium are added to the glass. The cobalt oxide absorbs strongly in the red (see Figure 6), while selenium absorbs in the blue. Thus, absorption is evened out over the visible spectrum and the glass appears colorless.

Other decolorizing systems exist, such as MnO_2. This material acts as an oxidizing agent and its own color (reddish-violet, due to Mn^{3+}) compensates for the greenish iron tint. Thus it serves two functions simultaneously.

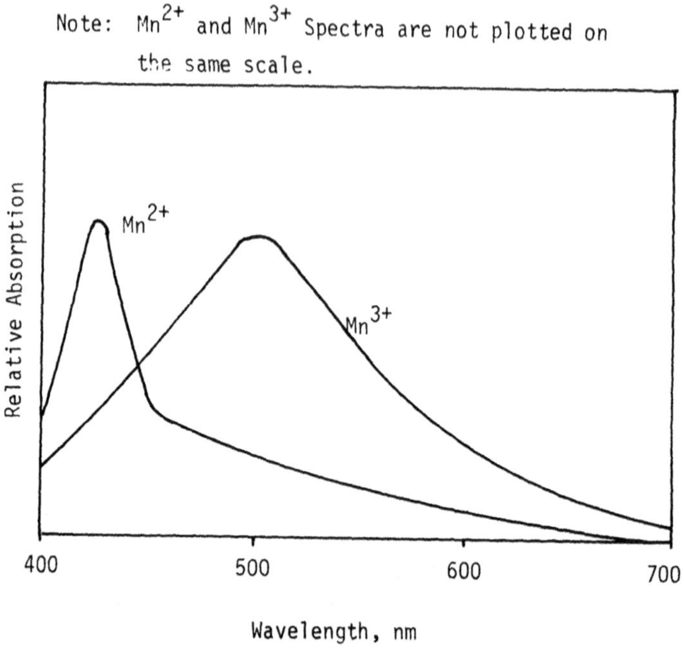

Figure 14. Absorption Spectrum of Mn^{2+} and Mn^{3+} in Glass.

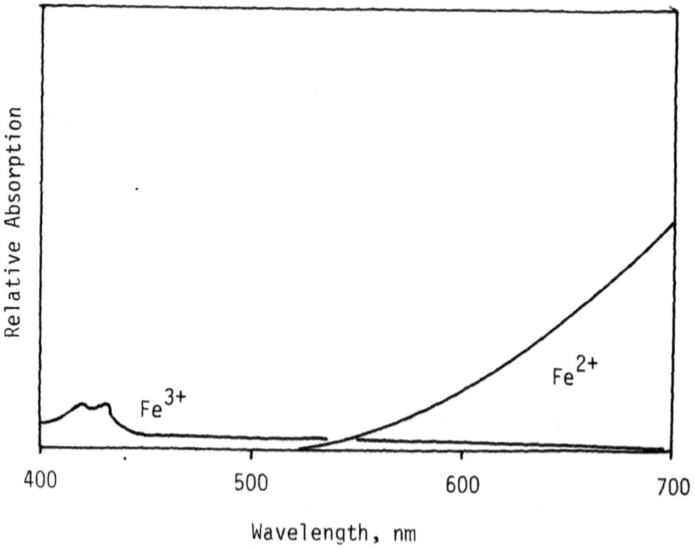

Figure 15. Absorption Spectra for Fe^{2+} and Fe^{3+} in Glasses.

Weyl's book refers to other decolorizing systems and the interested reader is invited to satisfy his curiousity by referring to it.

In addition to causing the need for decolorizing, iron is important in amber color, as mentioned previously. Iron also can be used, usually in combination with a slight addition of chromium oxide, to give a light blue-green (Georgia green) which is popular for many beverage containers.

Since cobalt colors have been discussed earlier, only nickel and copper colors remain to be discussed in the first transition series. Nickel has never played an important role in colored glass production. It absorbs almost the entire visible spectrum except under certain specific conditions which have been noted by Weyl.[4] Figure 16 shows the absorption due to Ni^{2+} in both tetrahedral and octahedral coordination. Since both coordinations exist in most silicate glasses, the total absorption extends across the entire visible spectrum.

Figure 17 shows the absorption for Cu^{2+} in glass. It results in a characteristic blue-green color. The production of this color in glasses offers no particular difficulties since, under oxidizing conditions, the Cu^{2+} ion is readily formed. However, under reducing conditions, Cu^{+} and metallic copper will also be formed. The latter grows into light scattering units which result in copper ruby glasses. This will be discussed in more detail in a later section.

RARE-EARTH ABSORPTION SPECTRA

As mentioned previously, rare-earth ions give colors because of splitting in their 4f electron shells. This situation is more complex (7 orbitals versus 5 in the d shells), but otherwise very analagous to that which occurs in the 3d shell of the transition metals. The other factor which distinguishes the rare-earths is that their 4f shell is deeply buried within the ion. This results in sharp absorption spectra which are insensitive to changes in glass composition and redox conditions.

The spectra of several of the rare-earths have no absorption in the visible range. Thus, Y, La, Gd, Yb,

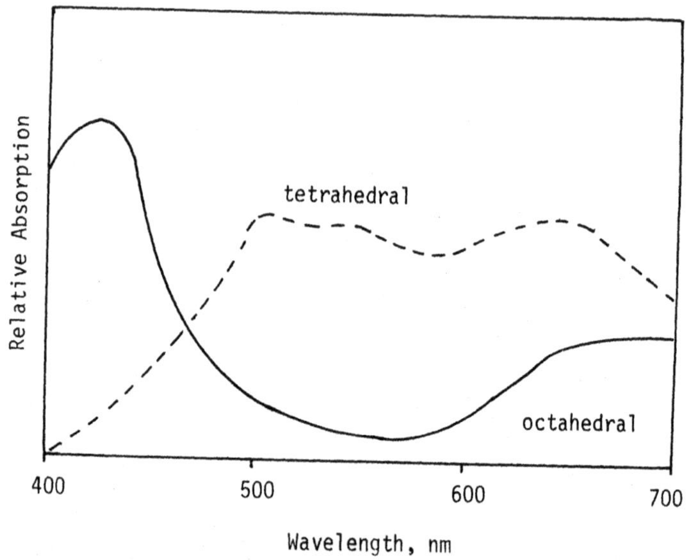

Figure 16. Absorption Spectra for Ni^{2+} in Tetrahedral and Octahedral Coordination in Glass.

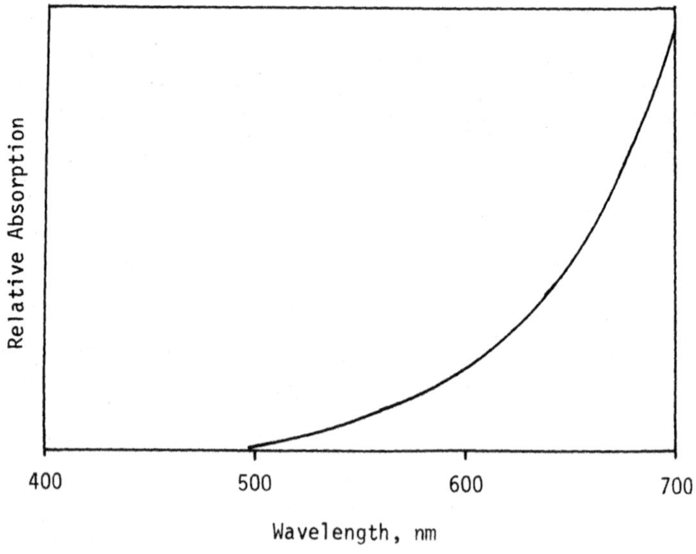

Figure 17. The Absorption Spectrum for Cu^{2+} in Glass.

and Lu are not represented in the following considerations. In addition, the element promethium will not be discussed since its half-life is so short that it does not exist in nature. Herring, et al[17] have covered most of the rare-earth colors in their comprehensive study.

Neodymium is one of the strongest rare-earth colorants and gives tones ranging from blue under fluorescent light to wine red in natural daylight or incandescent light. The transmission curve for a glass containing 0.2 mole percent Nd_2O_3 is shown in Figure 18.

Figure 19 shows the transmission curves for a glass containing 0.2 mole percent Pr_2O_3. Praseodymium glasses have a green color which the eye cannot distinguish from that of chromium-containing glasses, although their absorption spectra are quite different (see Figures 19 and 13). Because of this fact, chromium green glasses have largely precluded use of the more expensive praseodymium green colors.

Figure 20 shows the transmission spectrum of a glass containing 0.2 mole percent Er_2O_3. Erbium glasses are pink and represent the most stable, pink soda-lime glasses, since other pink or red colorants require carefully controlled redox and/or striking conditions. Although its one absorption band in the visible region is very strong, it is also very narrow, so erbium is not a particularly strong colorant.

Holmium glasses exhibit a strong blue absorption as shown in Figure 21, but since the human eye is not a very good detector in this region, the glasses have only a pale yellow color.

Cerium, samarium, europium, terbium, dysprosium, and thulium all give weak absorption bands in the blue and violet regions which cause pale yellow colors at sufficiently high concentrations. Since they are not of practical interest as colorants, their absorption spectra will not be presented.

DEFECT COLOR CENTERS

Colors may result from absorption of light by electrons which are not associated with any particular atom. For

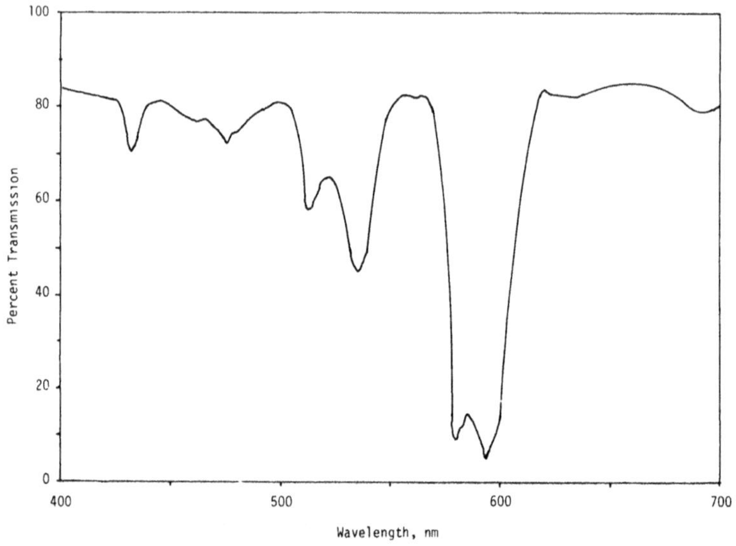

Figure 18. Transmission Curve for Glass Containing 0.2 Mol Percent Nd_2O_3.

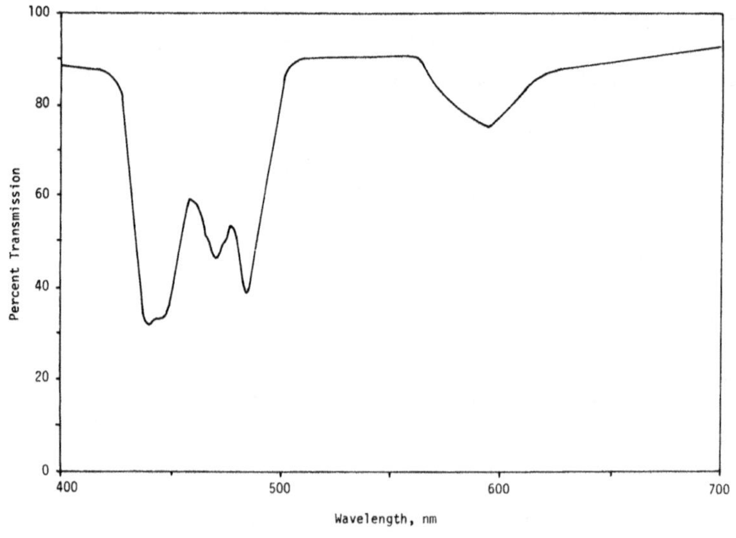

Figure 19. Transmission Curve for Glass Containing 0.2 Mol Percent Pr_2O_3.

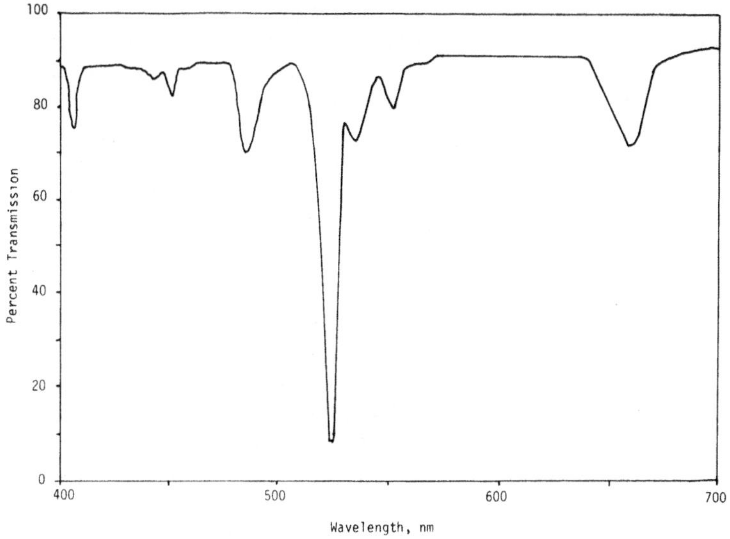

Figure 20. Transmission Curve for Glass Containing 0.2 Mol Percent Er_2O_3.

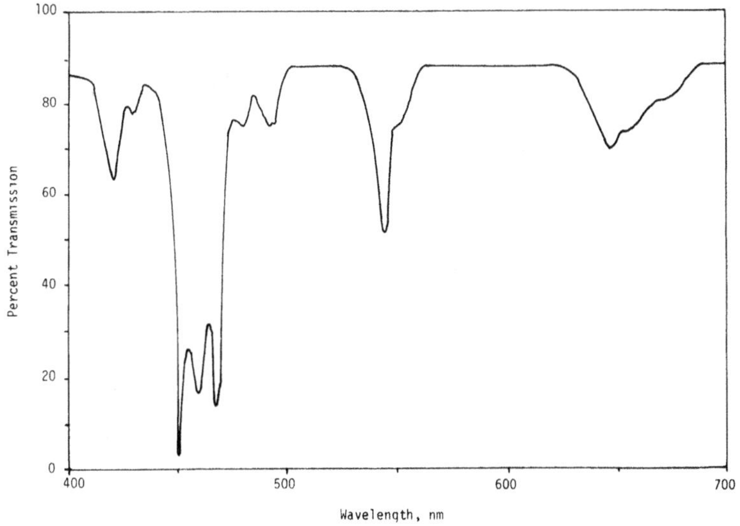

Figure 21. Transmission Curve for Glass Containing 0.2 Mol Percent Ho_2O_3.

example, bombardment of a glass by atomic particles or by high-energy x-rays will result in a general darkening of the glass. This is caused by anions being dislodged from their position in the structure by the powerful bombardment. In order to maintain the local charge neutrality, the proper number of electrons will be associated with the vacancy in the structure. Since these electrons will be more stable in some portions of the vacancy than others; usually close to a cation, a variety of energy levels are available to be occupied by such an electron. Thus, light can be absorbed by these electrons as they are shifted from one location to another within the vacancy. Since there is such a wide variety of energies available, the light is absorbed nearly uniformly throughout the visible spectrum. This usually results in a general darkening, rather than development of a specific color.

In most cases, the results of high energy bombardment can be effectively removed by heating the glass into its annealing range. This allows sufficient atomic mobility so that the defects can be annealed and the original structure restored.

SOLARIZATION

Long term exposure to the ultraviolet radiation in sunlight can result in another type of coloring phenomenon known as solarization. If certain multivalent ions or combinations of ions are present in the glass, their valence can be changed by ionizing radiation as follows:

$$Mn^{2+} + h\nu \rightarrow Mn^{3+} + e^-$$

$h\nu$ represents a photon of energy corresponding to the ultraviolet portion of the spectrum. The violet color which results in the case of manganese is due to the presence of Mn^{3+}. The electron generated by the solarization process will be trapped somewhere in the glass structure, often by Fe^{3+} in commercial glasses:

$$Fe^{3+} + e^- \rightarrow Fe^{2+}$$

Such a reaction serves to stabilize the solarized color center. If the electron could easily recombine with the Mn^{3+} ion, solarization would not build up a significant concentration of color centers. By removing the electron to a stable environment by reaction with Fe^{3+}, the Mn^{3+} color center attains a more permanent status. This two part reaction is the solarization mechanism responsible for the noticeable purple tint in much antique glassware.

Several other multivalent species can also result in solarization colors via analogous reactions:

$$Fe^{2+} + h\nu = Fe^{3+} + e^-$$

$$V^{3+} + h\nu = V^{4+} + e^-$$

The main consideration is that the rate of recombination of electrons with their original ionic species must be less than the rate of production of color centers by the solar radiation. Stabilizing reactions with Fe^{3+}, low temperatures, and high exposure to radiation all favor development of colors by the solarization mechanism.

Although the concentrations of ions capable of being solarized is low in most glasses, the color centers seem to have stronger absorption character than a similar ion formed by other means. A discussion of solarization and its control is given in Coloured Glasses by W. A. Weyl.[4]

COLORS DUE TO SCATTERING OF LIGHT

Whenever the medium through which light is passing is not optically homogeneous, some percentage of the light will be scattered. In the case of glass, this phenomenon can range from a problem which must be minimized (in optical glasses for example) to a commercially valuable asset (in opal or ruby glasses for instance).

Although glass-in-glass separations are known to occur in many silicate glasses, in no case has such a system been used for commercial glass colors. Such a scattering center would be far too difficult to control reproducably on a commercial scale. In all commercial glasses which owe some portion of their desired appearance to light scattering, the scattering centers are minute crystallites of one phase suspended in the glassy matrix. These crystallites range from metals to compounds; including gold, copper, fluorides, phosphates, sulfides, and oxides. The only essential is that the main glassy phase become supersaturated with some component. This phase then nucleates and grows crystals which are of proper size to scatter visible light. Control of the occurance and size of these crystallites is a complex problem. The driving force for nucleation of the second phase depends on the degree of supersaturation, which is a function of temperature as well as composition. The ability for nuclei of the second phase to form depends on the presence of nucleating sites, as well as upon temperature and time of heat treatment. Once nuclei are formed, whether or not they grow into crystallites of the proper size again depends on time, temperature, and the degree of supersaturation. Too little treatment and they will be too small; too long a treatment and they will be too large.

In addition, redox conditions of the melt are often important, for example, CdS crystallites cannot form under conditions which are either too oxidizing (no sulfide) or too reducing (Cd^{2+} reduced to metallic Cd), so the manufacture of colloidally colored glasses on a commercial scale is an economically hazardous undertaking. In order to do so successfully would require tight control of several composition dependent kinetic parameters in the same manufacturing operation; often two or more simultaneously. Kingery,[1] Weyl,[4] or Brophy, Rose, and Wulff[18] contain sections on nucleation and growth theory which point out the complexities of such systems.

In spite of the difficulties inherent in controlling light scattering systems, commercial production of opal glass is fairly common. It is often used to produce decorative tableware and cosmetic jars. This type of glass owes its white, milky color to uniform scattering of light throughout the visible spectrum. There are only two commercially important systems; namely, phosphate opals and fluoride opals. In both systems it is quite probable that liquid-in-liquid phase separation occurs in the melting operation followed by rapid crystallization of the phosphate or fluoride rich phase upon cooling. This "striking" takes place almost instantaneously and results in crystallites of alkali and alkaline earth phosphates or fluorides which scatter light throughout the visible spectrum. In most cases, the crystallites are mixed crystals of sodium and calcium phosphate or fluoride with calcium present in larger amounts than sodium.

Opal glasses can be made in a variety of colors in

addition to pure white by adding typical ionic colorants of the absorption type. For example, addition of a trace of cobalt oxide to opal glass will result in a blue opaque glass. Commercial use of such colored opal glass has not reached very large proportions because of limited markets.

The other prominant example of a glass colored by light scattering is colloidal ruby. Copper or gold metallic crystallites can be used to give such a colored glass. The number of scattering centers determines whether the color will be strong or weak, while the particle size determines the color. In addition, the ability of the particle to scatter light depends on the difference in refractive index between the glassy matrix and the scattering particle. In the case of metals, this presents no problem; however, in the case of some oxides, it might arise that the difference was too small to make an effective scattering center.

In the case of ruby glasses, blue light is scattered strongly by the suspended particles, allowing transmission of red light and development of the ruby color. Too large a particle size results in "muddy" color by scattering some of the necessary red light. Colloidal ruby colors are difficult to control because they must be "struck", i.e., reheated to develop the color. The problem arises because the effective growth temperature for the crystallites is higher than the nucleation temperature. Thus on initial cool-down, no nuclei are present as the glass passes through the growth range; thus no crystallites occur. On reheating, nuclei are formed in addition to those which formed on cool-down, so when the growth range is reached, crystallites readily occur with consequent color development. The process is still complex since the color depends on crystalline size which depends in turn on the number of nuclei, growth time and temperature, and degree of supersaturation. Successful operation of a continuous ruby glass furnace has never occurred in spite of a few commercial attempts. Control of the ruby color to a reasonable specification has not yet been found to be commercially feasible.

Weyl[4] mentions other colloidal systems, however, none have been used to any great degree in commercial applications.

VISIBLE COLORS DUE TO FLUORESCENCE

The ionic colors caused by light absorption are sometimes referred to as "subtractive" because the color results when part of the visible spectrum is removed by absorption. Fluorescent colors also result from electron transitions; however, in this case the electron emits a photon of visible light as it drops from a higher to a lower electron level. So one would be justified in referring to fluorescent colors as "additive".

The best way to understand fluorescence is to recall that in light absorption an excited state of the atom was produced which had only a limited average lifetime before it returned to the ground state. If the emitted energy is dissipated in the form of light, the process is known as luminescence. The electron transition energy difference and the light wavelength are related by equations (1) and (2), just as in the case of absorption. However, in this case, light of the appropriate wavelength is being emitted rather than absorbed.

In general, the term fluorescence is used to refer to luminescence when the absorption process and the emission process are practically instantaneous. The term phosphorescence is used to describe instances when a significant time lag exists between excitation and emission. This results in a phenomenon known as after-glow.

Figure 22 shows a schematic diagram of the electron transfer processes involved in fluorescence and phosphorescence. Absorption of light, often in the ultraviolet region, raises the electron in question from its ground state position, G, to a higher energy (excited) state represented by S_2 or S_3. Part of the electron's energy is almost always lost immediately by non-radiative decay leaving the electron in the lowest energy excited state represented by S_1. Fluorescence occurs when the energy level drops from S_1 to G by radiative decay occurring in about 10^{-8} seconds. This description serves to emphasize the fact that the fluorescent radiation is of lower energy (longer wavelength) than the exciting radiation.

Decay sometimes occurs from the lowest "stable" excited state to a metastable level, T, by a non-radiative process. The radiation emitted by further decay from T to G is

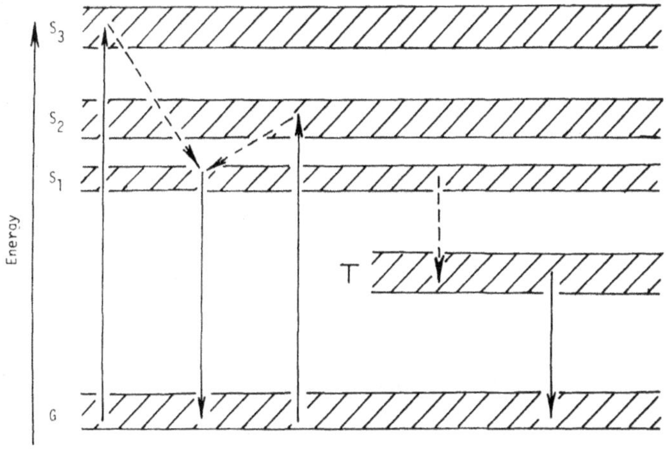

Figure 22. Schematic Energy Levels Showing Fluorescence and Phosphorescence Processes.

phosphorescence. Weyl[4] credits the existance of these metastable levels, T, to thermal vibrations. This accounts for the strong temperature dependence of phosphorescence.

The above discussion implies that just because an ion or other species in the glass results in a color center due to absorption is no reason to expect similar behavior with respect to fluorescence. In fact, many ions which have strong fluorescent character exhibit no absorption in the visible spectrum. Rather, their absorption is more likely to be found in the ultraviolet, with emission in the visible.

Weyl[4] has pointed out that fluorescent centers may be dissolved metallic atoms, a precipitated crystalline phase, or ionic constituents of the glass. Most of the interesting fluorescent centers are ions in the glass. Some of the same general comments made during the ligand field discussion of absorption centers apply to fluorescent centers as well. For example, changes in glass composition, ligand type, temperature, coordination number, valence, and concentration definitely effect the wavelength, intensity, and emission band width, just as in the case of absorption. Weyl[4] has presented figures which indicate the effects of temperature and glass composition on the fluorescence of uranium ions in glass.

Most often, fluorescence is characterized by both an excitation spectrum and an emission spectrum. Often a given ionic species in the glass will have more than one peak in its excitation spectrum with a similar, but not necessarily equal, number of peaks in its emission spectrum. The excitation spectrum is an absorption spectrum representing the wavelengths at which electrons are excited to higher energy levels. The emission spectrum represents the wavelengths at which fluorescent light is given off by electrons falling back to their ground levels.

Most of the rare-earth series have visible fluorescence bands; some possessing several. In those cases when more than one visible fluorescence band exists, the color can often be controlled by using the proper exciting wavelength. Herring, et al[17] have covered rare-earth fluorescence spectra giving both excitation and emission bands. Weyl[4] has also covered work up to about 1950 in this field. His work covers non-rare-earth fluorescing materials as well. In addition, Rindone[19] has covered fluorescence of many materials in a more recent treatment.

SUMMARY

In the preceeding presentation the physical and chemical processes which cause selective absorption or scattering of light have been given primary consideration. It should be remembered that color perception is a phenomenon which is experienced by individual people under specific lighting conditions. Both factors can contribute to different sensations by two different observers.

Absorption of light in the visible spectrum is controlled by electron transfer processes within individual color centers. In most cases this electron transfer takes place between split inner orbitals of transition-type cations. It can often be treated by ligand field theory. The glass composition, temperature, coordination number, and valence of the coloring cation, and several other factors can cause significant variations in absorption spectra.

Rare-earth absorption spectra are less sensitive to such extraneous influences because the color-causing electron transfer processes are taking place within deeply buried levels which are relatively isolated from the ion's surroundings.

Defects and solarization can also cause absorption centers which can cause or alter colors in glasses.

Light scattering systems include opal and ruby glasses. The color is determined by the wavelength and intensity of light scattered. This, in turn, is determined by the concentration and size of the scattering centers. Such colors are very sensitive to time and temperature of the nucleation and growth of the second phase, be it crystalline or glassy.

Fluorescent colors are caused by absorption of ultraviolet radiation with resulting re-emission in the visible range. Thus, such colors could be called "additive" just as absorption colors are referred to as subtractive. Most rare-earth ions have pronounced fluorescent bands.

This paper has been an attempt to present the basic concepts of color formation in glasses. Emphasis has been placed on why colors occur rather than an extensive review of literature in the field. Consequently, individual colors which may be of interest to some may have been omitted. For those who wish to explore individual colors further, Coloured Glasses by W. A. Weyl[4] is recommended. Since it was published nearly 20 years ago (1951), much work and some refinements in understanding have taken place in the interim.

ACKNOWLEDGEMENT

The author would like to thank the Brockway Glass Company, Inc. for permission to publish this work. Thanks are also expressed to the Glass Science Department at SUNY College of Ceramics, Alfred University, for the opportunity to make the presentation.

BIBLIOGRAPHY

1. W. D. Kingery, Introduction to Ceramics, John Wiley & Sons, New York (1960).
2. T. Bates, "Ligand Field Theory and Absorption Spectra of Transition-Metal Ions in Glasses", Chapter 5 in Modern Aspects of the Vitreous State, Vol. 2 (ed. by

J. D. MacKenzie) Butterworths, Washington, D. C. (1962).
3. F. A. Cotton and G. Wilkinson, Advanced Inorganic Chemistry, Interscience Publishers, New York (1962).
4. W. A. Weyl, Coloured Glasses, Society of Glass Technology (1951).
5. C. R. Kurkjian and E. A. Sigety, "Coordination of Fe^{3+} in Glass", Phys. Chem. Glasses 9 (3), 73 (1968).
6. W. D. Johnston, "Oxidation-Reduction Equilibria in Molten $Na_2O \cdot 2\ SiO_2$ Glass", J. Am. Ceram. Soc. 48, 184 (1965).
7. F. L. Harding and R. J. Ryder, "Amber Colour in Commercial Silicate Glasses", Canadian Ceramic Soc. Jour., 39, 59-63, (1970).
8. R. W. Douglas and M. S. Zaman, "The Chromophore in Iron-Sulfur Amber Glasses", Phys. Chem. Glasses 10 (4), 125 (1969).
9. K. Karlsson, "Absorption of Iron in Amber Glass", Glastek. Tidskr. 24 (1), 13-19 (1969) (in English).
10. D. Brown and R. W. Douglas, "Carbon-Sulphur Amber Glasses", Glass Technol. 6, 190-196 (1965).
11. S. M. Budd, V. H. Exelby, and J. J. Kirwan, "The Formation of Gas Bubbles in Glass at High Temperature", Glass Technol. 3 (4), 124 (1962).
12. W. R. Manring, "Comments on Reboil Phenomena in Amber Container Glasses", presented at the 30th Annual Conference on Glass Problems, No. 21-22, 1969, University of Illinois, Urbana, Illinois.
13. A. E. Pavlish and C. R. Austin, "Selenium-Ruby and other Glasses Colored by Selenium", J. Am. Ceram. Soc. 30 (1), 1 (1947).
14. H. Rawson, Inorganic Glass - Forming Systems, Academic Press, New York (1967).
15. P. Nath and R. W. Douglas, "Cr^{3+} - Cr^{6+} Equilibrium in Binary Alkali Silicate Glasses", Phys. Chem. Glasses 6 (6), 197 (1965).
16. A. Paul and D. Lahiri, "Manganous - Manganic Equilibrium in Alkali Borate Glasses", J. Am. Ceram. Soc. 49, 565 (1966).
17. A. P. Herring, R. W. Dean, and J. L. Drobnick, "The Use of Rare-Earth Oxides to Give Color or Visible Fluorescence to Soda-Lime Glasses", presented at the Am. Ceramic Soc. 70th Annual Meeting, Glass Division, Chicago, Illinois, April 24, 1968. Reprints available from Molybdenum Corporation of America.
18. J. H. Brophy, R. M. Rose, and J. Wulff, The Structure and Properties of Materials, Vol. II, John Wiley & Sons, New York (1964).

19. G. E. Rindone in Luminescence of Inorganic Solids (ed. by P. Goldberg) Academic Press, New York (1966).

The Fractology of Glass

V. D. Fréchette

SUNY College of Ceramics
at Alfred University
Alfred, N. Y.

INTRODUCTION

Fracture, the ultimate catastrophe in which the strength of a glass article is defeated, is no instantaneous event. It begins at a definable origin and it spreads with measurable velocity to its conclusion. The paths by which it propagates tell a complex story of static and dynamic stress configurations, weaknesses of design and material inhomogeneities. In short, it leaves a record of the event and it leaves it in a form which can be read. It is the present purpose to discuss the details of this record and how they can best be seen and interpreted.

The record is contained in the fracture surface, that is, it consists of all the details of the new surface generated by fracture. Such details are conveniently classified as transient and tendential. The tendential features are those which express long-range shifts in stress fields and they are to be seen in the profile of the fracture surface, i.e., the appearance when viewed from a direction parallel to the fracture surface. These tendential features are those in evidence when fragments of a broken container are reassembled and viewed perpendicular to the container surfaces.

The transient features are to be seen as relatively short-range perturbations in the fracture surface and are best examined by observations perpendicular to the surface under conditions of critical illumination that will be described. They record such subtle influences on the progress of fracture as sonic waves, material inhomogeneities, local stress gradients and velocity effects.

Both transient and tendential details pertain to the fracture once it has been initiated and is already catastrophic. Their study has the additional value of indicating the origin or nucleus of the fracture. If the fracture had not begun it would not, of course, have occurred at all; the origin is therefore the most important single feature and it can be located by examination of the fragments.

EXAMINATION PROCEDURES

Ability to detect the subtleties of the marking on fracture surfaces by optical means is contingent on the choice of optics and depends critically on the arrangement of illumination and on the orientation of the specimen. A number of arrangements have been suggested, including reflected light, transmitted light interferometry and decoration, as with HF fumes. The most effective for practically all purposes remains examination by light reflected from a well-defined source at such an angle that slight local undulations in place of the fracture surface are sufficient to spoil the light reflection so that these regions appear dark against a bright field or, alternatively, bright against a dark field. The effect can be intensified, if necessary, by silvering the surface to extinguish subsurface details and to enhance the brightness of surface reflections. It applies as well to examination with the naked eye or hand lens as to study with the highest magnifications of the microscope. The specimen may be conveniently hand-held for observation at low powers; at higher powers of magnification it is useful to support it in modeling clay in a hemisphere resting on a ring so that it may be rotated and tilted for best effect (Figure 1). Facility with specimen manipulation is frequently essential to correct interpreting of details. Indeed, the more complex situations frequently involve fracture-surface convolutions which make observations quite impossible from a single specimen setting.

Flexibility in choice of conditions for optical examination may be improved by provision of an easily adjust-

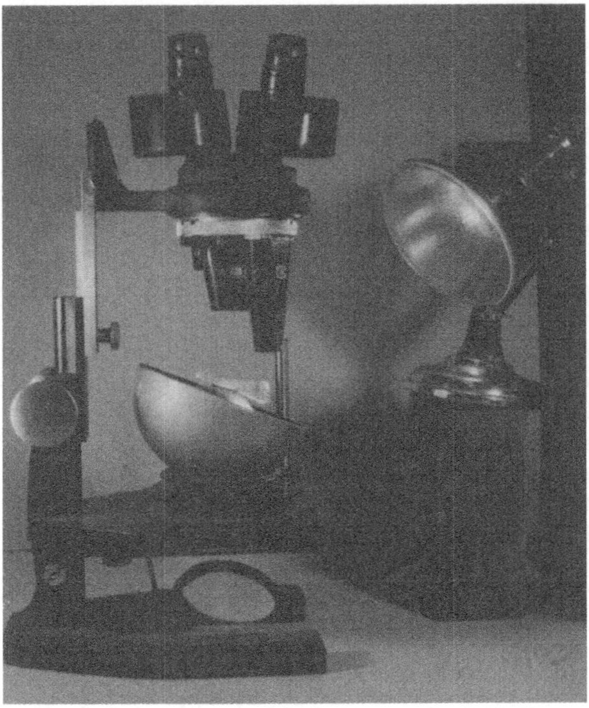

Figure 1. Convenient arrangement for studying fracture-exposed surfaces. The specimen orientation is adjusted by tilting and rotating the hemisphere as it rests on the ring.

able source of illumination. While movable lamps of various designs are commonly available, there is none that is really convenient for use with the sterescopic microscope. Light from a large source, cooled by passage through a flask of dilute copper sulphate and controlled by interposition of easily and quickly adjustable stops, is most effective. Analogous control of illumination must be arranged somewhat differently with the microscope at higher powers; no generalized solution is possible since it must be suited to the microscope design.

THE ANATOMY OF FRACTURE SURFACES

(A) The Origin

The key to many a fracture problem lies at the point of origin and the region immediately adjacent to it. The flaw which serves as the fracture nucleus may be an actual defect of manufacture or damage suffered subsequently or it may be merely one of the unfavorable atomic configurations which are inevitable in nonperiodic structures such as glasses.

Many fractures have a single origin from which all cracks are generated by extension and forking. But cases of multiple origins are not at all rare and indeed they are typical of certain conditions, particularly those of great violence.

Comparison of the origin flaw with other flaws in its vicinity may be enlightening. Where the origin flaw, or more specifically the origin flaw projected on the plane of the fracture which spreads from it, has in its neighborhood other flaws which, even though larger, have not served as fracture nuclei, it may be inferred that stresses must have peaked in its vicinity. The size of the initiating flaw is also significant. In cases where such a flaw is large, the stresses necessary to initiate catastrophe need not have been great. Conversely when the flaw constituting the origin is small it may be inferred that the stress causing fracture must have been correspondingly large. The conclusion is based, of course, on relationships calculated by Griffith, Inglis and others as already discussed under the heading of Strength of Glass.

The originating flaw may give some evidence of its own origin, whether occasioned by a bruise, scuff, scratch or percussion. The near vicinity should be examined for indications of the contributions which may have been made by glass defects such as stones, cords, striae, seeds, or manufacturing defects such as surface threads, mold marks, shear marks, or checks.

There may be more than one fracture origin if the event was an especially violent one. Hot glass articles quenched in water typically show many fracture origins when the thermal shock is well above the minimum necessary to just cause failure.

(B) The Mirror Region

While fracture may originate at a flaw somewhat inclined to the plane perpendicular to the principal tension, fracture spreads in this plane. It begins to spread slowly, accelerating at a rate that is determined by the magnitude of the stress and later also by the inertia of the specimen, since this inertia is responsible for the absorption of kinetic energy. At first the new fracture surface is smooth and rather flat and this stage is appropriately called the mirror region (Figure 2). Its diameter is inversely proportional to the square of the breaking stress.

Under conditions of persistent stress and consequent fracture acceleration, the mirror region becomes misty in appearance (Figures 2, 3). The effect is incompletely understood but it is thought to be the result of sonic waves generated along the fracture front by parting bonds; these sonic waves interreact with one another and with the stress field and they result in local deviations from the average fracture plane. The resulting roughness grows more

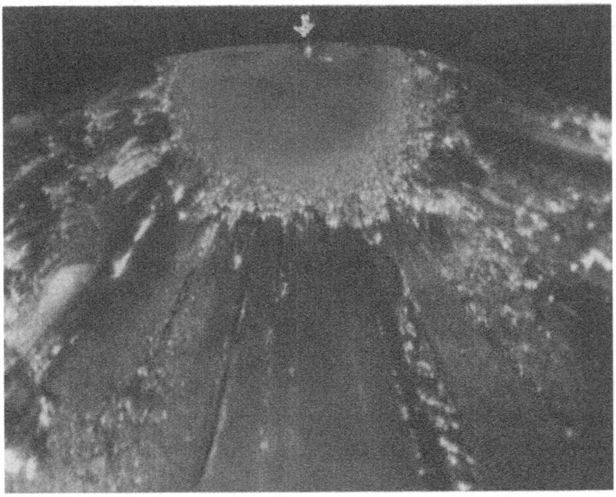

Figure 2. Fracture-exposed surface of glass rod broken by flexure. Origin (marked by arrow) is surrounded by mirror region, mist region and finally radial coarse hackle. 50X.

Figure 3. Details of the mist region formed as fracture attains high velocity. Fracture direction top to bottom. Incident dark-field illumination. 300X.

intense until finally it is terminated by bifurcation or forking to generate a pair of diverging fracture surfaces each of which is rough on a gross scale, a condition known as hackle.

(C) Hackle

Hackle markings consist of ridges running in the direction of fracture propagation. They arise from advance of the fracture front in a series of initially unconnected tongues which only later spread sidewards in a curving course to interconnect and complete the separation. The term hackle is well chosen, for individual members of the hackle pattern are strongly reminiscent of a feather, that portion generated by the advancing tongue representing the quill, flanked by nearly perpendicular markings (secondary hackle) resembling the vane of a feather.

Fast-spreading fractures appear to generate hackle because of independent advance at the front. Whereas slow fractures favor continuity through continual redistribution of stresses along the front, no such cooperation is possible when the fracture is traveling at velocities approaching the velocity of the sonic waves by which communication is maintained along the front. At such speeds continuity is lost as adjacent elements of the front deviate above and below the average plane of fracture. Later these isolated lance-like elements must spread laterally and curve towards one another in order to complete the separation of the specimen. In this process they may again form hackle, not because of high velocity but for a second reason.

Hackle is generated even in slow-moving fractures (Figure 4) when the direction of the principal tension rotates in the plane containing the fracture front. Since fracture in an isotropic medium occurs perpendicular to the local principal tension, such a rotation of stress implies a corresponding rotation of the fracture plane. At the same time continuity with the fracture surface already formed must be preserved; therefore the fracture front splits up into segments each of which rotates, as it advances, to become perpendicular to the new direction of principal tension (Figure 5).Thus the fracture spreads as a family of parallel but not coplanar elements, somewhat like the slats of a Venetian blind. Later, behind the fracture

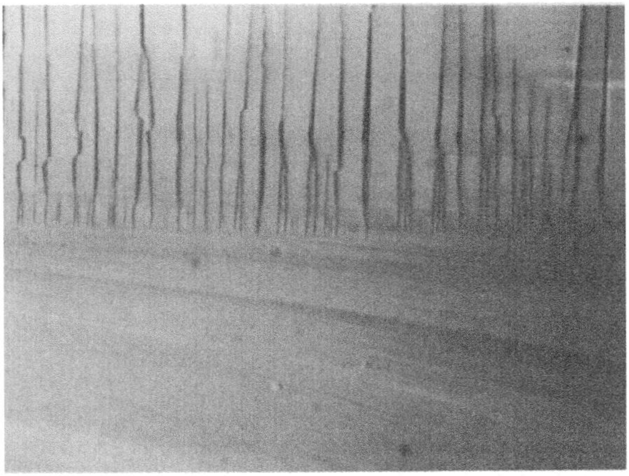

Figure 4. Wallner lines (curved) and hackle (jagged) on fracture-exposed surface of a container whose inside surface is at bottom, outside at top. The fracture front was far advanced at the inside surface indicating that it was in tension relative to the outside. Fracture direction was left to right. 70X.

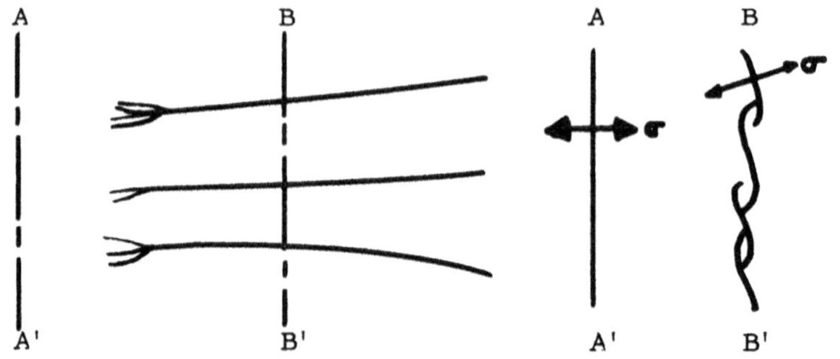

Figure 5. Generation of hackle by rotation of the principal stress. Fracture spread is from left to right. At Section AA' the fracture surface and principal stress are perpendicular to one another; at Section BB' the average fracture plane is no longer perpendicular to the (rotated) principal tension but local segments of it are. Separation (BB') is shown completed in three possible ways.

front, these extend laterally and curve towards one another to complete the parting.

The later joining of isolated tongues of the advancing fracture occurs by lateral extension of all of these portions in both directions. Overlap occurs and, between any two such tongues, there is a sort of competition as to which will be first to effect parting by intersecting the other (BB' in Figure 5). Both may be completed, in which case a splinter of glass is separated from the principal fragments of the specimen. This may involve considerable delay; long after the initial break of a glass specimen the tinkle of this splintering may be heard.

The pattern formed by hackle on the fracture surface has been compared also to river systems. Like river systems hackle ridges may first appear in profusion, joining one another much as tributaries flow together to form a river (the "river" in the fracture case consists of a hackle ridge of increased step height). The rivers tend to diverge from one another downstream. In their later stages, under new conditions of specimen cross-section or of stress configuration, they may also split to form "deltas", making the analogy complete.

(D) Rib Marks

Occasionally, during the spread of fracture, an arrest may occur as the result of a momentarily decreased stress field or a sudden change in the direction of the principal tension. When propagation resumes it is inevitable that the direction of principal tension will be altered and that the plane of the next segment of the fracture surface will be inclined with respect to the preceding. The resulting line of discontinuity is termed a rib mark. Its particular significance is that it marks the shape of the fracture front at the moment of arrest. In all fractures except those in a limited range of very slow velocities (about 10^{-4} meters per second) the shape of the front is convex in the direction of its propagation. The rib mark is useful in establishing, in post-mortem examination, the direction in which the fracture spread. It is also useful in inferring the stress configuration under which the fracture reached the point of arrest. The advanced portion of the rib mark identifies the region in which normal tension was maximum and the trailing portion marks that region where tension was least at the time that the front moved through.

(E) Wallner Lines

The spread of fracture past inhomogeneities, especially checks and scratches at the surface, initiates sonic waves radiating in all directions from such discontinuities. Where these sonic waves intersect the fracture front their displacement stress is added algebraically to the principal stresses. At any particular instant fracture occurs under the stress resultant, which is, in general, less or greater than the principal stress alone and is rotated slightly. As the fracture spreads under these conditions and as successive elements of the sonic wave train overtake it, the plane of fracture oscillates as it proceeds, generating on its surface a wave-like pattern termed Wallner lines. The phenomenon is a very common one and is therefore especially useful in inferring the shape of the fracture front from point to point throughout the fracture event. (Figure 4)

Interpretation of Wallner lines is based on the fact

that the origin of the sonic waves lies always on the older fracture surface, that is, the sonic waves are always coming from behind to overtake the spreading fracture. For qualitative purposes the sonic wave may be considered to intersect all points along the fracture front at the same instant, thus recording its shape at a single point in time. In fact, however, the sonic wave front is somewhat curved and it does not intersect the fracture front at a single instant; it overtakes the trailing regions of the fracture front first, of course, and the advance portions a little later. In the interval the advanced portions will have moved still farther ahead. The practical consequence is that the record left by the Wallner lines represents an exaggeration of the fracture front but this is not a practical difficulty. Near the origin of the sonic waves the Wallner lines give a very much distorted impression of the fracture front as a consequence of the considerable curvature of the sonic waves and because they radiate from a point not negligibly removed from the origin of fracture. This effect is prominent in Wallner lines appearing on the mirror region where careful analysis of the Wallner line shape can be used to calculate the velocity of fracture across the mirror region, the sonic velocity being known. Wallner lines may not appear on the surfaces generated by fractures moving very slowly, because the sonic vibration dies away quickly and is gone before the fracture has proceeded any great distance. Nor are Wallner lines easily seen on the surfaces exposed by very fast fractures because these generate surfaces are so rough that the comparatively subtle Wallner lines are lost in other detail.

(F) Effects at Inclusions

When the spreading fracture surface encounters an inclusion, or comes sufficiently close to one, it is likely to suffer a local change in plane as the result of either stored stresses in the vicinity or relaxation of principal stresses there. Once past the inclusion, the perturbation of the surface dies out. The effect of the inclusion sets in rather suddenly but it decays more slowly, leading to a tadpole-like shape tailing off downstream. (Figure 6) A more subtle effect resembles gull wings on either side of the discontinuity. While its formation has been attributed to generation of a sonic wave at the inclusion, thus classifying it as a Wallner line, that explanation is evidently incorrect because it may be shown that the "wings"

Figure 6. Fracture surface intersecting a bubble and generating a short-lived hackle step and full wing. Fracture spread top to bottom. 70X.

tilt in opposite senses. It is in fact the consequence of failure of the right and left elements of the fracture surface to meet in the wake of the inclusion. The mismatch redistributes stresses in such a sense as to restore the fracture into a single plane and so stress rearrangement, communicated with sonic velocity, causes a tilt in the fracture surface along a wing-shaped line. The line of mismatch and eventual tear-through, which is a special kind of hackle, trails behind the inclusion like a wake and dies out as the halves of the fracture finally regain continuity.

(G) Forking (Bifurcation)

Under conditions of accelerating fracture velocity a point is reached where the fracture plane forks or splits into two diverging planes. This forking occurs at something over half the (transverse) speed of sound in glass. Because of its necessary relationship to fracture acceleration and thus to the existence of energy in excess of that to just cause fracture, the frequency of forking can be taken as an indication of the applied stress and, as in the case of tempered or surface-treated glasses, of the stored energy. As examples it may be recalled that a window pane pinched in its frame may crack across without forking whereas the same pane struck by a hard-hit baseball breaks into myriad splinters as the fracture forks again and again. A tempered oven door window invariably "dices" when broken, i.e., breaks into many small, polygonal fragments, because of the stored energy imparted by the tempering process. (Figure 7)

Preston has given useful rough relationships between the angle of forking and the ratio between the two principal stresses in thin glass. (The stresses in question are those operating at the beginning of fracture and they alter, at the fracture front, as it proceeds.) Table I summarizes Preston's values.

INTERPRETATION

(A) Conditions of Fracture

Some conditions lead to fracture profiles so characteristic that a detailed study of the fracture-exposed surfaces may be unnecessary or may be limited to a few selected fragments for purposes of confirmation. The undulating path of fracture in glass panes broken by strong heating as, for example, by an incandescent lamp of high wattage, is so typical that it may be recognized at a glance (Figure 8). The reason for this wavy character is not clear, however.

Table I. Angle of Forking of Thin Glass

Specimen	Stress Condition	Ratio of Principal Stresses (σ_y/σ_x)	Maximum Angle of Forking
Tube	Torsion	-1	15°
Lath	Crossbending	0	45°
Container	Pressure	1/2	90°
Sheet	Central pressure	1	180°

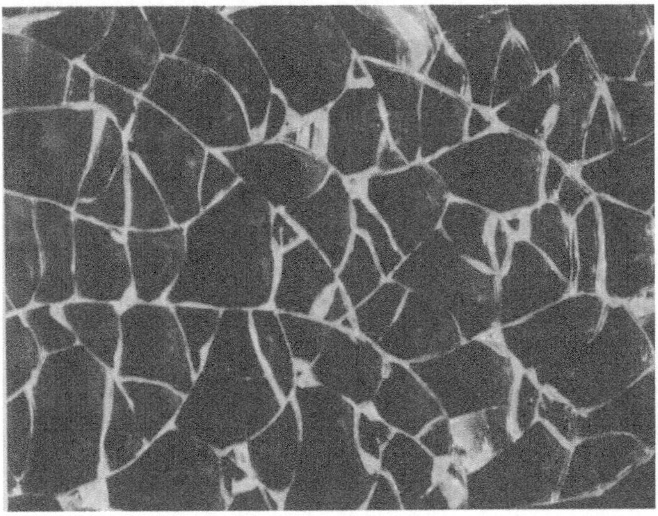

Figure 7. "Dicing" fracture of tempered plate glass. 1.5X.

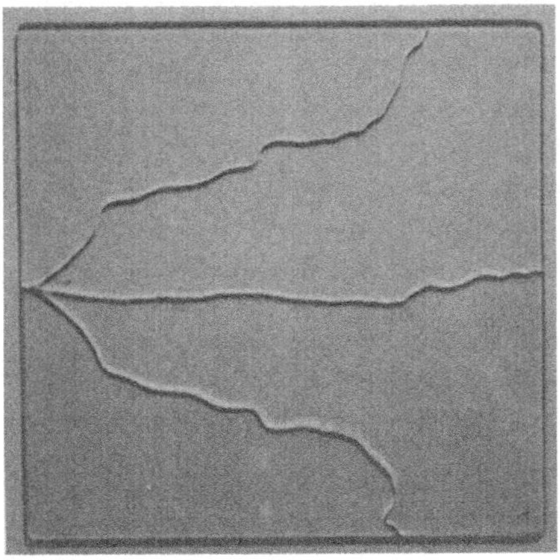

Figure 8. Face of glass pane broken by strong heating on one side. The wavy course of fracture is typical of thermally induced fracture. 1/4X.

Containers exploded by internal pressure yield a pattern (Figure 9) which is often confused with that of an external blunt impact (Figure 10). The impact case is easily recognized by observing the Wallner line shapes in the neighborhood of the origin. These indicate that the glass was under relatively high tension (Wallner lines advanced) at the inside bottle surface on the tangential cracks. The tendency of shell-like spalling from the inside surface near the origin is additional indication of impact. By contrast, exploded containers show nearly symmetrical Wallner lines on radial cracks, while tangential cracks show the Wallner lines advanced at the inside surface.

Filled glass containers dropped on their side often show three fracture origins, one on the inside surface opposite the point of impact and two more on the outside some distance on either side of it where the cantilever action exerted by the collapse concentrates tensile forces on the outside surface. Figure 10 shows this phenomenon.

Impact from a light projectile is easily recognized by the conoid generated, its truncated apex at the point of impact and its broad base at the back surface (Figure 11). The fracture originates in a ring about the area of impact, characteristically at several places, and flares outward as it travels into the glass. A ring of mist is followed by a collar of coarse hackle which may extend all the way to the back surface where the conoid may flare wide with generation of thin flakes (spall). Radial fractures, forming a star pattern, originate principally from the shoulder of the conoid at the beginning of the coarse hackle. More severe impact, especially if the glass is backed by a rigid support, may form additional conoids (Figure 12) and may fragment them. The entire sequence of events can be followed in detail from the markings on the fracture-exposed surfaces and by noting which cracks are continuous and which are discontinuous at intersections. In Figure 11, for example, it is clear that the apex end of the conoid is continuous across its intersections with radial cracks, showing that it formed first. From the beginning of the coarse hackle region, however, the conoid steps at the intersection with the radial cracks showing that once they started, the radial cracks extended faster than the conoid.

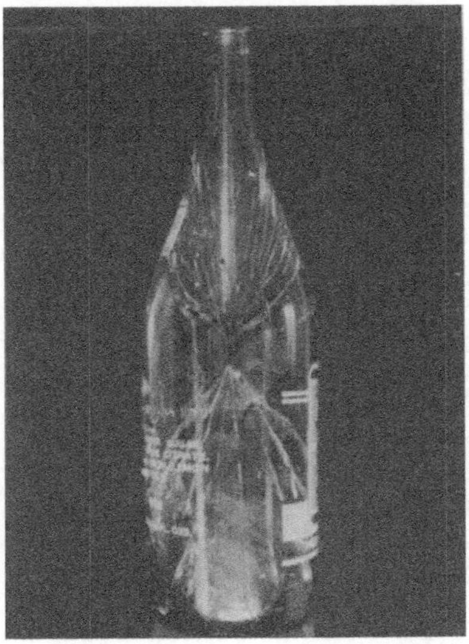

Figure 9. Bottle which exploded under internal pressure. Origin is on the vertical crack at the center on the outside surface.

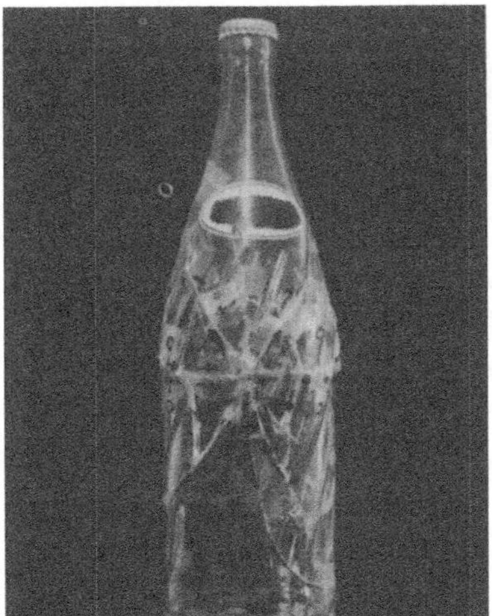

Figure 10. Bottle broken by tipping over on a hard surface while filled. Impact origin is at center on inside surface. Secondary origins lie on the horizontal center line on either side of it on the outside surface.

Figure 11. Conical fracture of plate glass struck by a BB shot. 3X.

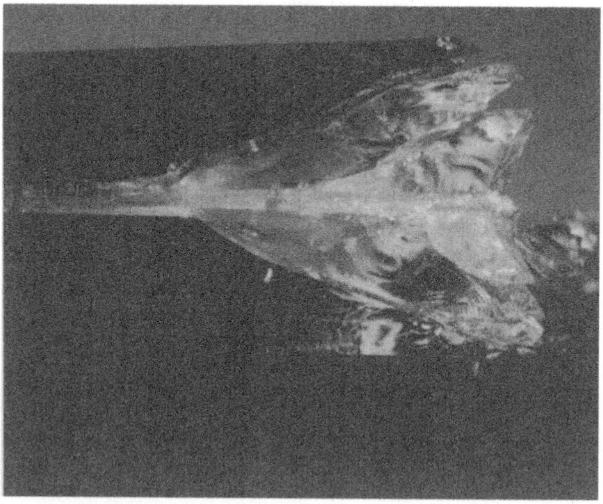

Figure 12. Plate glass backed with an aluminum plate and struck with a sharp projectile. The specimen has been folded back along a radial crack to show the fracture conoids. 3X.

(B) Ware Design

Valuable leads to sound design of glass ware may be obtained by locating the fracture origins of ware broken in simulative tests or better yet, ware broken in service. Where the fracture origins are found to be in a preferred part of the ware, redesign to distribute service stresses more favorably is indicated. Fractology does not itself provide the solution to the design problem but it does identify where the problem lies.

The bulge-ringed tumbler design was a consequence of the observation that straight-walled tumblers most often break in service by tipping over, the origin of fracture being on the rim, which is also the area most subject to the weakening effects of abrasion. Tempering the rim region of tumblers was another answer to recognition of the preferred location of break origins there.

Stress analysis is, of course, instructive in such cases, but it is not always possible to perform on complicated shapes or under ill-defined service conditions.

(C) Destructive Test Interpretation

The interpretation of laboratory destructive tests can be enormously facilitated by fractographic studies, especially in cases where the number of specimens is small. By finding the fracture origin and noting its size, character and location the test result may be evaluated as useful or spurious, and the quality of data enormously improved.

CONCLUSION

The principles of the fractology of glass involve interpretation of simple concepts, best mastered by practice on specimens broken under controlled conditions and beginning with specimens of simple geometry. They are most successful when combined with at least a qualitative knowledge of stress analysis and interpreted with a broad knowledge of manufacturing details and conditions of service.

BIBLIOGRAPHY

Murgatroyd, J. B., "The Significance of Surface Marks on Fractured Glass", J. Soc. Glass Tech., 26, 155T (1942).

Poncelet, E. F., "The Markings on Fracture Surfaces", J. Soc. Glass Tech., 42, 279T (1958).

Preston, F. W., "Angle of Forking of Glass Cracks as an Indicator of the Stress System:, J. Am. Ceram. Soc., 18, 175 (1935).

Pugh, S. F., "The Fracture of Brittle Materials", Brit. J. Appl. Phys., 18, 129 (1967).

Schardin, H., "Velocity Effects in Fracture", in Fracture-Proc. of Int. Conf. on Atomic Mechanisms of Fracture, Swampscott, Mass., 1959, John Wiley and Sons (1959), p. 297-330.

Shand, E. B., "Weakening Defects in Glass", Am. Ceram. Soc. Bull., 46, 111 (1967).

Shand, E. B., "Correlation of Strength of Glass with Fracture Flaws of Measured Size", J. Am. Ceram. Soc., 44, 451 (1961).

Shand, E. B., "Breaking Strength of Glass Determined from Dimensions of Fracture Mirrors", J. Am. Ceram. Soc., 42, 474 (1959).

Frechétte, V. D. and Cline, C. F., "Fractography of Ballistically Tested Ceramics", Am. Ceram. Soc. Bull., 49, 994 (1970).

Varner, J. R. and Fréchette, V. D., "Fracture Marks Associated with Transition-Region Behavior of Slow Cracks in Glass", Applied Physics Letters, 42, 1983 (1971).

The Strength of Glass

William C. LaCourse

SUNY College of Ceramics
at Alfred University
Alfred, N.Y. 14802

INTRODUCTION

The search for new high strength materials has always been one of the more active areas in the materials sciences, and while glass is not normally catagorized with high strength metallic alloys and crystalline ceramics, it does have particular properties which will warrant, or necessitate, its use in many new applications.

Although typical fracture strengths (In the present paper "strength", "modulus of rupture", etc., will be used interchangably to denote the applied stress required to cause fracture.) of commercial glass products can be as low as 1000-2000 p.s.i., there are several hidden factors which should be emphasized. Most importantly is the fact that these observed strengths are only 0.1% of the maximum observed strengths, and an even smaller fraction of the theoretical strength. Moreover, with the increasing importance of strength/weight ratios, particularly for aerospace and oceanographic (hydrospace) applications, glass becomes even more attractive. If consistent strengths of only 75,000 p.s.i. could be attained, the strength/weight ratio would be equivalent to those of several stainless steels. For other applications a combination of high strength and unique optical and corrosion resistant properties might well be exploited, and as will be seen in a later section, the fact that glass becomes stronger under compressive forces makes this material a leading candidate for undersea structural applications.

The present paper is concerned with the current understanding regarding: (1) factors which control the observed strength of glass, (2) various theories of strength, (3) what is being done to increase the design strength. Brief mention is made of possible avenues for future study.

THEORETICAL STRENGTH

Unlike metals which normally undergo a certain amount of deformation by plastic flow before failure, glasses behave in a typically brittle manner. That is, fracture occurs in the elastic portion of the stress strain curve, where the slope is sensibly constant and equal to Young's Modulus, E. This does not necessarily mean that plastic flow is not possible in glass. (Some workers suggest that glass does fail by plastic flow). It does suggest that, as in the case of brittle metallic alloys and crystalline ceramics, some other mechanism of failure becomes operable before reaching the critical stress required for plastic flow.

The lack of a low stress deformation mechanism is a direct consequence of the particular bond types and structure of glass. In metallic crystals bonding is non-directional and the resulting structures are therefore close-packed. In any close-packed material there are several crystal planes in which atoms will have little resistance to shear stresses, this being due to the small interatomic distances between equivalent sites in planes of closest packing, and to the fact that with metallic bonding, atoms have no preference for one particular lattice site. Consequently, if an atom moves from its equilibrium position under an applied stress, it progressively looses its attraction to that site and becomes increasingly attracted to the next closest site. When the atom reaches the mid point it is equally attracted to both sites. When one takes into account motion of dislocations, the critical stress to produce a permanent deformation is even less.

In covalent crystals structures are rarely close-packed due to the directional nature of covalent bonds, and the limited number of bonds which can be formed. Also, when covalent bonds are strained the attraction forces do not decrease and the stress must therefore be sufficient to break the bond before plastic flow can occur. Even dislocation movement is restricted in these crystals, and plastic flow is not normally observed at low temperatures.[1]

In glass, restrictions on plastic flow are even more rigid, since the lack of periodicity precludes the existence of planes of easy "slip". Plastic flow requires breaking bonds at one position, with subsequent movement of the atoms, and bond formation at some other site. It is this latter step which is most difficult in amorphous, non-metallic solids, and, as might be expected, fracture rather than flow, is the natural consequence of bond breakage in glass.

There have been several attempts to calculate the theoretical strength of glass from first principals,[2-4] and although results vary considerably, it is generally concluded that the theoretical strength (the stress required to cause fracture) is on the order of $2\text{-}5 \times 10^6$ p.s.i.. Most theories are based on the force-separation diagram for the Si-O bond in pure vitreous silica, shown schematically in Figure 1. Assuming the initial portion of the curve can be approximated by a sinosoidal function;

$$\sigma = \sigma_t \sin \left(\frac{\pi x}{\lambda/2} \right) \qquad (1)$$

where σ_t is the maximum stress required for bond breakage, and $\lambda/2$ is the range (approximate) of the interatomic forces. a_o, in Figure 1, is the equilibrium separation distance.

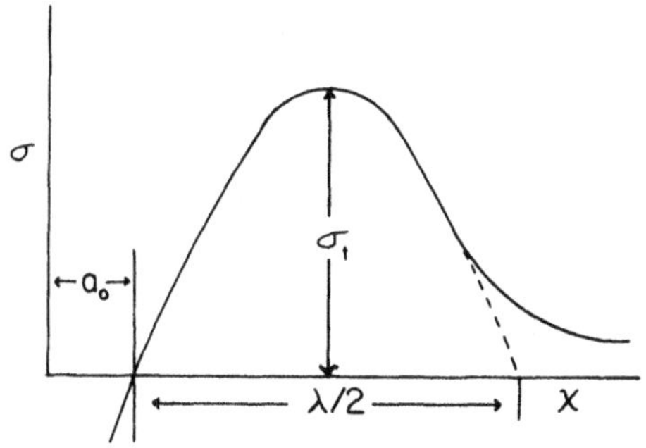

Figure 1. Force-separation diagram for calculating the theoretical strength of glass.

At small strains, where $\cos\theta \approx 1$,

$$\frac{\partial\sigma}{\partial x} \quad \frac{2\pi}{\lambda}\sigma_t \tag{2}$$

and for Young's Modulus

$$E = \frac{\sigma}{\Delta\ell/\ell} = \frac{\sigma}{x/a_o} \tag{3}$$

Differentiating (3) with respect to x, assuming E is constant, yields

$$\frac{\partial\sigma}{\partial x} = \frac{E}{a_o} \tag{4}$$

and equating Eq. 4 and 2 the following expression for the theoretical strength is obtained.

$$\sigma_t = \frac{E\lambda}{2\pi a_o} \tag{5}$$

The problem is now reduced to estimating a_o and λ and obtaining reasonable values for E. It is from this point that most differences in strength estimates arise. Intuitively, however, one would expect λ/a_o to be relatively constant for most silicate glasses, and therefore the theoretical stress should vary in direct proportion to Young's Modulus, E.

To obtain a numerical estimate, it is generally assumed that $\lambda/2$ is $\approx a_o$. With E on the order of 10.6×10^6 p.s.i.[5], σ_t is calculated to be about 3×10^6 p.s.i..

The theoretical strength can also be calculated in terms of surface energies by equating the total work done in causing fracture, to the total surface energy of the two surfaces created by fracture. Thus,

$$\sigma_t \int_0^{\lambda/2} \sin \frac{2\pi x}{\lambda} \, dx = 2S \tag{6}$$

where S is the surface energy. This leads to;

$$\sigma_t = \left(\frac{SE}{a_o}\right)^{1/2} \tag{7}$$

Although estimates obtained from eq.(7) are of the same order of magnitude as that quoted previously the method is generally less reliable since surface energies are not well known for most glasses.[6]

It is emphasized that these are rough extimates due to an inexact knowledge of the force-separation relation and to assumptions regarding the various parameters of eqs. (5) and (7). Hillig[7] has found that in silica, E is not constant but varies with the applied stress according to

$$E = E_o \ (1 + 3.0 \times 10^{-7}\sigma_a) \qquad (8)$$

and at the highest strengths measured, more than 1×10^6p.s.i., E was about 1.6 times the value at zero stress. This implies that the theoretical strength is a minimum of 1.6 times those estimates obtained assuming a constant E.

Hillig considers this dependence of E on applied stress, or more appropriately strain, to be a direct result of the random network structure of vitreous silica. Values of E at low strain result from "straightening out" previously linked Si-O-Si bonds, while higher strains are governed by distortions of SiO_4 tetrahedra and increases in Si-O bond lengths. An extreme example of this type of process is the stretching of rubber which involves molecular uncoiling at low strains and bond stretching at higher strains.

It should be noted that this behavior is not typical of most materials. As a matter of fact, from previous discussions of bonding in metallic systems one might expect the exact opposite behavior, and indeed this is the case. It is also found that most silicate glasses containing at least a few percent network modifiers do not show this behavior either.[8] On the basis of Hillig's arguments this observation is somewhat unexpected and the reasons for this behavior are unclear. The increased modulus for vitreous silica appears to be just one more anomolus property of this intriguing material.

Charles[6] has estimated the value of Young's Modulus for silicate glasses by assuming that the concentration of Si-O-Si linkages is the controlling factor in determining the modulus. Furthermore, using the experimentally determined modulus for fused silica (7.3 Kg/mm^2 = 10.6×10^6 p.s.i.), he finds that E for a series of soda-silicate glasses is a linear function of the concentration of Si-O-Si bonds. This of course suggests that the Si-O-Na bonds contribute little to observed strengths.

Other workers[9-11] have developed emperical formulas for the relation of E to composition. These formulas are generally of the type

$$E = C_1p_1 + C_2p_2 + \cdots\cdots C_np_n \qquad (9)$$

where p_n is the mole percent of oxide n and C_n is an empericallly determined parameter. For SiO_2, C_n is 10.6 irrespective of the glass composition, but coefficients for other oxides may be a function of composition. Those for alkali metal oxides, for instance, depend on the amount of SiO_2, Al_2O_3 and B_2O_3, and coefficients for divalent metal oxides such as MgO depend of the alkali concentration.

Table I shows both observed values of E, and those calculated using formulas developed by Phillips[9] and Capps and Blackburn.[10] Although agreement is quite good, it should be cautioned that relation composition to E is not equivalent to relation composition to observed strength, and indeed the latter relation has yet to be obtained, due

Table I. Observed and Calculated Values of Young's Modulus in Silicate Glasses[9,11]

Composition (mol %)								Young's Modulus. E p.s.i. x 10^6	
SiO_2	Na_2O	K_2O	Al_2O_3	B_2O_3	CaO	MgO	BeO	Calculated	Observed
100	----	----	----	----	----	----	----	10.6	-----
84	16	----	----	----	----	----	----	6.29	6.29
84	----	16	----	----	----	----	----	6.13	6.13
65	35	----	----	----	----	----	----	5.91	5.91
80	16	----	4	----	----	----	----	6.49	6.48
64	16	----	20	----	----	----	----	7.23	7.22
80	16	----	----	4	----	----	----	6.66	6.66
64	16	----	----	20	----	----	----	7.69	7.69
50	14.2	----	----	28.7	14.2	----	----	8.68	8.68

to complicating factors which will become evident in following discussions.

Assigning an intrinsic, or theoretical strength to glass is considered by some to be of questionable value.[12] It is argued that due to the random network structure there exists a distribution of Si-O bond distances and Si-O-Si bond angles. In such a structure many of the Si-O bonds will be "pre-stressed" in the sense that they are at distances greater than the average, and these bonds may be fractured at substantially lower stress levels. As these weakest bonds fail, surrounding bonds must support a greater load and the process continues, eventually leading to crack formation and fracture.

OBSERVED STRENGTHS OF GLASS

Although this is a valid argument it is not particularly relevant in the present discussion of glass strength since even under carefully controlled laboratory conditions, using specially prepared samples, the observed strengths are rarely at a level where this mechanism would be important. The reasons for the low observed strengths are many and varied, but some insight into the factors controlling strength can be gained by examining Table II which summarizes the results of several investigations. Pertinent information regarding surface conditions, and test conditions are included. Table III should be consulted for compositions of the various glasses.

One of the most striking features of this data is the obvious dependence of strength on the condition of the glass surface. This dependence is also shown in Figures 2 and 3. Although the range of values overlaps somewhat the general trend is toward lower strengths with increasing surface damage. The strength of "off the shelf" glass rods (as received from the manufacturer) is rarely over 20,000 p.s.i.,[13] and samples which have been purposely damaged (abraded) by sand blasting or grinding often fail at stresses below 5000 p.s.i.[14] The data presented in Figure 3 shows the effects of normal handling. As noted all samples of glasses receiving no special handling broke at stresses under 25,000 p.s.i. while samples handled in such a way as to minimize surface damage withstood loads of 300,000 p.s.i. and more, although a few still broke below 25,000 p.s.i.

Table II. Observed Strengths of Silicate Glasses.

Glass	SiO_2	B_2O_3	Al_2O_3	Na_2O	K_2O	CaO	MgO	BaO	PbO
Vitreous silica	100	-----	-----	-----	-----	-----	-----	-----	-----
X 8	70.5	0.5	2.6	16.3	1.2	5.7	2.9	-----	-----
R 6	67.7	1.5	2.8	15.6	0.6	5.6	4.0	2.0	-----
0088	70.5	2.25	2.3	11.9	5.3	4.3	3.0	0.1	-----
7740	80.6	12.6	2.0	4.2	-----	-----	-----	-----	-----
"E"	55.2	7.3	14.8	0.3	0.2	17.7	4.3	-----	-----
Lead	56.0	-----	1.3	4.6	8.0	-----	-----	-----	30.0
0080	72.0	-----	3.0	17.0	-----	5.0	3.0	-----	-----

Table III. Glass Compositions.

Glass	Diam. mm	Surface	Temp. °C	Atm.	Load Rate	Observed Strength (psi x 10^3) Minimum	Mean	Maximum	Ref.
SiO_2	1	pristine	25	air	17cm/sec.	80	650	800	15
"	.03	"	"	"	"	400	850	1400	15
"	"	"	"	Vac.	"	600	1100	1600	15
X 8	0.5	"	"	air	----	130	570	780	22
"	"	after 1 hour at 125°C	"	"	----	60	170	250	22
7740	"	pristine	"	"	----	100	425	680	23
"	4	etched	"	"	----	175	250	380	26
0080	.25	abraded	"	H_2O gas	.05 in./min.	14	22	29	20
"	"	"	-170	"	"	10	13	15	20
"E"	.0075	pristine	"	N_2	.39 in./min.	720	840	875	16
"	"	"	+200	"	"	200	220	250	17
R 6	1.3	slight abrasion	25	air	----	50	65	80	18
"	0.7	etched	"	"	----	250	310	380	18
soda-lime	rect.	abraded	"	"	800psi/sec.	6.5	7.5	8.0	14
"	"	"	"	"	1 psi/sec.	4.3	5.0	6.0	14
"	"	"	-170	N_2	800psi/sec.	7.2	18	27	14

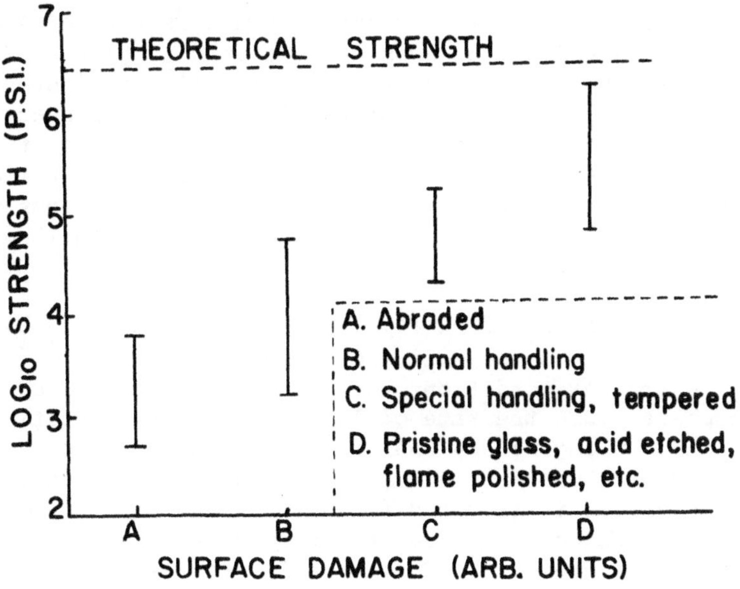

Figure 2. Ranges of observed strength as a function of surface damage.

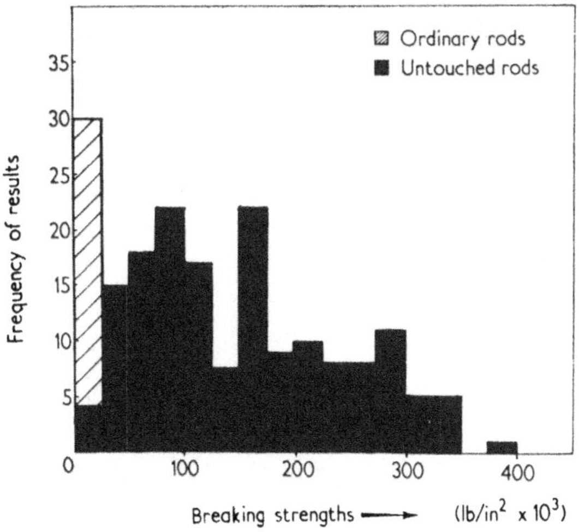

Figure 3. Comparison of strengths of "ordinary" and "untouched" 8 mm diameter glass rods.[13]

These large variations are a common feature of strength measurements and can be even more drastic in glasses whose mean strength is very high. Morley et al[15] for instance, report an average strength of 850,000 p.s.i. for specially prepared fused silica fibers. In this same series of tests, however, one sample fractured at 1,700,000 p.s.i. while another prepared under the same controlled conditions failed at only 100,000 p.s.i., more than an order of magnitude lower!

Another interesting aspect of glass strength, obvious from Table II, is the influence of test condtions;[13-24] 1. the ambient conditions such as temperature and/or atmosphere, 2. the rate of loading, 3. the duration of the load. Changing any one of these test conditions can alter results substantially, and if the surface condition is also varied it is not difficult to imagine the complications which can arise in interpreting results of strength measurements. Considerable progress in this area has been made in recent years however, and some of the more representative results are given in Figures 4-7.

Figure 4 shows the effects of temperature on "as drawn" and etched fibers.[18] For as drawn fibers which presumably

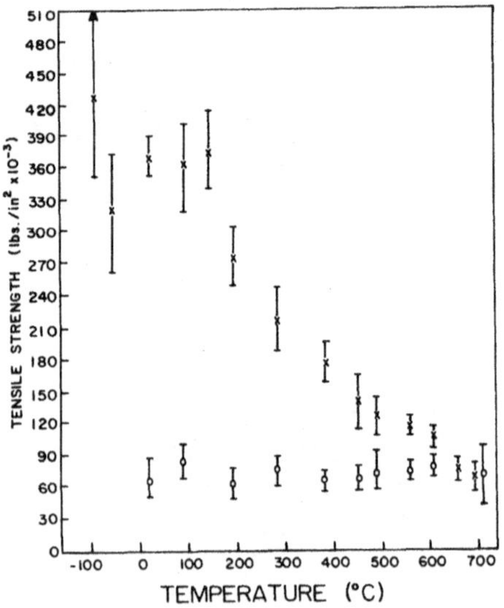

Figure 4. Breaking strength of as drawn and etched fibers as a function of temperature.[18] x = etched glass o = as drawn.

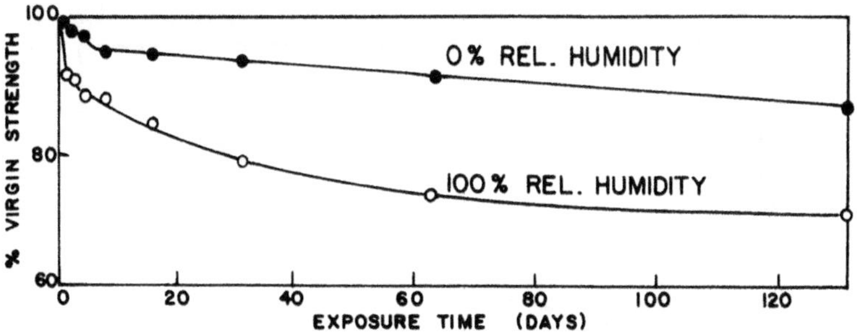

Figure 5. Decrease in strength of pristine glass on exposure to air.[59]

have a small amount of surface damage resulting from normal handling, there is little if any, change in strength with temperature. When the damaged surface layer is removed by etching in an HF solution, the low temperature strengths are found increased by a factor of 5-6, while high temperature strengths are equal to those of damaged glass. Similar strength improvements with etching have been reported by several authors.[25-27]

It should be mentioned that pristine glass (undamaged or etched) shows a decrease in room temperature strength after heat treatments above 50-100°C.[17,23,24] On the other hand, strengths of samples with abraded surfaces are enhanced with ageing, the magnitude depending on the ambient atmosphere, temperature, degree of surface damage, and the load applied during ageing. Figures 5-7 summarize most of the work on ageing effects.

Results shown in Figures 4-7 are indicative of the difficulties encountered by experimentalists and theoreticians alike. For example, it is clear that either testing samples at high temperatures, or simply testing at room temperature after heat treatment at elevated temperatures, produces qualitatively similar results .... a decrease in observed strength. In such cases the problem presents itself as to whether strengths are intrinsically less at higher temperatures, as implied by the top curve of Figure 4, or whether some irreversible, thermally activated mechanism is responsible as implied by Figure 7. This is a crucial point since if reduced high temperature strengths are not an intrinsic property of glass, but a consequence of some

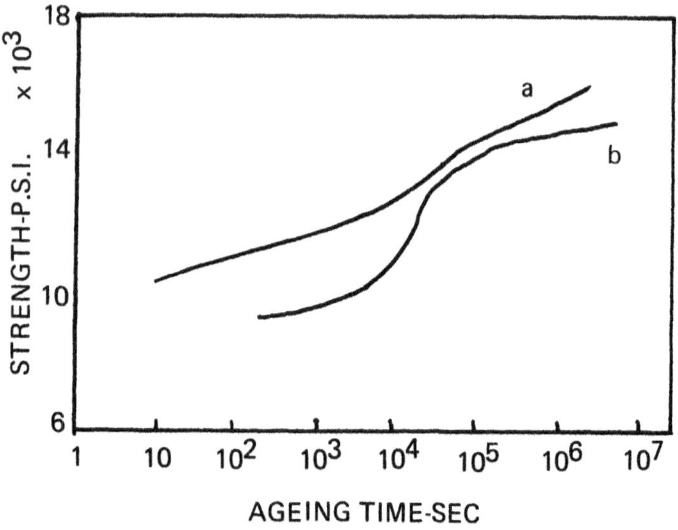

Figure 6. Increase in strength of glass damaged by emery cloth.[45] a) Aged in water. b) Aged in air at 100% relative humidity.

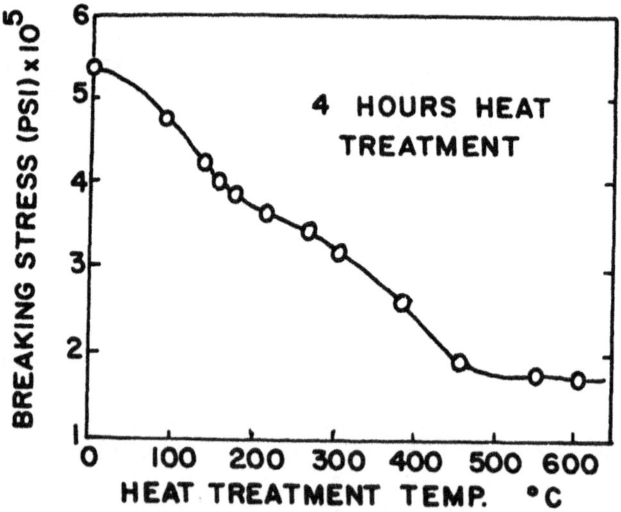

Figure 7. Variations in breaking strength with temperature of heat treatment.[59]

mechanism which operates at high temperature, then there is some possibility of improving high temperature strengths by suppressing this mechanism. This problem will be discussed further in later sections.

Figure 8, taken from the recent work of Ritter[28,29] demonstrates the dependence of observed strength on the rate of loading (dynamic fatigue). As noted, observed strengths increase with the rate of loading, but do not approach the instantaneous strength. The instantaneous strength can be considered as that which would be observed in the limit of infinitely fast loading, and is determined by measuring the strength at liquid nitrogen temperatures or in vacuum where no effect of loading rate is observed.[16]

In the limit of infinitely slow loading rates, (i.e., a constant load applied over extended periods of time) one obtains plots similar to those shown in Figure 9, where "time to fracture" is plotted against breaking stress.[20] The following features should be noted: 1) There appears to be a stress at which failure does not occur even after extended periods of time. 2) For the same stress, samples tested at higher temperatures fail in shorter times. 3) Moisture appears to speed up the process. These delayed

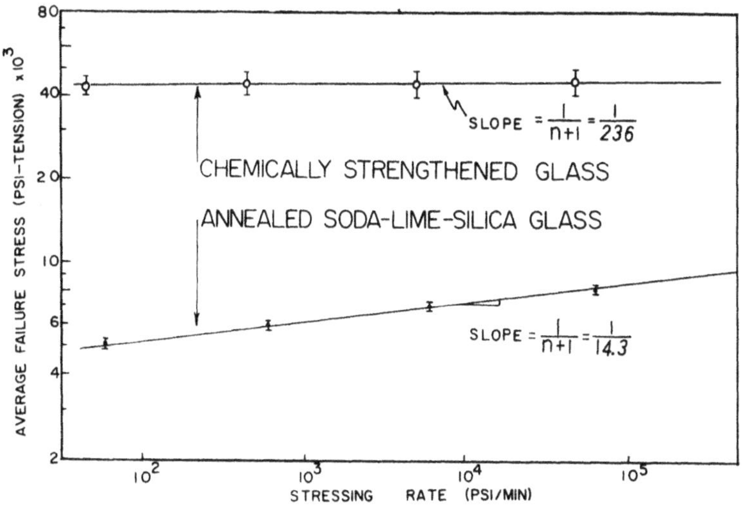

Figure 8. Strength of glass as a function of the rate of loading.[28]

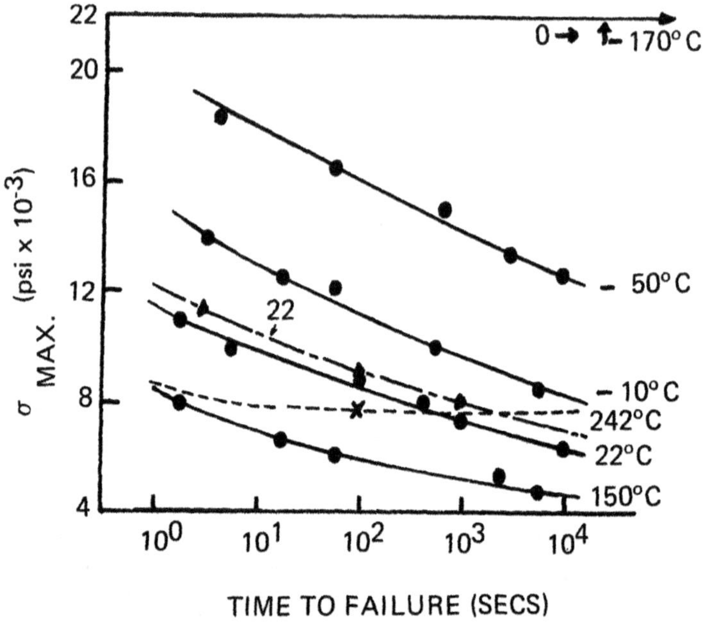

Figure 9. Static fatigue effects in an alkali silicate glass.[20]

fracture phenomena are generally termed "static fatigue" effects. In this paper the term will also include dynamic fatigue effects since the two are essentailly identical in nature.

Considering the several aspects of strength which have thus far been considered a brief recapitulation of the more pertinent results in order. The main points can be summarized as follows:

1. The theoretical strength of glass is on the order or 3-5 x 10^6 p.s.i.
2. Observed strengths can be 2-3 orders of magnitude less depending on the surface conditions.
3. In practically all measurement the scatter of results is large.
4. Glass exhibits static fatigue effects with a complex interdependence of strength on temperature, atmosphere, loading rate and degree of surface damage.

The static fatigue effects can be summarized in terms

of the following partial derivatives:

$$1.\quad \left.\frac{\partial S}{\partial T}\right)_{W,L,A,t} = \text{negative}$$

$$2.\quad \left.\frac{\partial S}{\partial W}\right)_{L,A,t,T} = \text{negative}$$

$$3.\quad \left.\frac{\partial S}{\partial L}\right)_{A,t,T,W} = \text{positive} \qquad (10)$$

$$4.\quad \left.\frac{\partial S}{\partial A}\right)_{t,T,W,L} = \text{negative}$$

$$5.\quad \left.\frac{\partial S}{\partial t}\right)_{T,W,L,A} = \text{negative}$$

where S is the breaking strength, T the temperature, W the water content of the atmosphere, L is the loading rate, A is the degree of surface damage, and t, the time of heat treatment. There are a few exceptions to this summary, for example, the increased strength of damaged glass on ageing in air, but these equations can be considered generally correct.

GRIFFITH MICROCRACK THEORY

Much of the puzzling behavior of glass can be understood by assuming the existence of structural flaw (e.g., cracks, voids, etc.) which can act as stress raisers, and thus cause premature fracture.

The mathematics of stress raisers was worked out over fifty years ago by Inglis.[30] He found that for the elliptical shaped crack of Figure 10, the stress concentration factor, defined as the ratio of the maximum stress developed at the crack tip to the applied stress, can be approximated by

$$\frac{\sigma_{max}}{\sigma_a} = 2\sqrt{\frac{\ell}{r}}$$

where r is the radius of curvature at the crack tip, and ℓ is the crack depth.

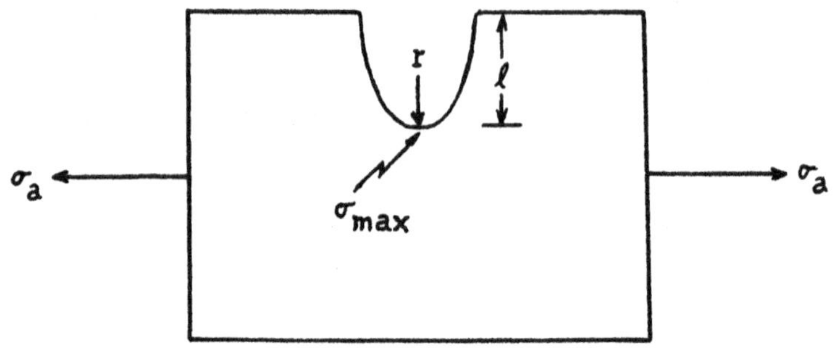

Figure 10. Assumed geometry for calculating the effective stress at a crack tip.

With $\ell = 10^{-4}$ inches and $r = 100$ Å, the stress concentration factor is on the order of 20. This means that an applied stress of only 200,000 p.s.i. produces a stress of 4,000,000 p.s.i. at the crack tip, which is on the order of magnitude of the theoretical strength, and may be sufficient to cause fracture in most glasses.

It should be noted that this mechanism does not apply to a material which fails by plastic flow. In metal, for instance, when the stress at the crack tip approaches some critical value, plastic flow via dislocation motion etc., can occur such that crack blunting occurs (increased r). A secondary effect, strain hardening, via dislocation entanglement, etc, may also occur, leading to a greater effective yield stress at the crack tip. This combination of "blunting and hardening" can arrest crack propagation at low stresses.

In brittle materials no mechanism of blunting or work hardening is available, and once the critical stress at the crack tip reaches the theoretical strength the specimen fractures.

In his classical work, Griffith[31] applied the concept of stress raisers to glass. By equating the elastic energy stored in the region of an elliptic shaped crack tip with the increase in surface energy as the crack grows, he obtained the following expression for the critical applied stress for crack propagation:

$$\sigma_a = \left(\frac{2ES}{\pi a} \right)^{1/2} \qquad (12)$$

where E is Young's Modulus for the glass, S is the surface energy and "a" is the crack depth. The Griffith criteria does not include the radius of curvature.

In his experimental work Griffith introduced cracks of macroscopic size and determined experimentally the constant:

$$\frac{\sigma_a}{(a)^{1/2}} = \left(\frac{2ES}{\pi} \right)^{1/2} = 240 \qquad (13)$$

Using this value for $(2ES/\pi)^{1/2}$ and noting the stress required to cause fracture in specimens which contained no observable flaws, Griffith estimates that undamaged glass contains a certain concentration of microscopic cracks (intrinsic?) for which "a" is on the order of 10^{-4} inches. These flaws were originally presumed to be distributed throughout the bulk of the glass although they are now considered to be purely a surface phenomena.

Hillig[32], in his critical review of the Griffith and Inglis criteria suggests that although the Griffith equation is a necessary condition for crack propagation, it is not sufficient. The Inglis criterion, Eq. (11), is both necessary and sufficient.

More recently Phillips[33] has suggested an interesting alternative to the Griffith Equation. He notes that during manufacturing and handling all glass is subject to a certain amount of surface damage, and he suggests that the resulting crack depth should be inversely proportional to Young's Modulus such that

$$\ell = d/E \qquad (14)$$

where d is a constant of proportionality and depends on the mechanism (e.g. sandblasting etc.) by which the surface is damaged.

Although the Griffith model of intrinsic flaws was widely accepted, for many years there was no direct experimental confirmation of their existence. Early attempts to identify these flaws by direct observation of the surface were inconclusive. Thus, in their early work, Andrade and Tsien[34] were able to produce visible cracks on borosilicate glasses by using a high temperature sodium vapor treatment, but it remained unclear as to whether these were the true Griffith cracks or merely artifacts resulting from the sodium vapor treatment.

This question remained unanswered until more than 20 years later when Ernsberger published results of his work on sodium vapor treatments and on a new ion-exchange technique.[35,36] For full details of this work the reader is referred to the original papers.[35-39] Only the main features will be discussed here.

As a result of his investigations Ernsberger concluded that the Andrade-Tsien cracks were not the true Griffith flaws. Periodic observation of the glass surface during the high temperature sodium vapor treatment revealed no cracks, but upon cooling after the treatment, a reproducible crack pattern formed. On heating, the cracks disappear and a completely different pattern forms during a second cooling process.

Although sodium vapor treatments indicated that the crack patterns developed by sodium vapor treatment were not true Griffith cracks, Ernsberger felt that they probably originated at such flaws. He was able to provide substantial evidence for this hypothesis by his work on ion-exchange.

In this method a surface tension is produced on soda-silicate glasses by a low temperature ion-exchange process in which Li diffuses into the glass and replaces a fraction of the Na ions. When the surface tries to contract to accommodate the smaller Li ions it is restricted by the interior glass and a surface tension results. The beauty of this method lies in the fact that by controlling the exchange process, surface tensions can be controlled to such an extent that crack growth does not occur during cooling. Crack propagation can be induced at room temperature by applying a stress or by moistening the surface, and therefore the origin of the cracks can be determined. (see Rossington, this book, for representative photographs.)

The major conclusions resulting from this study of crack formation under controlled conditions are as follows:

1. Cracks which appear after ion-exchange are not themselves Griffith cracks, but originate at pre-existing flaws, presumed to be the Griffith cracks.

2. These pre-existing flaws are not an intrinsic feature of glass, but result from mechanical damage, surface contamination during heat treatment, or devitrification sites.

3. Glasses having extremely high fracture strengths (etched glass or other pristine surfaces) do not contain Griffith cracks.

The mathematical discriptions of Inglis and Griffith, together with the experimental results obtained by Ernsberger regarding the origin of stress raisers, provide a solid basis for the understanding of many experimental observations described previously.

Specifically, the dependence of strength on the degree of surface damage (Figure 2) is fully accounted for by application of the Griffith or Inglis equations, where strength is directly related to the depth of surface cracks. Glass purposely damaged by sandblasting, grinding, etc. would be expected to contain rather deep flaws and therefore, a low strength. Similar arguments could be offered for "off the shelf" glasses. On the other hand glasses which have been given careful handling presumably contain fewer and less severe cracks leading to the increased strengths shown in Figure 3.

Results for pristine surfaces and etched glass are only partially accounted for. While it is true that strengths of glasses containing no observable Griffith cracks are much higher than for normal glass, they rarely approach the theoretical strength. It must be assumed that either some other factor controls at high stresses, or the ion-exchange technique is incapable of detecting very small flaws. This problem is discussed more fully in later sections.

Extreme variations in strength can be accounted for by simple statistical arguments. The probability that a glass sample will fail at some particular applied stress will be determined by the distribution of crack geometrics and concentration. The statistical approach will be discussed further in a later section.

Ernsberger's conclusions regarding possible flaw generation by surface contamination and devitrification, are at least partial explanations for effects of heat treatment and temperature on strengths of pristine glass. He was able to show that either process can lead to microcrack formation on previously undamaged surfaces. The wide variations in strengths of sensibly flawless glass may be related to similar effects, since most glass is exposed to contamination during manufacture.

The concept of stress raising surface flaws is therefore seen to be a powerful tool for understanding time independent properties of glass. However, in order to account for the various effects catagorized as static fatigue, it becomes necessary to postulate at least one further mechanism which may enhance the effect of stress raisers, and cause the time, temperature and atmosphere dependence of strength.

There are two possible mechanisms which can account, at least partially, for observed behavior: A) Surface Energy Lowering B) Stress Enhanced Corrosion.

A) Surface Energy Lowering: In the Griffith model (Eq. 12), the critical stress for crack propagation is directly proportional to $S^{1/2}$. It would, therefore, seem reasonable that any gas or fluid which decreases S in the region of the crack tip, would cause a corresponding decrease in the critical applied stress[40]. Assuming, for present purposes, that this is indeed the case, the temperature dependence of strength on long term loading could be explained by assuming that diffusion of water, or some other species, to the crack tip, controls the rate of crack propagation. Effects of humidity and vacuum are explained by the availability of the diffusion species.

Hillig[32] argues that this mechanism is inconsistent with the kinetics of static fatigue, and also presents some indirect evidence against such an effect. The failure time for "flawless" silica glass in dilute solutions of NaOH or HF is a sensitive function of both stress and concentration.[7] With dilute solutions it is argued that little change would be observed if surface energy were the controlling factor.

Despite these arguments there are scattered indications that this mechanism can be important in some instances.

The increase in crack velocities in the presence of certain non-corrosive gases is one such observation.[41] There does not appear to be sufficient information in the literature to ascertain the relative importance of the surface term, but its possible influence cannot be ruled out.

B) Stress Corrosion: The most widely accepted explanation for static fatigue was first given by Charles[19-21] and later refined by Charles and Hillig.[42,43] The basic assumption of this model is that the activation energy for corrosion of a surface is a function of the tensile stress acting on that surface. In the first approximation we have,

$$Q_a = Q_o + \sigma \left(\frac{\partial Q}{\partial \sigma} \right)_{\sigma=o} \qquad (15)$$

where Q_o is the activation energy in the absence of a net stress and σ is the effective stress at the point in question.

If the existence of surface flaws is taken into account one more term must be included because the surface energy of the solid is a function of its radius of curvature. This will also have an effect on the effective activation energy.

Including this surface energy term the corrosion rate, ν, normal to the reaction interface can be described by an Arrhenius equation of the form:

$$\nu = A \exp - \left[\frac{Q_o + F(1/r) + \sigma \left(\frac{\partial Q}{\partial \sigma}\right)_{\sigma=o}}{RT} \right] \qquad (16)$$

where F is the surface free energy and F(1/r) implies it is inversely proportional to the radius of curvature, r. At small radii of curvature it is seen that this term may very well control the rate. Also, the curvature is extremely important in the Inglis equation and will, therefore, influence the last term in Eq. 16. Crack geometry is therefore of extreme importance in a stress corrosion mechanism, with the radius of curvature being relatively more important than crack depth.[43]

It is emphasized that in the theory of rate processes,[44] the overall rate of a reaction involving a series of rate controlled steps is determined by the rate of the slowest step. The stress enhanced corrosion model described above implicitly assumes that chemical reaction at the crack tip is the slowest process and, therefore, will be the controlling mechanism. However, other possibilities exist. For example, there are several types of chemical or physical processes whose rate is controlled by one of the following steps, all of which may be important for the case in hand.

1. Diffusion of the reacting species to the reaction site.
2. Desorption of the reaction products.
3. Diffusion of the reaction products away from the reaction site.

In the first case, it is clear that if the reactant is not available at the surface, the rate can be no faster than the "arrival of the reacting species. A good example would be diffusion controlled crystal growth from a glass. The second case is applicable when the reaction product remains on the surface and decreases the concentration of reaction sites. Cases 2 and 3 can be combined to produce a situation in which the initial reaction rate is controlled by reaction at the surface, while for longer times the rate may be controlled by diffusion of reactants through the reaction products.

Another complication is a possible change in the rate controlling process with temperature. Due to differences in activation energies, the rate may be controlled by diffusion at one temperature, but by chemical reaction at some other temperature.

In their mathematical formulation of stress enhanced corrosion, Charles and Hillig develop relationships describing the change in crack geometry as a function of time, temperature and applied stress. The important results are shown schematically in Figure 11. The case of no applied stress, or stress independent corrosion, is shown in Figure 11a. Crack blunting results due to a sensibly uniform corrosion over the entire crack. The surface energy term will tend to accelerate blunting somewhat, since corrosion is enhanced at convex surfaces (the glass surface) and retarded at concave surfaces (when compared to a planar surface).

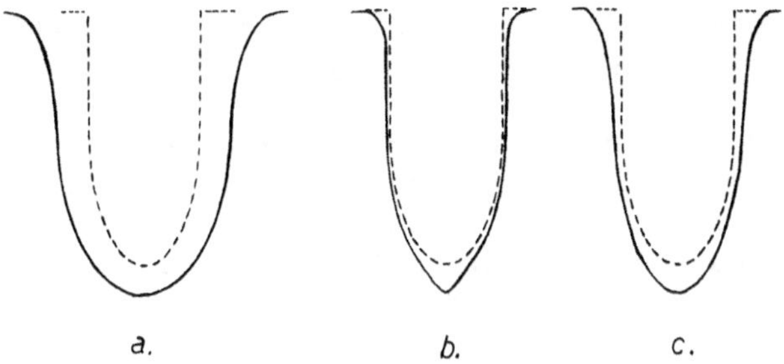

Figure 11. Crack geometries resulting from stress enhanced corrosion. a)Crack blunting. b) Crack sharpening. c) Radius of curvature remains the same. This case defines the "fatigue limit".[42]

Figure 11b. shows the surface resulting from corrosion in the presence of a large applied load. Since the stress is maximum at the crack tip, the last term in the exponent of Eq. 16 becomes most important. The corrosion rate is therefore maximum at the tip, leading to continuous sharpening and elongation.

The two limiting cases described above suggest the existence of some intermediate stress at which crack blunting, as a result of the surface energy term, and normal ageing, is exactly balanced by crack sharpening due to the applied stress. In this case, the resulting geometry is given in Figure 11c, and the applied stress leading to this result is called the fatigue limit. At stresses below the fatigue limit fracture is not observed since blunting is more favorable, leading to a decrease in the radius of curvature and the stress concentration factor.

Above the fatigue limit, crack sharpening progresses continuously with time, and the results of Ritter,[28] showing a decreased strength with decreasing rate of loading, are a direct consequence of this fact. As long as the rate of crack sharpening is finite, a decreased strength will be observed with finite loading rates.

Effects of temperature are also accounted for since stress corrosion involves a thermally activated chemical attack. Although there are complications due to possible changes in the rate determining processes, assuming no change in mechanism, one would expect to observe a linear plot of $\ln \nu$ vs. $1/RT$, with the slope equal to the activation energy. As shown in Figure 12 this is indeed the case for static fatigue in soda-lime silica glass.

The activation energy determined from the slope is about 19 kcal/mole which interestingly, is close to that for diffusion of Na^+ in this glass and suggests that this diffusion may be the rate controlling process in the static fatigue of this particular glass.

The slope change in the high temperature portion of Figure 12 implies a change in the rate controlling process, and is one example of several possible complications. In this particular case Mould[45] has suggested that desorption of the reacting species (water) occurs. The process is then limited by the amount of reactant which can get to the crack tip.

This of course brings in the effect of ambient atmosphere, and again the model is consistent with observed behavior. In vacuum measurements the observed increase in strength and decreased sensitivity to long term loading are attributable to the negligible water vapor content. Similarly, results shown previously in Figure 9 indicate a direct relation between water content and resistence to

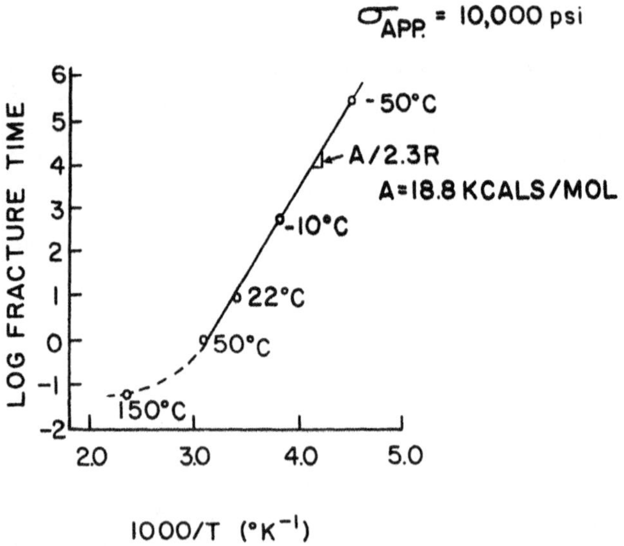

Figure 12. Temperature dependence of the fracture time for a constant applied stress.[20]

stress corrosion. Although no mention has been made of other reactants it should be clear that any corrosive media will tend to accelerate static fatigue effects. This can be an important design consideration, particularly for applications in which the glass will be exposed to corrosive vapors or liquids.

UNIVERSAL FATIGUE CURVE

Perhaps the acid test for any theory of static fatigue is its ability to satisfactorily account for the "Universal Fatigue Curves" (UFC) shown in Figure 13.

The existence of a UFC was first demonstrated by Mould and Southwick[24] in their comprehensive study of static fatigue. Similar curves, for any particular composition, can be obtained by performing the following experiments (For details see Mould and Southwick[24,46] and Mould[45,47]):

1. Samples are given a known, reproducible surface treatment, for instance grit-blasting, grinding with an abrasive of known grit size, or etching in a standard solution.

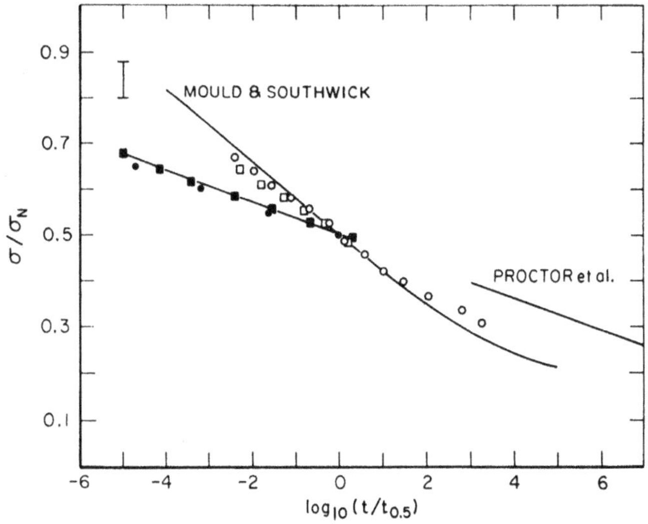

Figure 13. Universas Fatigue Curves for soda-lime-silicate glass (O,□) and vitreous silica (●,■).[81]

2. The instantaneous strength is obtained for samples with each surface treatment by testing at liquid nitrogen temperatures.

3. The dependence of tensile strength on load duration is obtained for a given temperature and atmosphere. This results in curves similar to those of Figure 9. Maintaining a constant temperature and atmosphere is critical, and is best done by immersing the sample in a water bath.

The UFC is now obtained by plotting

$$\frac{\sigma}{\sigma_N} = f\left(\frac{t}{t_{0.5}}\right) \tag{17}$$

where t is the "time to failure" for a load σ, σ_N is the load required to cause fracture at liquid nitrogen temperatures <u>for samples given identical surface treatment</u> and $t_{0.5}$ is the time to failure for a load

$$\sigma = .5\ \sigma_N \tag{18}$$

The surprising result is that for all glasses of the same composition, the points thus determined all lie on a single line (UFC) regardless of the type or severity of the surface damage. As shown in Figure 13, the UFC is unique for each particular composition. Each temperature and atmosphere also produce a unique curve.

The stress corrosion model satisfactorily accounts for the general shape of the UFC and also predicts a temperature atmosphere and compositional variation. It fails, however, to account for the dependence of $t_{0.5}$ on the liquid nitrogen strength.

$$t_{0.5} = A \exp\ (B/\sigma_N^{\ 2}) \tag{19}$$

A and B are independent of σ_N, but seem to be dependent on the <u>type</u> (not the severity) of surface damage.[24] It was found that glasses having a large initial crack depth and low initial strength weakened at a slower rate (when tested at a stress determined by Eq. 18) than did glasses with smaller cracks and larger initial strengths.

This behavior is not predicted by the stress corrosion model since there should be no basic difference in the chemical attack at the tip of a long crack and that in a shallow (but equally sharp) crack. This failure is not a sufficient condition to rule out the stress corrosion mechanism for static fatigue, as has been suggested.[48,49] The assumption of a stress enhanced corrosion may still be valid, but the rate controlling process may be diffusion of the reacting species to the crack tip. Mould and Southwick[24] have shown that this mechanism leads to the correct dependence of $t_{0.5}$ on σ_N. Combining this mechanism with preferential attack at the crack tip might possibly reconcile this discrepancy

There have been other examples in which stress corrosion does not appear to be applicable. In one case, Proctor et al [50] found that static fatigue, in high strength silica fibers and rods, is consistent with a stress corrosion model, with perhaps one notable exception. They found that pre-stressing glasses for short periods of time at some high load, actually increased the "time to failure" (compared to times for glasses with no pre-stressing) when these glasses were tested with decreased loads.

The important point here is that the pre-stress loads were obviously above the fatigue limit. (Although Proctor et al, do not quote a fatigue limit, glasses were observed to break at lower stresses. The pre-stress load is therefore above the static fatigue limit by definition!) This fact would appear to rule out crack blunting since this only occurs at stresses below the fatigue limit.(Figure 11)

There is also some evidence, admittedly sketchy at the present time, that contrary to the earlier conclusions of Gurney and Pearson,[51] glass may indeed exhibit a cyclic fatigue effect. Whitney et al[52] have shown that cyclic loading increases the strength. Other authors [53,54] have recently verified the effect, and although the stress corrosion model invoked, it is not clear that the results are consistent with such an effect, particularly if the load in considerably above the static fatigue limit, as noted previously. More work in this area is clearly called for.

COMPOSITIONAL EFFECTS

Before closing this section, a comment regarding the dependence of strength on composition would be in order. Ommission of this seemingly crucial variable in previous discussion has not been unintentional. There is simply too little reliable information available to make any valid conclusions. Other effects such as surface damage and static fatigue are probably orders of magnitude more effective in determining strength.

There have been a few studies[55] directed at determining the compositional dependence of strength, but in light of previous discussions, results must be considered suspect due to inherent experimental difficulties. An observed increase in strength may be due more to an increase in chemical durability (reducing stress corrosion) with composition or, as implied by Eq. 14, to an increase in abrasion resistance. Clearly any study of this nature must include an almost impossible control of experimental test conditions.

STRENGTH OF GLASS FIBERS

One feature of reported data, evident from Table II, is that glass fibers generally exhibit a greater modulus of rupture than does bulk glass, and the question presents itself as to whether this is an intrinsic feature of fibers. Early work did seem to indicate that strengths were directly related to fiber diameter[56,57], and several authors have suggested that there is an inherent structural difference, namely a preferential alignment of molecular units along the fiber axis, which accounts for the increased strengths.[58] Using statistical arguments, other workers have suggested that, due to a size effect, the probability of finding cracks of sufficient size to cause fracture, decreases as the size of the fiber decreases. Similar arguments can be used to account for an observed dependence of strength on fiber length.

Recent experimental evidence tends to discount the suggestion of inherently stronger fibers. Several years ago, Greene[12] had reported strengths on the order of 200,000 p.s.i. for bulk commercial glass. More recently, Proctor[13] has obtained room temperature strengths of up to 500,000 p.s.i. for 8 mm soda-lime glass, and 400,000 p.s.i. for an 8 mm borosilicate. These values compare favorably with a room temperature strength of about 550,000 p.s.i. for fine "E" glass fibers[16]. Moreover, work by Thomas[59] and Otto[60] has shown that high strength fibers do not generally exhibit a dependence of strength on sample diameter as long as care is taken to form the fibers under nearly identical conditions. There does appear to be a dependence on the forming conditions, however. Brannan[61] has also shown that fiber strengths are also isotropic, which implies that bonding is isotropic.

The latter conclusion, however, is not necessairily true for all diameters, and all drawing techniques. Some recent work by Bartenev[62] is interesting in this respect. It was reported that glass fibers from the Al_2O_3-B_2O_3-SiO_2 system produced by a special drawing technique, do show an increased strength with decreasing fiber diameter, although no effect on fiber length was detected. Moreover, removal of a surface layer, by etching, actually caused observed strengths to decrease from about 430,000 p.s.i. to 300,000 p.s.i.. Strengths of etched fibers were found to depend on the length.

Bartenev concluded that the greater strengths of unetched fibers was due to a greater strength of the structure itself. In the drawing technique used in this work, the fiber surface was maintained at a temperature near T_g, while the interior surface was significantly hotter. As a result, the surface layer underwent molecular orientation during the initial draw, and due to the low temperature, this oriented structure was presumably frozen in. The interior would also undergo a similar orientation initially, but the temperature is above T_g, and after the drawing tension is reduced, structural relaxation could occur.

This experiment is of considerable practical importance, but unfortunately there has been little subsequent work. No one seems to have tested for anisotropy in these particular fibers, although it would be expected.

The previous results are interesting, but fail to account for high strengths in non-oriented fibers, in which an intrinsic difference can be ruled out. One plausible explanation appears to be that fibers are more resistant to mechanical damage. Touching fiber ends together produces little change in observed properties, while a large decrease in strengths occurs for bulk glasses.[63] Simply touching silica rods with a finger can lead to drastic strength decreases.[26] It is therefore asserted that although exceptions are possible, fibers are not inherently stronger than bulk glasses. Differences in measured strengths are probably due to the greater resistance of fibers to surface damage resulting from handling and testing procedures.

COMPRESSIVE STRENGTH

With the current interest in undersea exploration glass has gained considerable attention as a possible structural material for deep submergence vehicles.[64,65] This is due in part to the fact that even abraded glass tends to be extremely strong in compression. (Note that compression stresses "close" rather than propagate cracks.) Other factors are the extremely attractive strength to weight ratio and the fact that most other materials, particularly metals, do not behave well under compressive loads. Hollow metal spheres, for instance,are less able to withstand shock with increasing hydrostatic pressure while glass spheres actually become more resistant.[64] This is of great importance for undersea applications.

Reasons for the poor behavior of metals can be understood from Figure 14. Yielding under an applied load is a natural consequence of the metallic bond and the periodic crystal structure. On the other hand, glass is covalently bonded and,moreover, possesses a random network structure, in which there are no slip planes, or dislocations. As a

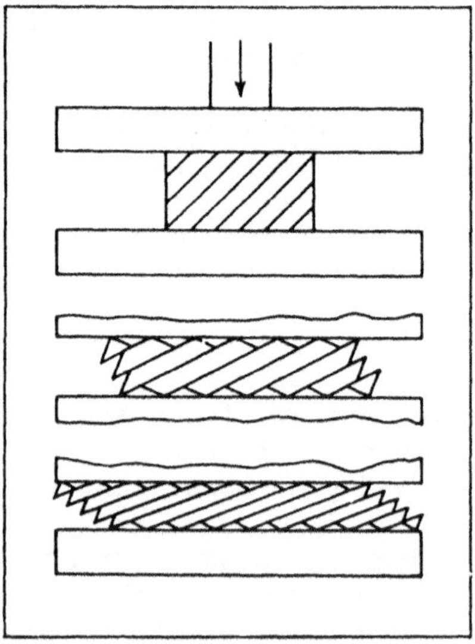

Figure 14. Effect of pressure on crystalline materials.[65]

result the glass tends to withstand compression forces until actual densification occurs.[66] (If buckling is suppressed.) Compressive strengths for hydrostatic loading can be up to 1.2×10^6 p.s.i. for fused silica and about 1.0×10^6 p.s.i. for borosilicate glass.[66]

OTHER THEORIES OF GLASS STRENGTH

Thus far only the theory of surface stress raisers has been discussed and for the most part plausible explanations of observed properties are possible within the general framework of this theory. There are other theories of glass strength, and, as will be seen, virtually all of them take into account the pre-existence of surface flaws, although in a few cases discussed below an assumption of pre-existing flaws is unnecessary.

Of the six models listed below only the plastic flow theory of Marsh[48,49] will be discussed in depth.

A. Statistical Models
B. Thermally Activated Flaw Initiation
C. Mobile Ion
D. Intrinsic Flaw
E. Levengood's Dislocation Model
F. Plastic Flow

A) As opposed to other models, the statistical approach to glass strength is based on the existence of microcracks, and on the concept that a "chain is only as strong as its weakest link." It is generally assumed that the surface is characterized by a random distribution of flaws. At each applied load one can determine the number of flaws per unit area which satisfy the Griffith (or Inglis) criteria for fracture.

These statistical models have been applied with considerable success to glasses which have undergone a considerable amount of surface damage, and are of some value in determinations of design strengths.

Application of such theories to high strength glasses has been less rewarding and in some cases actually led to erroneous interpretations. As a result of this approach, at one time it was generally agreed that the observed strength of glass should be a function of the area under maximum load. Thus, fiber strength should vary directly with the diameter and length, and glasses tested by bending should be weaker than those tested in tension.

It is now apparent, however, that while this line of reasoning has merit for glasses having a large number of flaws, it is questionable for high strength glass, which Ernsberger has shown may have no intrinsic, or pre-existing, flaws.[36] Furthermore, most statistical models are not capable of accounting for the various static fatigue data presented earlier.

B) Thermally Activated Flaw Initiation:[68,69] There have been several attempts to relate the temperature dependence of strength to a mechanism by which thermal motion of the lattice either creates a microcrack, or causes an existing one to grow. Such theories are generally based on a stress dependent rate process. Under zero stress, rates of bond breakage and formation are equal, but under an applied stress the activation energy for the forward reaction (dissociation) is reduced. The arguments are similar to those of Charles and Hillig[42] in their stress corrosion model.

It is argued that at some temperature and/or stress, the rate is such that there is a finite probability of enough bonds breaking in one particular area to create a crack, or to cause an existing crack to grow.

Hillig,[32] in his critical review of glass strength, suggests that although this type of a mechanism may be a "practical barrier" to attaining the theoretical strength of glass, it plays no role in determining observed strengths. Furthermore, there is now substantial evidence that much of the increase in strength below room temperature is due to the decreased effect of stress corrosion.[16] At temperatures below liquid nitrogen, the strength is almost independent of temperature, and recent data[66] discussed at length in a later section, indicates that virtually no portion of the strength-temperature curve is consistent with such an effect.

C) Ion Mobility: Cox[70] has recently suggested that both the static fatigue and fatigue free behavior of glass can be accounted for by assuming a localized bond weakening effect of mobile ions, presumably alkali. Cox calculates the weakening effect of a charged Na^+ ion in a neutral interstitial site. The Na^+ would normally be in a charged site around a non-bridging ion, but due to its mobility, there is a finite probability of finding it elsewhere. Its presence in the neutral site causes a bond weakening, due to elastic strain energy, and to electrostatic terms (oxygen is attracted and silicon repelled). When a sufficiently large number of bonds are weakened in a localized area, instantaneous flaw formation can result. Fatigue effects are explained by assuming a stress enhanced migration of Na^+ ions to the crack tip. Not enough experimental data is available so as to make a further discussion profitable at this time. However, some recent information[66] to be discussed in a later section, allows the possibility of such a mechanism.

D) Intrinsic Flaw Theories: The fact that glass always fails below the theoretical strengths, even though surfaces may be sensibly free of mechanically produced flaws, suggests the possibility of some intrinsic flaw, associated with the glass structure. It has already been pointed out that Griffith cracks are not intrinsic, but due to mechanical or chemical damage.

Gurney[71] notes that "E" glass fibers (and glasses in general) undergo a significant increase in density on annealing after a quick quench. This supports the idea of submicroscopic voids in the structure which can act as stress raisers. Rough calculations indicate that resulting stress concentration factors on the order of 3-10 are possible, which is consistent with observed strengths.

Evidence for some type of intrinsic flaw is also found in the work of Hillig[7] (see Figure 15) and others[67], who find that their high strength data can be accounted for if one assumes a bimodal distribution of (surface?) flaw sizes. Defects leading to failure at low stresses can be attributed to mechanical damage, with flaw sizes on the order of Griffith cracks, while high strengths may presumably be correlated with intrinsic flaws whose size may be as small as 10Å.

As evidence against a strength limited by internal voids or bubbles, it can be argued that fracture always originates at the surface. This is admittedly true for glasses failing at low applied stresses, but has not been proven for glasses

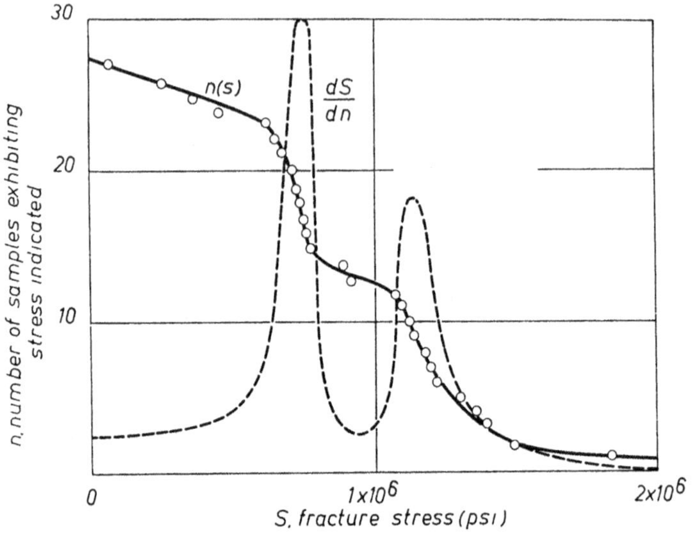

Figure 15. Fracture stress of vitreous silica at 78°K.[7]

failing at stresses approaching the theoretical strength. In fact, these high strength specimens may fracture so catastropically that the glass shatters into dust. In such cases it is clearly impossible to determine the origin of fracture.

E) Dislocations: Levengood[72-75] has suggested the intriguing possibility that glass contains dislocation-like defects, (it is emphasized that these defects are not claimed to be exactly like those found in crystals) which can be rendered visible by a special etching technique.[72] These defects are observed to "move" under an applied stress, in contrast to Griffith cracks which are found to be stationary.[75] It was further demonstrated that these defects form at lower stresses than do Griffith cracks and that once the defect concentration reaches a certain value (around 100/cm^2) they may "combine" to form a Griffith crack.

The concept of a moveable flaw is not generally associated with glass and the theory has not been widely accepted. Unfortunately there is little published information, besides the work discussed above, so an objective discussion of the model is not possible. Ernsberger[76] has apparently attempted to duplicate these experiments, with no success, and suggests the possibility that these flaws may be artifacts associated with the etching techniques used. This argument has been vigorously denied.[77]

F) Plastic Flow: The remaining theory to be discussed is based on the assumption that glass is capable of localized yielding by plastic flow. This concept had been considered for a number of years, but not until the work of Marsh[48,49] in 1963-64 did most researchers seriously consider the possibility. Considering the 40 odd years that brittle fracture was almost unanimously accepted, it is not surprising the plastic flow concept is meeting considerable opposition, with the controversy yet to subside.

Marsh suggests that the idea of glass as a purely brittle material is "incompatible" with observations of plastic flow in glass, and, in fact, fails to account for the behavior of glass under an applied load. A few of the problems mentioned by Marsh are: (there are others[49])

1. The maximum fatigue-free strengths observed in "flawless" glass are well below the theoretical strengths.

2. In macroscopic flaws the calculated stress at the crack tip is several times the theoretical strength, assuming purely elastic behavior.

3. The maximum propagation velocity of a crack front is a fraction of that expected from elastic theory.

4. Cracks fail to close on unloading.

5. Stress corrosion theories must presume crack sharpening to tip radii as low as 3Å to explain existing data.

6. Stress corrosion fails to account for the observed dependence of the fatigue rate on initial crack length.

Each one of these "discrepencies", and others noted in the original articles, are subject to debate. Number 6 has been discussed previously in the current article, and was not found to be inconsistent with the concept of stress corrosion. However, taken as a whole these objections have some validity.

Marsh suggests that these and several other discrepencies can be accounted for by assuming a plastic flow process in which the glass undergoes a permanent deformation under the influence of an applied stress. The flow process is thought to be closely related to the concentration of non-bridging oxygens, with these sites more able to react to the applied stress and to flow by a mechanism similar to dislocation motion in crystals. (It is interesting to speculate as to the possibility that motion of line defects composed primarily of non-bridging oxygens could be the process observed by Levengood[75]).

Others have suggested plastic deformation processes but proponents of brittle fracture had been able to discount these arguments since "flow stresses" calculated from indentation hardness measurements were much too low to account for observed strengths of glass. Marsh, however, considered that previous calculations were in error due to an erroneous assumption regarding the deformation process in hardness measurements.

He points out that as opposed to soft materials, in which indentations produce a "piled-up" region around the indentation, in hard materials, such as work hardened metals and glass, one must consider a pure "sinking in" mode in which the material flows out radially from the indentor. Taking this into account and recalculating flow stresses Marsh obtains extremely close agreement between observed and calculated strengths.

According to the plastic flow mechanism fracture occurs when local stresses exceed the flow stress. Catastrophic failure is consistent with this model since, as discussed in the introduction, there is no stabilizing mechanism at the yield point which can lead to work hardening. Thus, as in brittle metals, failure occurs instantaneously with very little additional elongation. Failure by plastic flow therefore does account for the low strengths of glass.

The second objection to brittle fracture is accounted for in the plastic flow model by assuming that crack blunting occurs via plastic flow reducing the stress concentration factor before the critical flow stress is reached. Also plastic flow implies a net movement of material which leads to a mismatch of bonds and thus explains the failure of cracks to close on unloading. The failure of the crack velocity to reach theoretical limits can be accounted for by noting that some fracture energy is dissipated during plastic flow.

Static fatigue effects are presumably due to a change in the yield stress with time (crack blunting via flow) and atmosphere. Water is also important in this model as it is presumed to "soften" the glass at the crack tip thus lowering the yield stress.

It would appear to be a little premature to pass judgment on the plastic flow mechanism as the jury is still deliberating. It does account admirably for many observations but some evidence is accumulating which leaves a reasonable doubt as to its general applicability. Wiederhorn and Townsend[78] have recently shown, for instance, that cracks introduced in microscope slides by a tensile stress do indeed close and regain up to 80% of their original strength. Crack healing would not be expected if plastic flow had occurred.

Ernsberger,[79] one of the stronger critics of this mechanism in glass, has suggested that indentation hardness is not related to plastic flow, but rather to a process of densification under an applied load. (For a thorough discussion of densification processes in glass the reader is referred to a recent review by Sakka and MacKenzie.[80] For present purposes it will suffice to say that densification, as opposed to plastic flow, is a reversible process in which volume is reduced.) He reasons that if densification is indeed the controlling mechanism, and the evidence tends to support this view, hardness measurements provide little information regarding the mechanism of failure under tensile stress, and the agreement between calculated and observed strengths from hardness values is purely coincidental.

RECENT DEVELOPMENTS

Major advances in the understanding of glass strength have been few and far between due to a large degree to a lack of reproducible results. There have been two recent developments, however, which will hopefully lead to more rapid advances in the future. The first is the development of a powerful technique whereby propagation of a crack can be controlled and when desired, followed optically over a wide range of velocities. The general validity and usefulness of the technique has already been proven in a variety of experiments.[78,81-83]

The method involves use of the double cantilever arrangement shown in Figure 16. Specimens are normally microscope slides in which a crack of known geometry has been introduced. The crack is made to propagate or stop by suitable control of the load. The method is well suited for studying fracture surface energies, and particularly for studying various static fatigue effects. Temperature, atmosphere, applied load and glass composition are easily varied and the relative effects of each can be determined, Universal Fatigue Curves can also be obtained.[81]

Thus far results of investigations using this technique are in general agreement with the Charles-Hillig model of stress corrosion, although certain effects at high velocities seem to be independent of the water content. Some evidence, both for and against a plastic flow mechanism has also evolved. Crack healing which can be observed by this technique[78] has already been mentioned in this respect.

Although a considerable amount of work has already been published, it is hoped and anticipated that the complicated

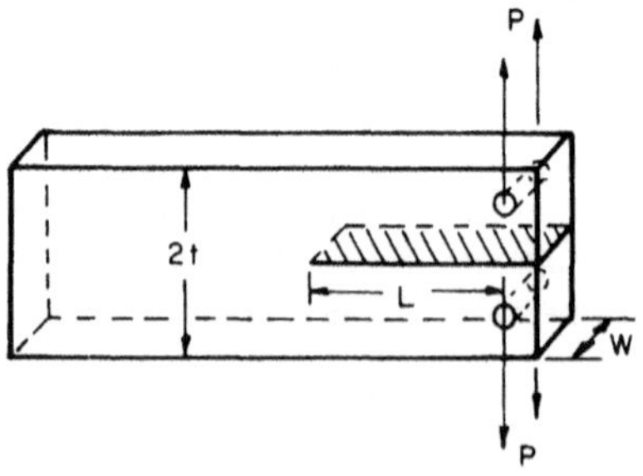

Figure 16. Double cantilever arrangement for studying fracture propagation in glass.[83]

mechanisms and dependencies giving rise to static fatigue will be further illucidated by continued application of this technique.

Perhaps even more significant is Ernsberger's recent work on a new technique for studying the strength of flawless glass surfaces. A discussion of his preliminary findings will provide a useful summation of the current views of glass strength.

In the Ernsberger method a small bubble is trapped in a capillary tube by sealing off the two ends. The glass is then heated to form a spherical bubble and to dissolve any remaining contaminents. The glass is then compressed slightly such that the bubble is flattened into a spheroid, and then annealed. By controlling the compressive step, the bubble geometry can be chosen to yield a particular stress concentration factor.

Figure 17 shows the resulting bubble geometry and stress distribution. With the bubble radius t, 26 times the tip curvature p, the maximum stress concentration factor is 7.3 at the crack tip, in a direction perpendicular

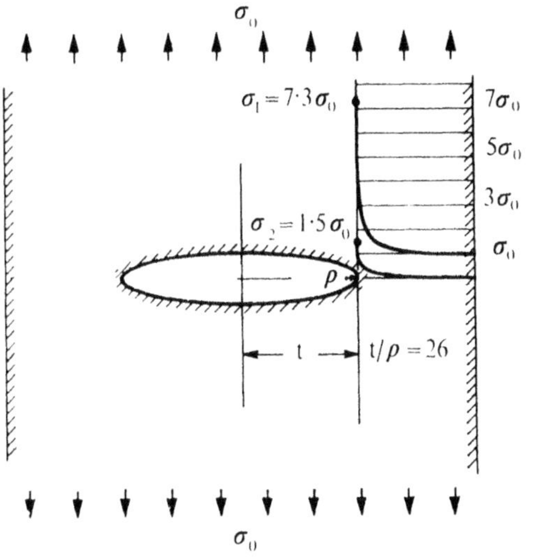

Figure 17. Geometry and stress distribution produced in the oblate bubble technique.[66]

to the equitorial plane. With this stress concentration factor the glass can always be made to fracture from the inside of the bubble. In this way the measured strengths are due to fracture from a pristine surface, which once formed is never in contact with ambient atmosphere and is, therefore, insulated from the corrosive action of water, and the possibility of mechanical damage. The surface should, therefore, be less susceptible to static fatigue, and since the surface is completely free of flaws resulting from mechanical damage, strengths should be reproducible and should approximate the theoretical limit.

The configuration shown in Figure 17 was used to study both compressive and tensile strengths. In the latter case, however, it was necessary to strengthen the outer surface by ion-exchange (see the following sections) to assure that fracture originated at the interior bubble.

Ernsberger reports the following results for soda-lime (Kimble R6, and Corning 0088) borosilicate (Corning 7740) and fused silica:

1. Failure in compression occurs either by shearing, in the case of R6 glass, or by densification in the case of 7740 and fused silica. Shear failure occurred at stresses between 600-650,000 p.s.i. and was the first observation of this mechanism in glass. Failure by densification occurred at significantly greater stresses, being about 1.1×10^6 p.s.i. in 7740 glass and about $1.2\text{-}1.3 \times 10^6$ p.s.i. in fused silica, at room temperature. At lower temperatures even greater strengths were observed. These are the highest values yet obtained, and may be significant in terms of possible undersea applications.

2. The room temperature tensile strength of R6 glass which contains about 68% SiO_2 is found to be 488,000 p.s.i. while that of 0088 glass which contains about 71% SiO_2 is 549,000 p.s.i. Ernsberger considers these differences to be statistically significant, and due to the difference in silica content. If so, this appears to be the first observation of a true compositional effect on glass strength. Measurements on fused silica would be interesting in this respect, but unfortunately no tensil strengths were reported, probably due to an inability to assure fracture from the interior bubble. It should be noted that the strengths measured here are actually greater than those measured

for glass fibers of similar composition[16] and can be considered further evidence against an intrinsically greater strength of glass fibers.

3. The effect of temperature on the strength of R6 glass is shown in Figure 18 and should be compared with Figure 4 given earlier. In all cases shown the temperature dependence was measured under conditions which minimized atmospheric effects, but Ernsberger's results seem to be most reliable. Ernsberger has shown that these effects could not be attributed to an irreversible degradation at high temperatures as was the case in the work of Brearley et al[22,23] and others.[27] This was demonstrated by testing samples at room temperature after heat treatment at 560°C, no decrease in strength was found. The lack of such an effect also leads to the conclusion that irreversible effects observed by others are due to surface contamination which react with the surface and create Griffith flaws. No contamination by foreign dust particles, etc. is possible in the present technique, although Ernsberger notes the possibility of a "self-contamination" effect in which the glass may produce its own surface contamination.

The general shape of the strength-temperature plots of Figure 18 is strong evidence against a direct effect of

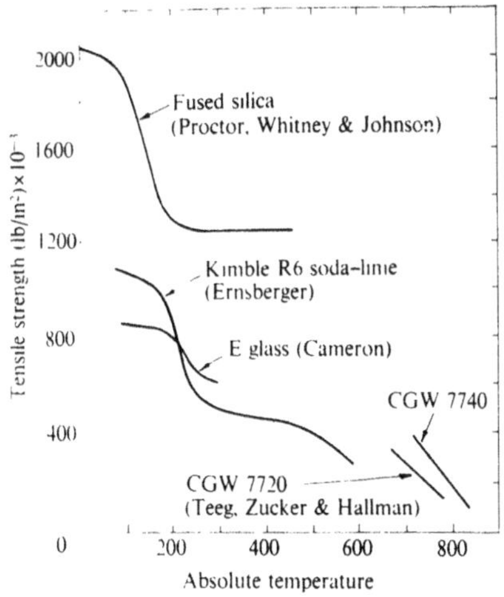

Figure 18. Effect of temperature on the breaking strength of glasses tested under conditions of minimum static fatigue.[66]

temperature on strength as implied in theories of thermally generated flaws. The temperature dependence appears to be minimal at low temperatures, and the sharp drop in strengths in the 100-300°K range, followed by a high temperature plateau, is also inconsistent with such a model. The general behavior appears to be more consistent with a model similar to that of Cox[70], discussed earlier, in which a temperature and stress dependent diffusion of some mobile species is controlling.

4. Somewhat surprisingly, static fatigue does play a role in the strengths discussed here, although the effects are small. A 10 fold increase in loading rate produced a 10% increase in observed strength, and loading below the critical stress did produce failure after extended times. Compared to previous work, however, these results indicate an improved static fatigue resistance. Ernsberger draws no concuusions, since the bubbles ambient conditions are not known. The possible influence of water cannot be ruled out. It is well known that virtually all glass contains water, and it is possible that a stress enhanced diffusion of "network" water, by the mechanism of Cox, could lead to observed behavior.

While the above work has not conclusively answered any of the questions associated with glass strength, for the first time a method is available which permits the study of truly flawless glass surfaces which, after forming, are <u>never</u> exposed to corrosive effects of the atmosphere or to mechanical damage. With proper design of experiments it is not unreasonable to assume that the various strength dependencies will begin to be unraveled. Further exploitation of this technique over the next several years will no doubt lead to more reliable data regarding the ultimate strength of glass, compositional effects, plastic flow, etc., and it would not be surprising if many current views of glass strength were correspondingly revised.

As a prerequisite to any future studies, it would appear that atmospheric effects (within the bubble) on static fatigue must be determined. Until this problem is clarified, measurements of compositional effects may still be suspect. As one final word of caution, it should be noted that, as of now, results from other laboratories have not been published. The acid test of this technique will be the ability of several laboratories to obtain quantitatively reproducible strengths for various compositions and test conditions.

STRENGTHENING GLASS

Previous sections have emphasized the fact that glass is intrinsically strong with observed strengths up to $1\text{-}2 \times 10^6$ p.s.i. For commercial applications, however, this high strength is rendered essentially useless by surface flaws and by static fatigue effects, and design strengths for most glass applications can be as low as 500-2000 p.s.i.

There are basically two approaches one could take in trying to attain higher _practical_ strengths:

1. Develop methods of producing and maintaining flawless glass.

2. Develop methods to neutralize the effects of existing flaws.

The first approach has yet to yield any significant results. The problem is not producing flawless surfaces since etching or special techniques such as fire polishing[50] have already been described. In practice, however, these are not productive, since glass surfaces are simply too susceptible to mechanical damage.

Several attempts have been made to protect surfaces from abrasion by treating with various coating materials, primarily organo-silicic compounds.[27,84] It was reported, for instance,that certain surface treatments could lead to increased strengths by offering increased abrasion resistance, and there was one report (since retracted) that certain of the organosilicic compounds could actually heal surface cracks by forming Si-O-Si bonds across the crack. (See Ref. 84).

Unfortunately most of the techniques utilizing the coating technique are relatively ineffective and results are non-reproducible. Symmers et al[27] find that in some cases these treatments actually lead to a deterioration of strength and little, if any, protection from mechanical damage. Even in cases where the _mean_ strength increased, scatter of the data was such that no _practical_ design improvements could be made.

All of the major advances in producing glasses with permanent, reproducible high strengths have resulted from the second approach. There currently exists several practical methods for rendering surface cracks less menacing, but they may all be classified under the general term "surface compression" techniques.

The idea behind surface compression is simple. Assume, for instance, that a glass has an intrinsic fracture strength on the order of 1,000,000 p.s.i. but that surface stress raisers reduce the effective strength to 10,000 p.s.i. If by some means the glass surface (and, therefore, surface cracks) is placed under a compressive stress of say, 50,000 p.s.i., the glass would then be able to withstand a tensile stress of 60,000 p.s.i., since before surface flaws experience a net tensile stress, the load must overcome the 50,000 p.s.i. surface compression forces.

There are two main catagories of strengthening via surface compression: 1. Physical tempering, 2. Chemical tempering.

PHYSICAL TEMPERING [85-87]

Glass strengthening by physical tempering is accomplished by heating the glass to some predetermined temperature, and then cooling quickly, usually by passing cool air over the surface. The upper temperature must be chosen such that the glass will not deform, but high enough to allow rapid relaxation of internal stresses during the initial cool. In practice this temperature is normally just below the softening point.

The physical basis of tempering can be understood with the help of the familiar volume-temperature relation shown in Figure 19. Above the glass transition temperature, T_g, volume relaxation occurs via two mechanisms; a) structural rearrangement, b) bond relaxation. Once the temperature has passed through T_g, however, the glass becomes rigid and any further volume change will occur only by changes in bond lengths.

An important feature of Figure 19 is the shift of T_g to larger volumes (lower densities) as the cooling rate is increased. Thus, when glass is cooled rapidly, as in tempering, the surface tends to follow curve A of Figure 19,

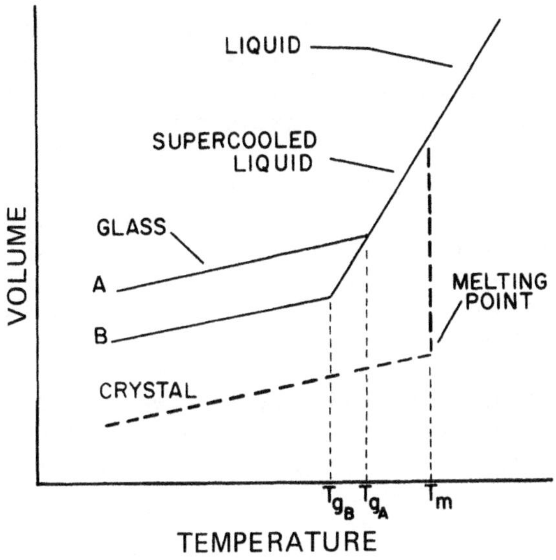

Figure 19. Volume-temperature relation for glass. A) fast cooling rate. B) slow cooling rate.

and solidifies in a low density configuration. The interior glass, however, cools much more slowly, following curve B, which leads to a greater density. The greater contraction of the interior with respect to the surface, tends to "pull" the surface into compression. Since the surface has cooled rapidly to a rigid state, no structural rearrangement is possible and the compressive stress is, therefore, frozen in.

Mathematical models of the tempering process have now reached a fairly sophisticated stage, but only a brief discussion of the more practical aspects of the theory will be given here.

One of the earliest and simplest models has been given by Bartenev.[85] He makes the simplifying assumptions that: a) above Tg the glass is perfectly fluid and below Tg perfectly elastic; b) instantaneous freezing occurs on cooling below Tg; c) a single Tg can be used to describe the cooling process; d) freezing begins only after the surface reaches ambient temperature, and e) cooling at the air interface is by covertion only.

These assumptions clearly represent a simplification of the process described previously, but the results are in surprisingly good agreement with experiment. The compressive stress is found to be a sensitive function of the heat transfer coefficient and of the maximum temperature before cooling. The theory predicts a rapid rise in surface compression at low initial temperatures, being zero at and below the instantaneous freezing point. As would be expected from assumptions b and c, the theory also predicts a temperature above which there is no further increase in surface compression.

Gardon[86] has reported very interesting work in which movies were made exhibiting the stress distribution in glass during the tempering process. The results of this study are shown schematically in Figure 20, where each "frame" represents the stress profile at a particular time during the tempering process. The surface is seen to be in tension initially but becomes subject to an increasing compressive stress. The final frame shows a parabolic stress distribution with the surface in compression and the interior in tension.

Similar stress distributions were obtained over a wide range of heat transfer coefficients. Results are

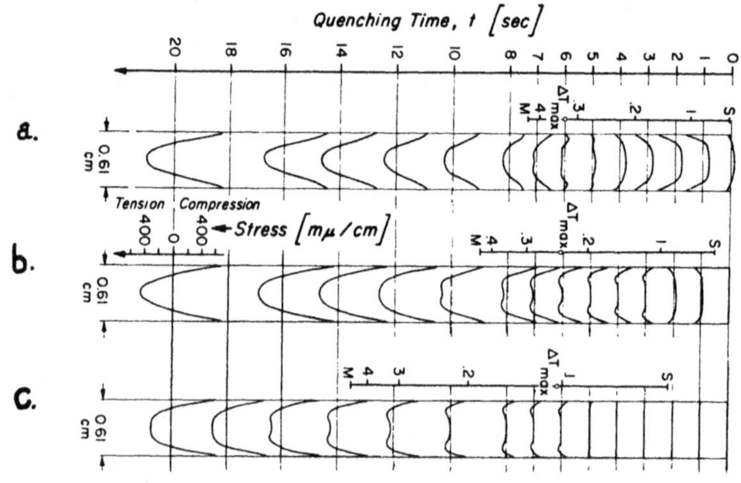

Figure 20. Stress distributions in glass during the tempering process. Initial temperatures were a) 616°C b) 648°C c) 738°C.[86]

summarized in Figure 21. As predicted by the Bartenev model there does appear to be a maximum, induced compressive stress for each heat transfer coefficient, h, but the predicted low temperature behavior is not observed.

In his treatment of the data Gardon included radiative heat transfer and removed Bartenev's restriction of having a single temperature T_g, describe the entire cooling process. In agreement with Figure 19, he assumes this value to be a function of the cooling rate.

The results of his calculations are shown in Figure 21. Agreement at low "h" values is good, but the model fails to predict observed leveling off of tempering at high initial temperatures and large "h". Much of this disagreement can probably be traced to assumptions regarding "instantaneous freezing" of the glass and to the non-ideal behavior of glass both above and below T_g. In more recent work, which will not be discussed here these assumptions have been removed,[87] and reasonable agreement with experiment is obtained.

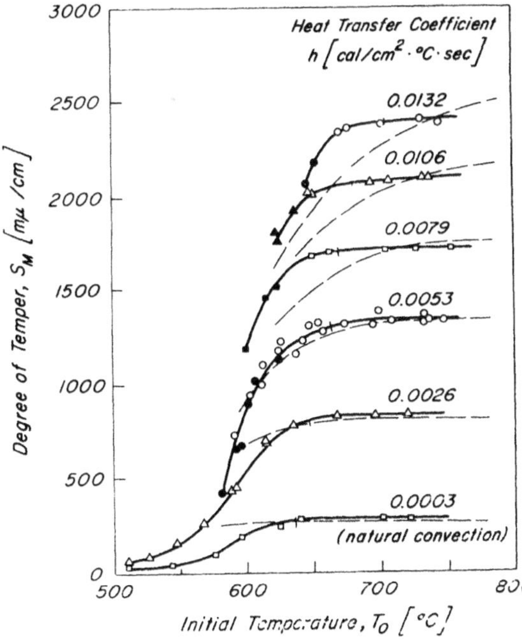

Figure 21. Degree of temper as a function of initial temperature and cooling rate. Dashed lines are the theoretical predictions of Gardon.[86]

In closing this section on physical tempering, mention should be made of the surprising results obtained by Vitman et al.[88] The strengthening process used was a "rapid quench" tempering, followed by etching in HF. The resulting glass exhibited strengths greater than glasses treated by only etching <u>or</u> tempering. It was originally felt that this result was due to a decrease in the surface damage on etching which more than compensated for the exposure of a new surface with lower compressive stress, but results obtained on glasses with extremely fast cooling rates indicate that the original hypothesis is incorrect.

Figure 22 represents the increase in strength due to quenching

$$\Delta\sigma = \sigma \text{ (quench + etch)} - \sigma \text{ (etch)}$$

for glasses of various thicknesses. The dotted line is the increase one would obtain if the only strengthening mechanism was the production of surface compression stresses. It is obvious that the results for large strength increases cannot be accounted for in this manner. The authors suggest that these results are indicative of a structural effect.

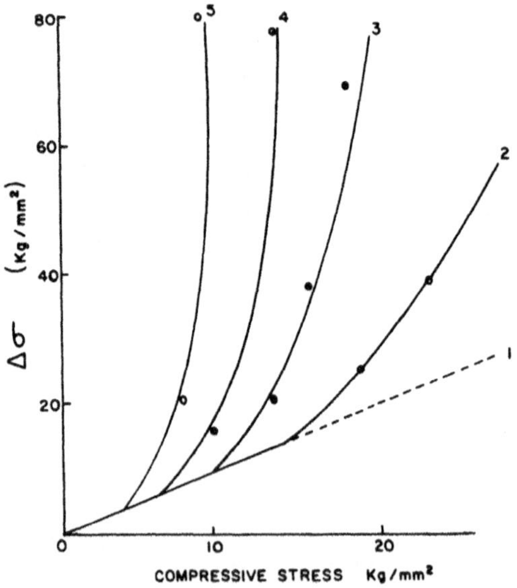

Figure 22. Strength increases after quenching and etching as a function of the induced compression stresses. Increases are larger for thinner glasses.[88]

More recently Boguslavski[89] has suggested that physical tempering is intimately connected with structural changes which occur during the heat treatment, as well as surface compression. Two specific structural changes were considered: a) Homogenization, b) Microplastic flow.

a) HOMOGENIZATION

A reduction in the microheterogeneity[90] of glass is favored by rapid quenching from the molten state. In the tempering processes, however, the glass is held at temperatures well below the liquidus and the only process which would occur is dehomogenization. (nucleation and growth etc.) This process must, therefore, be ruled out in the current discussion of strength <u>increases</u> although it may be of importance in other instances.

b) PLASTIC FLOW

It is a well known fact that the defect concentration of all solids rises rapidly with temperature, and specifically in glass, above the glass transition temperature, T_g. This being the case then the surface of a tempered glass which has experienced a rapid quench should contain a rather large concentration of non-bridging oxygens. Another way of stating this would be to say that glass at high temperatue is characterized by a depolymerized structure. The existence of non-bridging oxygens presumably allows a certain amount of plastic flow at the crack tip, which leads to crack blunting and a corresponding increase in strength.

This hypothesis is extremely interesting for it implies that by controlling the non-bridging oxygen content in the surface layer, by thermal and/or chemical manipulation it may be possible to produce a glass whose interior is brittle but strong, and a surface which is able to yield plastically to reduce stress concentrations. If this is true then the ideal glass may be more than a dream, but unfortunately the extensive work needed to verify these results has not been done.

CHEMICAL TEMPERING[91-104]

Chemical tempering normally denotes a process in which the surface compression results from a change in the chemical composition of the surface. A number of varied and sometimes complex processes are employed to produce this difference.

In the Schott overlay process, glass with a high thermal expansion coefficient is encased in low expansion glass. The process is carried out at elevated temperatures and when the composite cools, relatively larger contraction of the interior glass pulls the low expansion glass into compression. The main drawback is the fact that working temperatures of low expansion (hard) glasses are generally much higher than those of soft glasses, and deformation of the object can result.

The overlay process, which dates to 1891, is currently being used in a more sophisticated manner, to produce glass-ceramic dinnerware. In this process high expansion glass is formed into the desired shape and then crystallized. It may then be glazed with a low expansion glass with no danger of deforming the crystalline interior. The compression layer is produced in the surface as the piece cools after firing.

The overlay process may also be applied to fibers. Cooper and Krohn[91] have recently published data on cladded fibers produced by placing a solid rod of high expansion glass inside a tube of low expansion glass. The fiber is drawn from this composite. Although results showed considerable scatter, the mean strength of abraded fibers increased by 20,000 p.s.i. to about 42,000 p.s.i.

Other methods include dealkalinization of the surface layer which results in a reduced coefficient of expansion.[92] In this process the glass is heated in the presence of SO_2 or SO_3 which catalyses an ion-exchange between hydrogen (from atmospheric water) and alkali. Further heating dehydrates the surfaces leaving a silica rich, low expansion surface.

Surface crystallization can also be employed to produce the desired low expansion surface.[93,94] At first glance this would appear to be a highly useful technique since most crystalline materials have lower expansion coefficients than their glassy counter parts. The problem is that surface layers formed in this manner are generally weak and tend to flake. Also, the optical properties are impaired by the usually opaque crystalline surface.

Only one system, a lithium-aluminosilicate, has thus far exhibited useful properties.[93] Some glasses in this system have been treated to obtain strengths of more than 100,000 p.s.i. with little deterioration of optical quality.

ION-EXCHANGE TECHNIQUES

There are two basic ion-exchange processes which can be used to strengthen glass. The first, due to Hood and Stookey,[95] involves the exchange of larger cations in the glass for lithium ions. The exchange process is affected by placing the glass in a bath of molten lithium salts. Lithium ions from the bath diffuse into the glass, replacing sodium ions which are diffusing out. If the temperature of the bath is held <u>above</u> the strain point, relaxation of the surface network structure takes place to accomodate the smaller lithium ions, and no net stress is present before cooling. At the same time, however, the surface which is now high in lithium, has a lower coefficient of expansion than does the interior (soda-rich), and as the glass cools a net compression stress results in the same manner described previously.

A second ion-exchange process described by Kistler,[96] involves a reverse diffusion. That is, small ions from the glass surface are replaced by larger ions from a molten salt bath. This "crowding", or "ion-stuffing" mechanism is performed at temperatures <u>below</u> the strain point so that <u>no</u> structural relaxation is permitted. In this case surface compression is a direct result of the larger ions "expanding" the surface into compression.[100]

The process of ion-stuffing is not à priori obvious, except to the thermodynamicist, as the driving force for ion-exchange occurs mainly in the entropy term of the free energy expression. Diffusion of smaller ions from the glass is favored due to the increase in entropy of mixing, but this results in a net electrostatic charge on the glass which is compensated for by diffusion of the larger ion into the glass, despite its larger size.

The mathematics of ion-exchange have been worked out by several authors and the interested reader is referred to the original papers.[98,99] Only a qualitative discussion will be attempted here. The net compression stress induced in the surface as a result of ion-stuffing will depend on:[100]

1. The radius ratio of exchanging ions
2. The degree to which the exchange occurs
3. Stress relaxation due to structural rearrangement of the surface
4. Any change in the thermal expansion coefficient
5. Localized densification
6. The depth of the compression layer

The depth of the compression layer is more of a practical barrier than anything else. A thin layer will show a much smaller resistance to abrasion, due to an increased probability that a flaw will penetrate the surface layer. Figure 23 shows the difference between stress profiles produced by ion-stuffing and thermal tempering. Although greater compressive stresses can be obtained with a small maximum internal tension, the stressed inner layer may extend dangerously close to the surface. It is conceivable that if the tensile stress on the interior glass was sufficiently high, once a crack reached the strained region, spontaneous failure would result.

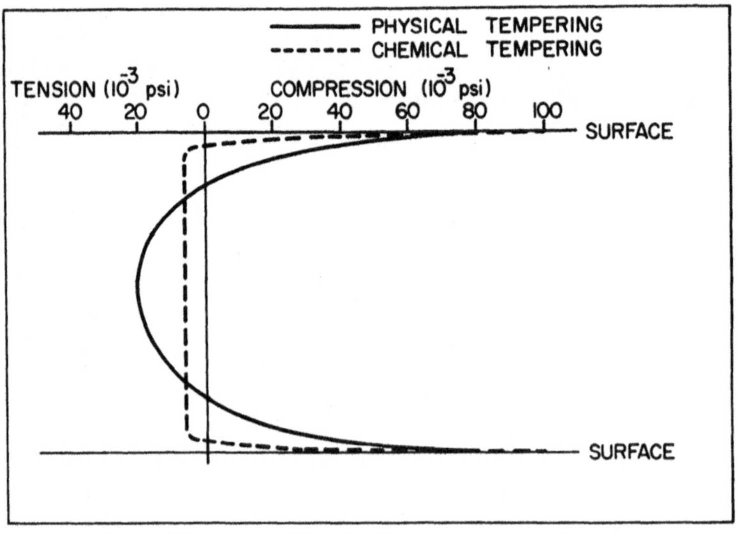

Figure 23. Stress distributions in glass resulting from thermal tempering and chemical tempering.[102]

In practice the compression layer must be on the order of 50 microns, and it is this design criterion which severely limits the flexability of the ion-exchange process for commercial applications. To illustrate this factor, consider the data of Table IV, where the maximum theoretical compression stress is shown for various combinations of ion-exchange. Values were calculated using[101]

$$S = \left(\frac{E}{1-\mu} \right) \times (\text{strain}) = 1/3 \left(\frac{E}{1-\mu} \right) \frac{\Delta V}{V}$$

where E is Young's Modulus, μ is Poisson's ratio and ΔV, the change in volume resulting from the ion-exchange. The important point is that the greater the volume change (radius ratio) the greater the compressive stresses.

It turns out, however, that the kinetics of ion-exchange is strongly dependent on the radius ratio of the diffusing ions, and that only the Na^+ for Li^+ exchange is economically feasible at the present time. Ward et al[103] calculate that at 375°C, 800 hours is necessary to develop the required 50 micron layer in a K^+ for Na^+ exchange. The production costs employing this exchange would clearly be prohibitive. Similar conclusions can be made for Cs and Rb exchanges.

Of course, the diffusion process is thermally activated and it would seem that a simple solution to the time problem would be to raise the temperature of the exchange bath. As shown in Figure 24, however, there is another effect which

Table IV. Calculated Compressive Stresses for Various Ion Exchanges.[101]

Glass Ion	Molten Salt Ion	Atomic Vol. Change cc/100g.	Induced Compressive Stress (psi)
Na	K	2.18	210,000
Li	Na	.92	90,000
Li	K	3.10	290,000
Li	Rb	4.40	320,000
Li	Cs	6.65	575,000

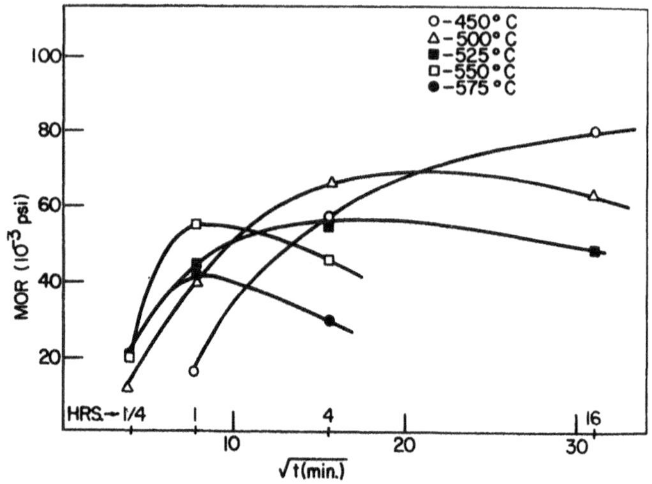

Figure 24. Effect of bath temperature on the strength of a chemically tempered glass. M.O.R. is Modulus of Rupture.[102]

must be considered; namely stress relaxation. Therefore, even though one can obtain a greater rate of ion-exchange, the surface structure is able to relax which lowers the induced surface compression. In this connection it should be noted that some relaxation may occur even at lower temperatures,and combined with possible densification effects (discussed by Burggraff[104]) and the lack of reliable data on the atomic volume differences between diffusing species, the calculations given in Table IV may be overestimating practical values by as much as 100%.

An important factor in determining the usefulness of a strengthened glass is its ability to maintain the strength over extended periods of time, and in a variety of conditions. Degradation by exposure to corrosive media and high temperature are perhaps the most important design problems.

Figure 25 shows the effect of aging chemically strengthened glass at 275°C for various times and Figure 26 shows the effect of increasing temperature. Both curves can be qualitatively attributed to a structural relaxation process and to a redistribution of surface ions. At lower

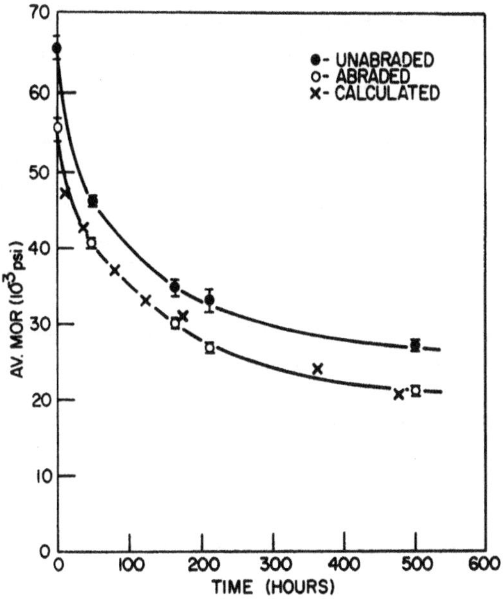

Figure 25. Effect of ageing chemically tempered glass. M.O.R. is the Modulus of Rupture.[102]

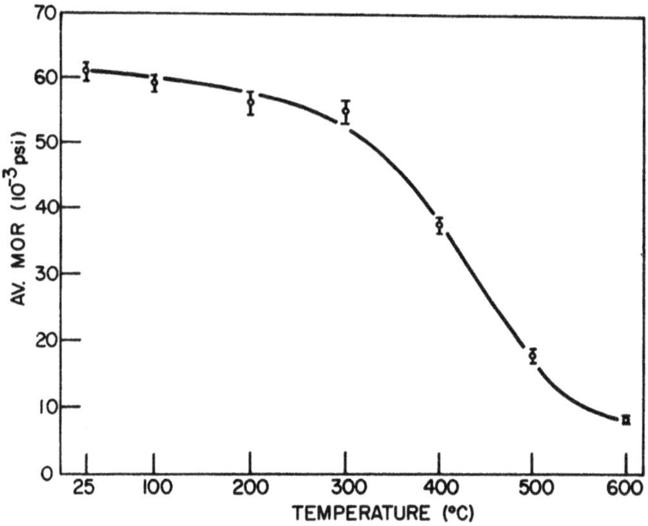

Figure 26. Effect of temperature on chemically tempered glass. M.O.R. is the Modulus of Rupture.[102]

temperatures where stress relaxation is prohibited, the latter process controls while at higher temperatures the former becomes more important.

Because of differences in diffusion kinetics, K^+- Na^+ exchanged glasses are about 20 times more stable at 300°C than are Na^+ - Li^+ exchanged glasses. Significant improvements might also be expected for Cs^+ - Li^+ and Rb^+ - Li^+, but the problem is in producing these exchanges in the first place. The very property which imparts desirable high temperature behavior is the prohibitive factor in their production.

The other factor to be considered is that of static fatigue. Ritter[29] has shown that chemically strengthened glass shows an almost negligible dependence on the rate of loading. This can be attributed to the residual surface compressive stress, which must be overcome before the crack tip exhibits a net tensile stress. This effectively reduces the rate of crack growth leading to increased strengths for any particular loading rate. One would expect that the fatigue limit should also be increased by an amount equal to the compressive stress, although this does not appear to be proven as yet.

Although strengths are generally higher for chemically strengthened glasses there are several other factors which must be considered for any specific application. In general chemically strengthened glass is more susceptible to abrasion and the time required for ion-exchange is orders of magnitude longer than for thermal tempering. Furthermore, as of now the range of compositions which can be employed is limited and they are generally more expensive. As a result it is often necessary to compromise certain other properties, as well as cost, to obtain the desired strengths.

The advantages over physical tempering include a significantly greater design strength and the fact that odd shaped glasses can be strengthened. Also, since the exchange process is carried out at lower temperatures there is no danger of distortions resulting from flow. It appears that whereas physical tempering is more flexible in terms of chemical composition, chemical tempering provides a greater degree of flexibility in overcoming physical design problems such as size and shape restrictions.

WHERE DO WE GO FROM HERE

The major conclusions to be drawn from the present survey are:

1. Glass is an inherently strong material, capable of sustaining loads up to 10^6 p.s.i., and more, in both tension and compression.

2. Surface defects reduce the maximum strength by as much as three orders of magnitude.

3. Griffith cracks are not an intrinsic feature of glass, but result from mechanical damage or surface contamination

4. If intrinsic flaws are present in glass, they need only be on the order of 10-100 Å to explain existing data. Such flaws may be associated with microscopic voids or a microheterogeneous structure.

5. Long term loading (static fatigue) can decrease observed strengths to 1/5 of the instantaneous strength.

6. Static fatigue results are in general agreement with the combined theories of Griffith cracks and a stress enhanced corrosion at the crack tip. However, other theories are able to account for these results.

7. The importance of atmospheric moisture in static fatigue has been clearly demonstrated.

8. Temperature appears to influence glass strength by an indirect mechanism, rather than by causing spontaneous crack formation. Indirect mechanisms include chemical attack, as in stress corrosion, and ionic diffusion.

9. The conflict between theories of brittle fracture and microplasticity has not been resolved. In their present formulations neither theory is able to account for all observed properties.

In obtaining satisfactory answers to many of the remaining problems, perhaps the most important consideration will be the proper design of experiments. Many of the early interpretations of strength measurements are suspect, due to inadequate specification and/or control of important parameters. Of course, many factors which control strength have become fully appreciated only in the past ten years, but with the new interpretations of Griffith crack origins, effects of heat treatment, and atmospheric effects, a complete specification of test conditions becomes imparative. Similarly, meaningful changes in fiber strengths can be produced by varying the drawing conditions, and the work of Porai-Koshits has shown the possibile importance of a microheterogeneous structure. It is clear that phase separation, on the atomic scale or of macroscopic dimensions, should have some effect on glasses whose strengths approach the theoretical limit, and possibly in abraded glasses under static fatigue conditions, if the phase separation causes preferential attack.

An example of where such problems might arise is in a study of the compositional dependence of strength. If all of these parameters have not been specified and controlled, it will be impossible to determine how composition affects strength. It is possible, for instance, to conceive of a hypothetical glass in which a change in composition leads to a decrease in the observed strength, while the calculated theoretical strength increases. The observed strengths in such a case, could be due to the fact that the compositional change produces a decreased abrasion resistance, an increase in the stress corrosion process, or phase separation, etc.. Without having a complete specification of thermal history, handling, and test conditions, absolutely no meaningful information can be extracted from such an experiment.

Perhaps the first order of business for future theoretical and basic investigations will be reconciliation, if possible, of the theories of plastic flow and brittle fracture. It is not clear how this might best be done, but it would be hoped that extensive work on the cantilever and oblate bubble techniques would produce some meaningful results.

Whether or not a plastic flow mechanism exists and operates is of more than academic interest, for it bears directly on static fatigue and on producing higher strength glasses. This is acutely evident in the work of Boguslavski.[89]

No doubt much of the progress in understanding glass

strengths will result from application of new techniques. The scanning electron microscope, with its depth perception, should be a valuable tool in future studies.[105] Also valuable, will be investigations of other glass systems, such as the phosphates[106], in which one can control the degree of structural continuity, and the chalcogenides[107], in which one deals with covalent-metallic bonding rather than ionic-covalent. Studies in both systems should lead to a clearer understanding of basic fracture phenomena in disordered systems.

REFERENCES

1. W. D. Kingery, Introduction to Ceramics, John Wiley & Sons, New York (1960).
2. I. Naray-Szabo and J. Ladik, Nature 188, 226 (1960).
3. I. Naray-Szabo, Pg. 327 in Symposium sur la Resistance Mechanique du Verre et les Moyens de l'Ameliorer, Union Scientifique Continentale du Verre, Charleroi, Belgium (1962).
4. E.U. Condon, Am. J. Phys. 22, 224 (1954).
5. S. Spinner, J. Am. Ceram. Soc. 37, 229 (1954).
6. R. J. Charles, Pg. 1 in Progress in Ceramic Science, Vol. 1, J. E. Burke, editor, Pergamon Press, N.Y. (1961).
7. W. B. Hillig, Pg. 295 in Symposium sur la Resistance Mechanique du Verre et les Moyens de l'Ameliorer, Union Scientifique Continentale du Verre, Charleroi, Belgium (1962).
8. F. P. Mallinder and B. A. Proctor, Phys Chem. Glasses 5, 91 (1964).
9. J. C. Phillips, Glass Tech. 5, 216 (1964).
10. W. Capps and D. H. Blackburn, Nat Bur. Stand. Rept. 5188 (1957).
11. M. L. Williams and G. E. Scott, Glass Tech. 11, 76 (1970).
12. C. H. Greene, Glass Tech. 7, 54 (1966).
13. B. A. Proctor, Appl. Mat. Res. 3, 28 (1964).
14. R. H. Kropschott and R. P. Mikesell, J. Appl. Phys. 28, 610 (1957).
15. J. G. Morley, P. A. Andrews and I. Whitney, Phys. Chem. Glasses 5, 1 (1964).
16. N. M. Cameron, Glass Tech. 9, 14 (1968).
17. N. M. Cameron, Ibid, Pg. 121.
18. J. E. Ritter and A. R. Cooper, Jr., Phys. Chem. Glasses 4, 76 (1963).
19. R. J. Charles, J. Appl. Phys. 29, 1549 (1958).
20. R. J. Charles, Ibid. Pg. 1554.

21. R. J. Charles, J. Appl. Phys. 29, 1657 (1958).
22. W. Brearley and D. G. Holloway, Phys. Cham. Glasses 4, 69 (1963).
23. W. Brearley, P. A. P. Hastilow and D. G. Holloway, Phys. Chem. Glasses 3, 181 (1962).
24. R. E. Mould and R. D. Southwick, J. Am. Ceram. Soc. 42, 582 (1959).
25. C. H. Greene, J. Am. Ceram. Soc. 39, 66 (1956).
26. B. Proctor, Phys. Chem. Glasses 3, 7 (1962).
27. C. Symmers, J. B. Ward and B. Sugarman, Phys. Chem. Glasses 3, 76 (1962).
28. J. E. Ritter, Jr., Phys. Chem. Glasses 11, 16 (1970).
29. J. E. Ritter, Jr., J. Appl. Phys. 40,340 (1969).
30. C. E. Inglis, Trans. Inst. Naval Arch. 55, 219 (1913).
31. A. A. Griffith, Phil. Trans. Roy. Soc. 221A, 163 (1920).
32. W. B. Hillig, Pg. 152 in Modern Aspects of the Vitreous State Vol. 2, J. O. Mackenzie, editor, Butterworths, London (1962).
33. C. J. Phillips, Amer. Sci. 53, 20 (1965).
34. E. N. Da C. Andrade and L. C. Tsien, Proc. Roy. Soc. A159, 346 (1937).
35. F. M. Ernsberger, Phys. Chem. Glasses 1, 37 (1960).
36. F. M. Ernsberger, Proc. Roy. Soc. A257, 213 (1960).
37. F. M. Ernsberger, Pg. 511 in Advances in Glass Technology, Plenum Press, N.Y. (1962).
38. F. M. Ernsberger, Pg. 57 in Progress in Ceramic Science, Vol. 3, J. E. Burke, editor, The Macmillan Co., N.Y. (1963).
39. F. M. Ernsberger, Glass Ind. 47, 300,481, and 542 (1966).
40. E. Orowan, Nature 154, 341 (1944).
41. W. C. Levengood and W. H. Johnson, J. Chem. Phys. 26, 1184 (1957).
42. R. J. Charles and W. B. Hillig, Pg. 511 in Symposium sur la Resistance Mechanigue du Verre et les Moyens de l'Ameliorer. Union Scientifigue Continentale du Verre, Charleroi, Belgium (1962).
43. R. J. Charles and W. B. Hillig, Pg. 682 in High Strength Materials, V. F. Mackey, editor, John Wiley and Sons, N.Y. (1965).
44. S. Glasstone, K. Laidler and H. Eyring, The Theory of Rate Processes, McGraw Hill Book Co., Inc., N.Y. (1941).
45. R. E. Mould, J. Am. Ceram. Soc. 43, 160 (1960).
46. R. E. Mould and R. D. Southwick, J. Am. Ceram. Soc. 42, 542 (1959).
47. R. W. Mould, J. Am. Ceram. Soc. 44, 481 (1961).
48. D. M. Marsh, Proc. Roy. Soc. A279, 420 (1964).

49. D. M. Marsh, Proc. Roy. Soc. A282, 33 (1964).
50. B. A. Proctor, I. Whitney and J. W. Johnson, Proc. Roy. Soc. A297, 534 (1967).
51. C. Gurney and S. Pearson, Proc. Phys. Soc. (London), 62B, 469 (1949).
52. I. Whitney, J. W. Johnson and B. A. Proctor, Nature, Lond. 210, 730 (1966).
53. A. L. Prantis, J. Am. Ceram. Soc. 52, 340 (1969).
54. E. B. Shand, J. Am. Ceram. Soc. 53, 53 (1970).
55. M. Watanbe, R. V. Caporali and R. E. Mould, Phys. Chem. Glasses, 2, 12 (1960).
56. F. O. Anderegg, Ind. Eng. Chem. 31, 290 (1939).
57. J. B. Murgatroyd, J. Soc. Glass Tech. 32, 291T (1948).
58. A. F. Prebus and J. V. Michener, Ind. Eng. Chem. 46, 146 (1954).
59. W. F. Thomas, Phys. Chem. Glasses 1, 4 (1960).
60. W. H. Otto, J. Am. Ceram. Soc. 38, 122 (1955).
61. R. T. Brannan, J. Am. Ceram. Soc. 36, 230 (1953).
62. G. M. Bartenev, Pg. 491 in Physics of Non-Crystalline Solids, J. Prins editor, North Holland Publishing Co., Amsterdam, (1965).
63. C. R. Morelock, private communication to W. B. Hillig (Ref. 32).
64. J. E. Ritter Jr., Glass Ind. 49,603 (1968).
65. F. M. Ernsberger, Pg. 191 in Research into Glass Vol. 1 PPG Industries, Pittsburgh (1967) (Reprinted from Machine Design, April 1965).
66. F. M. Ernsberger, Phys. Chem. Glasses 10, 240 (1969).
67. J. Cornelissen, H. W. J. H. Meyer, A. M. Kruithof, and H. C. Hamaker, Pg. 489 in Advances in Glass Technology Plenum Press, N.Y. (1962).
68. J. C. Fisher, J. Appl. Phys 19, 1062 (1948).
69. E. F. Poncelet, Pg. 201 in Fracturing of Metals, American Society for Metals, Cleveland (1948).
70. S. M. Cox, Phys. Chem. Glasses 10, 226 (1969).
71. C. Gurney, Proc. Roy. Soc. A282, 24 (1964).
72. W. C. Levengood, J. Appl. Phys. 30, 378 (1959).
73. W. C. Levengood and T. S. Vong, J. Appl. Phys. 31, 1104 (1959).
74. W. C. Levengood, J. Appl. Phys. 32, 2525 (1961).
75. W. C. Levengood, Inter. J. Fracture Mech. 2, 400 (1966).
76. F. M. Ernsberger, J. Am. Ceram. Soc. 52, 404 (1969).
77. W. C. Levengood, Ibid, page 403.
78. S. M. Wiederhorn and P. R. Townsend, J. Am. Ceram. Soc. 53, 486 (1970).
79. F. M. Ernsberger, J. Am. Ceram. Soc. 51, 545 (1968).

80. S. Sakka and J. D. Mackenzie, J. Non-Cryst. Sol. 1, 107 (1969).
81. S. M. Wiederhorn and L. H. Bolz, J. Am. Ceram. Soc. 53, 543 (1970).
82. S. M. Wiederhorn, J. Am. Ceram. Soc. 50, 407 (1967).
83. S. M. Wiederhorn, J. Am. Ceram. Soc. 52, 99 (1969).
84. P. C. Fletcher and J. J. Tillman, J. Am. Ceram. Soc. 47, 382 (1964).
85. G. M. Bartenev, J. Tech. Phys. 19, 1423 (1949)(Russian).
86. R. Gardon, Proc. VII Inter. Cong. on Glass, International Glass Commission, Charleroi, Belgium (1965).
87. E. H. Lee, T. G. Rogers and T. C. Woo, J. Am. Ceram. Soc. 48, 480 (1965).
88. F. F. Vitman, I. A. Boguslavski, and V. P. Pukh, Sov. Phys. Solid State 4, 1582 (1963).
89. I. A. Boguslavski, Glass and Ceramics 21, 562 (1964).
90. E. A. Porai-Koshits, O. A. Goganov and V. I. Averjanov, pg. 117 in Physics of Non-Crystalline Solids, J. A. Prins, editor, North Holland Publishing Co., Amsterdam (1965).
91. D. A. Krohn and A. R. Cooper, J. Am. Ceram. Soc. 52, 661 (1969).
92. J. S. Olcott, Science 40, 1189 (1963).
93. J. S. Olcott, and S. D. Stookey, pg. 400 in Advances in Glass Technology, Plenum Press, New York (1962).
94. H. M. Garfinkel, D. L. Rothermal and S. D. Stookey, Ibid, pg. 404.
95. H. P. Hood and S. D. Stookey, U. S. Pat. 2,998,675 (1961).
96. S. S. Kistler, J. Am. Ceram. Soc. 45, 59 (1962).
97. M. E. Nordberg, E. L. Mochel, H. M. Garfinkel and J. S. Olcott, J. Am. Ceram. Soc. 47, 215 (1964).
98. R. H. Doremus, J. Phys Chem. 68, 2212 (1964).
99. R. H. Doremus, pg. 6 in Modern Aspects of the Vitreous State Vol. 2, J. D. Mackenzie, editor, Butterworths, London (1962).
100. C. L. Quackenbush, M. S. Thesis, Alfred U. 1969.
101. S. D. Stookey, pg. 669 in High Strength Materials, V. F. Zackay, editor, John Wiley and Sons, New York (1964).
102. H. M. Garfinkel, Glass Ind. 50, 28 and 74 (1969).
103. J. B. Ward, B. Sugarman, and C. Symmers, Glass Tech. 6, 90 (1965).
104. A. J. Burggraaf, Phys. Chem. Glasses 7, 169 (1966).
105. T. R. Bott and A. J. Barker, Glass Tech. 9, 42 (1968).
106. A. E. R. Westman, pg. 63 in Modern Aspects of the Vitreous State, Vol. 2, J. D. Mackenzie, editor, Butterworths, London (1960).
107. B. T. Kolomiets, Phys. Stat. Solidi 7, 359, 372 (1964).

Surface Chemistry of Glass

D. R. Rossington
College of Ceramics
Alfred University
Alfred, N.Y.

INTRODUCTION

When two immiscible phases, such as a gas and a solid, are brought into contact, the concentration of the gas phase is found to be greater at the interface than in its bulk. This process is termed adsorption, and is due to the fact that the atoms in a solid surface are subjected to unbalanced forces of attraction normal to the plane of the surface.

Adsorption is to be distinguished from absorption, which involves the bulk penetration of the structure of a solid by a gas and is governed by laws of diffusion.

Adsorption is a spontaneous process and is accompanied by a decrease in the free energy of the system (i.e ΔG is negative). There is also a decrease in the entropy of the system (ΔS is negative), because the gas molecules are passing from a vapor phase with three-dimensional mobility to a surface phase with two-dimensional mobility or no mobility at all.

Thus, from the equation:

$$\Delta G = \Delta H - T\Delta S$$

It is seen that ΔH ,the heat of adsorption must also be negative. Therefore, adsorption is an exothermic process and the magnitude of ΔH helps to define the type of forces responsible for the process.

If adsorption takes place through the interaction of Van der Waals forces of physical attraction, such as those responsible for the deviation of gases from ideal gas laws and the liquefaction of gases, the process is known as physical adsorption. Because physical adsorption involves the interaction of weak forces, the magnitude of ΔH is low, usually being of the order of 0.5 to 10 kcal/mole. Physical adsorption is usually non-specific, in that most gases will physically adsorb on any type of surface, and will normally only take place at temperatures near the boiling point of the gas at the pressure under consideration. A physically adsorbed species, which is still essentially a molecule, may be removed by raising the temperature of the adsorbent.

If adsorption takes place through the formation of a chemical bond between the adsorbent and adsorbate, or some interaction of the electron charge clouds of the atoms involved, this process is known as chemisorption. Because the forces of chemical attraction in this case are larger than the physical Van der Waals forces values of ΔH for chemisorption may range from 5 to 150 kcal/mole or more. Chemisorption is usually a very specific process and will only take place on an atomically clean surface.

In very general terms, solids may be divided into two main groups, crystalline and amorphous. In a crystal surface the atoms are arranged in a regular and repeating symmetrical pattern. Crystalline materials have a sharp melting point and often cleave in preferred directions, exposing a particular crystal plane in the surface.

Glasses, on the other hand, are amorphous materials, although not all amorphous materials are glasses. Very broadly, a glass is an amorphous solid which may be melted to a viscous liquid by heating and then cooled again to a solid state without crystallization. The surface of glasses, therefore, is much less regular and more difficult to define, than the surface of other types of solids.

ADSORPTION AND REACTIONS ON GLASS SURFACES

Glass surfaces are normally assumed to be inert and, in fact, this is one of their advantages in the use of glass as a container material, with no contamination of the contents.

However, in many cases molecular interactions of a glass surface are very important and most surfaces, when treated in particular ways, can be extremely reactive. The following sections will deal with some examples of physical, chemical and biological aspects of the activity of glass surfaces.

When dealing with glass on a molecular level, adsorption studies of gas molecules on the surface are frequently used to provide information of processes occurring at the surface. Many studies have been made with porous glasses of very high surface area and by extrapolation it is possible to construct a model of the active groups existing on a flat glass surface.

One of the main problems when dealing with any aspect of glass is that its structure is very poorly defined. Figure 1 shows a representation of pure silica, a continuous three dimensional random polymer. In practice, of course, glasses contain other elements. Alkali metal oxides on boric oxide are added to reduce the melting temperature and viscosity by breaking some of the Si-O bonds. Calcium or aluminum oxides are added to control other properties such as expansion, durability, etc. Thus, Figure 2 may be used to represent the structure of a glass body. Most commercial glasses contain at least 60% silica and, therefore, a study of the surface chemistry of glass is best obtained by first finding out as much information as possible about pure silica surfaces, and then treating glass as an impure form of silica.

One of the most useful techniques for studying the nature of pure silica surfaces is infrared spectroscopy. The details of this are given in another chapter and will not be repeated here, although some of the findings will be discussed.

The adsorption of a gas molecule by a solid surface may take place via one of two mechanisms, physical adsorption or chemisorption, as mentioned in the introduction.

The energy released upon physical adsorption (the heat of adsorption) is low, being of the order of 0.5 to 10 kcal/mole

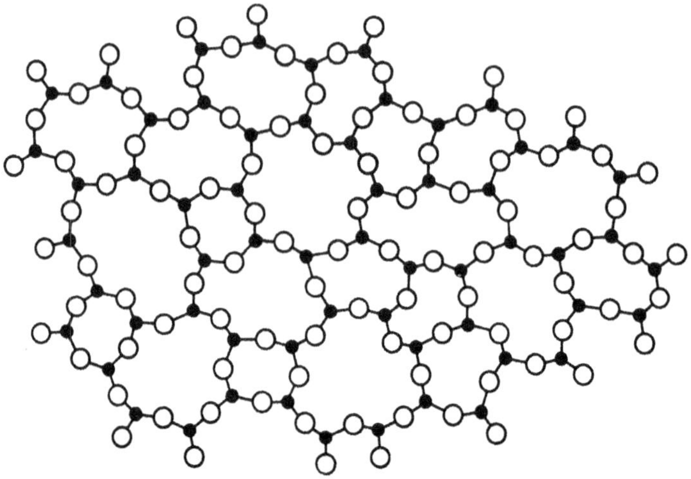

Figure 1. Pure silica. A continuous three-dimensional polymer.

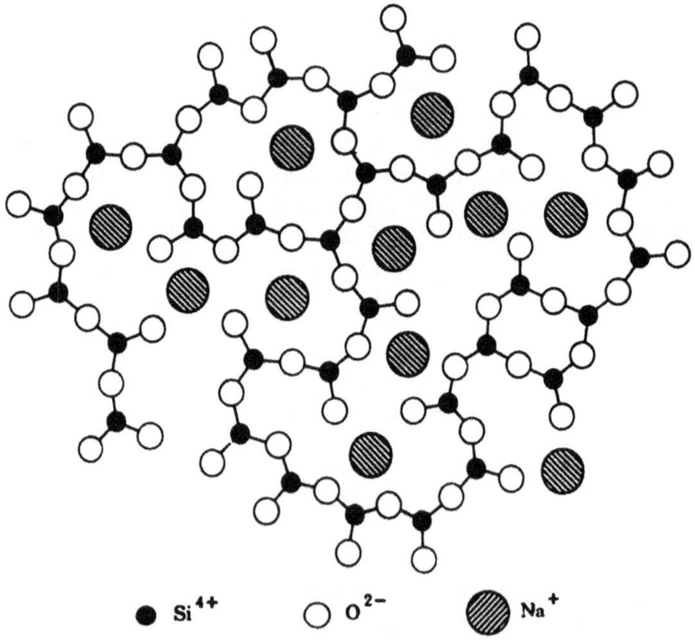

Figure 2. Schematic representation of a sodium silicate glass.

Figure 3 shows the physical adsorption of a carbon monoxide molecule on a surface site, the dotted line representing the Van der Waals bond.

Chemisorption, on the other hand, involves the formation of a chemical bond between adsorbate and adsorbent, i.e., some electron transfer or sharing takes place. Thus, chemisorption is usually very specific and will only take place on an atomically clean surface. Heats of chemisorption may range from 5 to over 100 kcal/mole. Figure 4 represents the chemisorption of CO on a surface, the full lines representing normal chemical bonds.

Infrared spectroscopy has shown the effect of increasing temperature on removing adsorbed water from a silica surface. This is shown in Figure 5 where it may be seen that at room temperature a silica surface consists of surface hydroxyl groups and molecular water physically adsorbed. As the silica is heated the physically adsorbed water and some of the hydroxyl groups are removed, leaving behind surface hydroxyl groups of two types, those that are close enough to be hydrogen-bonded and those that are sufficiently far apart so as not to interact with one another.

Up to about 400°C the dehydroxylation of a silica surface is reversible, but above this temperature a more drastic change in the surface takes place and subsequent water

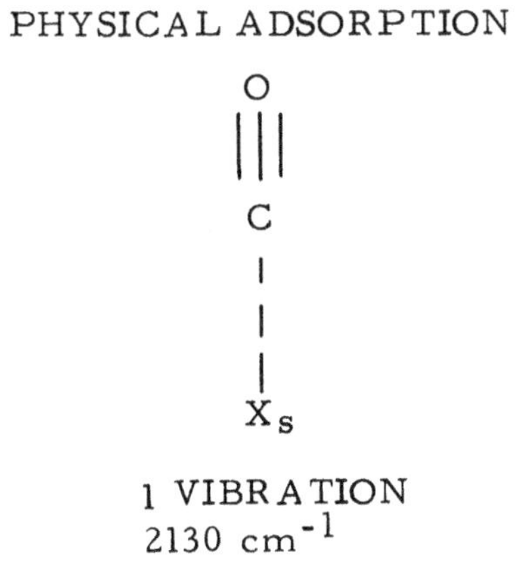

Figure 3. Physical adsorption of CO.

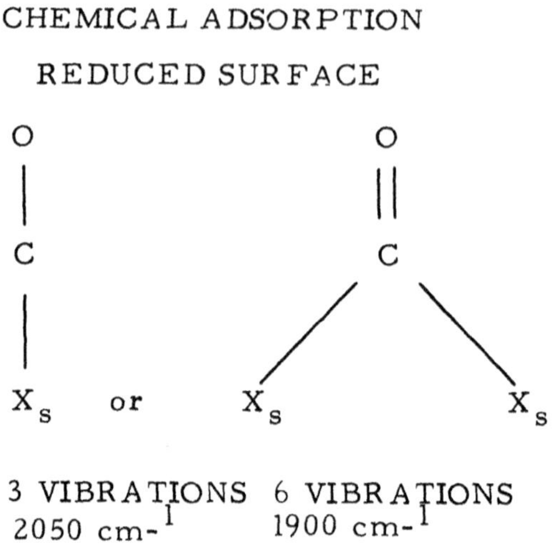

Figure 4. Chemisorption of CO. X_s represents a surface adsorption site.

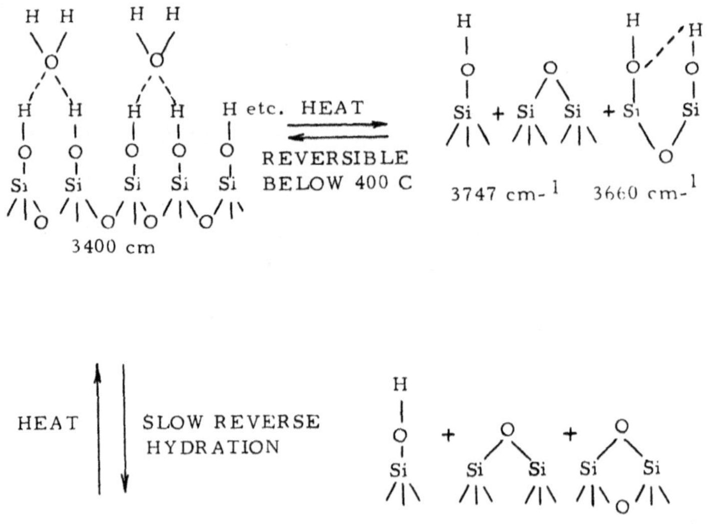

Figure 5. Dehydration of a silica surface.

adsorption becomes more difficult. After heating to 850°C a silica surface will not absorb water without special treatment to reform an hydroxylated surface.

A knowledge of the specific surface area of silica samples is necessary in order to study processes occurring at the surface. The surface area is usually determined by a method of gas adsorption using an equation developed by Brunauer, Emmett and Teller.[1]

Figure 6 shows some results by Taylor, Hockey and Pethica[2] on the surface area of silica powder heated to various temperatures. Sample RA2 was a high surface area silica heated in water for 5 hours at 95°C, filtered and dried in air at 115°C. Sample RA2/700 was heated to 700°C for 48 hours to obtain an annealed surface, then rehydrated in the same way as RA2. It is seen from Figure 6 that heating at temperatures below 750°C results in a stable surface whose area does not change with prolonged heating. At temperatures above 940°C sintering occurs and there is a continuous decrease in surface area with time.

Figure 7 shows the decrease in surface concentration of hydroxyl groups with heating in air. The concentration was determined by measuring the weight loss due to the removal of hydroxyl groups. Figure 8 shows similar results for heating in vacuum. These determinations and those of many other workers show that the previous thermal history of a glass surface can have a great effect on its subsequent behavior.

The thermal history of a glass has a vital effect on its adsorptive properties. Sewell and Morgan[3] recently studied the adsorption of methanol on vitreous silica and on a soda-lime-silica glass. The composition of the glass is shown in Table I.

Table I. Composition of Glass[3]

SiO_2	72.7	Wt. %
Na_2O	12.9	
CaO	8.2	
MgO	3.8	
Al_2O_3	1.3	
K_2O	0.6	
SO_3	0.35	
Fe_2O_3	0.15	

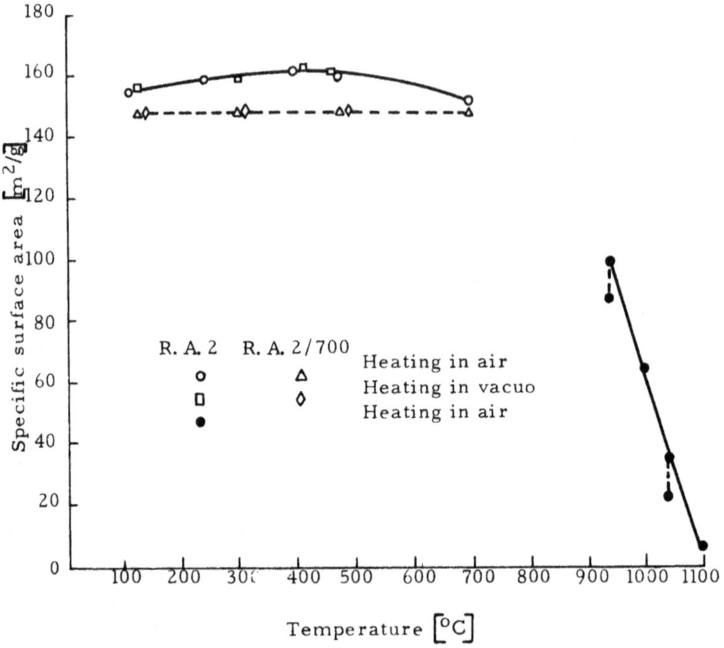

Figure 6. Specific surface areas of silica as a function of temperature (from ref. 2).

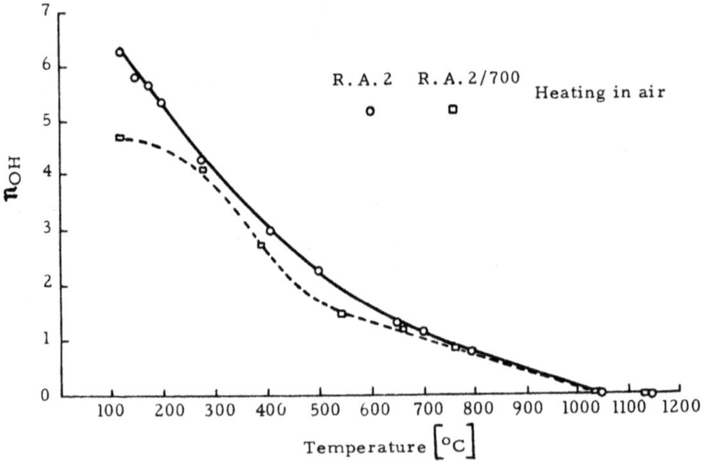

Figure 7. Surface hydroxyl population per 100 A^2 of surface (N_{OH}) as a function of temperature. Heating in air. (from ref. 2).

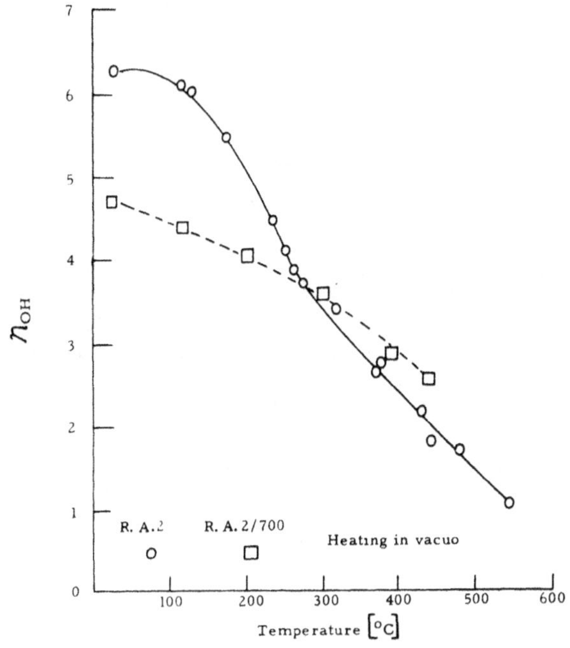

Figure 8. Surface hydroxyl population per 100 A^2 of surface (N_{OH}) as a function of temperature. Heating in vacuo. (from ref. 2).

The experiments were carried out at low pressures and adsorption isotherms determined by a gas chromatographic technique. At the pressures used the surface coverage by adsorbed methanol was low in order to keep adsorbate-adsorbate interactions to a minimum. Isotherms were determined in the temperature range 30°C to 80°C, in which region physical adsorption of methanol occurred. No evidence of chemisorption of methanol was detected. Figure 9 shows a typical set of adsorption isotherms.

The Clausius-Clapeyron equation may be expressed in the form

$$\frac{d \ln p}{d(1/T)} = \frac{-q}{R} \qquad (1)$$

where p = pressure required for a constant current of adsorption at temperature T

q = heat of adsorption

R = gas constant

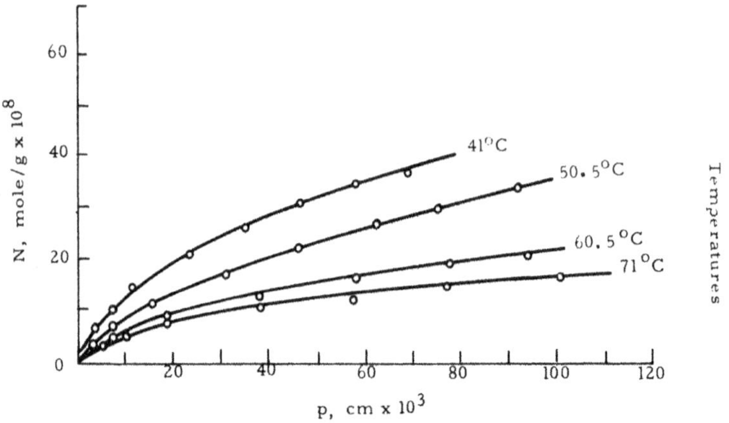

Figure 9. Adsorption isotherms for methanol on silica grains after degassing at 100° C for 3 hours. (from ref. 3).

Therefore, by taking constant amount of adsorption (i.e., a horizontal line in Figure 9) and observing the different values of p and T, and plotting in the form of ln p versus 1/T and a line of slope -q/R may be obtained.

Figure 10 shows such values obtained for the heat of adsorption of methanol on the two glasses, as a function of degassing temperature.

Sewell and Morgan interpreted their results in terms of the energy required for different types of bond formation. The heat of adsorption in the 4 kcal/mole range (after high temperature degassing) are consistent with the energy required for the formation of a single hydrogen bond for each adsorbed molecule, which is in the range 3-6 kcal/mole.

The high heats of adsorption, in the 10 kcal/mole range, are associated with low temperature degassing and are considered to be caused by the formation of multiple hydrogen bonds for each adsorbed molecule. Figure 11 shows a diagram of the adsorption after low and high temperature degassing. These models are consistent with the known relationship between surface hydroxyl concentration and degassing temperature.

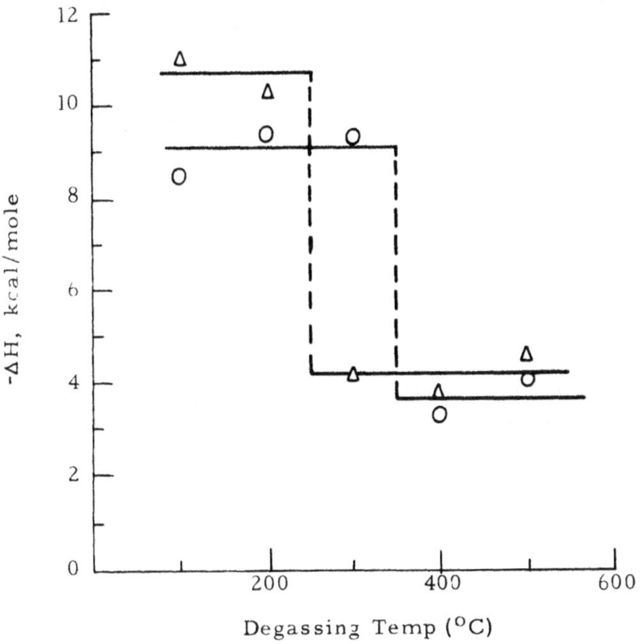

Figure 10. Isosteric heats of adsorption for methanol on so lime-silica glass (O) and silica (Δ). (from ref. 3).

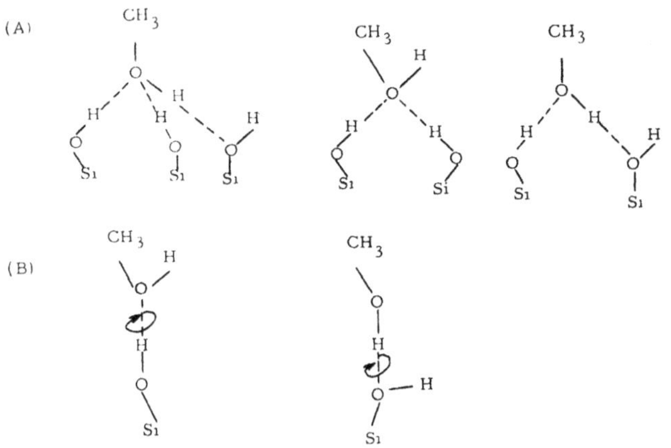

Figure 11. Model for methanol adsorption on silica and soda-lime-silica glass. Configurations for (A) low-temperature degassing and (B) high-temperature degassing. (from ref. 3).

The degassing treatments did not greatly affect the magnitude of adsorption in the low pressure regions. All of the isotherms measured at 40°C on silica were similar, regardless of degassing temperature. Taking into account the variation of heat of adsorption with degassing temperature, this could mean that there are fewer active sites for adsorption at the high surface hydroxyl concentrations. This, in turn, could be due to certain directional bond requirements for the formation of the configurations shown in Figure 11.

Sewell and Morgan considered that the behavior of glass and silica as adsorbants towards methanol are similar, although the amount of adsorption on glass is approximately half that on silica under identical conditions. This difference may be due to polarization in the silanol groups caused by the other constituents present in the soda-lime-silica glass.

The solubility of silica, in the form of porous glass and silica gel, in nitric acid solutions has been investigated by Elmer and Nordberg,[4] who showed that the solubility at constant temperature decreased with increasing strength of free HNO_3. The rate of solution of silica was also found to depend upon acid concentration, the initial rate being a maximum at 0.8N. The solubility at the three temperatures studied, may be expressed in the form:

95°C	$\log C = -0.168N - 0.332$
65°C	$\log C = -0.167N - 0.584$
36°C	$\log C = -0.184N - 0.796$

where C = concentration of silica in mg. per ml of acid solution N = normality of HNO_3 solution.

Elmer and Nordberg used their results to explain several processing phenomena encountered in industrial processes. For instance, in the manufacture of high-silica fiber glass it was found necessary to use strong acid solutions to leach the glass. Dilute acid solutions produced weak fibers, due to the fact that in such solutions the amount of silica which goes into solution is greater than in strong acid. Thus, in a weak acid more glass is dissolved from the silica skeleton before the solution becomes saturated with silica.

The influence of reaction products can greatly affect

surface reactions on glass. Wang and Tooley[5] have made a detailed study of the case of water interaction with a soda-lime-silica glass. The reaction products from this interaction have been identified as sodium metasilicate, calcium metasilicate and metasilicic acid.[6]

Wang and Tooley investigated the reaction of glass with each individual reaction product and came to the conclusion that the rate-controlling processes in the overall reaction between water and glass were the diffusion of cations and the breakdown of the glass network. They also presented a qualitative description of the mechanism of the reaction.

The use of glass as a component in vacuum systems and as a substrate for thin film deposition requires that its surface be made as clean as possible. A cleaning process for borosilicate glass has been developed by Tichane[7] which is claimed to be superior to a chromic acid wash. The efficiency of the process was determined by examination of electron photomicrographs of glass surfaces after different treatments. The recommended treatment for the glass in question (Corning Code 7740 - 81% SiO_2, 4% Na_2O, 13% B_2O_3, 2% Al_2O_3) is:

1. Prepolish with precipitated calcium carbonate (when possible).
2. Mild etch for 10 minutes with 1% sodium hydroxide solution at 95°C.
3. A 2-minute cleaning with 5% hydrochloric acid at 50°C.
4. Final rinse with distilled water at room temperature.

Reactions at glass surfaces may sometimes be severe enough to cause corrosion of the glass. Simpson[8] developed a method of measuring the surface durability of glass by exposing glass samples to alternate cycles of "fogging" and "clearing" by temperature cycling between 55°C and 50°C with a relative humidity approximately 100%. This was carried out in a haze test oven[8] and the time-temperature variation is shown in Figure 12.

The increase in haze (scattered light) on the glass surface was measured by a photoelectric cell. The relative amount of scattering is a direct measure of the surface deterioration.

The surface reactions involve a complex series of inter-

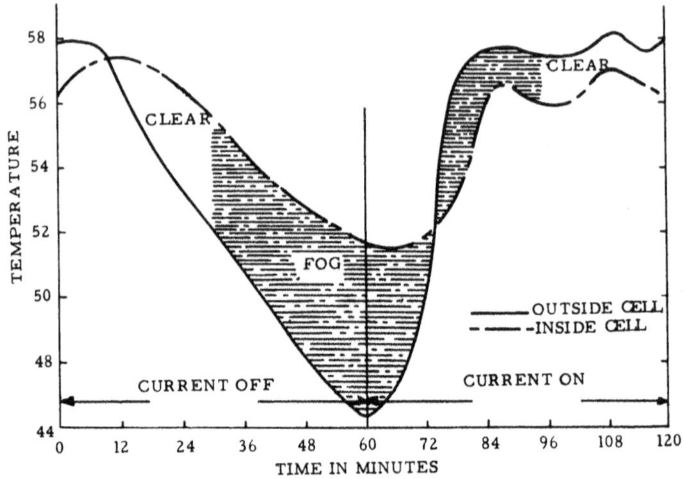

Figure 12. Temperature variations during cycling. (from ref. 8).

actions which may be expressed in a simplified form[9] as:

1. Adsorption of a water film on the glass surface.
2. Selective diffusion and activation of the surface of a leached solution involving alkali silicates.
3. The leached solution undergoes a hydrolysis process which may be expressed as:

$$Na_2O \cdot SiO_2 + 2H_2O \rightarrow 2NaOH + H_2SiO_3$$

The alkali hydroxide thus formed becomes a corrosive agent on most glasses and later reacts with CO_2 in the atmosphere to form sodium carbonate with the separation of silica in some form. In the original paper[8] Simpson gives the results of durability studies for many different glass compositions.

An example of the difference in reactivity between different forms of silica is shown in Figure 13. This graph[10] shows the activity of quartz, cristobalite and silicic acid for the hydrogen-deuterium equilibrium reaction

$$H_2 + D_2 \rightarrow 2HD$$

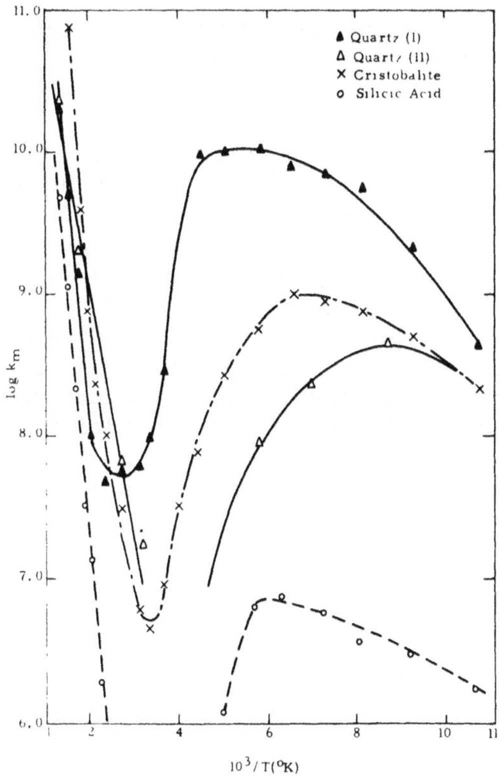

Figure 13. Temperature dependence of H_2-D_2 equilibration reaction on silicas. (from ref. 10).

The temperature range represented in Figure 13 is 77^oK to 750^oK. k_m is an absolute rate constant which corrects the experimentally determined first order rate constant for such variables as surface area of the sample, pressure or reaction gas, volume of reaction vessel. The units of k_m are molecules reacting per cm^2 by surface per second.

The decrease in activity in the intermediate temperature region (near 200^oK) is due to the desorption of hydrogen from the surface in a change-over from a physically adsorbed state at low temperatures to a chemisorbed state at high temperatures.

For all temperatures the order of reactivity is:

quartz $\approx$ cristobalite > silicic acid

The low temperature activity is considerably higher for quartz (A-II) which has been baked out in a reducing atmosphere.

In the low temperature region the sites responsible for surface activity will depend upon the nature of the material. For oxides such as silica and alumina, the requirement of lowest energy produces a surface of oxygen ions[11] which are bonded to hydrogen to satisfy the chemical valences. The condensation of these hydroxyl groups during thermal activation results in a surface with high strain energy and increased catalytic activity.[12] This activity is not decreased by adsorbed oxygen so that the active sites on these oxides may be regarded as dehydration centers.[13]

The surface of silica may be represented by a series of silanol groups:

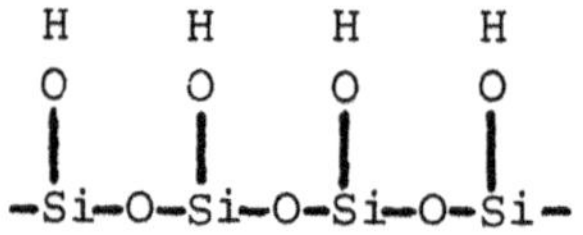

Initially there is considerable hydrogen bonding between the groups. Thermal activation may produce an ionic surface

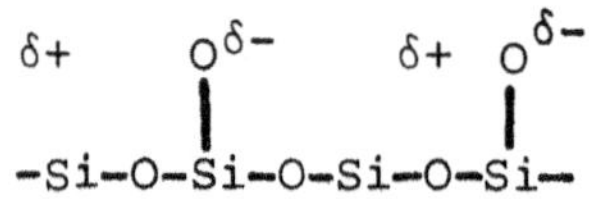

according to a similar model for alumina surfaces.[14] Alternatively, the active sites may be paired by oxygen bridges:

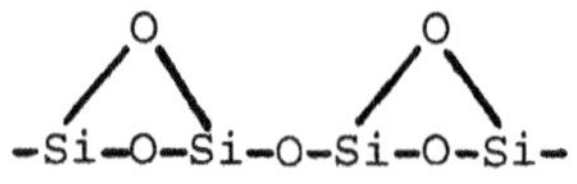

This form has been suggested by Iler[15] for silica gel. Such a surface would catalyze reactions such as hydrogen-deuterium equilibration.

Each of the silicas catalyzed hydrogen-deuterium equilibration at low temperatures. The reaction had a small activation energy, 200 to 2000 cal, which is less than the usual lower limit for chemisorption. These activation energies may represent the energy required to form an activated complex between hydrogen retained on the surface after bakeout and hydrogen held in a Van der Waals layer.

Quartz and cristobalite which were baked out in a vacuum under identical conditions showed similar low-temperature activities. Quartz which was baked out longer in a reducing atmosphere showed considerable increase in low-temperature activity. Luminescence studies by Ewles and Heap[16] on silica which had been heated in hydrogen at temperatures as low as 573°K showed that defects caused by the removal of surface oxygen, had appeared on the surface.

In the high temperature region above 473°K the rate of equilibration can be expressed by the Arrhenius equation:

$$k_m = A \exp (-E/RT)$$

A = frequency factor
E = activation energy
R = gas constant
T = temperature

Values of the parameters A and E are given in Table II for three forms of silica.

Table II. Arrhenius Parameters for High-Temperature Equilibration

Silica Form	A(molecules cm^{-2} sec^{-1} x 10^{-10}	E_a(kcal $mole^{-1}$)
Quartz	7.25	13.65
Cristobalite	8.90	12.30
Silicic Acid	5.25	13.25

FLAWS IN GLASS SURFACES

In the case of brittle materials, particularly glass, the great difference between the experimental and theoretical values for tensile strength has been attributed for many years to stress concentration at minute flaws. This interpretation was given in the classic work of Griffith fifty years ago.[17]

These Griffith flaws may be as narrow as 5A, and normal methods of microscopy, including electron microscopy, do not have sufficient resolution for their detection. However, in 1960, and in subsequent work, Ernsberger[18,19] developed a technique for the detection of strength-impairing flaws in glass.

Full details are presented in the original papers, but in essence the technique was developed from earlier work by Andrade and Tsien[20] who showed that crack patterns on the surface of hard glass could be observed after treatment with sodium vapor at temperatures up to 500°C. Until 1960 it was believed that the Andrade-Tsien cracks were produced during this high temperature treatment of the glass. However, Ernsberger showed that such cracks appear during the cooling phase. It was also shown that the component cracks in the network appear in succession, and that they arise from thermal contraction of the surface film.

The treatment of glass surfaces by exposure to sodium vapor causes a considerable change in physical properties due to the high concentration of alkali ions in the surface. For example, the surface layer increases in volume and the softening point is lowered.

Although a detailed examination of Andrade-Tsien cracks can provide useful information concerning flaws in glass, Ernsberger's great contribution was the development of an ion exchange technique for detailing the Griffith microcracks. The chief disadvantage of the sodium vapor method to reveal Griffith cracks is that only the deepest flaws survive the high temperature sodium vapor treatment.

When a glass is exposed to a molten salt containing lithium ions, an ion exchange occurs. Sodium ions diffuse out of the glass and are replaced by lithium ions. The film of glass affected by the ion exchange tends to shrink, due to the difference in ion size.

$$\text{radius ratio } \frac{Li^+}{Na^+} = 0.63$$

The tendency to shrink is opposed by the underlying glass composition and the result is tensile stress in the surface film of ion-exchange glass.

If Griffith microcracks are present in this stressed film they serve as origins for the propagation of visible cracks. The growth of the crack is limited to two dimensions for they cannot penetrate deeper than the tension state of the surface. This is determined by the depth of ion exchange, which is of the order of 2 microns for a 2 hour ion exchange treatment at 200°C.

The time and temperature for the ion exchange have to be carefully controlled. If the treatment continues for too long a time cracks will form during the exchange leading to spurious results. Under proper conditions the crack growth does not take place during the ion exchange or as the glass cools. Growth only begins after the treated surface is dipped into water, due to the reduction in surface energy of the glass by penetrating water.[5,6] This enables the study of crack propagation under isothermal conditions.

Figure 14 shows a typical crack pattern developed by the Ernsberger technique. This glass slide had been initially given a short etch in aqueous HF in order to reduce the number of flaws thus avoiding mutual interference of propagating cracks.

The arrow in Figure 14 points to the S-shaped step between two straight and parallel branches of a crack. The point of inflection in the step is a crack origin invisible in the absence of a bending stress.

Ernsberger considers that this ion exchange technique seems capable of revealing the position of strength impairing flaws, although it is not known whether it reveals all such flaws. He also considers the limit of resolution of cracks is about 2 microns, i.e., two parallel microcracks can be detected individually unless they are closer than 2 μ.

The density of microcrack distribution on a surface could be determined by simply counting all of the cracks. This is difficult and tedious and is overcome by confining the ion exchange to several small areas on the glass and taking a statistical sampling of the cracks formed. Ernsberger found that by protecting a large part of the glass surface with a polymer film, and exposing large numbers of small areas to the molten salt treatment a valid statistical

Figure 14. Shrinkage cracks produced by treatment of soft glass with a droplet of $LiNO_2$ + KNO_3 eutectic at 200° C, with a supplemental uniaxial stress. (from ref. 18).

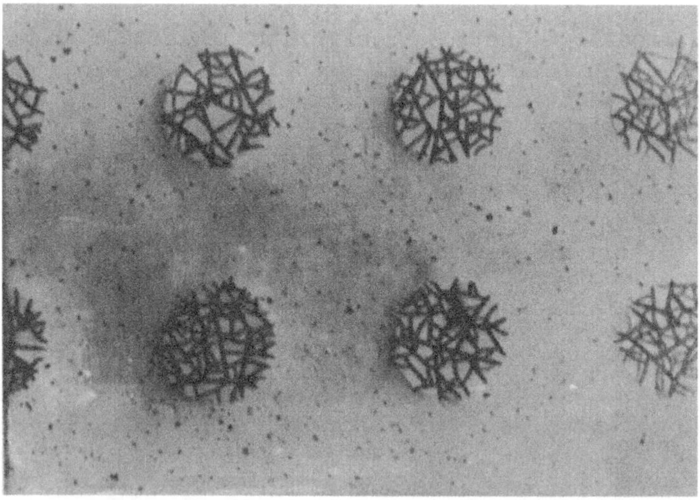

Figure 15. Crack patterns produced by photoresist-limited ion-exchange treatment of glazing quality plate glass. Diameter of area, 0.4 mm. (from ref. 19).

sample of cracks may be obtained.

Figure 15 shows a typical crack pattern produced by this method, which was obtained after 15 minutes ion exchange at 250°F. The circular areas exposed are 0.4 mm diameter and the number of microcracks averages 70 per area, or approximately 56,000 per $cm.^2$

For mirror-quality glass the number of cracks is drastically reduced, being approximately 700 per $cm.^2$

The number of microcracks in plate glass is reduced by etching for 1 minute with 5% HF as shown by comparing Figures 16 and 17. After 3 minutes etching the original highly flawed surface is almost free of strength impairing microcracks.

Many other photographs of varying treatments for glass are given in the original paper. Very briefly the main conclusions drawn by Ernsberger are that plate glass has a high and uniform density of microcracks, whereas sheet glass has a lower density and less uniformity. Polishing plate glass reduced the density of microcracks. Three strength impairing surface defects were identified by controlled experiments with initially flawless sheet glass surfaces:

1. grinding fissures
2. microabrasions
3. devitrification nuclei developed by contamination on the glass surface

SURFACE TENSION OF MOLTEN GLASS

All normal methods of evaluating surface tension can be used with glass systems, some of the more commonly used being the maximum bubble pressure method, and the sessile drop method. The variation of surface tension with temperature of the melt is very dependent upon the glass composition. Figure 18 shows the surface tension of glasses in the $CaO-Al_2O_3 - SiO_2$ system. The compositions of the glasses are given in Table III.

The results in Figure 18 were obtained by Barrett and Thomas[21] using the maximum bubble pressure technique. In this method the maximum pressure sustained by a bubble just before release from a tube immersed in a liquid is given by:

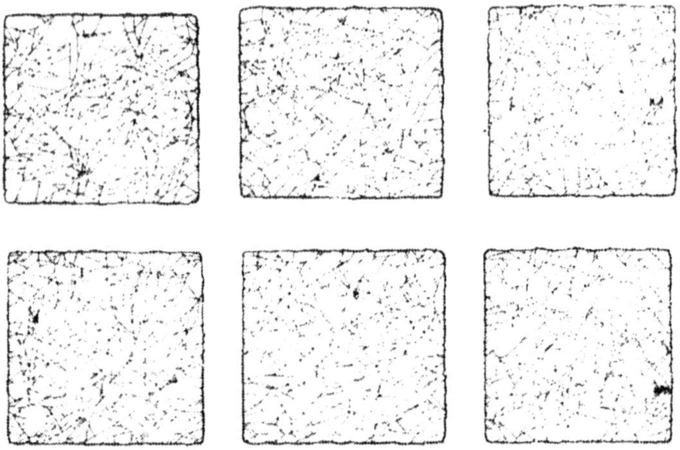

Figure 16. Effect of etching in reducing concentration of microcracks. Plate glass, unetched. Areas 0.25 cm^2. (from ref. 19).

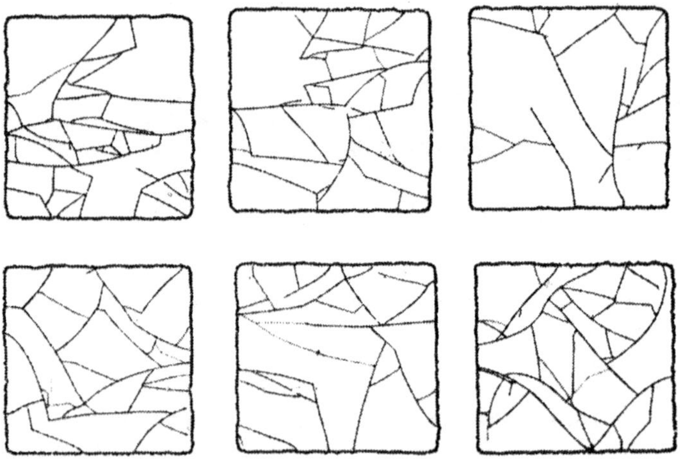

Figure 17. Effect of etching in reducing concentration of microcracks. Plate glass, etched 1 min. in 5% aqueous HF. Areas 0.25 cm^2. (from ref. 19).

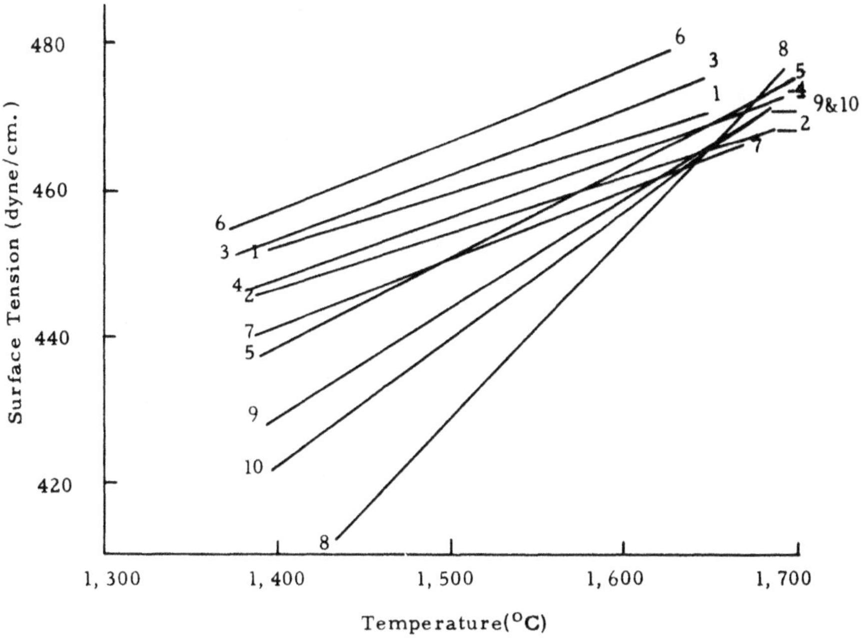

Figure 18. Surface tension of glasses in the system $CaO-Al_2O_3-SiO_2$ as a function of temperature. (from ref. 21).

Table III. Percentage Composition of Glasses[21]

Glass	CaO	Al_2O_3	SiO_2
1	39	19	42
2	35	10	55
3	34	30	36
4	30	25	45
5	30	10	60
6	29	40	31
7	25	20	55
8	25	10	65
9	23	15	62
10	15	20	65

$$P_{max} = \frac{2\gamma}{r} + gh\,(\rho - \rho^{o})$$

where γ = surface tension of liquid
r = radius of tube
h = depth of tube outlet beneath liquid surface
ρ = density of liquid
ρo = density of air (or surrounding phase)

The bubble can sustain the maximum pressure at the position of minimum radius. The applied pressure is balanced by the pressure difference across the curved surface plus a hydrostatic pressure depending on the depth of the tube outlet.

The rapidly increasing viscosity limited practical application of this method to temperatures above approximately 1300°C. Referring again to Figure 18, it is seen that within the temperature range studied the surface tension is a linear function of temperature. Such a relation, without any discontinuities, indicates the absence of any molecular association or "compounds" in the melt. It is also seen that the surface tension increases with alumina and silica contents and that with the exception of glasses with the highest alumina content (30-40%, glasses 3 and 8) they all have approximately the same value for the surface tension in the region of 1650°C.

In the sessile drop method the shape of a drop of liquid on a solid surface is determined, usually by an optical method. A sessile drop configuration is shown in Figure 19 and at equilibrium Young's equation applies.

$$\gamma_{SL} + \gamma_{LV}\cos\theta = \gamma_{SV} \qquad (2)$$

where γ_{SL} = surface tension of the solid-liquid interface
γ_{LV} = surface tension of the liquid-vapor interface
γ_{SV} = surface tension of the solid-vapor interface
θ = contact angle

Cases where $\theta<90^{o}$ are usually referred to as "wetting" and where $\theta>90^{o}$, as "non-wetting".

From equation 2, if values for γ_{SL} and γ_{SV} are known, an optical determination of the contact angle θ enables γ_{LV}, the surface tension of the liquid, to be calculated. Butler

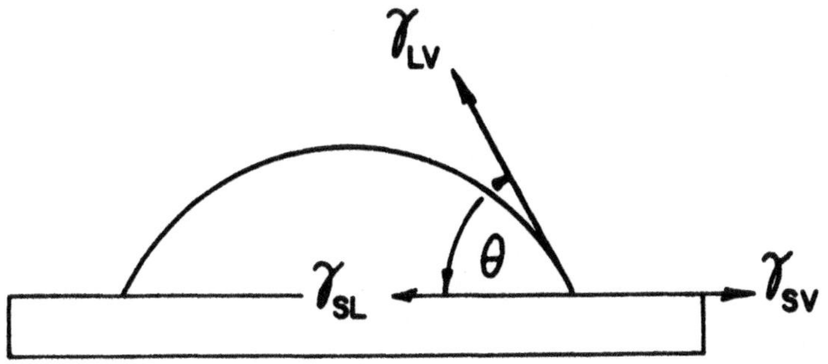

Figure 19. Sessile drop configuration.

and Bloom[22] have recently presented a computer program for calculating the shape of a sessile drop by fitting the coordinates of the periphery of a drop image to a theoretical drop contour. The program also provides size and shape parameters to enable the calculation of the contact angle.

One advantage of the sessile drop method is that it enables any variation of surface tension with time to be studied. From Young's equation it may be seen that if γ_{LV} will result in an increase in cos θ, and, therefore, a decrease in θ.

These considerations are very important in wetting problem of glazes etc., on refractory surfaces. Figure 20 shows the results of Towers[23] for the surface tension of a melt on polished alumina at 1415°C.

BIOLOGICAL APPLICATIONS

It seems increasingly certain that in future years glass will be of great importance in biological applications. Already the areas of bio-ceramics and bio-engineering are becoming disciplines in their own right and much work is currently underway in the field of bone implant and transplant materials.

Although in some cases the fact that glass is inert to many substances is an advantage, in biological applications it is often extremely reactive. For instance, a normal glass surface causes blood to clot rapidly (within a few minutes), and thus could not be used in such possible areas

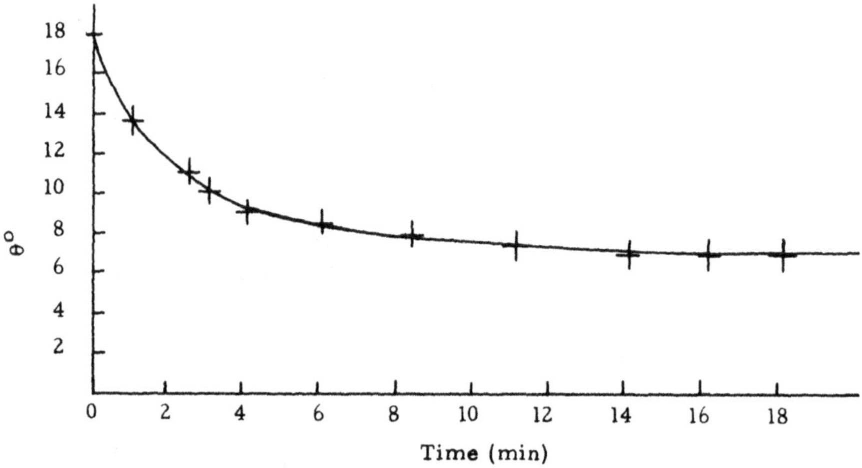

Figure 20. Variation of surface tension with time. Melt composition 30.6 CaO· 37 Al_2O_3· 32.4 SiO_2 on polished alumina support. (from ref. 23).

as heart valves, heart-lung machines, etc. The surface bonds of a glass are free to attract the various chemical groups often found in living organisms such as amino groups ($-NH_2$), hydroxyl (-OH), methyl ($-CH_3$) and others.[24]

Alkalies attack a glass network by an etching process involving the breaking of Si-O bonds. The rate of removal of surface glass is proportional to the alkalinity and is steady with respect to time. Acids attack glass by the exchange of hydrogen ions of the acid with other anions, for example, sodium. This process is diffusion controlled and, therefore, its rate decreases with increasing time.

Biological interactions with glass are considered to be largely chemical in nature, and can range from simple bonding interactions to corrosion reactions. Recently some workers[25] at Corning Glass Works have developed a method for permanently coating glass with heparin in order to prevent the clotting of blood. Heparin is a long chain polysaccharide molecule and is a universal agent for preventing blood clotting. The coating of heparin on glass was achieved by an ionic linkage using an amino silane. This heparin/silane/glass system presented a durable high concentration of heparin at the glass surface which resisted blood clots. The authors estimated that a concentration of 3μg heparin/cm^2 was sufficient to be non-thrombogenic. Such a concentration, from the dimensions of the heparin molecule (approximately 10 x 200A), corresponds to a monolayer coverage under the extreme condition of all the molecules standing on end. This seems unlikely and, therefore, multilayer coverage is probable. Table IV shows the results obtained comparing the clotting behavior of blood in contact with Corning 7740 glass.

Table IV. Heparinated Tubes [25]

	Time (min.)		
	Tube 1	Tube 2	Tube 3
1. First signs of fibrin (air-blood interface)	46-51	50-62	71-80
2. Patially clotted (from air interface downward)	54-59	56-66	83-88
3. Complete clotting	60-66	62-75	90-103
Controls			
1. First signs of fibrin (top and wall of tube)	5-7.5	5.5-8.0	6.0-8.0
2. Partially clotted	6-8	6-8.2	6.2-8.5
3. Complete clotting	6.5-9.0	6.5-9.0	6.8-10

The effect of glass surface on living organisms can be extremely subtle. For instance, it is known that when tuberculosis vaccine is kept in hand-blown borosilicate glass bottles, its effectiveness is decreased compared to vaccine stored in machine-blown bottles.

The generally accepted theory of cells bonding to glass is that of a calcium bond between the living cell and the glass, as shown in Figure 21. However, some workers now believe that potassium plays an important part in this process. The ratio of sodium to potassium is low within a cell, but high in the medium surrounding it, suggesting that the transport of potassium across the cell wall may be involved, as shown in Figure 21. Further evidence that calcium may not be as important as originally considered in cell bonding is in the fact that solutions of substances which effectively remove calcium by complexing still support the growth of certain types of cells. Such cells will only grow on suitable glass substrates. They die in ordinary lime or borosilicate glasses and grow best in glasses in which potassium has been introduced by ion exchange.

As well as chemical composition the structure of glass can have a significant effect on its behavior towards living matter. Silicosis, which involves the growth of fibrous tissue in the lungs, is caused by breathing crystalline quartz dust. On the other hand, there is evidence that breathing vitreous glass dust, which is just another physical form of silica, has no effect.

There are thus many ways in which glass and biological systems can interact. Most are not yet known or understood and this whole area of study is just at its threshold. In the future a more complete understanding of these interactions will result in more useful applications of glass to biological problems.

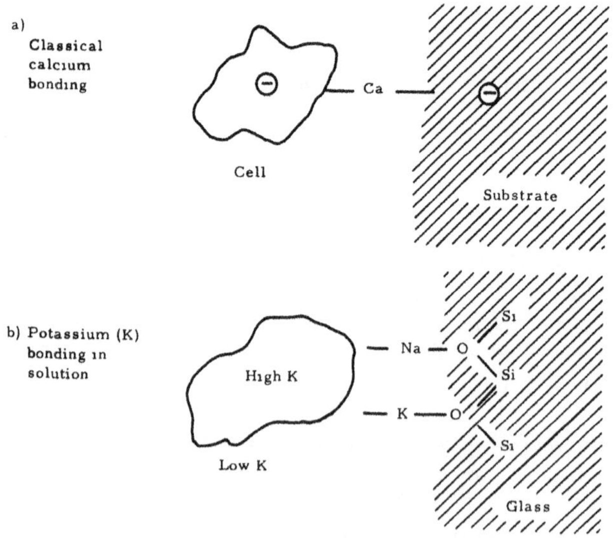

Figure 21. Cell bonding to glass.

APPENDIX

The form of the adsorption isotherm for the physical adsorption of gases on many finely divided solids is usually as shown in Figure 22, and the equation derived by Brunauer, Emmett and Teller[1] is:

$$\frac{P}{V(P_o-P)} = \frac{1}{V_m c} + \frac{c-1}{V_m c} \cdot \frac{P}{P_o} \qquad (1)$$

where V = volume of gas adsorbed at equilibrium pressure P (reduced to STP)

P_o = saturation vapor pressure of the gas at the temperature of adsorption

c = constant, related to the heat of adsorption

V_m = volume of gas molecules required to completely cover the surface of the solid with a layer one molecule thick (reduced to STP)

Thus by measuring the volume of gas adsorbed at various equilibrium pressures and plotting the results as $\frac{P}{V(P_o-P}$ versus $\frac{P}{P_o}$ a straight line of slope $\frac{c-1}{V_m c}$ and intercept $\frac{1}{V_m c}$ is obtained, and so the value of V_m may be calculated.

The B.E.T. equation is valid (i.e., equation 1 produces a straight line) in a pressure range $.05 < \frac{P}{P_o} < 0.35$.

The surface area of the sample is obtained from the value of V_m by the use of the equation:

$$\text{surface area} = \frac{NV_m}{V}$$

where N = Avogadro's number

= area covered by one molecule of adsorbed gas in the plane of the surface

V = volume of one mole of gas

Nitrogen adsorption and krypton adsorption at -196^oC are often used for determining surface areas. Values for nitrogen are:

P_o = 760 cm at -196°C

= 16.2×10^{-16} cm^2

Krypton:

P_o = 0.21 cm at -196°C

= 20.2×10^{-16} cm^2

Parameters for many other gases are given by McClellan and Harnsberger.[26]

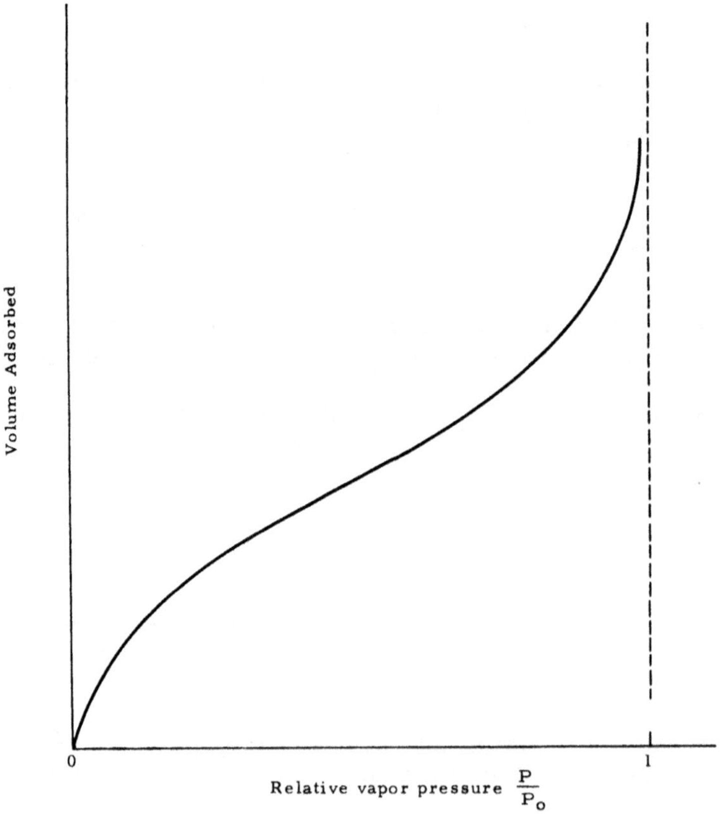

Figure 22. Type II multilayer adsorption isotherm.

REFERENCES

1. S. Brunnauer, P. H. Emmett and E. Teller, J. Am. Chem. Soc., 60, 309 (1938).
2. J. A. E. Taylor, J. A. Hockey and B. Pethica, Proc. Brit. Ceram. Soc., No. 5, 133 (1965).
3. P. A. Sewell and A. M. Morgan, J. Am. Ceram. Soc., 52, 136 (1969).
4. T. H. Elmer and M. E. Nordberg, J. Am. Ceram. Soc., 41, 517 (1958).
5. F. Fu-Yen Wang and F. V. Tooley, J. Am. Ceram. Soc., 41, 521 (1958).
6. F. Fu-Yen Wang and F. V. Tooley, J. Am. Ceram. Soc., 41, 467 (1958).
7. R. M. Tichane, Bull. Am. Ceram. Soc., 42, 441 (1963).
8. H. Simpson, Bull. Am. Ceram. Soc., 30, 41 (1951).
9. H. Simpson, J. Am. Ceram. Soc., 36, 143 (1953).
10. J. Harris and D. R. Rossington, J. Am. Ceram. Soc., 51, 511 (1968).
11. W. A. Weyl, Research (London) 3, 230 (1950).
12. E. B. Cornelius, T. H. Millikan, E. A. Mills and A. E. Oblad, J. Phys. Chem., 59, 809 (1955).
13. V. C. F. Holm and R. W. Blue, Ind. Eng. Chem., 44, 107 (1952).
14. H. Pines and J. Ravoire, J. Phys. Chem., 65, 1859 (1961).
15. R. K. Iler, The Colloid Chemistry of Silica and Silicates, Cornell University Press, 1955.
16. J. Ewles and C. N. Heap, Trans. Faraday Soc., 48, 331 (1952).
17. A. A. Griffith, Phil. Trans. Roy. Soc., London, A221, 163 (1920).
18. F. M. Ernsberger, Proc. Roy. Soc., A257, 213 (1960).
19. F. M. Ernsberger, Adv. in Glass Technology, p. 511, Plenum Press, New York, 1963.
20. E. N. da C. Andrade and L. C. Tsien, Proc. Roy. Soc., A159, 346 (1937).
21. L. R. Barrett and A. G. Thomas, J. Soc. Glass Tech., 43, 179 (1959).
22. J. N. Butler and B. H. Bloom, Surface Science, 4, 1 (1966).
23. H. Towers, Trans. Brit. Ceram. Soc., 53, 180 (1954).
24. P. B. Adams, New Scientist, p. 25, Jan. 2, 1969.
25. L. S. Hersh, H. H. Weetall and I. W. Brown, J. Biomed. Mat. Res., 3, 471 (1969).
26. A. L. McClellan and H. F. Harnsberger, J. Coll. Interface Sci., 23, 577 (1967).

The Corrosive Nature of Molten Glass

Vernon L. Burdick

State University College of Ceramics at
Alfred University
Alfred, N.Y.

INTRODUCTION

One of the outstanding characteristics of molten glass is its ability to chemically and physically attack the refractories which serve to contain the glass. The physical attack usually consists of penetration of voids, cracks and joints by the molten glass or in some cases refractories may be abraded. Chemical corrosion is much more deleterious to the life of the refractories since it involves the dissolution of some or all of their components. Additionally, this chemical interaction is potentially a source of defective glass product.

Defects known as stones are often traced to refractory materials wherein the interparticle bond has been dissolved by the glass and refractory particles are carried into the batch and remain undissolved. Other effects are noted when the dissolved refractory significantly changes the glass composition. These may be manifested as inhomogeneities such as cords or knots, or as devitrification tendencies. Secondary crystallization may also occur near the glass-refractory interface. All of these possibilities add up to a sizeable problem which exists throughout the glass industry.

Since it is recognized that chemical corrosion of refractory materials by molten glass is the most important factor governing the life of a glass melting unit, this paper will consider the study of several aspects of the corrosion problem. These topics include corrosion processes and kinetics, testing for corrosion resistance and selection of refractories for corrosion resistance.

CORROSION PROCESSES

Two basic processes are involved in corrosion. These are: (1) a chemical reaction at the glass-refractory interface and (2) the removal of reaction products from the interface. While this seems like a small number of processes to consider, the complexity arises when several important variables are considered. These include: (1) the temperature dependence of reaction rates, melt viscosity and melt density; (2) the type and velocity of fluid flow; (3) the composition dependence of melt viscosity and density; surface tension effects and (4) the solubility limits of various components in the melt.

Some of the complications are illustrated by examining the profiles of corroded samples. Several distinct profiles have been studied by Busby and Eccles[1] and by Vago and Smith.[2] Figure 1a shows a tapered profile produced by an upward convection current which has brought fresh glass to the lower portion of the sample. The current can be caused by a density gradient alone, where the reaction products are lighter than the bulk glass. This would occur, for example, when alumina or silica is dissolving in a lead glass under free convection conditions. In contrast, Figure 1b shows a tapered profile resulting from a downward current. Fresh glass is presented to the top of the immersed portion and a heavier reaction product is formed, which causes the downward current. This condition would prevail, for example when alumina dissolves in sodium silicate glass.

Features other than the taper below the glass line are also important. At or near the glass line, Figures 1b, 1c and 1d show respectively a cut, a groove and a neck. The first two have been attributed to surface tension differences between the bulk glass and glass containing dissolved refractory.[2] When the surface tension of the glass at the glass-refractory interface is decreased by the dissolution currents should flow up and away from the dissolving surface. The driving force for this is the tendency to lower the overall surface free energy by spreading the glass of lower

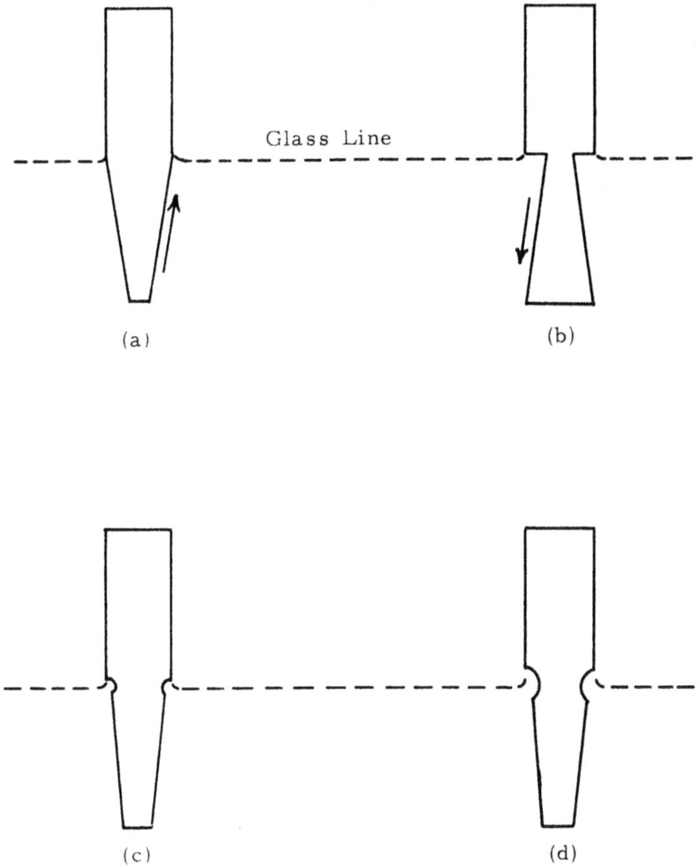

Figure 1. Corrosion profiles. (a) Produced by density current flowing upward. (b) Produced by density current flowing downward. (c) Shows a groove at the glass line from surface tension currents. (d) Shows a neck at the glass line from a combination of surface tension currents and vaporization.

surface tension over the bulk glass. This accounts for the cut at the glass line. (Figure 1b) On the other hand, if the surface tension is increased by the dissolution of the refractory currents should flow toward the refractory at the glass line and then down into the bulk glass. The impingement of fresh glass on the refractory at the glass line can account for the groove shown in Figure 1c. It is well to note that these currents act in conjunction with other currents produced by density or thermal gradients. The resulting flow patterns can be complex, thus more complicated profiles can be expected. Surface tension currents can reinforce convection currents thereby increasing the overall rate of corrosion. Additionally, the two types of currents can interact near the glass line to increase the velocity of fluid flow or to produce a turbulent pattern thereby increasing the necking effect.

The neck profile (Figure 1d) is attributed[2] to a combination of currents produced by an increase in surface tension during dissolution and the presence of vaporizable components in the glass which attack the refractory above the meniscus level. Without an increase in surface tension, the attack by corrosive vapor can produce a cut similar to that shown in Figure 1b.

For samples completely immersed in molten glass, the diffusion of reaction products through the moving boundary layer (produced by density currents) becomes important. Several investigations[1-7] have shown that the diffusion process determines the dissolution rate of refractories in molten glass. The taper below the glass line can be explained by the combination of density currents and diffusion through the layer of glass enriched by reaction products.

Corrosion of horizontal surfaces also plays an important role in determining refractory life. The upward drilling phenomenon is most destructive of surfaces facing downward. The attack occurs at surfaces formed at cracks and joints as well as at the top of the throat of a glass tank or other built in surfaces. Busby[8] has outlined three conditions which are usually fulfilled when upward drilling occurs. First, the density of the glass enriched by refractory dissolution must be greater than that of the bulk glass. Under this condition, the reaction products can be removed by density currents. A second condition requires that the surface tension of the glass must be increased by the dissolution process. This provides a driving force for removal of reaction products in order to reduce surface energy. A third condition which accelerates upward drilling is the presence of gas bubbles at the surface of the refractory. This effect involves a disruption of the diffusion boundary layer thereby allowing the dissolution reaction to proceed more rapidly.

Model experiments on upward drilling have been performed[8,9] using the system ice-alcohol-water. In this system, at temperatures below the freezing point of ice, the water-alcohol solution is less dense than water and the surface tension of the solution is increased by dissolving ice. This fulfills the conditions outlined above. Upward drilling occurred wherever there was a bubble present at the lower surface of the ice. Apparently[8] there is an intense disturbance of the solution boundary layer in the vicinity of the bubble where the phases air, alcohol, ice, water and solution meet.

A practical system of soda-lime glass with zirconia-alumina-silica refractories behaves just as the model system. Alumina and zirconia increase the density and surface tension of the glass upon dissolution of the refractory. Severe upward drilling is known to occur in this system. In contrast, when lead glass is used, the density is decreased by dissolving the refractory. Thus, reaction products are not removed by density currents and upward drilling is inhibited.

KINETICS OF CORROSION

The kinetics of corrosion are usually treated on the basis of the two basic processes of chemical reaction at the glass-refractory interface and the diffusion of the reactant to, or of, the products from the interface through the boundary layer.[5,6,8,9]

(A) Chemical Reaction Control

The rate of corrosion is limited by the surface chemical reaction whenever this step is the slowest process. The reaction rate can be expressed as:[5b]

$$\frac{dQ}{dt} = K \cdot \frac{A_c}{A_o} \cdot C_m \qquad (1)$$

where Q = quantity of refractory dissolved/unit area ($moles/cm^2$),
t = time (sec),
A_c = actual (chemically active) area of refractory (cm^2),
A_o = apparent (macroscopic) area of refractory (cm^2),
C_m = concentration of refractory in the melt ($moles/cm^3$),
K = rate constant.

It is evident that the rate of the reaction will be increased by an increase in chemically active surface area. Any surface roughening such as chemical etching or physical abrasion will produce this effect.

(B) Diffusion Control

If the chemical reaction at the glass-refractory interface is rapid enough to keep the interfacial glass saturated, then the corrosion rate will be controlled by the rate of removal of reaction products by diffusion. The rate of corrosion can then be expressed by the Noyes-Nernst equation:[10]

$$\frac{dQ}{dt} = k\ (C_s - C_m) = \frac{D}{\delta}\ (C_s - C_m) \qquad (2)$$

where C_s = saturation concentration of refractory in the melt ($moles/cm^3$),
k = rate constant = D/δ,
D = diffusion constant $cm^2/sec.$,
δ = effective boundary layer thickness (cm).

The effective boundary layer thickness, δ, as defined as: (see Figure 2)

$$\delta = (C_s - C_m)/(dC/dx) \qquad (2a)$$

where $\frac{dC}{dx}$ = the concentration gradient at the interface,

x = the distance from the interface.

Thus, it is evident that the corrosion rate is directly dependent upon the concentration gradient, the diffusion constant and the boundary layer thickness. The rate of corrosion is highly influenced by the velocity with which

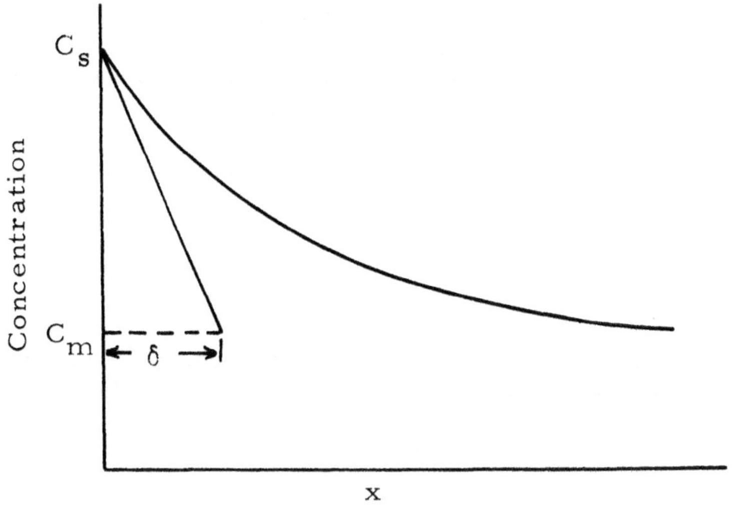

Figure 2. Definition of the boundary layer thickness.

the melt moves over the refractory. At high velocities, δ is reduced, thus the rate is high. Under steady state conditions, δ reaches an equilibrium value which is controlled by the magnitude of density gradients, the viscosity of the melt, the average diffusion coefficient and the size of the bath.[5b] Wagner[12] has dealt with these variables in terms of the dimensionless groups known as the Grashoff and Prandtl numbers. For a flat vertical surface where natural convection and diffusion are rate controlling factors, the dissolution rate is given by:

$$\frac{Q}{t} = 0.545 \frac{D(C_s - C_m)}{y} \cdot (G_r \cdot P_r)^{1/4} \quad (3)$$

where y = distance from the leading edge of the corroding surface (cm),

G_r = Grashoff number = $\frac{gy^3}{v^2} \frac{\Delta\rho}{\rho}$,

P_r = Prandtl number = v/D,

v = kinematic viscosity (cm^2/sec),
ρ = density of bulk melt (g/cc),
Δρ = density difference between saturated and bulk melt,
g = gravitational constant (cm/sec^2).

For a plate of height, h, the average rate of dissolution was given as:

$$\frac{Q}{t} = 0.726\ D\ (C_s - C_m) \cdot \left(\frac{g\Delta\rho}{vDh\rho}\right)^{1/4} \qquad (4)$$

Combining equations (2) and (4), the boundary layer thickness can be expressed as:

$$\delta = 1.38 \left(\frac{vDh\rho}{g\Delta\rho}\right)^{1/4} \qquad (5)$$

Under forced convection where the melt is stirred and laminar flow takes place over a plate, the velocity, U, replaces Δρ as a variable. The rearranged expression for the corrosion rate is:[5b,8]

$$\frac{Q}{t} = \frac{D(C_s - C_m)}{3.09} \left(\frac{U}{vy}\right)^{1/2} \left(\frac{v}{D}\right)^{1/3} \qquad (6)$$

and $$\delta = 3.09 \left(\frac{vy}{U}\right)^{1/2} \left(\frac{D}{v}\right)^{1/3} \qquad (7)$$

where: y = distance from the leading edge of the plate

Cooper[13] has modified the Hoyes-Hernst relationship to account for the recession of the refractory surface during corrosion. This amounts to a substitution of the quantity $[(C_s - C_m)/(1 - C_s\bar{V})]$ for the quantity $(C_s - C_m)$ in the foregoing equations. Further derivations have been made to allow the analysis of the corrosion of vertical cylindrical surfaces under natural and forced convection conditions.[3] Many glass-refractory systems have been investigated and most of these are diffusion controlled. Considerable success has been achieved in applying the foregoing relationships to characterize the corrosion kinetics, and in some cases diffusion constants have been calculated which agree reasonably well with those obtained by other methods. A selected list of glass-refractory systems which have been studied is given in Table I.

Table I. Selected List of Refractory Corrosion Studies

Reference Number	Refractory Material	Glass	Temperature °C	Conditions
1	Al_2O_3-Single Crystal	Soda Lime, Barium Crown, Lead, E	1400-1500	Free Convection
3	Al_2O_3-Single Crystal	Calcium-Aluminum Silicate	1340-1550	Molecular Diffusion Free and Forced Convection
4	Al_2O_3-Single Crystal and Polycrystalline, Mullite, Silica, Anorthite	Calcium-Aluminum Silicate	1350-1500	Free and Forced Convection
5a & b	Al_2O_3-Single Crystal and Polycrystalline	Calcium Aluminum Silicate, Sodium Silicate	1280-1650	Free and Forced Convection
6	SiO_2	Sodium Silicate Potassium Silicate	1000-1400 1000-1200	Free and Forced Convection
7	Al_2O_3, MgO, $MgAl_2O_4$ Single Crystal	Sodium Silicate	1200-1500	Free Convection
11	Al_2O_3-Single Crystal	Calcium Silicate, Sodium Silicate Magnesium Silicate	1300-1660	Free Convection
14	Quartz	B_2O_3	1130-1380	Free Convection
15	Zircon	E, Soda Lime Silicate	1450	Forced Convection

(C) Temperature Dependence of Corrosion Rates

The temperature dependence is usually evaluated in the steady state on the basis of the rate process equation:[5b]

$$\frac{Q}{t} = A \exp \ (-E/RT)$$

where A = constant,
E = apparent activation energy for the process (cal/mole),
R = gas constant (cal/mole/°K),
T = temperature (°K).

In the case of chemical reaction control, E would represent the energy of activation for the reaction. During diffusion control E would represent the activation energy for diffusion. If one process is definitely in control, the magnitude of E can sometimes be used to help identify the mechanism of corrosion. Often the corrosion process is complex and caution must be used in applying activation energy values. It is likely that many of them are composite values. References included in Table I have also considered the temperature dependence of corrosion.

LABORATORY CORROSION TESTS

Unfortunately much of the information generated by basic studies of corrosion cannot be directly applied to the improvement in furnace design and manufacture of refractories. Although the basic information is necessary, the actual performance of refractories is modified by such parameters as porosity, grain size distribution and phase assemblage. Thus, a continual comparative analysis of the corrosion resistance of refractories must be carried out. The Refractories Committee of the Society of Glass Technology of Great Britain has evaluated eleven corrosion tests.[15,16] They recommended that two types of tests were suitable for corrosion resistance measurements; namely, the static finger test and the rotating finger test. The criteria used in evaluating the tests were: (1) The test should place samples in the same order of merit as is found in practical applications: (2) The test should simulate the conditions of practical applications; (3) The time required for the test should be short; and (4) The apparatus should be well controlled and inexpensive to run.

For the static finger tests, samples can be either rectangular bars or cylinders which have been cut from larger samples. Their sizes can range from 6 x 0.8 x 0.8 in. to 1.2 x 0.4 x 0.4 in.[16] From three to five test pieces including a standard are often used. The tests can be run by placing each of the samples in a small crucible of test glass (150 cc), or they can all be placed in a large crucible containing several pounds of glass. The recommended temperature control limits are ± 8°C. After the heat treatment is carried out at appropriate temperatures and times, the amount or corrosion can be determined in several ways. These include: (1) Measuring the glass line attack with dividers or a micrometer; (2) Measuring the penetration and loss of area by planimetering a photograph of a silhouette of the sample; (3) Measuring loss by a traveling microscope; (4) Measuring the loss in volume by a mercury volumeter; and (5) Measuring photographs of sections cut through the samples.

Samples for the rotating finger test are usually cut to about 5 x 1 x 1 in., placed in a ceramic holder and rotated in the molten glass at speeds ranging from 2-14 rpm. The samples can be held from 2-3 in. from the axis of rotation. The test can be carried out for a certain number of revolutions or for a set time period. In this test, volumetric measurements can be used to assess the amount of corrosion since the corroding portions of the sample are completely immersed.

In using these tests some of their shortcomings should be recognized. One definite problem is to obtain representative samples. Circular cores drilled through the center of full sized block have been recommended.[16] Additionally, the effects of upward drilling cannot be adequately evaluated from the static finger test.

At least one example exists[15] where the tests have failed to predict the behavior of a refractory in a practical application. Samples of zircon-sillimanite were tested and the results showed that this refractory should perform as well as or better than sillimanite. A trial in a soda-lime tank showed the opposite effect. Apparently, the test results were invalid because the test samples were not stirred close enough to the crucible wall where appreciable disturbance occurs.

There is also a need to interpret the results of corrosion tests with regard to the knowledge of the system being tested. That is, the properties of the glass, the reaction products and the refractory material should be considered.

SELECTION OF REFRACTORIES

Refractories for the glass industry are comprised of a small number of oxides including SiO_2, Al_2O_3, Cr_2O_3, ZrO_2 and MgO. A wide variety of products is available through the various chemical and mineralogical combinations of these materials. Additionally the microstructure of the refractories varies with the nature of the starting raw materials as well as with the chemistry and heat treatment. The degree of heat treatment provides a convenient means of classifying the refractories i.e., as fusion cast or bonded.

There are essentially four different types of fusion cast refractories including aluminosilicate, alumina, chrome-alumina and alumina-zirconia-silica. Typical composition of these are listed in Table II.[17,18] These materials are frequently used in units producing soda-lime

Table II. Compositions of Fusion Cast Refractories (Oxide Wt. Percent)

Oxide	AZS*	Chrome-Alumina	Fused Alumina Alpha	Fused Alumina Alpha Beta	Fused Alumina Beta	Alumina Silica
Al_2O_3	45-51	60	99.3	94.8	94.5	74
ZrO_2	33-41					1
SiO_2	10-15	1.8	0.1	1.1	0.1	19
Cr_2O_3		27				
Alkaline earth, alkali and others	1-1.6	10.5	0.6	4.1	5.4	6

*AZS = Alumina-zirconia-silica

glasses. In a typical application for the melter section of the tank which may run at 2800°F, it is found that the chrome-alumina type has the best corrosion resistance. Although this factor is the most important regarding refractory life, the strong coloring potential of chromium ions in the glass precludes the use of this material in many cases. The second best material, the alumina-zirconia-silica type, is chosen for glass contact because its corrosion resistance is higher than the alumina or alumino-silicate types and it lacks coloring ions.[18]

The alumina-zirconia-silica refractories contain alpha alumina and zirconia as crystalline phases which are embedded in a glassy matrix. The glassy phase which may reach as high as 18%, is the most susceptible to corrosive attack. By increasing the zirconia content at the expense of silica, the amount of glass can be reduced thereby increasing the corrosion resistance. It seems that the glassy phase should be entirely eliminated, however, it is a necessary ingredient from another standpoint. The crystalline constituents differ in thermal expansion coefficient and are anisotropic. Furthermore, the monoclinic-tetragonal inversion in zirconia is accompanied by a drastic volume change. The glassy phase is needed to absorb the strain produced by the thermal expansion of the crystalline components during thermal cycling.

In the refining portion of the operation, the temperature is lower, thus corrosion is less severe. An additional requirement is for refractories having a low potential for producing seeds. For these purposes, the fused alumina materials are preferred. They have a very low glass content, thereby having less tendency for producing seeds and stones. Their corrosion resistance is only slightly less than the alumina-zirconia-silica type at the lower temperatures. The uses of the aluminosilicate type have largely been taken over by the other more corrosion resistant types. They are still used in areas where corrosive action is not severe such as lower side wall areas and cooler portions of the tank.

Bonded refractories for glass contact include clay, mullite, alumina zirconia-silica, zircon, alumina and chromic oxide materials. Typical compositions of these are shown in Table III. Zircon refractories serve in soda-lime glass tank applications as well as in the more acidic

Table III. Composition of Bonded Refractories (Oxide Wt. Percent)

Oxide	Clay	Mullite	Bonded AZS	Zircon	Alumina	Chromic Oxide
Al_2O_3	25-50	60-76	65-70	0.3	94-99	-
SiO_2	45-70	38-20	10-20	33	0.1-4.0	-
ZrO_2			20-23	65		
Cr_2O_3						95-98

boro-silicate glass tank. In the soda-lime application where it is used as a bottom paver, the surface of zircon materials is attacked by the alkali and a concentration of zirconia is formed at the interface. This forms a highly resistant layer and greatly reduces corrosion. Chromic oxide is the most corrosion resistant of these materials but again the coloring potential is the limiting factor. Additionally, this material vaporizes when used above the glass line. It is used for critical areas such as in throats of boro-silicate tanks.

Bonded alumina-zirconia-silica refractories have much the same advantages as the fused material. They have good corrosion resistance as compared to the mullite type and the absence of coloring ions. The bonded material is not as dense as the fused material, thus it cannot at present replace the fused material in critical applications. A summary of typical uses of the fused and bonded materials is given in Table IV.

With the limited number of materials available to produce refractories for the glass industry and with the ever increasing demand for better refractory life, development of improved refractories will probably center around the use of higher purity raw materials and control of microstructure. Both can yield better corrosion resistance by eliminating or isolating and protecting phases which are susceptible to attack.

Table IV. Typical Uses of Refractories for Glass Contact

	Soda-Lime Glass	Sodium Silicate Glass	Borosilicate Glass
Melter:			
Sidewall:	AZS Clay	Clay Bonded AZS	Zircon, AZS Chrome
Bottom:	Zircon AZS, Clay	Zircon Clay	Zircon Clay
Refiner:			
Sidewall:	Fused Alumina		Zircon
Bottom:	Zircon, Fused Alumina, Clay		Chrome, AZS
Forehearth Channels:	Fused Alumina Zircon, Mullite Clay	Bonded AZS Clay	Zircon Chrome, Chrome Alumina

REFERENCES

1. T. S. Busby and J. Eccles, "A Study of the Solution of Single Crystals of Corundum in Molten Glass," Glass Technology, 5 (3), 115-123 (1964).
2. E. Vago and C. E. Smith, "The Corrosion of Refractory Materials by Molten Glass," Paper #62, Proc. Seventh International Congress on Glass, Brussels (1965).
3. A. R. Cooper, Jr. and W. D. Kingery, "Dissolution in Ceramic Systems: I, Molecular Diffusion, Natural Convection and Forced Convection Studies of Sapphire Dissolution in Calcium Aluminum Silicate," J. Am. Ceram. Soc., 47 (1), 37-43 (1964).
4. B. N. Smaddar, W. D. Kingery and A. R. Cooper, Jr., "Dissolution in Ceramic Systems: II, Dissolution of Alumina, Mullite, Anorthite and Silica in Calcium Aluminum Silicate Slag," J. Am. Ceram. Soc., 47 (5), 249-254 (1964).

5a. L. Reed and L. R. Barrett, "The Slagging of Refractories: I, The Controlling Mechanism in Refractory Corrosion," Trans. Brit. Ceram. Soc., 54, 671 (1955).

b. L. Reed and L. R. Barrett, "The Slagging of Refractories: II, The Kinetics of Corrosion," Trans. Brit. Ceram. Soc., 63, 509-534 (1964).

6. Klaus Schwerdtfeger, "Dissolution of Solid Oxides in Oxide Melts. The Rate of Dissolution of Solid Silica in Na_2O-SiO_2 and K_2O-SiO_2 Melts," J. Phys. Chem., 70 (7), 2131-2137 (1966).
7. D. Barham and L. R. Barrett, "The Dissolution of Magnesium Aluminate Spinel in Sodium Silicate Melts," Trans. Brit. Ceram. Soc., 67 (2), 49-56 (1968).
8. T. S. Busby, "The Solution of Refractories by Molten Glass", Paper #61, Proc. Seventh International Congress on Glass, Brussels (1965).
9. F. W. Preston and J. C. Turnball "The Physics of Upward Drilling", Am. J. Sci. 239, 92-105 (1941).
10. W. D. Kingery, "Corrosion of Refractories by Liquid Melts," Symposium on High Temperatures, Stanford Research Institute (1956), pp. 126-132.
11. N. McCallum and L. R. Barrett, "Some Aspects of the Corrosion of Refractories," Trans Brit. Ceram. Soc. 51, 523-544 (1952).
12. C. Wagner, "The Dissolution Rate of Sodium Chloride with Diffusion and Natural Convection as Rate Determining Factors", J. Phys. Coll. Chem. 53, 1030-33 (1949).

13. A. R. Cooper, Jr., "Modification of Noyes-Nernst Equation", J. Chem. Phys. 38, 284-85 (1963).
14. J. R. Hutchins III, "Dissolution Kinetics in Viscous Systems Where Diffusion and Free Convection are Rate-Controlling," Glass Technology, 7 (2), 42-53 (1966).
15. E. Vago and C. F. Smith, "The Corrosion of Zircon Refractories by Molten Glasses: I, The Effect of the Porosity of Zircon Refractories on Their Resistance to Corrosion by a Calcium Alumino-Borosilicate Glass," Glass Technology, 2 (6), 218-26 (1961): II, "The Effect of the Concentration of Zirconium Silicate Dissolved in a Calcium Alumino-Borosilicate Glass and in a Soda-Lime Silicate Sheet Glass on the Relative Rate of Corrosion of Zircon Refractories in These Glasses," Ibid., pp. 227-234: III, "The Effect of Varying the Speed and Duration of Stirring Tests in a Calcium Alumino-Borosilicate Glass," Ibid., 4 (4), 129-33 (1963).
16. Refractories Comm. Soc. Glass Tech. Report, "Corrosion Tests for Evaluating the Solution of Tank Block Materials by Molten Glass," Glass Technology, 2 (6), 235-37 (1961).
17. E. A. Thomas, "Refractories for Glass Melting", Chas. Taylor Sons Co. Bulletin 360 (1968).
18. R. W. Brown, "Fused Cast Refractories for Glass Furnaces," Refractories Journal 44 (6) 200-206 (1968).

Diffusion Processes in Glass

A. R. Cooper

Case Western Reserve
Cleveland, Ohio 44106

INTRODUCTION

(A) Diffusion Processes in Glass

Diffusion, that is, the motion of one atom relative to other atoms, plays an important, yes, a vital role in many processes important to glass science and glass technology. To indicate a few, one might first mention that the process of glass making itself, in every case except for that of a single component glass like silica, involves diffusion, as glasses typically originate as discrete particles of material of chemical compositions quite different from that of the ultimate glass. Thus, the motion of the atoms from a grain of silica, say relative to those of a grain of limestone, is necessary to produce the homogeneous product we require for modern glass. At the same time, diffusion processes participate in the development of cords during the glass making process, as it requires diffusion for material from refractories to become imperfectly dissolved in glass, and it also requires diffusion within the glass melt for composition variations to develope near the surface of the melt due to selective evaporation of one or more species. Otherwise, the region where the composition was depleted in the evaporating species would be confined to a single molecular layer, or two.

Diffusion is vital in the precipitation or crystallization of species whose composition is different from that of the bulk composition. Diffusion is closely connected to the process of electrical conduction in typical oxide glasses. Diffusion provides a mechanism for mechanical loss, that is a discrepancy from perfect elastic behavior. Perhaps this consequence of diffusion is most important in practical glass technology as it permits residual stresses developed by thermal tempering or ion exchange to decay at temperatures substantially lower than would be possible by truly viscous relaxation. Diffusion is important in the peculiar behavior of mixed alkali glasses and their slow change in specific volume at temperatures near to room temperature. The demonstrated capability of strengthening glasses substantially by exchanging ions of large size with those of smaller size is a classic case of diffusion. Finally, the decomposition of a glass into two phases of different chemical composition is controlled by diffusion processes, where the diffusion coefficient has a negative value. Thus, an understanding of diffusion processes is of vital importance to the glass technologist and glass scientist.

The following is a complementary viewpoint. Granted that, as indicated above, glass technology and science depend on understanding of diffusion phenomena, it is interesting to ask whether glass is particularly useful as a model material in order to get better fundamental understanding of the diffusion process itself. It seems to me that the answer here is clearly positive. With the advent of the electron microprobe and its ability to measure concentrations in a region of the order of a few cubic microns, the possibility of doing careful diffusion studies at high temperatures, quenching to room temperature and measuring the diffusion profiles has been almost completely realized. Glasses are especially useful materials in this regard as they can be quenched from high temperatures without fear that phase transformations or other changes in the overall structure will occur to distort the composition distribution. At the same time glasses, particularly silicate glasses, have wide solubility ranges for a number of different oxides. Thus, it is possible to examine in a well chosen glass system the effects of ionic radii, ionic charge, ionic polarizability on diffusion in a particular host glass. Further, the high solubility allows multicomponent diffusion studies to be conducted with a much greater flexibility than is possible in other systems.

It is hoped that these remarks have indicated that the relationship of diffusion processes to glass technology and science is almost ideal. Understanding of diffusion processes is vital to the development of the discipline of glass technology, and glassy materials are uniquely well adapted to permit serious studies of fundamentals of diffusion. The prediction of self-diffusion coefficients in glasses from first principles is not yet possible.[1] An ordering of the magnitude of coefficients, based on the concept that the lower the charge the more mobile is a diffusing species, with all other factors kept nearly constant, is usually successful.[2] Somewhat less widely applicable is the notion that for ions of the same charge, the smaller the ionic radius, the greater the mobility.[3]

The following will be an attempt to look fairly seriously at an important process in glass technology, i.e., chemical strengthening of glass, and to use this process to provoke our interest in some of the fundamental topics regarding diffusion processes in glass.

(B) Ion Exchange Strengthening of Long Circular Cylinders

Typically, because of their higher mobilities in glass, alkali ions, most frequently sodium and potassium, are exchanged at a glass surface in order to take advantage of the fact that these ions occupy a different amount of volume in the glass structure, that is to say, the oxides of these ions have different partial molar volumes. Naturally, because of the well known sensitivity of glass to flaws on the surface, it is desirable to exchange larger ions (from a molten salt, perhaps) with smaller ions in the glass, producing a compression of the surface region and hence a compression on the outer portion of the cylinder.

For simplicity, we will consider the large ions be be A and the small ions to be B. We will also concern ourselves with a long circular cylinder of radius $r = b$, and length ℓ, as shown in Figure 1. A salt bath, with an excess of A ions, surrounds a glass with an excess of B ions. Let us first focus on diffusion processes in the glass and write the Fick's law expression for the flux density j of species A (mols A/ Cm^2/sec) as a function of a diffusion coefficient D, molar density ρ, and the change in the mol fraction, x, of species A, with radius r.

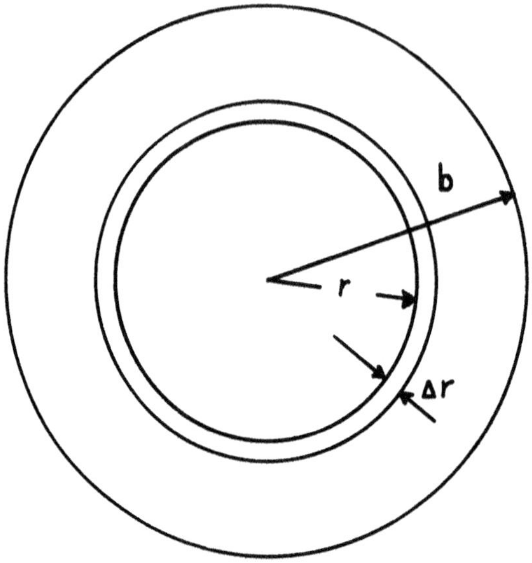

Figure 1

$$j = -D_{AB}\rho \frac{\partial x}{\partial r} \qquad (1)$$

Flux density is typically a vector, but since we are considering only concentration changes in the r direction we recognize that flux density is a number, positive values being associated with fluxes in the outward radial direction. It soon becomes apparent from examining equation (1) and considering our problem that we do not have the necessary information to solve equation (1), that is to say, we neither know j, nor do we know the concentration gradient. To put this in a more usable form, let us consider the following balance which will result in a derivation of Fick's second law for the radial symmetry or our problem. We consider a ring between radius r and $r + \Delta r$. Equation (2) gives the balance of species A.

$$2\pi\ r\ell j(r) - 2\pi\ r\ell j(r + \Delta r) = \rho \frac{\partial x}{\partial t} (2\pi r \Delta r \ell) \qquad (2)$$

On the left hand side is the difference between the flux into the element and the flux out of the element. On the right hand side we have the product of the change in concentration with time and the volume of the element, i.e., the change in the mols of species A in the element per unit time. Simplification yields;

$$\frac{rj\,(r) - rj(r + \Delta r)}{\Delta r} = r\rho\frac{\partial x}{\partial t} \qquad (3)$$

Taking the limit as $\Delta r \rightarrow 0$, use of the definition of a derivative, and substitution of equation (1), with D_{AB} constant gives us

$$\frac{1}{r} D_{AB} \frac{\partial}{\partial r} \left(r \frac{\partial x}{\partial r} \right) = \frac{\partial x}{\partial t} \qquad (4)$$

which is Fick's second law for a material with cylindrical symmetry. This is an equation which we can solve, provided we can introduce proper boundry and initial conditions.

For our case, it is desirable to make some simplifying assumptions regarding boundary conditions, that we can check later on.

$$\text{i. } x\,(b,\, t) = x_b$$

$$\text{ii. } x\,(r<b,\, 0) = x_o \qquad (5)$$

$$\text{iii. } \frac{\partial x}{\partial r} = 0 \text{ at } r = 0$$

The assumption that the surface of the glass immediately takes on the concentration of A that is in equilibrium with the salt bath, gives boundary condition (i). Surface mol fraction, x_b, is established by the mol fraction of species A in the salt, by the Nernst distribution coefficient[4], and by the total alkali mol fraction of the glass. Boundary condition (ii) is only the statement that the concentration is uniform throughout the glass before ion exchange begins, and condition (iii) is just a recognition of the radial symmetry of the problem.

For a given diffusion equation, especially with constant diffusion coefficient, and for usual sets of boundary conditions, such as those above, it is often possible to find the appropriate solution.[5,6] For our case the mol fraction, $x(a)$, at $r = a$ is given by;

$$\frac{x(a) - x_o}{x(b) - x_o} = 1-2 \sum_{n=1}^{\infty} \exp\,(-\beta_n^2\, Dt/b^2) \frac{J_o(a\beta_n/b)}{\beta_n J_1\,(\beta_n)} \qquad (6)$$

where x_o and $x(b)$ are the original and surface concentrations, J_o and J_1, Bessel functions, β_n is the nth root of the equation $J_o(\beta) = 0$, D the appropriate diffusion coefficient for ion exchange, and time,t. Graphical representation of equation (6) can be found for various values of Dt/b^2.[6]

As shown by Crank[5], integration of equation (6) leads to the following expression (most appropriate at short times) for the fractional exchange; the ratio of potassium exchanged at time t, M(t), to that exchanged at infinite time, M(∞):

$$\frac{M(t)}{M(\infty)} = \frac{4}{\sqrt{\pi}}\left(\frac{Dt}{b^2}\right)^{1/2} - \left(\frac{Dt}{b^2}\right) - \frac{1}{3\sqrt{\pi}}\left(\frac{Dt}{b^2}\right)^{3/2} \cdots \qquad 7)$$

A plot of $M(t)/M(\infty)$ vs. (Dt/b^2) is available.[5]

Equation (6) is useful for comparing theoretical concentration profiles with those measured for ion exchanged glasses.[7] Either equation (6) or equation (7) allows calculation of D from the results.

Measurements, for example, by electron microprobe, can be used to determine whether or not the assumed boundary condition leading to the solutions shown above are correct.

Figure 2 indicates schematically the concentration profiles to be expected for three cases: 1) the resistance to transport is entirely contained within the glass; 2) the phase boundary reaction occurs instantaneously, and there is some resistance to diffusion within the salt bath; and, 3) for the case where the phase boundary reaction is not instantaneous. Recently[7] it has been shown that for $K \rightarrow Na$ exchange in alumino-silicate glass, after a short period the kinetics are controlled by diffusion in the glass, with $D = 5 \times 10^{-11}$ cm^2/sec at 400°C.

Knowing the concentration distribution, it is a straightforward matter to calculate the stress at any radius, a, in the cylinder. The axial stress, P(z), is of primary interest and is related to the distribution of mol fraction A as follows;

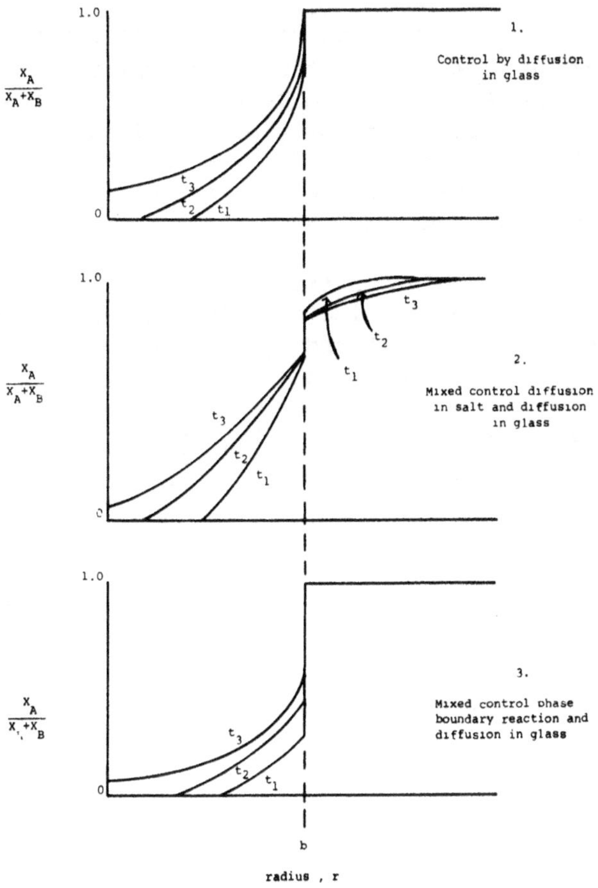

Figure 2

$$P_z(a) = \frac{BE}{1-p} \frac{2}{b^2} \{ \int_0^b xr\, dr - x(a) \} \qquad (8)$$

Equation (8) is taken directly from thermal stress theory because of the complete correspondence of stresses developed by concentration differences and stresses developed by temperature differences. In this equation E is Young's Modulus, p is Poisson's ratio, and B, the dilation coefficient, (analogous to thermal expansion) is defined as;

$$\frac{1}{3V} \left(\frac{\partial V}{\partial x}\right)_{P,T}$$

where V is the Volume.

Equation (8) displays the conditions necessary to obtain maximum surface stress. First consider the factor BE/1-p. Young's Modulus, E, and Poisson's ratio,p, are relatively independent of composition for typical silicate glasses.[9] values of the dilation coefficient, B, are less well documented. Thus we are motivated to relate B to more familiar thermodynamic quantities.

Starting with the definition of B we divide it into two terms. Using the premise that ion exchange is occuring only between species A and B, and the mols, n_i, of all other species are constant:

$$B = \frac{1}{3V}\left\{\left[\left(\frac{\partial V}{\partial n_A}\right)\left(\frac{\partial n_A}{\partial x}\right)\right]_{n_B,n_C,\cdots P,T} + \left[\left(\frac{\partial V}{\partial n_B}\right)\left(\frac{\partial n_B}{\partial x}\right)\right]_{n_A,n_C,\cdots P,T}\right\} \quad (9)$$

Now the condition of equimolar exchange of A and B gives:*

$$\left(\frac{\partial n_A}{\partial x}\right)_{n_B,n_C\cdots P,T} = -\left(\frac{\partial n_B}{\partial x}\right)_{n_A,n_C\cdots P,T} = N,$$ the total number of mols, which along with the definition of partial molar volume;

$$\bar{V}_i \equiv \left(\frac{\partial V}{\partial n_i}\right)_{n_j,n_k\cdots P,T}$$

yields

$$B = \frac{N}{3V}\left(\bar{V}_A - \bar{V}_B\right) = \rho/3\left(\bar{V}_A - \bar{V}_B\right) \quad (10)$$

where ρ is molar density.

Not surprisingly, the dilation coefficient is seen to depend on the difference between the partial molar volume of the two alkali oxides. This factor depends on both the base composition and on the particular alkali ions. Larger values of B may explain the higher stresses generally obtained when alkali-alumino-silicate[10] glasses are used for exchange. A value of B ≈ (.02) mol fraction^{-1} has been calculated[7] for an alumino-silicate glass.

---

* Remembering that x is the mol fraction of A

Now let us consider the terms within the bracket, { }, in equation (9). It is simply the difference between the mol fraction of A at a particular point, a, and the average mol fraction, $\langle x \rangle \equiv \frac{2}{b^2} \int_0^b xr\,dr$, of A. It is clear that the magnitude of this difference is enhanced by increasing i) the total alkali mol fraction of the glass; ii) the surface mol fraction, x(b), of exchanging alkali ions, and by minimizing the bulk mol fraction, $x(\infty)$, in the glass.

If the goal is maximum surface stress, then the less diffusion the better, because diffusion will increase the average mol fraction, $\langle x \rangle$, and diminish the difference.

However, if some strengthening beneath the surface plane is desirable to reduce fracture sensitivity to flaws, then diffusion is necessary to produce compression throughout the flawed region. Naturally, the higher the value of D, the less will be the time required for diffusion.

(C) Phenomenological Treatment of Diffusion

An equivalent description[11] of the flux density of a species is given by the product of the density of the species and its velocity, v, relative to some frame of reference.

$$j = \rho x v$$

where ρ is the overall molar density, and hence ρx is the molar density of species A. Now, defining mobility, β, as the velocity of a species per unit force, f, on a particle of that species, we can rewrite our expression for flux density:

$$j = \rho x \beta f$$

As we shall see, there are a number of different sources of force on an ion, but we will first focus on the one that is simply related to the concentration gradient. Recognizing that chemical potential, μ, is the partial free energy per atom of a species, and that the spatial derivative of chemical potential is the force on a particle of that species, and finally that the force will act in the direction of decreasing chemical potential, we can obtain the following equation;

$$j = -\rho x \beta \frac{\partial \mu}{\partial r} \quad (11)$$

Recall, that although we have not indicated with a subscript for x, β, μ, and j above, all of these quantities refer to species A. Comparison of equation (11) and equation (1) allows us to obtain D_{AB} readily[12] as shown below, where γ is the activity coefficient of species A, k is Boltzmann's constant and T is absolute temperature:

$$D_{AB} = \beta \left(x \frac{\partial \mu}{\partial x}\right) = kT\left[1 + \frac{\partial \ell n \gamma}{\partial \ell n x}\right] \beta \quad (12)$$

Notice that the diffusion coefficient contains a part solely related to the thermodynamics, $\frac{\partial \mu}{\partial x}$, or $kT\left[1+\frac{d\ell n\gamma}{d\ell n x}\right]$ and a part that is only concerned with kinetics, the mobility, β. Equation (12), it must be emphasized, relates to cases when the only influence on species A is the force caused by its own concentration gradient. This limitation is fulfilled perfectly when we consider the self diffusion coefficient of a species, that is the results of an imaginary experiment where we could in some manner identify without influencing the behavior of, for example, half of the sodium atoms in a piece of glass. If we watched how those sodium atoms exchanged with other sodium atoms in the glass, their mixing would give us a measure of the self diffusion coefficient of sodium. In this case, the only influence effecting the exchange of marked and unmarked sodiums is the concentration gradient of marked and unmarked sodiums. The closest that we can come to such an experiment is to use different isotopes of the same species and assume that both isotopes will behave almost identically with regard to diffusion. We also expect the system would show ideal thermodynamic behavior with respect to changes in isotopic ratio. Thus the change in log activity coefficient with the change in log mol fraction would vanish (or equivalently the change in chemical potential with the change in log mol fraction would be simple kT). Thus, the self diffusion coefficient, for example of species A, can be written simply as follows:

$$D_{AA} = kT\beta_A \quad (13)$$

Now we are prepared to return to the case of ion exchange strengthening and ask how the interdiffusion coefficient,

D_{AB}, depend on the individual mobilities or self-diffusion coefficients of the exchanging ions. To do this we must consider a specific model. The simplest is to assume that the alkali ions exchange in a relatively rigid glass structture. That is to say, that except for the sodium and potassium ions, the glass does not participate in the exchange process. There may be forces acting on the ions in addition to those caused directly by the gradients of chemical composition. Specifically, we assume that the major additional force is due to a local electrical field, E, which may develop if one species is more mobile than the other. This field would naturally be in the direction to retard the more mobile species. Adopting a sign convention for electrical fields such that the force on an ion is equal to the product of the field and the charge on the ion, the following expressions are obtained for the flux of A and B:

$$j_A = -\beta_A \rho x_A \left(\frac{\partial \mu_A}{\partial r} - eEz_A \right)$$
$$j_B = -\beta_B \rho x_B \left(\frac{\partial \mu_B}{\partial r} - eEz_B \right) \qquad (14)$$

Note that z is the valence of the particular ion. Hence, ez is the electrical charge of an ion. We have necessairly dropped the previous simple convention, and use a subscript to identify the specific ions.

Requiring that there be no build-up of space charge is equivalent to requiring that the flux of A and B sum to zero, since the charge on both ions is identical (unity in the case of alkali ions). Together with the Gibbs-Duhen equation, $x_A \partial \mu_A + x_B \partial \mu_B = 0$, this allows us to solve eq-

tions (14) and obtain, $eE = \left[\frac{\beta_A - \beta_B}{x_A \beta_A + x_B \beta_B} \right] \left(x_A \frac{\partial \mu_A}{\partial r} \right)$

Note that E is proportional to the mobility difference and to the chemical potential gradient.

Substituting back into the flux density for A, we get the following expression:[13]

$$j_A = -\rho_A \left[\frac{(x_A + x_B)\beta_A \beta_B}{x_A \beta_A + x_B \beta_B} \right] \left(\frac{\partial \mu_A}{\partial r} \right) \qquad (15)$$

Now we can introduce the relations between self diffusion coefficients and mobility (equation (13)) and the relations between change in chemical potential and change in concentration (from equation (12)) to obtain an expression for the interdiffusion coefficient for ion exchange in terms of the thermodynamics of the system and the self diffusion coefficients of the exchanging species as follows:

$$D_{AB} = \frac{(x_A + x_B) D_{AA} D_{BB}}{x_A D_{AA} + x_B D_{BB}} \left(1 + \frac{d\ln\gamma_A}{d\ln x_A}\right) \qquad (16)$$

Notice that the function of the self diffusion coefficients typically yields a result that lies between the value of each self diffusion coefficient and in most cases is fairly close to the value of the lower self-diffusion coefficient. This shows that if rate of ion exchange is an important factor, it is important to have the lower diffusion coefficient as large as possible.

In a more detailed consideration of the forces, we might have recognized that in addition to the forces due to a concentration gradient and the forces due to a local electrical field, there might also be forces due to the strain energy that the diffusion process itself is producing. If the strain energy is primarily due to hydrostatic pressures, then the coefficient necessary to account for its effect on a particular ion is the partial molar volume of that ion. Thus, if we were to call the gradient of the strain energy, that is the strain field, S, the expression for the flux density for A would need to be written as follows:

$$J_A = -\beta_A \rho_A \left(\frac{\partial \mu_A}{\partial r} - eEz_A - S\bar{V}_A\right) \qquad (17)$$

This is a much more complicatedly coupled diffusion equation and to my knowledge, it has not yet been treated. However, ion exchange experiments where an external strain field was applied have indicated little effect from the strain on the overall kinetics.[7] Thus, we may conclude ignoring the effect of the last term in equation (17) is not too serious.

(D) Stress Relaxation

Once an ion exchanged cylinder is removed from the salt bath, it can achieve a reduction in its own free energy by eliminating the chemical concentration gradient, eliminating the electrical field and eliminating the strain field. In other works, immediately after removal from the salt bath, the ion exchanged cylinder tends to return to a homogeneous stress-free state. This tendency is virtually prevented by taking the glass to a much lower temperature to reduce the values of the self-diffusion coefficients. It is clear, however, that since the diffusion-time product, Dt, is what is decisive (see equations (6) and (7)) we are required to have enormous decreases in the self-diffusion coefficient, because, while the surface exchange process may take on the order of several hours, ion exchange strengthening is expected to persist for perhaps a hundred years. Thus, the diffusion coefficient will have bo be of the order of 10^6 times less at the operating temperature than it was at the temperature where the ion exchange strengthening took place. Before turning to the matter of the temperature dependence of the self-diffusion coefficient, however, let us consider quickly a couple of other consequences of this so-called fundamental approach to diffusion that we have been pursuing.

(E) Electrical Conductivity

The electrical conductivity, σ, is given by:

$$I = \sigma E \qquad (18)$$

where I is the electrical current (flux density of charge) and E, the electrical field. When charge is transported by ions, we can define the electrical current to be the product of the flux density of the ion, times the charge on the ion, ze. From equation (14), in the absence of a chemical potential gradient, we can readily write an expression for the flux density of an ion relative to the electrical field as follows:

$$I_A = j_A z_A e = \beta_A \rho_A z_A^2 \, e^2 E$$

It is clear that if more than one ionic charge carrier is present, the expression for the electrical conductivity is obtained from the sum of the contributions of each species

$$\sigma = \frac{e^2}{kT} \Sigma \, D_{ii} \rho_i z_i^2 = \frac{e^2}{kT} \frac{D_{AA} \rho_A z_A^2}{t_A} \qquad (19)$$

where t_i is the transport number of species i, (the fraction of total charge carried by species i) and we have also used equation (13). Equation (19) for ionic conductivity is commonly called the Nernst-Einstein equation. Interestingly enough, equation (19) seems to underpredict the electrical conductivity[14] in an alternating field for simple single alkali glasses. An explanation for this discrepancy based on jump correlation is due to Haven and Stevels.[15] Following the ideas for diffusion in crystals, it suggests that diffusing alkali ions are correlated in their motion and hence have a diminished diffusion coefficient compared to that of the charge carrier. Later Haven and Verbeck[16] showed that an interstitialcy mechanism where in addition to correlation the longer jump of the charge compared to the ion contributes to the discrepancy. More recently L. W. Barr has shown[17] that measurement of self diffusion coefficients of various sodium isotopes is consistent with the interstitialcy mechanism. Further, the anomalous behavior of mixed alkali glasses with respect to diffusion and internal friction[18] can also be reconciled to the idea that diffusion jumping involves the coupled motion of two alkali ions.

An alternate viewpoint is that there are two types of alkali motions in a glass: 1) where the alkali re-orients itself with respect to the nearest non-bridging oxygen ion, but does not leave this non-bridging oxygen ion, and 2) the alkali ion jumps from one non-bridging oxygen ion to another. Both of these processes would contribute to electrical conductivity in an alternating field, but only the latter would contribute to self diffusion coefficient measurements.

(F) Stokes-Einstein Relations

The Stokes-Einstein expression for self diffusion coefficient of a species follows directly from the approach we have been adopting. If we consider a sphere moving through a viscous liquid, the velocity per unit force is just established by Stokes' solution for the flow around the sphere as *

$$\beta = \frac{1}{6\pi\eta R_s}$$

*The coefficient may be either 1/4 or 1/6 depending on the boundary conditions at sphere interface.

where R_S is the radius of the sphere and η is the viscosity. Substituting the relation between self-diffusion coefficients and mobility, we obtain the so-called Stokes-Einstein relations for self-diffusion coefficients;

$$D_{AA} = \frac{kT}{6\pi\eta R_A} \qquad (20)$$

where R_A is the radius of the diffusing species. While the Stokes-Einstein law has been reasonably successful for estimation of diffusion coefficients of ions in aqueous solutions, it predicts a result that is often 10 orders of magnitude too low for the diffusion coefficient of alkali ions in rigid glasses. Thus, while the Stokes-Einstein equation is of little use to us in terms of prediction, the much higher diffusion coefficient than predicted reveals that alkali diffusion must occur by an easier process than viscous flow. Presumably viscous flow involves breaking of silicon-oxygen bonds, while alkali diffusion only requires alkali-oxygen bonds to be severed.

(G) An Atomistic Approach to Alkali Diffusion in Glass

The diffusion coefficient can be related with complete generality to the mean free path, λ, i.e., the average distance of successive jumps of the diffusing ion, and the time, Γ, which is the average interval between the beginning of each successive jump, in the following way:

$$D = \frac{\lambda^2}{3\Gamma} \qquad (21)$$

Consistent with the idea that diffusion occurs only when an alkali ion exchanges the non-bridging oxygen ion to which it is bonded, we might quite appropriately estimate the distance, λ, as being the average distance between alkali ions bonded to adjacent non-bridging oxygens in the glass. A value of 5 Å would seem reasonable, according to this model, while a maximum of jumps of over 7 Å would be allowable.

The problem now remains to calculate the value of Γ. Typically, the starting point for this calculation is to assume that the frequency, ν, of atomic vibrations in a rigid glass is given by

$$h\nu = kT \tag{22}$$

where h is Planck's constant. The interval between successive jumps, Γ, would be equal, as indicated by the next equation, to the frequency of atomic vibrations, times the probability, p, that a vibration produces sufficient energy to allow a jump to take place.

$$\frac{1}{\Gamma} = \nu p = \frac{kT}{h} p$$

Introducing these concepts, we obtain an expression for D_{AA} that depends on the probability of an alkali (A) ion jumping from a non-bridging oxygen to a neighboring non-bridging oxygen;

$$D = \frac{\lambda^2}{3} \frac{kT}{h} p \tag{23}$$

Adopting the concept of an activated state as the state along the path of the diffusion jump where the ion has its maximum free energy, we find that the probability of an ion making a jump in a given fluctuation of the glass structure is, according to Boltzmann's statistics:

$$p = \exp -(\Delta F^{\dagger}/kT)$$

$\Delta F^{\dagger}$, the activation free energy, can be decomposed into the activation (internal) energy $\Delta H^{\dagger}$, and the activation entropy change, $\Delta S^{\dagger}$, which leads to the result that the diffusion coefficient is given by;

$$D = \frac{\lambda^2}{3\Gamma} \frac{kT}{h} \exp (\Delta S^{\dagger}/k) \exp -(\Delta H^{\dagger}/kT) \tag{24}$$

Neither the activation entropy nor the activation energy have been calculated with very great success for glasses. Thus the prediction of self diffusion coefficients from first principles is not a well established practice at this time, although Anderson and Stuart[1b] have made estimates of the activation energy. The activation energy, $\Delta H^{\dagger}$, is of particular importance to the utility of the ion exchange strengthening process, because all the information regarding temperature dependence of the diffusion coefficient is contained in this parameter. Typically, activation energies for alkali diffusion in glass run from 20 to 30 kcal/mol.[19] Comparing this value with the coulombic energy associated with an Na^{+}—$O^{=}$ bond, we see that activation energies are of the order of 1/10 the bond energy.

To make explicit the temperature dependence of alkali self diffusion coefficients, let us tabulate the fractional change in D at various temperatures for different values of $\Delta H^{\dagger}$. Table I gives the ratio of the diffusion coefficient at several use temperatures, 300,400, and 500°K compared to the diffusion coefficient at a common exchange temperature of 700°K for various typical values of $\Delta H^{\dagger}$. It is notable that for the long time conditions described earlier, i.e., $D/D_{700}<10^{-6}$, none of the activation energies are sufficiently high to prevent substantial stress relaxation at 500°K. At 400°K (only 100° above room temperature), only the activation energy of 30,000 cal/mol gives a sufficiently small ratio to guarantee long time (100 years) persistence of stress. However, at room temperature, it is clear that even the lowest activation energy glass would maintain its strength for long periods.

Table I. Temperature Dependence of D.

	$H^{\mp}$, kcal/mol		
	20	25	30
$D_{500^\circ K}/D_{700^\circ K}$	$10^{-2.6}$	$10^{-3.3}$	$10^{-3.9}$
$D_{400^\circ K}/D_{700^\circ K}$	$10^{-4.6}$	$10^{-5.8}$	10^{-7}
$D_{300^\circ K}/D_{700^\circ K}$	$10^{-8.5}$	10^{-11}	10^{-13}

(H) Limitations

By focusing on the question of ion exchange strengthening of glass and the diffusion process controlling it, we have necessarily not been able to consider many other aspects of diffusion processes in glass. Primary among these are processes where other flows, such as free convection and forced convection, influence a diffusion controlled phenomena; e.g., kinetics of precipitation or dissolution, or the homogenization of glass by stirring. Nor have we covered the area of multicomponent diffusion so common in glass at higher temperatures, where we must concern ourselves, not only with the exchange of two species, as discussed above, but where more species participate in the diffusion process. This adds additional complexity to the problem, but it is believed that pursuing the principles suggested tn this discussion allows these problems to be treated in a straigntforward manner.[20-22]

REFERENCES

1.a. R. H. Doremus, Pg. 1 in Modern Aspects of the Vitreous State, Vol. 2, Ed. J. D. Mackenzie, Butterworths, London (1962).
b. O. L. Anderson and D. A. Stuart, J. Am. Ceram. Soc., 37, 573 (1954).
2.a. G. H. Frischat, J. Am. Ceram. Soc., 52, 625 (1969).
b. T. B. King and P. J. Koros, Pg. 80 in Kinetics of High Temperature Processes, Ed. W. D. Kingery, John Wiley, New York (1959).
3.a. K. H. Stern, J. Phys. Chem., 72, 2256 (1968).
b. A. K. Varshneya, M. S. Thesis, C. W. R. U., Cleveland, Ohio.
4.a. J. L. Barton, Verres et Refr., 22, 268 (1968).
b. J. P. Lacharme, C. R. Acad. Sci. Paris, 267, 559 (1968).
5. J. Crank, The Mathematics of Diffusion, Oxford University Press, Fair Lawn, N. J. (1956).
6. H. S. Carslaw and J. C. Jaeger, Conduction of Heat in Solids, 2nd edition, Oxford University Press, Fair Lawn, N. J. (1959).
7. A. R. Cooper and D. Krohn, J. Am. Ceram. Soc., 52, 665 (1969).
8.a. B. Boley and J. Weiner, Theory of Thermal Stresses, John Wiley, New York (1960).

b. O. Richmond, W. C. Leslie, and H. A. Wreidt, Trans. A. S. M., 57, 294 (1964).

9. F. Tooley, Handbook of Glass Manufacture, Ogden, New York (1953).

10. M. E. Nordberg, E. L. Mochel, N. M. Garfinkle and J. S. Olcott, J. Am. Ceram. Soc., 47, 215 (1964).

11. R. B. Bird, W. E. Stewart, E. N. Lightfoot, Transport Phenomena, John Wiley, New York (1960).

12. P. Shewmon, Diffusion in Solids, McGraw Hill, New York (1963), Pg. 125 ff.

13. J. H. Heasley, and A. R. Cooper, J. Am. Ceram. Soc., 49, 280 (1966).

14. J. V. Fitzgerald, J. Chem. Phys., 20, 922 (1952).

15. Y. Haven and J. M. Stevals, IV. Congs. Intern. du Verre, Paris, (1956).

16. Y. Haven and B. Verbeck, J. Phys. Chem. Glasses, 6, 38 (1965).

17. L. W. Barr, Inter. Conf. On Non-Cryst. Materials Sheffield, England, (1970).

18a. J. E. Shelby and D. Day, J. Am. Ceram. Soc., 52, 169 (1969).

b. D. Day and W. E. Steinkamp, Ibid. pg. 571.

19a. V. I. Malkin and B. M. Mogutnov, Dokl. Akad. Nauk SSSR, 141, 1127, (1960).

b. J. R. Johnson, Ph.D. Thesis, Ohio State University, (1950).

c. A. R. Cooper and A. K. Varshneya, J. Am. Ceram. Soc., 51, 103 (1968).

20. A. R. Cooper, Pg. 217 in Advances in Glass Technology, Plenum Press, New York (1962).

21. A. R. Cooper, Glass Tech., 7, 2 (1966).

22. P. K. Gupta and A. R. Cooper, To be published, Physica (1971).

Electrical Properties of Glass

L. L. Hench and H. F. Schaake

University of Florida
Gainesville, Florida 32601

INTRODUCTION

The purpose of this review paper on the electrical properties of glass is threefold. First, the theoretical basis for understanding the electrical behavior of glass will be reviewed. Second, a synthesis of the understanding of composition and structure effects will be presented; and third, important problems in this area will be discussed. The theoretical review will first consider a general theory of conductivity, followed by more specific discussions of ionic conductivity and electronic conductivity. Dielectric properties of glass will also be considered, including theory, polarization mechanisms, and the current status of understanding composition, thermal history, microstructural and ultra-structural effects. Dielectric breakdown and switching glasses will be briefly considered.

Because of the immense scope of the subject of electrical behavior of glass, a thorough historical survey of this field will not be attempted. The reader is referred to Owen[1], Mackenzie[2], Pearson[3], Morey[4], Volf[5], Doremus[6] and Weyl[7] for authoratative historical reviews in this area. A detailed bibliography on this subject has also been compiled for the period 1957-1968 by the International Commission on Glass.[8]

INTRODUCTORY THEORY OF ELECTRICAL CONDUCTIVITY

In order for a current to flow in a glass, charge must somehow be transported. Electrons may move through the lattice in one of two ways: either by moving with nuclei, in which case ionic conductivity (σ_i) occurs, or by becoming detached from one atom, and moving to another, in which case electronic conductivity (σ_e) is present. It is also possible, of course, that both electronic and ionic conduction processes occur simultaneously. The transfer number, $t_x = \sigma_x/\sigma_T$, expresses the fraction of total conductivity contributed by each type of charge carrier.

The conductivity of any material is defined as the ratio of the current density to the electric field required to produce the density, i.e.,

$$\sigma = \frac{j}{E} \tag{1}$$

While in the most general case σ is a second rank tensor* since both j and E are vectors, because a glass is isotropic in its properties, we need only consider it as a scalar. The current density is the charge crossing a unit

---

*A "zeroeth" rank tensor is a scalar, while a first rank tensor is a vector. A second rank tensor either (1) relates one vector linearly to another, such as is in the case of the conductivity, where an applied electric vector field induces a vector current density which is proportional to the applied electric field (i.e., doubling the applied field without changing its direction results in a doubling of the magnitude of the current density) but which is not necessarily in the same direction as the electric field; or (2) relates a second rank tensor (such as strain) to a scalar such as temperature. An example of the latter is the thermal expansion tensor. Second rank tensors are generally represented by 3 x 3 matrices (3 rows and 3 columns) and are manipulated by matrix algebra.

area per second. Thus if the charge on a carrier, q, is the product of its valence times unit charge e, q = Ze, its velocity is v, and there are n identical carriers per unit volume, the current density will be:

$$j = nqv \tag{2}$$

We now inquire as to how the velocity, v, is related to the electric field, E. For electrons in a conduction band, the classical free electron model can be used with an argument that the electron will be accelerated by the field, and therefore its velocity at any time t will be changed from its initial random value, v_o, by an amount:

$$\Delta v(t) = (\text{acceleration x } t) = \frac{qE}{m} t \tag{3}$$

where m is the mass of the electron (or more properly, the effective mass, in which factors introduced by quantum mechanics are taken into account). But the electron will undergo this acceleration only for an increment of time before it collides with an atom and its random velocity is restored. If the mean time between collisions is τ, the average velocity of an electron is independent of time and is given by

$$\langle v \rangle = \langle v_o + v(t) \rangle = 0 + \frac{qE\tau}{2m} \tag{4}$$

Combining equations 1, 2 and 4 we find that

$$\sigma = \frac{nq^2\tau}{2m} \tag{5}$$

Equation (5) is generally written as

$$\sigma = nq\mu \tag{6}$$

where μ is the mobility defined by:

$$\mu = \frac{q\tau}{2m} \tag{7}$$

While equation (7) is based on a simple model, a full-scale quantum statistical mechanical treatment yields similar results. Another case of interest is that of ions which hop from site to site. A relation between the ordinary diffusion constant, D_i, and the mobility can be derived from the Boltzmann transport equation,[9] yielding the so-called Einstein relation:

$$\mu = \frac{e}{kT} D \tag{8}$$

The essential argument in deriving equation (8) is that the electric field only slightly perturbs the diffusion process leading to a slightly larger number of ions moving in the direction of the field than against it (or vice versa, depending on the sign of the charge of the ion). The 1/T dependence of equation (8) is very slight, and is usually swamped out by the exponential temperature dependence of the diffusion coefficient, $D = D_o \exp [-Q_D/RT]$. Consequently, a general expression for ionic conduction will be of the form:

$$\sigma = nq \frac{e}{RT} D_o \exp - [Q_D/RT] = \sigma_o \exp - [Q_D/RT] \tag{9}$$

For many years since the classical experiments of Warburg[10] which established that ions moved in glass under an electric field it was tacitly assumed that all glasses were ionic conductors. During the intervening years, details of ion transport mechanisms that led to relations such as equation (9) have become well accepted.[11] Series of glasses involving compositional variations have been investigated and have led to an understanding of the effect of network modifying ions on σ_o and Q_D. Relatively little has been done to correlate electrical properties with the recent information of glass microstructures involving both liquid phase separation and nucleation and crystal growth.

Within the last two decades the phenomenon of electronic transport on oxide glasses containing transition metal ions and glasses based on the chalcogenide series of elements has also been well established.[1,2,3,6] More recently extensive controversy has developed concerning the mechanisms of electronic motion in amorphous solids.[6] The theoretical understanding has been extensively hampered by a lack of a variety of conductivity data on glasses with well documented microstructures. A variety of compositional ranges in both oxide and chalcogenide systems have been studied but accompanying phase equilibria, metastable phase separation, or glass transition data has usually not been available.

It is the author's hope that the following sections

will reflect the current state of understanding in both the ionic and electronic transport areas of glass. Examples of applications of the theoretical treatments presented will indicate the direction that theory-properties-structure correlations are headed.

IONIC CONDUCTIVITY - THEORY

Ionic conductivity involves the long range migration of ionic charge carriers through the glass under the driving force of an applied electric field. The charge carriers will be the most mobile ions in the glass and therefore for silicate glasses are usually +1 valence cations moving in an immobile SiO_2 based matrix. The electric force on the cation serves to perturb its random thermal motion by increasing the probability of a transition in the direction of the applied field. This behavior can be described easily by considering the two schematic potential energy wells shown in Figure 1. The potential barrier W represents that maximum barrier in the path of least resistance in the glass.

Consider that the ions move in one dimension parallel to the x axis jumping over the potential barrier W. The

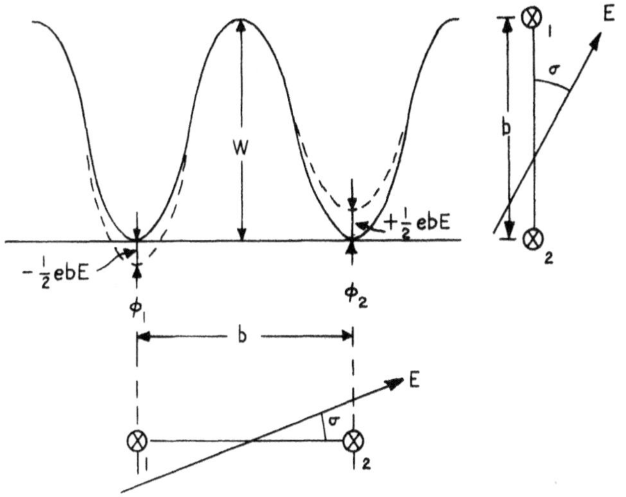

Figure 1. In site energy configuration in a glass. Solid lines indicate energy without field; dashed lines indicate energy with field E applied.

probability that an ion will move to either the right or the left is:

$$p = \alpha \frac{kT}{h} \quad \exp\,[-W/kT] \qquad (10)$$

where α is an accommodation coefficient related to the irreversibility of the jump; kT/h is the vibrational frequency of the ion in the well, k is Boltzmann's constant, and T is absolute temperature. When a field E is applied, the ion coordinations are slightly distorted within the glass structure and the potential barrier to the ion motion is slightly shifted as shown in Figure 1. The field will lower the potential barrier on the side in the direction of the field and raise it on the other side by an equivalent amount. If the distance between the two wells is (b) the potential on the right side will now be smaller by an amount, $1/2\ zeEb = 1/2\ Fb$, where F is the force on the ion and z is the valence of the ion. Therefore, the probability of motion to the right is:

$$p+ \;= \frac{1}{2}\,\alpha\,\frac{kT}{h}\,\exp\,-\;[W - \frac{1}{2}\,Fb]/kT \qquad (11)$$

or

$$p+ \;= \frac{1}{2}\,p\,\exp\,[+Fb/2kT]$$

The probability of motion to the left is thus:

$$p- \;= \frac{1}{2}\,\alpha\,\frac{kT}{h}\,\exp\,-\;[W + \frac{1}{2}\,Fb]/kT \qquad (12)$$

or

$$p- \;= \frac{1}{2}\,p\,\exp\,[-Fb/2kT]$$

Consequently, the positive transitions will be more frequent than the negative and therefore there will be an average drift velocity to the right in the direction of the field. The mean velocity of the drift motion is:

$$\bar{v} = b(p+ \; - \; p-) = \frac{1}{2}\,bp\left[\exp + \frac{Fb}{2kT} - \exp - \frac{Fb}{2kT}\right]$$

$$\bar{v} = bp\,\sinh\frac{Fb}{2kT} \qquad (13)$$

It is worthwhile to repeat that this result is based on the electric field merely directing the random diffusion of ions that occur spontaneously at any temperature, T_i. The field does not pull the ions out of the wells. In fact, as long as the field strength is much smaller than kT, i.e., 1/2 Fb <<kT, then sinh can be replaced by the argument and the drift velocity can be expressed as:

$$\bar{v} \simeq \frac{b^2 pF}{2kT} \qquad (14)$$

In the case where very large field strengths exist, the first term in sinh will dominate and:

$$\bar{v} \simeq \text{Const. } \exp\,[bF/2kT] \qquad (15)$$

At room temperature bF is small compared with kT up to field strengths of 10V/cm. Therefore, the current will be proportional to the field strength in agreement with Ohm's law.

$$j = nze\bar{v}; \quad n = \#\text{ of ions/cc},\ q = ze \qquad (16)$$

so

$$j = \frac{nze\, b^2\, pF}{2kT} = \frac{nz^2e^2b^2pE}{2kT} \qquad (17)$$

and substituting:

$$p = \frac{\alpha kT}{h} \exp\text{-}[W/kT] \text{ and } W = \frac{\Delta F_{DC}}{N} \qquad (18)$$

where ΔF_{DC} is the change in free energy for dc conduction in units of kilocalories/mole and N is Avagadros number $= 6.023 \times 10^{23}$ atoms/mole.

yields:

$$j = \frac{n\alpha z^2e^2b^2E}{2h} \exp\text{-}\left[\frac{\Delta F_{DC}}{RT}\right] \qquad (19)$$

Therefore the electrical resistance of glass

$$R_{(ohms)} = E/j = \frac{2h}{n\alpha z^2e^2b^2} \exp+[\Delta F_{DC}/RT] \qquad (20)$$

or $\quad \log R = \log \frac{2h}{n\alpha z^2e^2b^2} + \Delta F_{DC}/RT$

or conversely, $\log G_{(mohs)} = \log \frac{n\alpha z^2e^2b^2}{2h} - \Delta F_{DC}/RT \qquad (21)$

which is similar to the Rasch-Hinrichsen law of electrical resistivity of glass discovered empirically a number of years ago,[12] $\log \rho = A+B/T$. That a variety of glasses do behave according to eq. 21 is shown in Figure 2. (note σ^{-1} $(\text{ohm} - \text{cm}^{-1}) = G(\text{ohm}^{-1})$ (thickness) (area^{-1}).

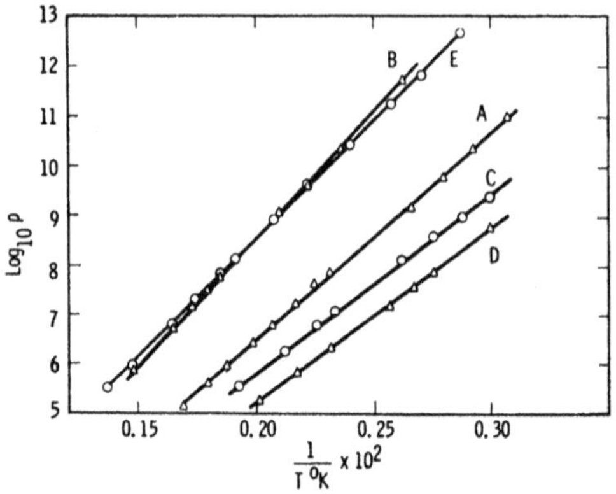

Figure 2(a). Resistivity of some ionic glasses.[19] A.) $18Na_2O \cdot 10CaO \cdot 72SiO_2$, B.) $10Na_2O \cdot 20CaO \cdot 70SiO_2$, C.) $12Na_2O \cdot 88SiO_2$, D.) $24Na_2O \cdot 76SiO_2$, E.) Pyrex.

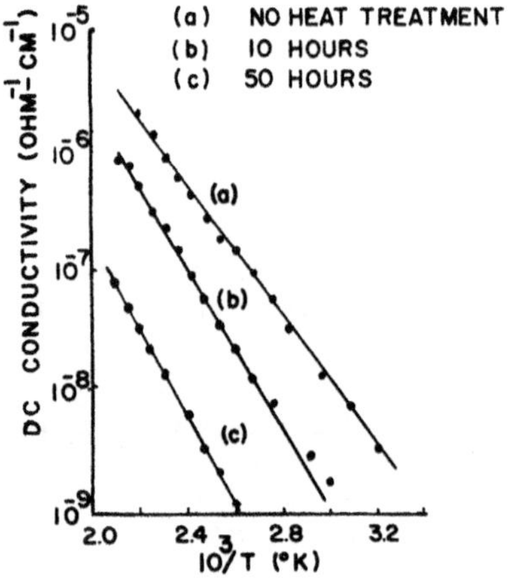

Figure 2(b). Conductivity of $30Li_2O \cdot 70SiO_2$ glass.[18]

The free energy of activation for dc conduction is also given in Table I for several glasses together with diffusion activation energies. The similarity between the two activation energies gives additional credence to the formulation of conductivity theory based on field directed diffusion of charged ions. It is also possible to distinguish between enthalpy and entropy changes involved in the dc conduction by using the Second Law of Thermodynamics as:

$$\Delta F_{DC} = \Delta H_{DC} - T\Delta S_{DC} \quad (22)$$

which yields a final expression for the conductance:

$$\log G = \log\left[\frac{n\alpha z^2 e^2 b^2}{2h} \exp\frac{\Delta S}{R}\right] - \frac{\Delta H_{DC}}{RT} \quad (23)$$

Therefore, the enthalpy and entropy can be calculated from the slope of the log σ vs 1/T plot respectively. Some interpretations of ΔS values have been given, although they are largely speculative in nature.[1,13]

IONIC CONDUCTIVITY - ELECTRODE POLARIZATION

Ion transport within the glass under a dc potential eventually leads to a build-up of charge at the glass-electrode interface if the ions are not replenished at the electrode. This phenomena is termed electrode polarization and it can drastically affect the magnitude of dc conduction measured. Three regions of ionic conductivity can be seen on Figure 3 which is an experimental measurement of the conductivity or current after a voltage of 1 volt had been applied for the times indicated.[14] The conductivity at very short times increases due to various dielectric con-

Table I. Activation Energies for Diffusion and DC Conductivity.

Composition*					E_a ($\frac{Kcal}{mole}$)	E_a ($\frac{Kcal}{mole}$)
Na_2O	CaO	Al_2O_3	SiO_2	GeO_2	Diffusion (Na^+)	Conduction
33.3	---	---	66.7	---	13 - 14	14 - 16
25.0	---	---	---	75.0	17 - 18	16 - 18
** 15.7	---	12.1	72.2	---	16.4	15.6
** 11.0	---	16.1	72.9	---	15.6	15.1
** 15.9	11.9	---	72.2	---	22.0	20.8
** 14.5	12.3	5.8	67.4	---	20.2	19.5

* Mole %

** R. W. Heckman, J. A. Ringler, E. L. Williams, Phys. Chem. Glasses, 8, 145 (1967).

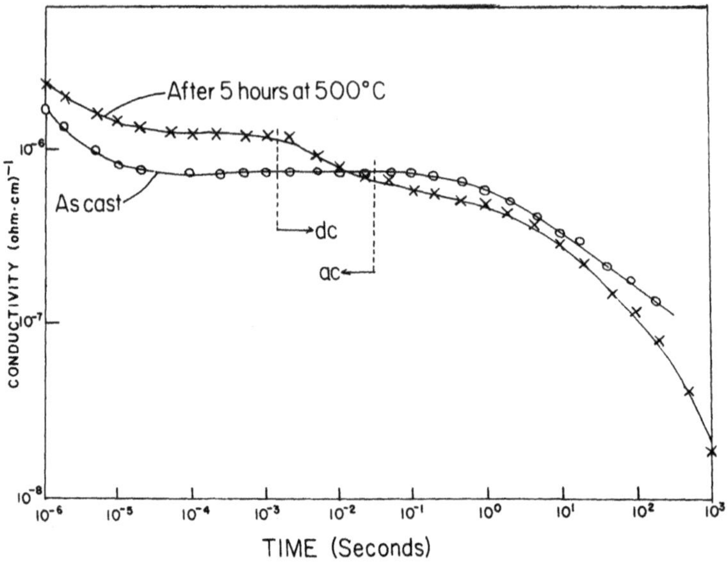

Figure 3. Conductivity as a function of time for a $30Li_2O \cdot 70SiO_2$ glass, as cast and heat treated.[14]

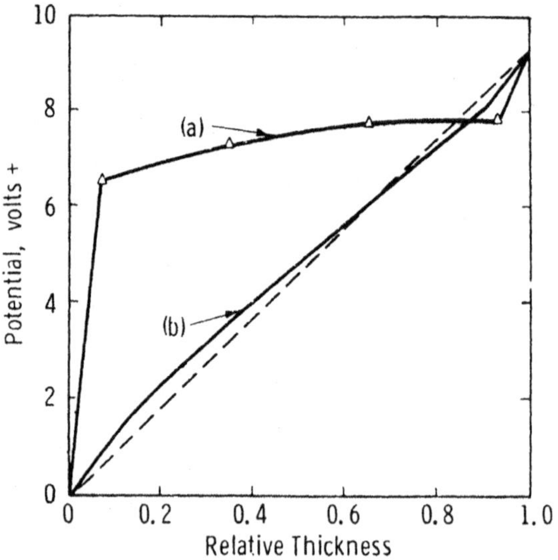

Figure 4. Potential distribution in glass.[15] A.) An alkali-lead-silicate after 10 hours at 382°C, B.) An alkali-free glass after 18 hours at 460°C. The field strength is 160-180 v/cm.

tributions discussed in a following section. At long time intervals the conductivity decreases indefinitely and only at intermediate intervals is the conductivity time independent. It is within this time independent region that conductivity can be correctly termed σ_{DC}.

The reason for the continued decrease in σ at long times is that the mobile cations pile up at the electrodes which results in an induced back field. The back field concentrates the potential drop in the electrode region as shown by Proctor and Sutton[15] in Figure 4. Since the potential across the bulk of the sample is decreased to a fraction of the applied potential, the force, F, on the mobile ions in the bulk is decreased proportionally and the conductivity decreases as predicted from equation 17. The back field builds up indefinitely as time progresses, resulting in a continuous decrease in glass conductivity.

It is thus evident that dc conductivity is dependent on the time interval at which the measurement is made and care must be taken to conduct measurements in the time-independent region of behavior. This may be especially difficult at higher temperatures when the conductivity is in the range of $>10^{-5}ohm^{-1}cm^{-1}$ because electrode polarization occurs so rapidly. (<10 sec.)

There are three possible solutions to the problem of accurate experimental measurement of ionic σ_{DC} posed by electrode polarization. The first is to conduct σ vs. t curves such as Figure 3 for each temperature to ensure that measurements are made in the time independent regime. Secondly, it is possible to use non-blocking electrodes which provide a source and sink for the mobile ions, e.g., sodium amalgam electrodes for Na containing glasses.[16] However, it should also be pointed out that a brief investigation of Li-Cd alloy electrodes for Li_2O-SiO_2 glasses[17] indicated that the alloy composition must be adjusted such that the chemical potential across the interface is equal before the electrode is completely non-blocking.

The third approach is to calculate the current flow required to produce the back potential necessary to decrease the conductivity of glass to a given fraction of its true non-blocked value. Kinser and Hench[14] have shown that the back potential, ϕ_b, due to the polarization charge per unit area, q, is simply:

$$\phi_b = \frac{qd}{2\varepsilon\varepsilon_o} \qquad (24)$$

where ε is the permittivity of the glass, d is the thickness, and ε_o is the permittivity of free space. The charge, q_b, to form the back potential can be related to the current flow as:

$$q_b = \int_o^t j dt = jt \qquad (25)$$

The current, j, required to form the back charge is directly related to the applied potential, ϕ_a, and the conductivitv of the glass, as $j = \phi_a \sigma$, and therefore on substitution into eq. 25 yields:

$$q_b = \phi_a \sigma t \qquad (26)$$

So, if there is a critical response time for the measuring equipment, t_x, and an acceptable level of error in the conductivity measurement of 0.1, for example, then the ratio ϕ_b/ϕ_a is equal to 0.9 and the limit of conductivity that can be accurately measured within t_x for the glass is:

$$\phi_b/\phi_a = \frac{0.9(2\varepsilon\varepsilon_o)}{d\ t_x} = \sigma_x \qquad (27)$$

Since σ increases rapidly with temperature, equation(27) implies that there will be a critical temperature, T_c, where the conductivity error becomes greater than 10% when T_c is exceeded. Such a behavior is shown in Figure 5. By using equation(27) and an estimated value of ΔF_{DC}, this problem can be eliminated.

IONIC CONDUCTIVITY - COMPOSITION EFFECTS

A number of investigators[1,19,20,21] have investigated the change in electrical conductivity of alkali-silicate glasses associated with the replacement of silica by alkali ions. In nearly all cases, a rapid increase in conductivity is observed until a range of 20-30 mole% modifier is reached, after which the effect becomes progressively less. The range of conductivity is from 10^{-16}ohm^{-1}cm^{-1} to 10^{-7}ohm^{-1}cm^{-1} at room temperature. The relative effect of alkali type on conductivity is $Na^+ > Li^+ > K^+$. The addition of only parts per million of Na^+ ions is sufficient to drastically increase the conductivity of fused silica and the conductivity is directly related to the Na^+ content.[22]

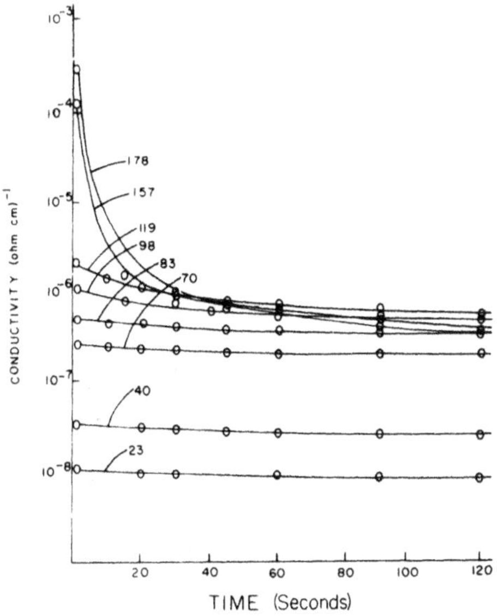

Figure 5. Conductivity as a function of time for a $33Na_2O \cdot 67SiO_2$ glass at various temperatures (in °C).[14]

Additions of divalent modifiers to a soda silica glass have consistently been observed to decrease the electrical conductivity[1,4,19] with the relative effect being $Ba^{++} > Pb^{++} > Sr^{++} > Ca^{++} > Mg^{++} > Zn^{++} > Be^{++}$. Additions of a 20 weight % Ca to an 18 % Na_2O.82 SiO_2 glass will decrease the room temperature conductivity to 10^{-15}ohm^{-1}cm^{-1} whereas Zn^{++} additions up to 40% will be of no influence. The effectiveness of the divalent oxides in decreasing the conductivity has been found[23] to increase as the radius of the metal ion is increased. This correlation is shown in Figure 6. The interpretation of the effect is that the divalent ions block the conduction paths of the alkali ions and as the divalent ion size increases, the blocking becomes more effective. It is also possible that the divalent and monovalent ions coordinate with mutual oxygen ions, thus maximizing the charge distribution in the glass.

Partial substitution of one alkali metal oxide for another in silicate, borosilicate and borate glasses produce such large increases in electrical resistance that this behavior is termed "the mixed alkali effect." This effect is illustrated for three systems in Figure 7 taken from

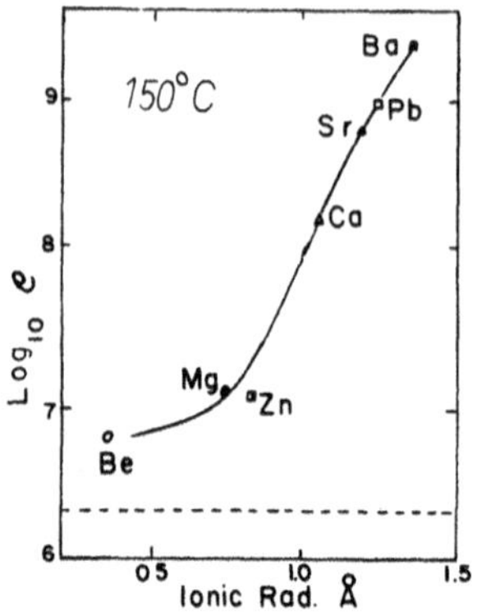

Figure 6. The effect of divalent ion radii on resistivity of $20Na_2O \cdot 20MO \cdot 60SiO_2$ glass.[1] Dashed line is the resistivity of $20Na_2O \cdot 80SiO_2$.

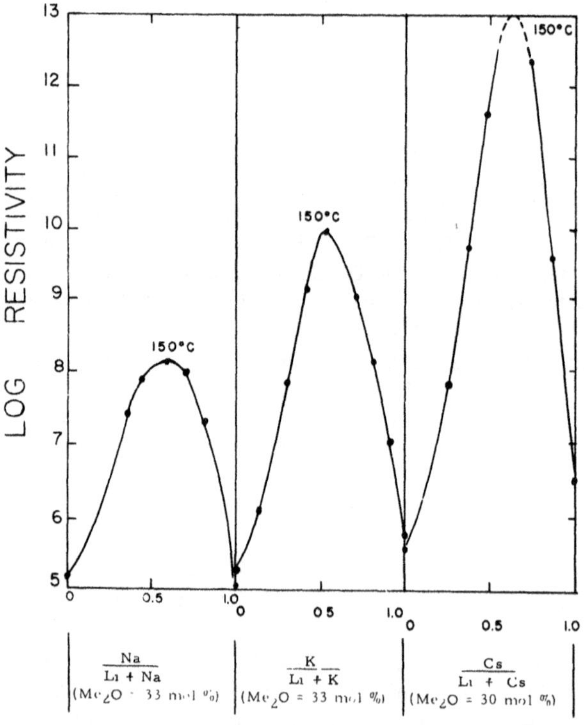

Figure 7. Resistivity of silicate glasses containing mixed alkali ions.[24]

Charles.[24] Early work[25] interpreted the mixed alkali effect largely in terms of the motion of "coupled ions." The evidence suggesting the formation of a cation complex was that the activation energy for conduction also increases by almost a factor of two as the substitution occurs. Charles has recently shown[24] that the substitutions induce phase separation in the Li-Na-Cs cation glasses investigated. His conclusion,therefore, is that the smaller, more mobile ions are isolated in the dispersed phase with the larger ions in the matrix phase thus decreasing the total dc conductivity. However, in order for the conductivity to be suppressed to the magnitude demonstrated, the larger cations must also be largely coordinated within the dispersed phases yielding a nearly alkali free silica rich matrix which determines the dc conductivity.

Substitution of Al_2O_3 for SiO_2 in sodium silicate glasses is especially interesting.[26] As the Al_2O_3 content increases,the electrical conductivity is also increased and the activation energy for conduction is reduced. This behavior appears to be associated with the aluminum ion entering the network in fourfold coordination reducing the number of non-bridging oxygen ions. The non-bridging oxygen coordination shell about the sodium ions must be decreased, which could serve to increase the sodium ion mobility.

In light of Charles' observations on the mixed alkali effect it would indeed be interesting to compare the structures of the soda-silica glasses with and without the Al_2O_3 substitutions.

IONIC CONDUCTIVITY - STRUCTURE EFFECTS

With the exception of Charles' studies[24,27] there have been relatively few attempts to obtain correlations between composition-structure and the ionic conductivity of glass. It has also been recognized for many years that thermal treatments in the annealing-transformation range significantly alter the dc conductivity of glasses.[28,29] However, there has been relatively little work done to establish structural interpretations of thermal history variations in electrical properties either.

Charles[27] examined the electrical properties and structural differences in lithia-silica glasses resulting from different quenching rates. His results indicated that the changes in electrical properties could be related to the morphology of the phase separation. A rapid quench of a low, 7.5 mole %, lithia-silicate glass changed the phase separated structure from that of a dispersed phase to a tubular, interconnected phase, which was perhaps spinodal in origin. The dc conductivity was increased by nearly an order of magnitude by the rapid quenching due to the presence of the interconnected Li_2O-rich phase.

A series of investigations concerned with structural interpretations of electrical conductivity changes accompanying thermal treatments on Li_2O-SiO_2 and Na_2O-SiO_2 glasses has recently been completed by Kinser and Hench.[17,18,30,31,32] These studies have shown that the dc conductivity of both types of alkali silicate glasses over the compositional ranges investigated decreases by over two orders of magnitude concurrent with the appearance of the equilibrium crystal phases in the glass. This behavior is illustrated in Figure 8. The major change in the conductivity is associated with the presence of only a few percent of crystals in the glass. Consequently, it appears that the more energetic cations, the ones that usually participate in the thermally activated dc conduction, are the first removed from the glass during the crystallization

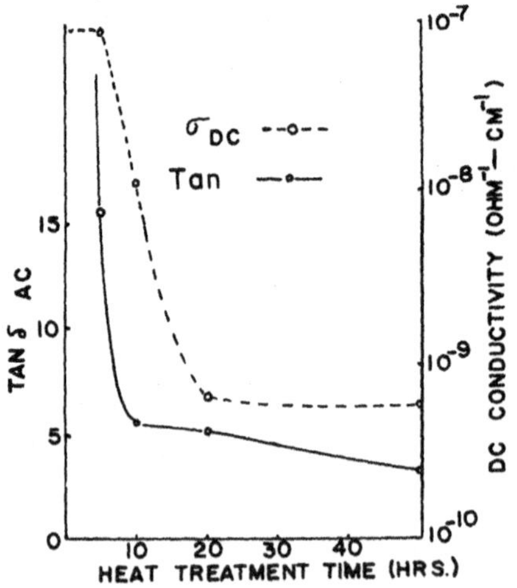

Figure 8. Conductivity and loss tangent for a $30Li_2O \cdot 70SiO_2$ glass as a function of heat treatment at 500°C.[18]

process. Since crystallization is diffusion controlled in these glasses[34] it is reasonable that it be the same ions that respond first to the free energy gradient for crystallization as would respond most easily to the dc potential.

IONIC CONDUCTIVITY - MOLTEN SILICATES

The determination of the electrical conductance of molten glasses has been largely prompted by an interest in the structure of glasses and the manner in which the structure changes upon melting. However, the recent interest in the production of glass by means of electric furnaces provides a commercial impetus for understanding the nature of the electrical conductance of the molten glasses. In particular, it is of interest to understand the effect of composition on the conductance of glasses in the molten state. Obviously, from a commercial standpoint, the effect of temperature on electrical conductivity is of extreme importance also.

There have been several investigations of the electrical conductivity of molten silicates which contain alkali oxides.[35-39] Different workers disagree about which of the three systems, Li_2O-SiO_2, Na_2O-SiO_2, or K_2O-SiO_2 is the most conducting. The recent authoritative work of R.E. Tickle[35] has shown that the relative conductivities of the alkali oxide silicate glasses depend upon the level of concentration of the alkali ions. For concentrations of R_2O greater than 22%, the order of equivalent conductances was found to be $Li_2O > Na_2O > K_2O$, but below about 22% R_2O, the order was $Na_2O > Li_2O > K_2O$. These results appear to clarify the controversy existing between the studies conducted by Endell and Hellbrugge,[36] Brockris et al.,[37] and Urnes.[38]

A comparison of resistivity isotherms at 1400°C is made for various concentrations of Li_2O-SiO_2 glasses measured by different workers,[35-38] and is shown in Figure 9.

The resistivity of a series of soda silica glasses[35] is shown as a function of reciprocal temperature in Figure 10 over a temperature range of 900°C to 1600°C.

The numbers on the curves denote the mole % of Na_2O in the binary glasses. The curvature of the plots is clearly seen, and the departure from the simple exponential temper-

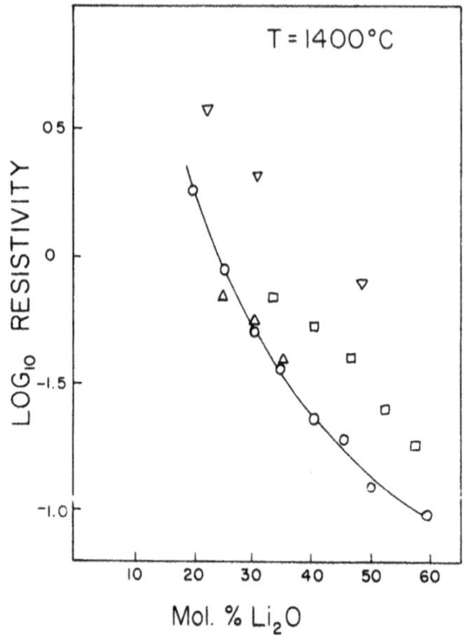

Figure 9. Resistivity of molten lithia-silica glasses as a function of Li_2O content.[35-38]

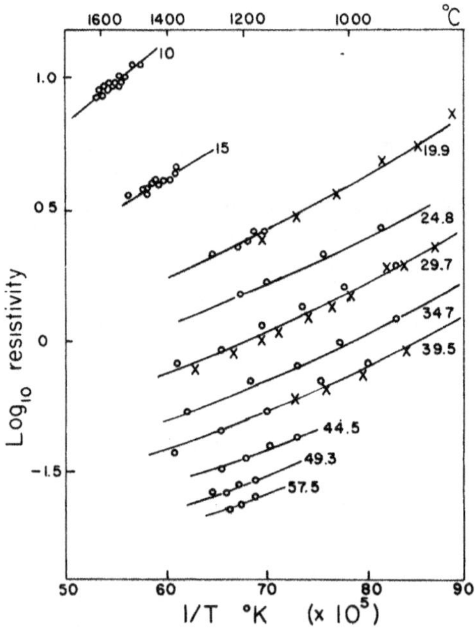

Figure 10. Resistivity of molten $x(Na_2O)\cdot(100-x)SiO_2$ glasses as a function of temperature.[35]

ature dependence of the Rasch-Hinrichsen law is pronounced. It is significant that the low temperature and high temperature data match well, and can be represented by a single curved line for each of the compositions studied. The resistivity of soda silica glasses decreased markedly with an increase in the percent of Na_2O present.

The conductivity of the alkaline earth silicate glasses is, in general, about an order of magnitude lower than that of alkali containing glasses of equivalent concentration. The general order of equivalent conductances for the various alkaline oxides is CaO>MgO>SrO >BaO. The differences between the resistivities of the high and low members of this series is very small with the resistivity of calcium containing glasses being approximately one-half that of barium containing glasses of equivalent concentration.

The change in conductivity of calcium oxide silicate glasses[36] is shown as a function of reciprocal temperature in Figure 11. The numbers on the curves represent the following percentages of CaO in the binary glasses:

Glass #	1	2	3	4	5	6	7	8
% CaO	55	50	30	45	40	35	20	30

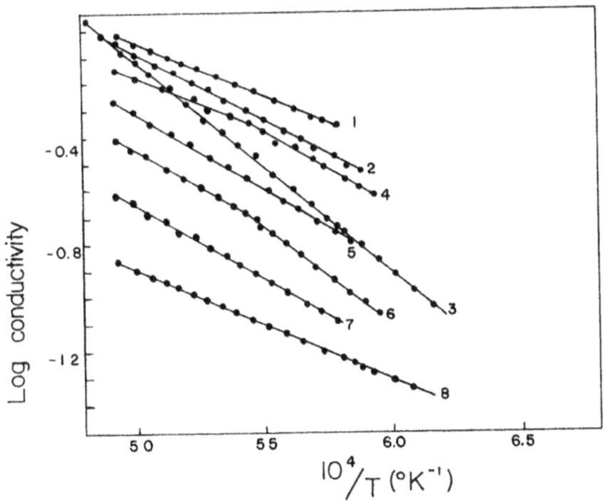

Figure 11. Conductivity of molten CaO-SiO_2 glasses as a function of temperature. (Glass #1 is 55% CaO, #8 is 30% CaO, remainder is given in text.)[36]

Binary liquid silicates containing iron and manganese modifier ions possess electrical conductivities nearly equivalent to that of the alkali silicate systems. Binary liquid silicates containing a second network former such as Al_2O_3 or TiO_2 possess electrical conductivities that are as much as three to four orders of magnitude lower than those of the alkali silicate liquids.

The electrical conductivity of glasses and liquid melts containing two alkali modifier ions has received considerable attention in the literature because of the so-called mixed alkali effect mentioned above. Tickle has established that the mixed alkali effect increases as the temperature is decreased. The conductivity of glass systems containing alkali modifier ions at high temperatures is nearly equivalent to linear additions of the conductivities of the two end members, while at lower temperatures, the conductivity is decidedly less. The mixed alkali effect may decrease the conductivity by as much as a factor of 4 at a temperature of 800°C for soda potassium silicate glasses. At 1200°C the decrease in conductivity in the mixed soda potassium glass is only a factor of 2.

Understanding the effects of composition and temperature on the electrical conductivity of the glasses of the alkali-alkaline earth-silicate systems is of the greatest importance from a commercial viewpoint. Most commercial glasses are based upon combinations of alkali and alkaline earth modifiers with silica. It is indeed unfortunate that there has been only meager systematic investigations of the electrical conductivity of glasses in these systems. Recent publications by a team of Russian investigators have been concerned with the electrical conductivity of glasses in the system Na_2O-RO-SiO_2.[40] The RO modifiers that have been studied in the ternary systems include CaO, MgO, PbO, ZnO, BaO, BeO, and CdO. The temperature range covered has been from 900°C to 1400°C. The investigations show that in the range of 1100°C to 1400°C, the conductivity of these glasses obey the well-known exponential relation, $\log \sigma = A - B/T$, where A and B are constants, and T is the absolute temperature.

Perhaps the most significant result of this research has been the finding that the conductivity is proportional to the concentration of sodium oxide present in the ternary systems. This result was found for all of the ternary glasses investigated. A plot of the variation of resistivity of glasses of the systems, Na_2O-RO-SiO_2, with the alkali ion concentration at 1200°C is shown in Figure 12. It can be seen from this figure that the resistivity is directly a function of concentration of sodium present in the glasses regardless of the nature and concentration of the bivalent cation present. The temperature dependence and sodium ion concentration dependence of ternary glasses in the Na_2O-RO-SiO_2 system can be expressed by the following equation:

$$\log \sigma = 1.508 - 0.0204C - \frac{4836-128C}{T}$$

where C = Na_2O content in weight percent and

T = °K

The above equation has been found by the Russians, Kostanyan et al.,[41] to be valid over a temperature range of 1100°C to 1450°C for the following composition ranges: 12-20 Na_2O, 0-6 MgO, 0-11 CaO, 0-11 Al_2O_3 and 68-78 SiO_2.

It must be pointed out that these results apply to glasses with high sodium concentration ($Na_2O > 10$ mole %) and relatively low bivalent cation concentration. It has been stated by Kostanyan et al., that when the sodium ion concentration decreases to less than 10% or the bivalent cation concentration increases to greater than 10%, the relationship of conductivity to concentration becomes more complex.

It is tempting to extend the results of Kostanyan et al., to predict that the concentration of alkali modifier cations will be the controlling factor in the electrical conductivity of ternary glasses containing potassium or lithium ions as well as the soda containing glasses. This conclusion would seem to be reasonable considering the fact that the relative conductivities of the alkali glasses are nearly comparable and are nearly an order of magnitude greater than the conductivities of the alkaline containing binary silicate glasses. In other words, the bivalent cations appear to have very little influence on the mobility of the sodium ions in the liquid melt and, consequently, one could conclude that they would have relatively little influence on the monovalent lithium and potassium ions as well. However, it

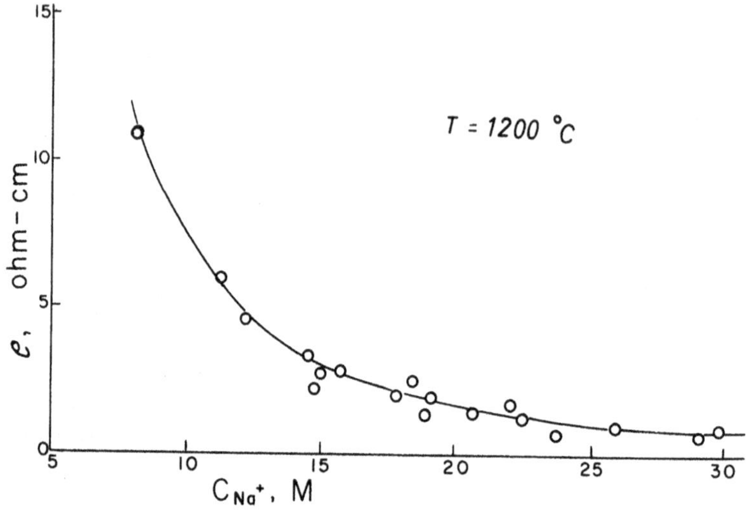

Figure 12. Dependence of resistivity of Na_2O-RO-SiO_2 glasses on a wight per cent of Na^+ ions. (Data points represent MgO, CaO, BaO, BeO, and CdO containing glasses with 68-78% SiO_2).[40]

must be cautioned that the effect of bivalent cations on the mixed alkali effect is at present unknown.

It is also of interest to note that the electrical conductivity of sodium borate glasses is also a function of sodium ion concentration. Sodium borate glasses possess approximately 50% higher values of conductivity than sodium silicate glasses of equivalent alkali concentration.

A change in the conductivity will be noticed if either of the two terms on the right hand side of equation (23) are changed. The first term may be changed if either the ion vibrational frequency (see also equation(10) or the connectivity of hopping sites (i.e., the number of nearest neighbor equivalent hopping sites) is altered. Smaller effects may be seen also if the transfer number, Z, or the hopping distance, b, are changed. The second term may be altered if the height of the energy barrier is changed. This energy consists of two terms: (1) the electrostatic energy required to move a point charge q = Ze to the midpoint between the two sites, and (2) the elastic energy required to stretch the lattice so that the ion will "pop" through to the new site. It is uncertain which of these factors is the principal cause of the difference between the borate and silicate glasses.

Martin and Derge[42] have conducted an extensive investigation of the effects of temperature and composition of calcia alumina silicate slags from alumina percentages of 5% - 20%, and calcia percentages from 35% - 50% over a temperature range of 1450°C to 1600°C. Only compositions above the liquidus surface were investigated. The results show a maximum in conductivity occurring in the region, 38% to 43% SiO_2, 44% to 50% CaO, and 10% to 15% Al_2O_3. This peak is close to the composition of the ternary eutectic in the CaO-Al_2O_3-SiO_2 system which is located at approximately 41% SiO_2, 47.3% CaO, and 11.7% Al_2O_3. The conductivity was observed to be a minimum at concentrations of 46% SiO_2, 37% CaO, and 17% Al_2O_3. Other than these regions of maximum and minimum conductivity, there appears to be a general trend for the conductivity to increase with increasing basicity of the melt.

ELECTRONIC CONDUCTIVITY

This present section is a review of the phenomena of electronic conduction in amorphous semiconductors. An attempt will be made to unify the discussion around Anderson's theorem and hopping phenomena based on Holstein's small polaron model.

Several omissions in this area are rather glaring; however, to have done justice to such topics as Hall conductivity, optical properties, and thermopower would have necessitated a great expansion into fields where there is even less agreement than in the mechanism of electric conduction.

BAND STRUCTURE

The present state of knowledge of the electronic structure of amorphous materials may best be summarized as primitive. The only calculations which have been performed are those on one-dimensional models, principally a disordered Kronig-Penney model (Figure 13). Disorder is invariably assumed to be of a strictly random nature, which is appropriate to a gas but not to a glass.

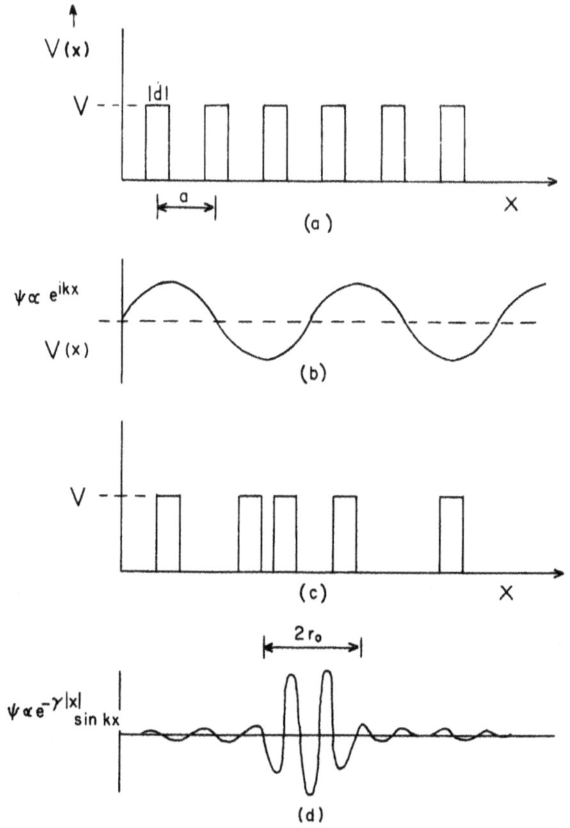

Figure 13. Ordered and disordered Kronig Penney model. A.) ordered, B.) qualitative solution to Schrödinger equation, C.) disordered, D.) qualitative solution to Schrödinger equation.

This limited development is a result of two other conditions: (1) The position of the atoms in a real amorphous material has not been mathematically described, and (2) as a result, no proven methods exist for the calculation of a band structure in a glass. It is worthwhile exploring these problems a little further.

The most reliable and detailed information on the atomic structure of glasses which is currently available is the radial distribution function obtained by macroscopic diffraction methods. Accurate experimental techniques for obtaining the RDF have been developed only in the past ten years.[43] It is impossible, however, to reconstruct the three-dimensional distribution of atoms from the RDF; any given RDF can result from an essentially infinite number of three-dimensional structures. Furthermore, when more than one type of atom is present, the assignment of peaks in the RDF to specific atomic correlations is impossible without additional information. Qualitative information on the three-dimensional structure can be obtained from the RDF by regarding the atoms as nearly hard spheres, and then using geometrical arguments to construct the lattice.

For a particularly useful example, let us consider the work done on amorphous germanium.[44] It is found from the area under the first peak of the RDF, which corresponds to nearest neighbor correlations, that the crystalline coordination and spacing are preserved in the amorphous state. This suggests that the fundamental building block in amorphous Ge is a Ge_4 tetrahedron. These tetrahedra may be connected by having common corners. It has been suggested that the fourfold coordination of Ge may be satisfied on an extended scale if the tetrahedra are oriented in a staggered or an eclipsed configuration. Repeating only the staggered configuration leads to the diamond crystal lattice, while repeating only the eclipsed configuration (with some small distortion) leads to a regular dodecahedron each of whose sides consist of a ring of five atoms. The lack of a periodic structure is attributed to the appearance of these elements with five-fold symmetry. By mixing the staggered and eclipsed configuration, it is found that a structure may be obtained which produces the actual RDF fairly well.[45,46] More recent work suggests that rather than the eclipsed configuration, the tetrahedra assume a configuration midway between the staggered and eclipsed configurations.[47]

We are left, therefore, with only the calculations performed on one-dimensional models.[48] We shall review only the results of these calculations.

Starting with a perfectly periodic structure, we find that all states are the normal band states (Figure 13a,b). As a small amount of disorder is introduced, the number of band states decreases and localized states are formed having energies at the top and bottom of the band. As the amount of disorder increases more states are split off from the band and form localized states. A definite energy, E_c (Figure 14), is found below (or above) which all states are localized. Finally, if the disorder is sufficient, all

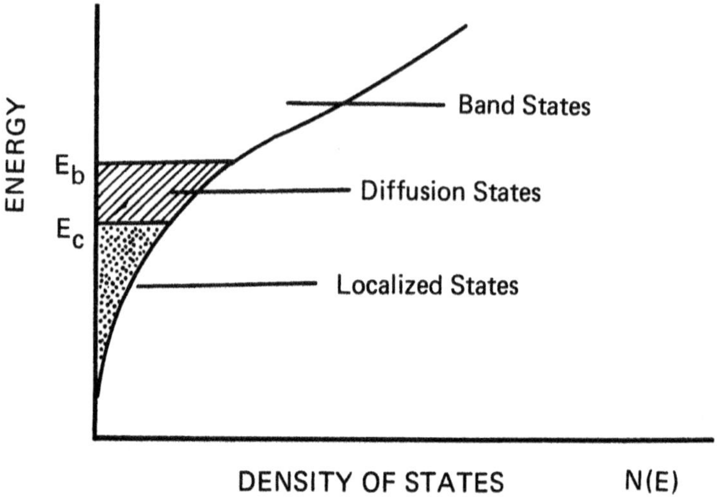

Figure 14. Density of states (conduction band) in a partially disordered lattice.

states will be localized (Figure 13c,d). This latter result was first proved by Anderson in 1958.[49]

Anderson considered the tight-binding model, that is, one in which the electronic wave functions may be approximated reasonably well as a sum of atomic or molecular wave-functions. He found that if the quantity

$$\frac{J_{ij}}{\Delta E} \tag{28}$$

is sufficiently small (of the order 1/5), then all states are localized. Here J_{ij} is defined as:

$$J_{ij} = \sum_{n \neq i} \int \psi_i^* V_n \psi_j \, dt \tag{29}$$

with ψ_i the wavefunction of the ith atom or molecule, V_n the potential of the nth atom or molecule and ΔE is the mean spread of energy levels of the electron on each of the atoms or molecules isolated from each other.

The essential reason for this result is that due to the scattering of the wave from the various centers in which the phase undergoes a random shift, there will be, at best, few sites in the lattice where the scattered waves interfere constructively. If the energy of such a site is just right, then the phases can be adjusted and the wave-function can blow up. The probability of finding such a site increases much less rapidly with increasing distance than the amplitude of the wave-function decreases, provided that Anderson's criterion is met. The wave-function therefore goes to zero at infinite distances.

Now let us consider a partially disordered solid. Considering first the sites which have the smallest energy, we find that these are the least numerous (at least for random disorder), since they fall at the extremity of the energy distribution. Hence, electrons localized on these sites must be the most localized (i.e., the radius of the electron cloud is the smallest), since the probability of finding another site of anywhere near the proper energy is smallest because of the low density of these sites. This condition is analagous to the impurity problem in a semiconductor. If we have a semiconductor with a single impurity (a donor for simplicity) the energy of the donor state decreases as the charge increases, and the radius decreases.

Now let us suppose that we are in an energy region E near E_c, the critical energy for electron localization, where band states begin. Again we can identify critical sites. That is, the probability for finding an electron is large in a certain region, and we can identify a certain site which causes this modulation of the wave-function. By this, we mean that if the energy of the site is changed appreciably, the state disappears from our range of energy interest.

We now proceed to calculate the wave-function of an electron on each site, with all other sites so altered as to be outside the range of interest. We can use the wave-function so generated as a new set of "molecular" functions and apply them in Anderson's theory.[49] We will find that if $\left\langle \frac{J}{\Delta E} \right\rangle$ is sufficiently small, our states will be localized. Here J is the overlap integral, defined in equation (29), using our "molecular" wave-functions. By considering sites above and below E_c (the Anderson parameter of those above E_c will be exceeded), we can pin down E_c.

We need to consider one additional topic before leaving the structure problem, that of the effect of short range order on the electronic structure. As we have previously mentioned, disorder in any real glass is not completely random. If we again limit our attention to the conduction band of an n-type semiconductor, it may be possible by limiting the type of disorder to split off a relatively high density of states to below the conduction band, and hence form fairly delocalized wave-functions on possibly even band states. As an extreme example of this, we can imagine starting with a monatomic crystal, and then fractionally increasing every other nuclear charge, and decreasing the remaining charges by the same amount. The mean deviation of such a process would be large; however, the result will be to eventually split the conduction band in two. It is not inconceivable that some related structure will arise in a glass. We have seen that in silicon and germanium, there are probably eclipsed and staggered configurations of tetrahedra, with comparatively small distortions of these. If the electronic energies of electrons in these configurations differ moderately, there may result a splitting of the energies into two groups about which the deviation is small. We shall consider such a possibility in a future section.

TRANSPORT

We next will consider the problem of transport in a disordered lattice. Three regimes have been recognized: band conduction, as in crystalline materials, for electrons with energy above E_C; thermally activated hopping among localized states for electrons with energy below E_C; and a diffusion mechanism for electrons with energy just slightly greater than E_c.[50]

The hopping mechanism may be understood by considering a localized electron on one site, and a second site nearby having a slightly different energy. (In order for the electron to hop it will need to gain or lose energy from thermal vibrations.) Due to its localized nature, however, the electron will polarize the lattice. Movement of the lattice distortion at high temperature (in the classical region) also requires energy.

Let us first consider the so-called non-adiabatic hopping. The classical activation energy for a hop when the energy of both sites is the same has been derived from electrostatics by Mott.[51] First, we compute the binding energy of the localized electron; that is, the energy lost by the system when the electron polarizes the lattice.

The energy of the electron in polarizing the lattice will be lowered by

$$E_e = -\frac{e^2}{r_o}\left(\frac{1}{k_\infty} - \frac{1}{k_s}\right) \tag{30}$$

where k_s is the static dielectric constant (more precisely, to avoid many polaron effects, the dielectric constant measured at 10^{10} Hz or so), k_∞ the infinite frequency dielectric constant, and r_o the radius of the electronic state. The energy required to polarize the lattice is

$$E_\ell = \frac{1}{2}\,\frac{e^2}{r_o}\left(\frac{1}{k_\infty} - \frac{1}{k_s}\right) \tag{31}$$

So that the binding energy, the sum of these two terms will be

$$-E_b = E_e + E_\ell = -\frac{1}{2}\,\frac{e^2}{r_o}\left(\frac{1}{k_\infty} - \frac{1}{k_s}\right) \tag{32}$$

Note that since the binding energy is inversely proportional to its radius, the localized states further from the conduction band will have a greater binding energy.

In order for the electron to move classically, the electron on each of the two sites must have the same energy. The minimum energy required to do this is depicted in Figure 15.

Let the electronic energy of the electron on site 1 be 0 and on site 2 be ΔE (Figure 15a). Then when we allow the electron to polarize the lattice on site 1, the electronic energy is reduced by $2E_b$, while the lattice stores an elastic energy E_b, which means that the total energy of the system is reduced by E_b (Figure 15b). In order for the electron to jump to site 2, the electronic energies must be made equal. The minimum energy required to do this is shown in Figure 15c. In determining this energy, we have assumed that the change in the electronic energy is pro-

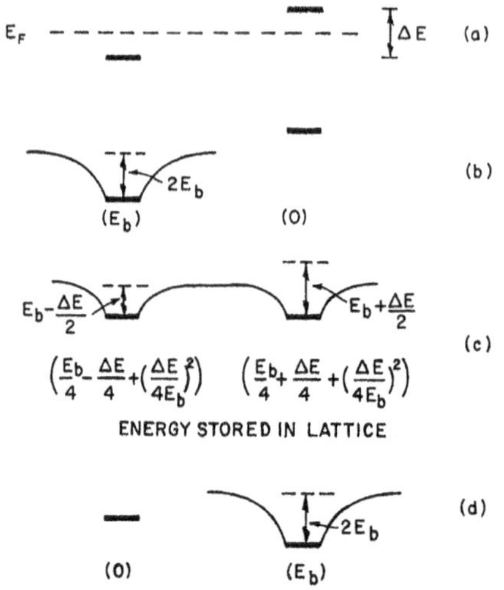

Figure 15. Localized electron hopping model. A.) Site energies, B.) lattice polarization on site 1, C.) configuration at mid-hop, D.) configuration at conclusion of hop.

portional to the displacement of the ions, while the energy stored in the lattice is proportional to the square of this displacement. If we calculate the difference in energy between this configuration, and the ground state in Figure 15b, we find that it is

$$E_H = \frac{\Delta E}{2} + \frac{E_b}{2} + \frac{(\Delta E)^2}{8E_b} \tag{33}$$

This will be the classical activation energy for hopping.

With the energy on the two sites equal, the electron will be able to tunnel from site 1 to site 2. The probability for tunneling will be proportional to:

$$\exp(-2\alpha d) \tag{34}$$

where α is the tunneling constant and d the distance between the two sites. Once the electron has moved, the system will return to the configuration shown in Figure 15d, and the transition will have been accomplished.

The overall transition rate can be written as:

$$W = \nu \exp(-2\alpha d) \exp\left[-(\Delta E + E_b + \Delta E^2/4E_b)/2kT\right] \quad (35)$$

where ν is a phonon frequency. For a quantum mechanical treatment, see Schaake and Hench.[52,53]

The third term in the activation energy will be significant if $\Delta E > 2E_b$. It merely reflects the fact that if the electron is loosely bound to the lattice the probability of its gaining large amounts of energy is small.

Equation (35) is valid only over a limited range. In particular, consider the case when two sites are close together (adiabatic hopping).[54,55] Under these conditions the electron may tunnel back and forth between the two sites when their energies are similar; the energy of the electron and hence the half-way configuration (Figure 15c) is lowered by a resonance process, yielding for the transition rate:

$$W = \nu \exp\left[-\left(\frac{\Delta E}{2} + \frac{E_p}{2} + \frac{\Delta E^2}{8E_b} - \frac{1}{2} J\right)/kT\right] \quad (36)$$

The criteria for this equation is that $J > h\omega_o$ where ω_o is the optical phonon frequency. J is as defined in equation (29).

When the temperatures are sufficiently low, neither of the above equations is valid. This results from the fact that a single phonon process may be favored over a multiphonon process such as considered above. A calculation of the transition rate for single acoustical phonons in the non-adiabatic case has been made by Schaake,[55] who finds

$$W = \frac{E_1{}^2 \Delta E J^2}{\pi \rho h^4 c^5} \exp\left[-2S(T)\right] \coth(-\Delta E/kT) \quad (37)$$

where E_1 is the deformation potential, ρ the density, c the velocity of sound, and $\exp[-2S(T)]$ is a complex function of temperature which decreases more rapidly with increasing T as the binding energy is increased.

Physically, this process would correspond to the

electron gaining or losing the necessary energy from a single phonon, with the electron and its distortion then tunneling to the new site.

Where the binding energy is very small, such as impurity states in silicon, the exponential term goes to unity, and the results become identical to Miller and Abrahams[56]. Also of interest is the fact that equation (35) is valid only if

$$\frac{E_b}{2h\omega_o} \coth \frac{h\omega_o}{2kT} >> 1 \qquad (35)$$

If the binding energy is very small (again as in Si), equation (37) will form a good approximation through an extensive temperature range, although interactions with single optical phonons will need to be considered also.

A consideration of the single phonon process for the adiabatic case has not been made; it appears, however, that a calculation should make the dependence on the distance between the two sites, which is implied by the overlap integral $J \propto \exp[-2\alpha d]$, disappear.

A third hopping region, also adiabatic, may exist. If the sites are extremely close, then the electron may desire to localize itself on two sites, with the lattice polarization then surrounding both sites. Under these conditions, transitions would take place between bonding and anti-bonding orbitals, and the transition rate activation energy should be given by an expression in the high temperature region such as

$$W_H = \nu \exp[(-\frac{\Delta E}{2} - \frac{\Delta E^2}{8E_b})/kT] \qquad (39)$$

Where ΔE is the energy difference between the bonding and anti-bonding orbitals. The range of validity of this expression has not been determined.

Let us briefly summarize this previous discussion. For hopping of an electron between two sites, we find that the transition rate will increase as the sites come closer together, other factors being the same, and at some critical distance, levels off. The transition rate increases exponentially with temperature, with the activation energy also increasing. If the binding energy is small, this increase may not come until quite high temperatures. Finally, in the high temperature region, the activation energy will be smaller for closer sites than for more distant sites.

Turning now to transport in the band states, we find comparatively little theoretical work done on them. It would appear that electrons in them would obey laws similar to conventional semiconductors with, however, a much lower mobility due to the large amount of scattering caused by the disorder.[57]

The band state just above the critical energy for localization, E_c, has been investigated by several authors.[50,52,57] In this region, the electronic wave-function will be highly modulated, with comparatively large probabilities for finding the electron in some regions and small probabilities in the regions connecting them. In this diffusion region a mobility equation such as

$$\mu = \frac{ea^2}{kT} \nu_{el} \tag{40}$$

where a is the distance between the maxima in the probability. ν_{el} is an electronic frequency (i.e., the frequency of the electron in a maxima region).

An additional factor must be taken into consideration in these diffusion states; if the dielectric constant of the material is high, and localization small, then there may be appreciable interaction between the lattice and the electron, leading essentially to polaron states, which we shall discuss shortly. From this, we will be able to conclude that as the temperature increases, E_c will increase, i.e., more states become localized.

Up to now, our discussion has been based on the assumption that conduction in the crystalline phase is according to conventional semiconductor theory. At least in the transition metal oxides, there is no agreement as to this.[58,59] A brief review of polaron theory is therefore in order.

Consider a narrow band semiconductor. A localized wave-function can be constructed by summing over a number of band states. The expectation energy of such a packet

inadequate specification and/or control of important parameters. Of course many factors which control strength have become fully appreciated only in the past ten years, but with the new interpretations of Griffith crack origins, effects of heat treatment, and atmospheric effects, a complete specification of test conditions becomes imparative. Similarly, meaningful changes in fiber strengths can be produced by varying the drawing conditions, and the work of Poraj-Kosic has shown the possible importance of a microheterogeneous structure. It is clear that phase separation on the atomic scale or of macroscopic dimensions, should have some effect on glasses whose strengths approach the theoretical limit and possibly in abraded glasses under static fatigue, if the phase separation causes preferential attack.

An example of where such problems might arise is in a study of the compositional dependence of strength. If all of these parameters have not been specified and controlled, it will be impossible to determine how composition affects strength. It is possible, for instance, to conceive of a hypothetical glass in which a change in composition leads to a decrease in the observed strength while the calculated theoretical strength increases. The observed strengths in such a case could be due to the fact that the compositional change produces a decreased abrasion resistence, an increase in the stress corrosion process, or phase separation, etc. Without having a complete specification of thermal history, handling, and test conditions absolutely no meaningful information can be extracted from such an experiment.

Perhaps the first order of business for future theoretical and basic investigations will be reconciliation, if possible, of the theories of plastic flow and brittle fracture. It is not clear how this might best be accomplished, but it would be hoped that extensive work on the cantilever and blate bubble techniques would produce some meaningful results.

Whether or not a plastic flow mechanism exists and operates is of more than academic interest for it bears directly on static fatigue and on producing higher strength glasses. This is acutely evident in the work of Boguslavski.[89]

No doubt much of the progress in understanding glass

where a_{mn} is the site separation distance, and W_{mn} the ition rate. Decause of the exponential dependence of W_{mn} on distance, due to the overlap integral, σ_{mn} will rapidly decrease with increasing distance. The conductivity of such a network in the low temperature regime has been calculated by Miller and Abrahams.[56] They find that;

$$\sigma_{tot} \propto \exp(-\Delta/kT) \qquad (42)$$

where Δ is the mean spacing of the energy levels at the Fermi level; that is to say, only hops in the vicinity of the Fermi level, with energy differences kT or less, contribute to the conductivity.

Several additional points need to be considered. Because of the exponential dependence of the overlap integral, the conductivity will also be proportional to $\exp(-2\alpha d)$, where d is the mean spacing of the sites in the region of the Fermi level. As the temperature is increased, we eventually enter the high temperature region, and the conductivity should become proportional to:

$$\sigma \propto \exp[-(\Delta + E_b + \frac{\Delta^2}{4E_b})/2kT] \qquad (43)$$

In other words, the activation energy will increase as multiphonon processes become important. Schnakenburg[62] has estimated that the transition between these two regions will come at about half the Debye temperature. This is only true for the case of tight binding, the more general rules stated proviously will always apply. At very low temperature, $\Delta \to 0$, as pointed out by Mott[51], since hops between distant neighbors with very small energy differences will dominate the expression.

Finally, if the semiconductor is not degenerate, that is, if the Fermi level is located in a band gap, then an additional factor of $\exp(-E_g/2kT)$ will enter into the conductivity expression, reflecting the population of carriers at the bottom of the conduction band. The meaning of Δ for a rapidly changing density of localized states with increasing energy, however, is complicated.

The ac conductivity of this model has been considered by Pollack and Geballe[63]. Each hop will contribute to the conductivity through the factor;

$$\frac{\omega^2\tau}{1+\omega^2\tau^2} \quad \text{with} \quad \tau = \frac{1}{W} \tag{44}$$

Jops can occur between distant neighbors with a large τ, and between near neighbors with a small τ. On summing all possible hops it is found that

$$\sigma_{AC}(\omega) \propto \omega^{0.8} \tag{45}$$

The above conductivity, due to non-adiabatic hopping, is additive to the dc conductivity. An additional contribution may come from adiabatic hopping. Since all such hops occur with approximately the same transition rate, if $\omega < W$, we will find a contribution:

$$\sigma_{AC}(\omega) \propto \omega^2 \tag{46}$$

The temperature dependence occurs through τ, and through a site population factor, $\exp(-(E-E_F)/kT)$.

If the glass is heterogeneous, that is, consists of two or more phases, then additional mechanisms are possible. These have been considered recently by Schaake.[64] The conductivity will be given by a complex mixture equation. If the conductivity of the isolated (precipitated) phase is greater than the matrix phase at high temperatures and smaller at low temperatures, then the activation energy of the composite is found to obey a simple mixture formula, when the conductivities are equal, and to go to the activation energy of the matrix phase at high and low temperatures.

Two new ac effects may be introduced by the heterogeneities. Maxwell- Sillars losses occur when the ratio of the dielectric constants and conductivities of the two phases differ. Since the distribution of the electric field at high frequencies, is determined primarily by the conductivities, if the ratios of these constants differ, a relaxation region going from one regime to the other will be observed. If the concentration of the precipitate is very small, the real ac conductivity will obey an equation such as equation (44).

The second mechanism occurs if the precipitated phase has greater conductivity than the matrix phase, and if it is surrounded by a barrier through which the carriers must tunnel. If these conditions are met then the dc path conductivity will be dependent on frequency, since avove a critical frequency, the carriers will be less and less able to complete their tunneling through the barrier. An appropriate expression is

$$\sigma_{dc\ path}(\omega)=\sigma_{composite}+(\sigma_{matrix}-\sigma_{composite})\ (1+\omega^2\tau_o^2)^{-1} \qquad (47)$$

where τ_o is the tunneling time. This equation predicts a decreasing conductivity with increasing frequency when $\omega \gtrsim 1/\tau_o$.

The complex part of the conductivity has not been considered, since for each of these mechanisms it is implied to follow from the Kramers-Kronig relations.

MATERIALS

The problem of applying the previous theory to the experimental results will concern us in this section. There is no agreement as to the conduction mechanisms in any material at the present time; indeed, different experiments appear to give conflicting results. One aspect, common to most amorphous materials, is that the conductivity is insensitive to small amounts of cation impurities (several per cent), in marked contrast with the situation in crystalline semiconductors, where impurities, in the parts per million range, cause orders of magnitude change in the conductivity. This is frequently attributed to the ability of the glass to satisfy the bond requirements of the impurities.[48] It may also be a result of the effects of the impurities being swamped by defects produced by the amorphous state.

The most frequently studied amorphous materials which exhibit semiconducting properties fall into one of three categories: elemental glasses (Ge, Si, As, Te, C, B, Sb, Se)[57], chalcogenide glasses[65] (compounds containing S, Se, and/or Te), and transition metal oxide glasses (V, Ti, and Fe oxides in particular).[66]

Some recently reported conductivity data on several amorphous semiconductors is shown in Figures 16 and 17. The high temperature dc activation energy in the elemental

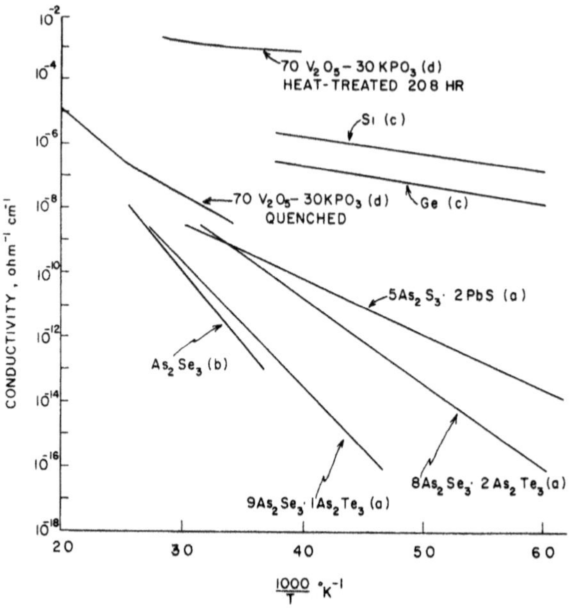

Figure 16. D. C. conductivity of some amorphous semiconductors. A.) A. E. Owen and J. M. Robertson, J. Non-crystalline Solids, 2, 40 (1970). B.) J. E. Lang, Ph. D. Thesis, Rensselaer Polytechnic Institute, July, 1968.

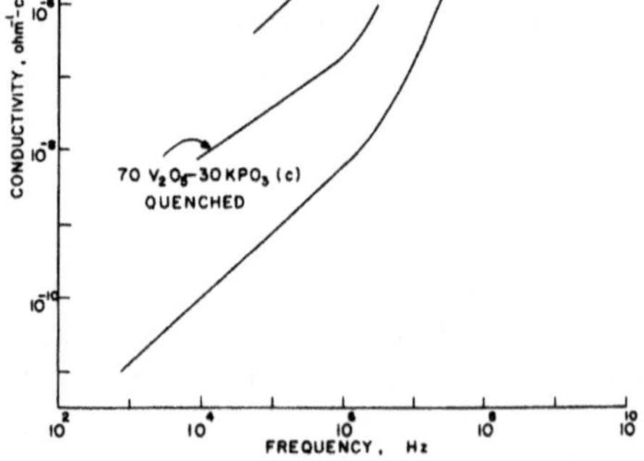

Figure 17. A. C. room temperature conductivity of some amorphous semiconductors. A.) A. E. Owen and J. H. Robertson, J. Non-crystalline Solids, 2, 40 (1970).
B.) H. K. Rockstad, J. Non-crystalline Solids, 2, (1970).
C.) H. F. Schaake and L. L. Hench, J. Non-crystalline Solids, 2, 292 (1970).

and chalcogenide glasses is attributed to be due, mainly, to excitation of carriers across the forbidded gap, as these activation energies correspond reasonably well to half the optically measured gap energy. In amorphous Ge and Si, the measured activation energy drops with decreasing temperature; this drop is generally attributed to a hopping of carriers in localized states near the center of the gap. These states arise from overlap in the localized tails from the top of the valence band and the bottom of the conduction band

The ac conductivity in the elemental and chalcogenide glasses, where it has been fully explored, frequently is found to have two regions; one proportional to ω and the other, at higher frequencies to ω^2 (see data for As_2S_3, Figure 17). The linear region is evidence of non-adiabatic hopping. The ω^2 region is probably due to adiabatic hopping to nearby sites.

In the transition metal oxides there is little agreement as to whether the conduction mechanism in the crystalline material is by normal band conduction or by motion of a polaron.[58] We shall limit our discussion primarily to V_2O_5-P_2O_5 glasses.

Optical measurements indicate a band gap width of 2-2.5 ev. The high temperature dc activation energy is found to be from 0.2 to 0.6 ev, indicating extrinsic conduction, i.e., conduction by carriers from impurities.

Mott argues that the activation energy for conduction in the glass is due to the binding energy of the polaron.[51] This seems questionable in that according to the previous sections, all hops would have this activation energy, and an ac relaxation peak with this temperature dependence should be observed. In the V_2O_5-P_2O_5 glasses the ac conductivity shows linear and quadratic regions, but it is very slightly dependent on temperature at most.[52] This is indicative of a very small binding energy (of the order of a few hundredths ev at most) of the adiabatic hopping of the polaron for most of the localized states. If we go to a heterogeneous structure, where we have a matrix in which the binding energy is high, and controls overall conduction, we would need to introduce considerable disorder in the matrix to accomplish such localization. This in turn would cause an appreciable amount of the activation energy to be due to the term Δ (equation 43).

Since the activation energy also decreases to a few hundredths of an electron-volt at low temperatures, this latter explanation is unsatisfactory. Furthermore, since the ac conduction is only slightly dependent on temperature, the majority of the activation energy cannot be attributed to the thermal population of localized states. Therefore, it appears that the semiconductor is degenerate, and that the activation energy is due to thermal activation over barriers, or due to regions in the glass having band conduction states a few tenths of an electron-volt above the Fermi level. At low temperatures, hopping in the neighborhood of the Fermi level would dominate in these regions. These two cases would be particularly attractive if a large number of states are split off of the conduction band as previously described. Optical absorption evidence at this time is inconclusive on this.[67]

The thermopower results favor a hopping mechanism.[51] The effect of a heterogeneous electronic structure as postulated above on the thermopower has not been considered theoretically, however.

Heat treatment of a quenched 30% KPO_3 - 70% V_2O_5 glass has been found to produce a marked phase separation.[68] The electrical properties of these glasses are found to obey the properties discussed under the conduction mechanisms of heterogeneous glasses.

The low frequency ac conductivity (Figure 17) of the heat treated glass is found to fit a Debye relaxation process with a single well defined relaxation time. This relaxation time is found to have an activation energy equal to the activation energy of dc conduction. This is strong evidence of Maxwell-Wagner losses (see the Dielectrics section of this paper). Presumably, these losses mask the non-adiabatic hopping which is probably also present.

At high temperatures, the dc activation energy has a well defined break. At temperatures above this break, the dc path conductivity is found to decrease with increasing frequency, decreasing to a value at 4 MHz equal to that expected by a continuation of the low temperature activation energy. This is in agreement with equation (47), where the break in the dc activation energy is attributed to the onset of the conductivity in the isolated phase exceeding the conductivity in the matrix phase, while the decreasing conductivity with increasing frequency is due to a tunneling process through a barrier surrounding the isolated phase.

DIELECTRIC PROPERTIES

A. Importance

Dielectric properties are of special importance when glass is used either as a capacitive element in electronic applications or as insulation. The dielectric constant, dielectric loss factor, and dielectric strength usually determine the suitability of a particular glass for such applications. Variation of dielectric properties with frequency, field strength, and other circuit variables influence the performance of glass. Environmental effects, such as temperature, humidity, and radiation also influence the use of glass in dielectric applications. Consequently, it is necessary to examine in this section the dielectric theory of glass in terms of materials response, circuit response, and environmental response.

Glass as an insulating material has some definite advantages over plastics which are major competitors. Glass can also be molded to a variety of shapes, although flexibility still remains a problem in most instances where this feature of insulation is required. However, glass possesses superior electrical properties, is absent from creep or deformation under stresses at room temperature, and resists environmental changes particularly at high temperature where plastics oxidize, gassify, or decompose. Glass also can be used to form gas-tight seals with metals and other ceramic components, and therefore becomes an integral part of the electronic device.

B. Theory

Dielectric properties comprise the non-long-range conducting electrical characteristics of glass. Dielectric responses result from the short range motion of charge carriers under the influence of an applied electric field. The motion of the charges leads to the storage of electrical energy and the capacitance of the dielectric. Consequently, as a first step in examining the dielectric behavior of glass, let us briefly consider the definition of capacitance. Capacitance is a measure of the ability of any two conductors in proximity to store charge when a potential difference is applied across them.

$$C = Q/V = \frac{\text{Coul}}{\text{Volt}} = \text{Farad (F)} \quad \begin{array}{l} Q = \text{Coulombs} \\ V = \text{Volts of potential difference} \end{array} \tag{48}$$

The capacitance of a vacuum capacitor is determined purely by the geometry. It can be shown from elementary electrostatics[69] that the charge density on the plates, Q, is proportional to the area (A) in meters and the electric intensity applied (E = V/d), where d is the distance between the plates (in meters). The proportionality constant is defined as ε_o, the permittivity of free space, and is equal to 8.854×10^{-12} Coulombs^2/m^2 or F/m. Thus, the capacitance of a parallel plate capacitor shown in Figure 18a will be equal to:

$$Q = qA = \pm\varepsilon_o\, EA = \varepsilon_o (V/d) A \tag{49}$$

$$C = Q/V = \frac{\varepsilon_o (V/d) A}{V} = \varepsilon_o\, A/d \quad \begin{array}{l} A = m^2 \\ d = m \\ C = F \end{array} \tag{50}$$

When a material is inserted between the plates the capacitance is increased.

The dielectric constant (k) is defined as the ratio of the capacitance of a condenser with a dielectric between the plates to that with a vacuum between the plates.

$$k = C/C_o = \frac{\varepsilon\, A/d}{\varepsilon_o A/d} = \varepsilon/\varepsilon_o\ ; \tag{51}$$

where ε is the permittivity of the material, also in units of C^2/m^2 or F/m.

Thus the dielectric constant of a material is the ratio of the permittivity of the material to the permittivity of free space. It can be seen in Table II that typical values of the dielectric constant of insulating glasses are between 5 and 10.

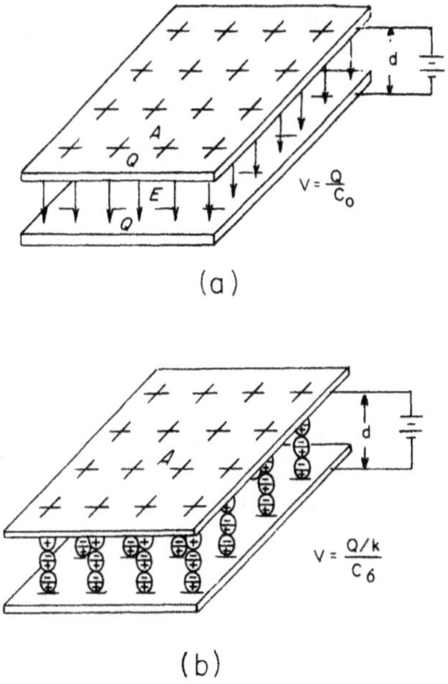

Figure 18. Charge on a parallel plate capacitor. A.) with vacuum as the dielectric. B.) with a dielectric material between plates.

Table II. Dielectric Properties of Insulating Glasses at Room Temperature.

Glass	Frequency (Hz)	Dielectric Constant
SiO_2	10^2-10^{10}	3.78
Corning 7040 (Na_2O, K_2O, SiO_2 B_2O_3)	10^2 10^6	4.84 4.73
Pyrex (Na_2O, SiO_2, B_2O_3)	10^2 10^6	5 02 4 84
Corning 0080 (Na_2O, CaO, SiO_2)	10^2 10^6	8 30 6.90
Corning 1770 (CaO, Al_2O_3, SiO_2)	10^2 10^6	6 25 6.00
9% Na_2O - 91% SiO_2	10^2 10^6	6.4 5.4
20% Na_2O - 80% SiO_2	10^2 10^6	10.8 6.6
12 8% Li_2O - 87.2% SiO_2	10^2 10^6	9.94 4.95
12 8% Na_2O - 87.2% SiO_2	10^2 10^6	8.09 5.66
12 8% K_2O - 87.2% SiO_2	10^2 10^6	7.53 6.09

The charge stored in a glass is on the order of micro-Coulombs or micro-micro-Coulombs. So, the permittivity of glasses is ordinarily in the range of microfarads/m or picofarads/m (micro-microfarad/m).

Equation(51) shows that the presence of the material between the plates of a condenser increases the ability of the plates to store charge. The reason for this is a result of the material containing charged species, which can be displaced in response to the field applied across the material. The displaced charges within the material comprise dipoles with a moment, $\mu = Qd$ (coul-m) where d is the separation distance. The electric dipoles, shown in Figure 18b, are oriented with respect to the applied field. The effect of the orientation is to "tie up" charges on the plates of the condenser and thus neutralize part of the applied field. The charge which is not neutralized by dipoles within the material, called "free charge", equal to Q/k, produces an electric field and voltage towards the outside; $V = \frac{Q/k}{C_o}$. Therefore, a smaller external field is required to maintain the same surface charge because some of the charge is held by the polarization in the dielectric.

There are four primary mechanisms of polarization in glasses. Each mechanism involves a short range motion of charge and contributes to the total polarization of the material. The polarization mechanisms include: electronic polarization (P_e), atomic polarization (P_a), orientational polarization (P_o), and interfacial polarization (P_i). A schematic of the mechanism of operation of each of these major types of polarization as well as two special cases is given in Figure 19.

Electronic polarization is due to the shift of the valence electron cloud of the ions within the material, with respect to the positive nucleus. This mechanism of polarization occurs at very high frequencies (10^{15} Hz*) which are in the ultraviolet optical range. The mechanism of polarization gives rise to a resonance absorption peak in the optical range, as shown in Figure 20. The index of refraction of the glass will depend on the electronic polarization occurring within the glass.

---

*Hz = Hertz = cps = sec^{1}

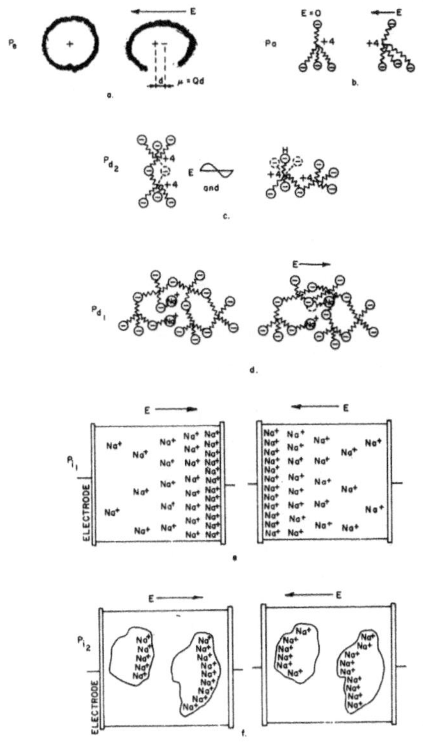

Figure 19. Schematic of polarization mechanisms in a glass. A.) electronic. B.) atomic or ionic. C.) high frequency oscillatory dipoles. D.) low frequency cation dipole. E.) interfacial space charge polarization at electrodes. F.) interfacial polarization at heterogeneities.

At frequencies in the infrared range (10^{12} - 10^{13} Hz), Figure 20, atomic or ionic polarization is occurring. Atomic polarization is the displacement of positive and negative ions in a material with respect to each other, such as is shown in Figure 19b for a SiO_4^{4-}-tetrahedron. A resonance absorption occurs at a frequency characteristic of the bond strength between the ions. If there are several types of ions in a glass or a distribution in bond strength, the infrared absorption will be quite broad.

In the sub-infrared range of frequencies, orientational polarization contributes to the dielectric properties of glass. Orientational polarization, also referred to in some texts as dipolar polarization, involves the perturbation

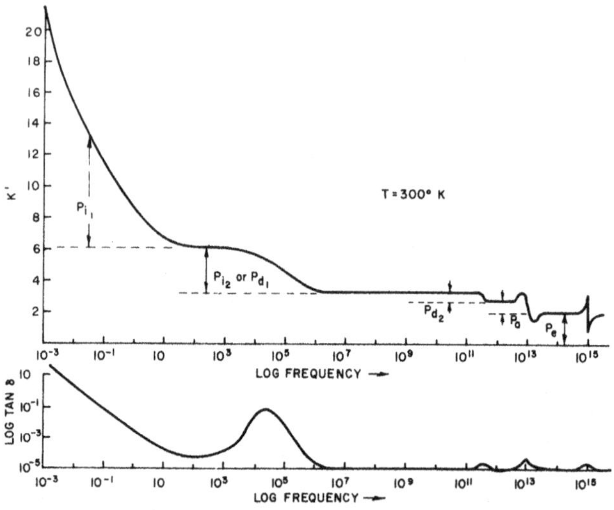

Figure 20. Frequency dependence of the polarization mechanisms in glass. Upper: contribution to the charging constant (representative values of k'). Lower: contribution to the loss angle (representative values of tan δ).

of the thermal motion of ionic or molecular dipoles, producing a net dipolar orientation in the direction of the applied field. Mechanisms of orientational polarization can be generally divided into two categories: molecules containing a permanent dipole moment may be rotated against an elastic restoring force about an equilibrium position. This effect is especially important for a variety of liquids and gases and polar solids, such as ice and many plastics. Two models of this type of polarization in glass are shown in Figure 19c. The left hand side of the figure depicts the oscillation of a Si-O-Si bond about an equilibrium position under a sinusoidal ac field. Since such a bond will possess a dipole moment when it is asymmetrical, the oscillation produces an orientational polarization. The frequency of relaxation of such a mechanism is very high, 10^{11} Hz, at room temperature. A similar mode of orientational polarization due to the oscillation of an OH^- group about an equilibrium position is shown on the right hand side of Figure 19c. The dipole moment is again due to the asymmetry of the Si-OH configurations in the random glass network. Oscillation of these moments also occurs in the range of 10^{11} - 10^{12} Hz.

In several sources,[1,70] these mechanisms are termed "Stevels deformation polarization."

The second mechanism of orientational polarization is an especially important contribution to the room temperature dielectric behavior of glass. It involves the rotation of dipoles between two equivalent equilibrium positions. It is the spontaneous alignment of dipoles in one of the equilibrium positions which gives rise to the non-linear polarization behavior of the ferroelectric materials that is responsible for dielectric constant values of 10^4 or more in such materials. In linear glass dielectrics, which we are concerned with in this section, orientational polarization occurs largely as a result of motion of charged ions between the interstitial positions within the ionic structure of a glass. Figure 19b shows a schematic of orientational polarization involving the oscillation of a sodium ion between two equivalent positions. Such oscillations occur continuously, the applied field makes the jumps in a direction parallel to the field more probable. Since an appreciable distance is involved in such an ionic transition, the polarization occurs at a frequency range of 10^3 - 10^6 Hz, at room temperature. Because this mechanism involves the same mobile cations that contribute to the dc conductivity, several authors refer to it as "migration losses".[1,70]

The last polarization mechanism, interfacial or space charge polarization, occurs when mobile charge carriers are impeded by a physical barrier that inhibits charge migration. The charges pile up at the barrier, producing a localized polarization of the material. When an ac field is of sufficiently low frequency, less than 10^{-3} Hz, a net oscillation of charge can be produced between barriers as far apart as 1 cm (Figure 19e), producing a very large capacitance and dielectric constant (Figure 20). If the barriers are an internal structural feature (Figure 19f), or the density of charges contributing to the interfacial polarization is sufficiently large, the frequency range of sensitivity for interfacial polarization may extend into the kilocycle range (Figure 20). In such a case it may be impossible to distinguish the frequency response of an orientational mechanism, such as P_{o1}, and an interfacial mechanism such as P_{i_2}.

Of the four polarization mechanisms contributing to the dielectric properties of materials, this paper will discuss only orientational and interfacial polarization since they influence the circuit characteristics of glass. Infrared and electronic polarization in glass are more appropriately discussed in the optical properties papers in this symposium.

Now that we have a qualitative understanding of the physical basis of polarization, let us consider in detail (1) the effects of polarization on the circuit behavior of glass, and (2) a quantitative description of orientational and interfacial polarization models.

CIRCUIT DESCRIPTION OF A GLASS DIELECTRIC

If a sinusoidal potential $V = V_o \exp[i\omega t]$ is applied to the dielectric, the charge must vary with time, as shown in equation 52, which constitutes a charging current

$$Q = CV;\ \text{so}\ I_c = dQ/dt = C\,dV/dt = i\omega CV = \omega CV_o \exp[i(\omega t+\pi/2)] \quad (52)$$

The charging current in an ideal dielectric thus leads the applied voltage by $\pi/2$ radians (90°). This relationship is seen vectorially in Figure 21.

In addition to the charging current, associated with storage of electric charge by the dipoles, a loss current must also be considered for real dielectrics. The loss current arises from two sources: (1) The long range migration of charges, dc ohmic conduction, and (2) the dissipation of energy associated with rotation or oscillation of dipoles. The latter contribution to the dielectric losses is a consequence of the charged particles having a specific mass and, therefore, an inertial resistance to being moved. Electrical energy from the field is lost in the overcoming of this inertia during polarization. The ac conduction from the inertial resistance and the dc conduction both are in phase with the applied voltage and thus a loss current can be written as:

$$I_\ell = (\sigma_{DC} + \sigma_{AC})V \quad (53)$$

where σ is the conductivity in units of mho/cm or $(\text{ohm-cm})^{-1}$

The total current for a real material is thus:

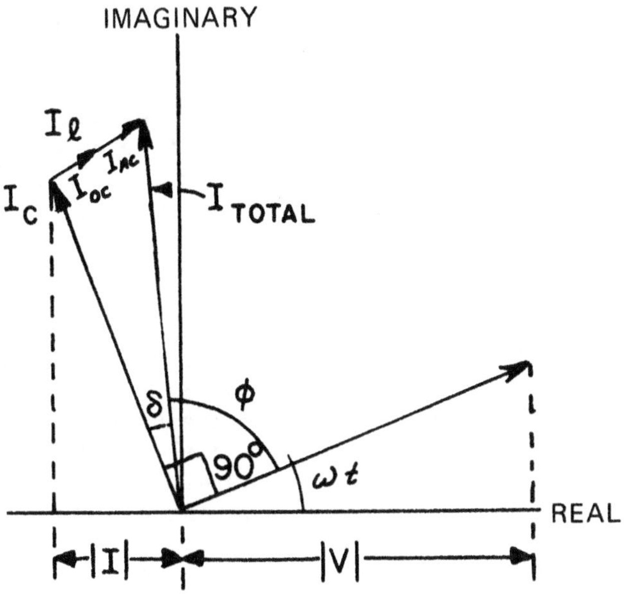

Figure 21. Vector diagram of charging, loss and total currents in a glass dielectric.

$$I_T = I_C + I_\ell = (i\omega C+\sigma_{DC}+\sigma_{AC})V \qquad (54)$$

As shown in Figure 21 the total current in a real dielectric is a complex quantity which leads the voltage by an angle $(90^\circ - \delta)$ where δ is called the loss angle.

An alternative way of expressing the concept of a real dielectric possessing both charging and loss processes is to use a complex permittivity to describe the material:

$$\varepsilon^* = \varepsilon' - i\varepsilon'' \text{ and } k^* = \varepsilon^*/\varepsilon_o = k' - ik'' \qquad (55)$$

Thus, the total current in the dielectric can now be expressed in terms of the single material parameter, k^* since:

$$C = k^*C_o \text{ so } Q = CV = k^*C_oV \qquad (56)$$

and $I=dQ/dt=C\ dV/dt=k^*C_oi\omega V=(k' -ik'')C_oi\omega V_o \exp i\omega t$

and thus: $$I_T = i\omega k'\ C_oV + \omega k''C_oV \qquad (57)$$

The first term on the right hand side of eq. (57) describes charge storage in the dielectric and k' is thus called the charging constant or often just dielectric constant. The second term represents the dielectric losses and ε" and k" are referred to as the dielectric loss factor and relative loss factor, respectively. The loss tangent or loss angle, or dissipation factor

$$\tan \delta = \varepsilon''/\varepsilon' = k''/k'$$

represents the relative expenditure of energy to obtain a given amount of charge storage. It is the "interest rate" so to speak. The product, k' tan δ, is sometimes termed the total loss factor and provides the primary criterion for evaluating the usefulness of a dielectric as an insulator. To minimize k", the losses in the insulator, it is desirable to have a small dielectric constant and most important a very small loss angle. The inverse of the loss tangent, Q = 1/tan δ, is used as a figure of merit in high frequency insulation applications. In dielectric heating the critical parameters are the dielectric constant, k', and the dielectric conductivity, $\delta_T = \omega k''$.

RELATION OF DIELECTRIC CONSTANT TO POLARIZATION

In order to obtain a quantitative understanding of the dielectric properties of glasses it is necessary to establish a relationship between the complex dielectric constant, k*, and the polarization in the glass. This can be done by considering the total electric displacement field, D, in the glass. As shown in Figure 18b, D will be the sum of the electric field established if there was not a dielectric in the condenser plus the polarization field within the material, i.e.,

$$D = \varepsilon_0 E + P = \varepsilon^* E \tag{58}$$

Thus the total electric displacement in the material is related to the external field E by the complex permittivity of the material, ε*. Consequently, the polarization can be expressed as:

$$P = E(\varepsilon^* - \varepsilon_0) \tag{59}$$

since: $k^* = \varepsilon^*/\varepsilon_0$ then $P = E(\varepsilon_0 k^* - \varepsilon_0) = \varepsilon_0(k^*-1)$

and rearranging: $k^*-1 = P/\varepsilon_0 E$ or $k^* = 1+P/\varepsilon_0 E$ (60)

and defining: $P/\varepsilon_0 E \equiv \chi$, the electric (61)
susceptibility we obtain: $k^* = 1 + \chi$ (62)

Equation (60) provides the relationship desired between the dielectric constant and the total polarization in the glass. However, it would be even more useful to have a relationship between k* and the fundamental polarizability of charge mechanisms contributing to the total polarization, P.

Such a relationship can be obtained through the following arguments. P is equal to the total dipole moment induced in the material by the electric field. Thus:

$P = N_i \bar{\mu}_i$ where N_i is the number of dipoles of type i and μ_i is their average dipole moment. (63)

Now, the average dipole moment of the charged particles is proportional to the local electric field (E') which acts on the particle:

$\bar{\mu}_i = \alpha_i E'$ where α_i is the polarizability or average dipole moment per unit local field strength and has units of $C^2 sec^2/kg$. (64)

Thus the total polarization is:

$$P = N_i \alpha_i E' \quad (65)$$

For gases with little molecular interaction the locally acting field E' is the same as the external field, E. However, for solids, including glasses, polarization of the surrounding medium affects the magnitude of the local field. Mosotti was the first to derive the local field contribution by the integration of the normal component of the polarization vector over the surface of a spherical cavity in the material.[71] The result obtained is:

$$E' = E_{(applied)} + P/3\varepsilon_0 \quad (66)$$

so since $N_i \alpha_i = P/E'$ from equation (65) then:

$$N_i\alpha_i = \frac{P}{E_a + P/3\varepsilon_o} \qquad (67)$$

and by substituting equations (61) and (62) in that order we obtain:

$$N_i\alpha_i = \frac{1}{\dfrac{1}{(k^*-1)\varepsilon_o} + \dfrac{1}{3\varepsilon_o}} \qquad (68)$$

Further rearrangement yields:

$$N_i\alpha_i = \frac{3\varepsilon_o^{\,2}\ (k^*-1)}{\varepsilon_o\ (k^*+2)} \qquad (69)$$

so:

$$\frac{k^*-1}{k^*+2} = \frac{1}{3\varepsilon_o}\ N_i\alpha_i \qquad (70)$$

This result is the famous Clausius-Mosotti equation which describes the relation between the complex dielectric constant of a material and the number of polarizable species N_i, and the polarizability of the species α_i. As we saw earlier, there are four major classes of polarizable species in glass, α_e, α_a, α_d, and α_i. Thus:

$$\frac{k^*-1}{k^*+2} = \frac{1}{3\varepsilon_o}\ [N_e\alpha_e + N_a\alpha_a + N_d\alpha_d + N_i\alpha_i] \qquad (71)$$

ORIENTATIONAL POLARIZATION THEORY

As was mentioned earlier, the physical theory of resonance absorption which explains the contribution of α_e, and α_a to k* in the optical and infrared region, will not be considered in this discussion.

It is important to discuss orientational polarization, however, since it markedly influences the insulation and capacitive applications of glass and is strongly affected by the composition, structure, and thermal history of glass. For several other theoretical treatments the reader is referred to Frohlich[72] and Daniels.[73]

Let's consider a bistable dipole model such as shown in Figure 19d. A change in the coordination of the Na^+ ion from the position at left to that at right involves a change in energy as represented in Figure 1. There is a random oscillation of Na^+ ions between these positions at any temperature above 0^oK with the probability of a jump, ω, being exponentially related to the temperature and the energy barrier W:

$$\text{No Field} \quad \omega = A\ e^{-W/kT}; \quad T = {}^ok \qquad k = 1.37\times10^{-16} erg/{}^oK \tag{72}$$

However, with an electric field applied the potential energy of the two sites will become unequal by an amount;

$$\phi_1 - \phi_2 = e(bE) = ebE \cos\theta \tag{73}$$

where b is distance separating the potential wells and θ is the angle between the field vector and the jump vector. Thus this model is equivalent to a turn of 180° of a dipole of the moment:

$$\mu_d = 1/2\ Zeb \quad \text{where } Z = \text{valence of the ion} \tag{74}$$

Let us assume that there are N bistable dipoles per unit volume with N being small enough that there will be no dipolar interaction. Also assume that $\cos\theta = 1$ for all dipoles, $\omega_1 = \omega_2$ without a field applied, and $W >> kT$.

The probability of jumps from 1 to 2 can thus be written as:

$$\omega_{12} = A \exp -\left[\frac{W+\mu E}{kT}\right] \tag{75}$$

$$\text{or} \quad \omega_{12} = A \exp - [W/kT] \exp - [\mu E/kT] \tag{76}$$

Since $\mu \approx 10^{-18}$ esu, then as long as E is less than 10^5 esu where dielectric breakdown begins to occur, $\mu E/kT$ will be much less than unity. So if the last term in eq. (36) is expanded as

$$e^{-x} = 1 - x + x^2/2\,! + \ldots, \text{ we obtain:}$$

$$e^{-\mu E/kT} = 1 - \mu E/kT \tag{77}$$

which on substitution into eq. (76) yields:

$$\omega_{12} = A \exp - [W/kT]\ (1-E/kT) = \omega(1-\mu E/kT) \tag{78}$$

The probability of a jump of the ion in the opposite direction will be:

$$\omega_{21} = A \exp-[W-\mu E/kT] \quad (79)$$

which yields the following expression from the above argument eq. (77):

$$\omega_{21} = \omega(1+\mu E/kT) \quad (80)$$

Under equilibrium conditions the average population of charges in wells 1 and 2 will not change with time. Consequently, those going into well 1 must come out of well 2 and vice versa.
Thus:

$$N_1\omega_{12} = N_2\omega_{21} \quad (81)$$

Thus:

$$N_1\omega(1-\mu E/kT) = N_2\omega(1+\mu E/kT) \quad (82)$$

which upon rearranging yields:

$$N_1 - N_2 = (N_1+N_2)\ \mu E/kT \quad (83)$$

And since the number of wells occupied per unit volume is constant: $N_1 + N_2 = N$ (84)
the polarization per unit volume is:

$$P = (N_1-N_2) \quad (85)$$

Therefore: $P = \dfrac{N\mu^2 E}{kT}$ (86)

and since $\mu = 1/2Zeb$, $P = \dfrac{Z^2Ne^2}{4}\left(\dfrac{b^2E}{kT}\right)$ (87)

Consequently, the dielectric constant of a glass containing neighboring positions of such bistable dipoles depends on the number of dipoles, oscillation length, and temperature as:

$$k=1+\frac{P}{\varepsilon_0 E} = 1 + \frac{Z^2Ne^2b^2}{4\ kT} \quad (88)$$

TIME DEPENDENCE OF ORIENTATIONAL POLARIZATION

In order to describe the ac behavior of the dielectric properties of glass it is necessary to discuss the time dependent response of the above bistable dipole model. The change in number of dipoles in site 1 is equal to the outflow to site 2 minus the inflow from site 2; thus:

$$\frac{dN_1}{dt} = -N_1\omega_{12} + N_2\omega_{21} \tag{89}$$

and since $N_1+N_2 = N$(constant)

$$\frac{dN_2}{dt} = dN_1/dt \tag{90}$$

and thus:

$$\frac{d(N_1-N_2)}{dt} = 2\ dN/dt \tag{91}$$

By using expansions for $\mu E/kT \ll 1$, as in eq. (77), we can obtain for the rate of change of occupancy:

$$1/2\left(\frac{d(N_1-N_2)}{dt}\right) = N_1\omega_{12} + N_2\omega_{21} = -N_1\omega(1-\mu E/kT) + N_2\omega(1+\mu E/kT) \tag{92}$$

By rearranging this result is obtained:

$$1/2\left(\frac{d(N_1-N_2)}{dt}\right) = -\omega(N_1-N_2) + \omega(N_1+N_2)\ \mu E/kT \tag{93}$$

which is a differential equation for N_1-N_2. Since $P = (N_1-N_2)$, eq. (93), is also a differential equation describing the time dependence of P:

$$1/2\left(\frac{dP}{dt}\right)\frac{1}{\mu} = -\omega/\mu(P) + \omega N\ \mu E/kT \tag{94}$$

which upon rearranging yields:

$$1/2\omega\left(\frac{dP}{dt}\right) + P = N\mu^2E/kT \tag{95}$$

Equation (95) is a relaxation equation with a relaxation time $\tau = 1/2\omega$ characteristic of the rate of relaxation.

$$\text{So,} \quad \tau \frac{dP_D}{dt} + P_D = N\alpha_D E = P_S \tag{96}$$

where α_D is the dipolar polarizability, μ^2/kT, and the subscript D has been assigned to designate that it is the dipolar polarization that is time dependent. In this sense P_S is the static or zero frequency value of the polarization. Equation (96) shows that when E changes with time, P, at a given moment, will generally differ from P_S. As the oscillation time increases there will be a trend towards an equilibrium value of P_S at a rate of change of dP_D/dt.

Integration of eq. (96) will lead to a solution of the time or frequency dependence of the dipolar contribution to the dielectric constant. However, several steps of simplification are desirable. For example, since α_e and α_a occur very rapidly; i.e., when $\tau > 10^{-11}$ sec, a high frequency polarization contribution can be defined as;

$$P_\infty = P_e + P_a \tag{97}$$

and at frequencies of 10^{11} sec^{-1} or greater the dielectric constant will depend only on P_∞ as:

$$k_\infty \simeq 1 + \frac{P_\infty}{\varepsilon_o E} \quad n^2. \quad \text{(n is the index of refraction of the glass)} \tag{98}$$

Consequently, at frequencies of 10^2 sec^{-1} to 10^{11} sec^{-1}, the low frequency or static value of the dielectric constant (k_S) can be expressed as:

$$k_S - 1 = \frac{P_D + P_\infty}{\varepsilon_o E} \tag{99}$$

By inserting equation (98) into (99) and simplifying, the dipolar polarization can be written in terms of the relaxed static dielectric constant, k_S, and the unrelaxed high frequency constant k_∞.

$$k_S - 1 = \frac{P_D + (k_\infty - 1)\,\varepsilon_o E}{\varepsilon_o E} \tag{100}$$

$$(k_S - k_\infty)\,\varepsilon_o E = P_D = P_S - P_\infty \tag{101}$$

Upon inserting eq. (101) into eq. (96) one obtains the differential equation describing the change of the dielectric constant from k_S to k_∞ resulting from the dependence of the dipolar polarization:

$$\tau \frac{dP_D}{dt} + P_D = P_S - P_\infty = (k_S - k_\infty)\varepsilon_o E \qquad (102)$$

The general solution of the above equation is most easily obtained by using complex variables; i.e., let $E^* = E = E_o \cos \omega t$. If we assume that relevant physical laws hold for E^* as for E we can express the time dependent orientation polarization as:

$$P^*_D = \varepsilon_o k^* E^* \qquad (103)$$

Substitution of eq. (99) into eq. (97) and solving for P^*_D yields:

$$P^*_D = k_e^{-t/\tau} + \frac{(k_S - k_\infty)\varepsilon_o}{1+i\omega\tau} E_o \exp(i\omega t) \qquad (104)$$

The first term on the right hand side of the equation describes the time dependent decay of the dc charge on the capacitor. The second term describes the ac behavior of the polarization when a field of magnitude E_o and frequency ω is applied.

Since the electronic and atomic polarization are frequency independent in the range of interest they can be separated from the time dependent expression of k^* by the following definition:

$$k^* - k_\infty = \frac{P^*_D}{\varepsilon_o E^*} \qquad (105)$$

Now by substitution of eq. (101) into the above definition:

$$k^* - k_\infty = \frac{(k_S - k_\infty)\varepsilon_o E^*}{(1+i\omega\tau)\varepsilon_o E^*} \qquad (106)$$

or $$k^* = k_\infty + \frac{k_S - k_\infty}{1+i\omega\tau}$$

And, recalling that:

$$k^* = k' - ik''$$

we can separate eq. (106) into real and imaginary parts as:

$$k' = k_\infty + \frac{k_S - k_\infty}{1+\omega^2\tau^2} \tag{107}$$

and

$$k'' = (k_S - k_\infty)\left[\frac{\omega\tau}{1+\omega^2\tau^2}\right] \tag{108}$$

and

$$\tan\delta = \frac{k''}{k'} = \frac{(k_S - k_\infty)\,\omega\tau}{k_S + k_\infty\,\omega^2\tau^2} \tag{109}$$

Eqns.(107-109) are the desired frequency dependent relationships of the charging and loss constants and the loss tangent. These equations are known as the Debye equations and they yield the graphical relationships shown in Figure 22.

At low frequencies the charging constant is frequency independent at a value which reflects the contribution of $P_e+P_a+P_d$. As the applied frequency $\omega = 2\pi f$ approaches a value equal to $1/\tau$, k' passes through an inflection and

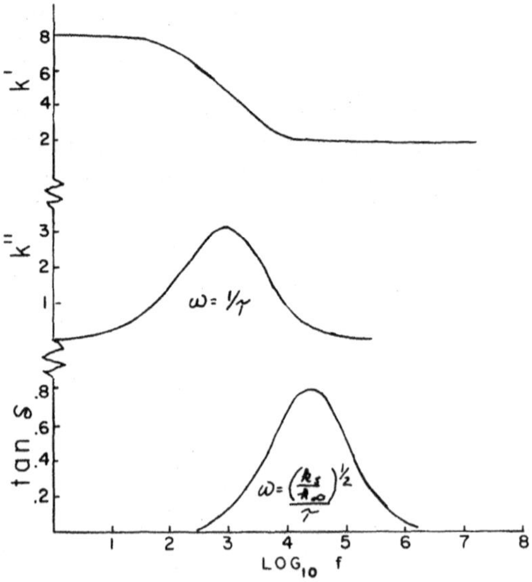

Figure 22. Frequency variation of dielectric parameters due to a Debye relaxation process.

at a higher frequency becomes asymptotic to k_∞ which depends only on P_e+P_a. When $\omega = 1/\tau$, the oscillating charges are coupled directly with the oscillating field and absorb a maximum in electrical energy and k" goes through a maxima as a result. The magnitude of the loss peak will be $(k_S-k_\infty)/2$ and thus is directly dependent on the number of oscillating charges and their distance of motion as given in eq. (87). Since P_D is inversely proportional to temperature, eq. (87), the magnitude of the peak height should vary with the reciprocal of the absolute temperature of the glass.

Figure 22 and eq. (109) also show that the loss tangent goes through a maximum when plotted as a function of log frequency. However, the loss tangent peak is displaced to a higher frequency by the quantity $\sqrt{k_S/k_\infty}$. Consequently, it is often possible to observe a maxima in tan δ where the k" maxima is at a frequency too low to be measured.

TEMPERATURE DEPENDENCE OF ORIENTATIONAL POLARIZATION

The development of the bistable dipole model also provides a basis for understanding the temperature dependence of the dielectric properties of glass. Equation (77) expressed the probability of a transition of the oscillating charge in terms of Boltzmann statistics as $\omega = A \exp[-Q/kT]$. It was also shown in equation (96) that ω is directly related to the relaxation time of the orientational process, $\tau = 1/2\,\omega$. Consequently, the relaxation time is exponentially temperature dependent:

$$\tau = \frac{1}{2A} \exp[+Q/RT] \qquad (110)$$

where Q is an activation energy for the relaxation process in units of kilocalories per mole. It is also possible to define an intrinsic relaxation time for the relaxation process as:

$$1/2\ A \equiv \tau_o \qquad (111)$$

τ_o will be proportional to a term $\left(\frac{\alpha kT}{h}\right)^{-1}$

which includes a frequency factor, $\frac{kT}{h}$, describing the average frequency of oscillation of an harmonic oscillator in an ensemble of temperature, T. τ_o also includes the term α which is an accommodation coefficient that is related to the irreversibility of the oscillations.

By combining equation (110) and the Debye equations it is possible to describe the temperature variation in the location of the tan δ loss peak and the ε'' loss peak. Since tan δ_{max} occurs when

$$\omega = \sqrt{\frac{k_S}{k_\infty}}/\tau \text{ and } \omega = 2\pi f$$

then $$f_{max}(\tan\delta) = \frac{\sqrt{k_S/k_\infty}}{2\pi\tau_o \exp[+Q/RT]} \quad (112)$$

or $$f_m = f_o\sqrt{k_s/k_\infty}\exp[-Q/RT] \quad (113)$$

and similarly: $$f_m(\varepsilon'') = \frac{1}{2\pi\tau}$$

so: $$f_m(\varepsilon'') = f_o \exp[-Q/RT] \quad (114)$$

As shown in Figures 23 and 24, tan δ_{max} will vary with temperature in the manner predicted by equation (113). The slope of the log f_m versus $1/T\ {}^oK^{-1}$ plot in Figure 24 will be equal to $-Q/R$ as given in equation (113). Thus, an experimental measurement of the dielectric loss peaks as a function of temperature enables calculations of relaxation activation energies to be made.

There are several other important parameters which can also be obtained from dielectric data. τ_o can be calculated from data such as shown in Figure 24. Values of 10^{-13} sec to 10^{-14} sec are indicative that the relaxation process is ionic in origin. However, very low values of τ_o are suggestive that the relaxation process is electronic space charge polarization.

Since the quantity, $k_S - k_\infty$, is directly related to the dipolar polarization, P_D, in the glass, the magnitude of tan δ_{max} or ε''_{max} can be used to calculate the number of oscillating charges and their distance of oscillation by employing equation (87). The inverse temperature dependence of P in equation (87) also provides somewhat of a test for the presence of dipolar polarization in the glass. When the measuring temperature increases, the polarizability of the dipoles decreases and the loss peak height should decrease. However, the rapid increase in σ_{DC} with increasing

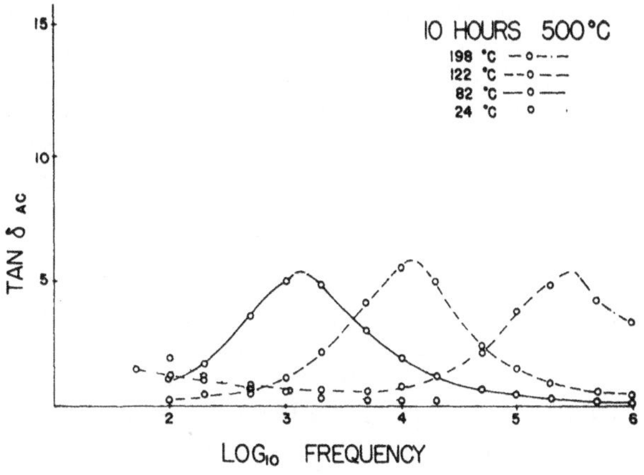

Figure 23. Shift of the dielectric loss peak due to increasing temperature.[18]

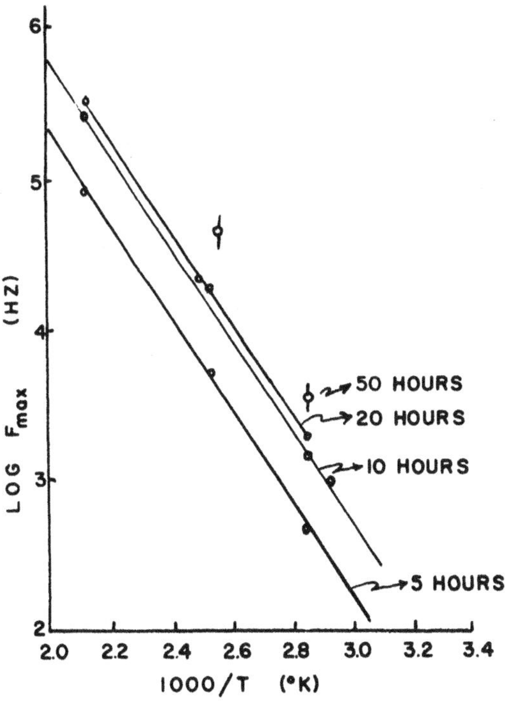

Figure 24. Logarithmic temperature dependence of the frequency location at the dielectric loss peak.[18]

temperature in glass usually results in the ac loss peak being swamped in magnitude by dc losses at temperatures of one-fourth the softening point. Consequently, analysis of a decrease in the height of tan δ peaks with temperature is often not possible. If temperature independence can be observed it is suggestive that an interfacial polarization may be operating.

Dielectric loss data in glasses can also be employed as a measure of the distribution of ion energies or configurations in the glass structure. Often, an appreciable broadening in the Debye peaks is observed. The broadening is due to a distribution in relaxation times which may be a consequence of a distribution of Q or a distribution in the oscillation distances between equivalent sites, b. There are a variety of techniques in the literature[73] which can be used to analyze distribution of relaxation times. Often a log normal distribution is a satisfactory description:

Let

$$\frac{\varepsilon''}{\varepsilon_s'-\varepsilon_\infty'} = \int_0^\infty \frac{\omega\tau}{1+\omega^2\tau^2}\, g(\tau)\, d\ln\tau \qquad (115)$$

where $g(\tau)$ is a weighting function which gives the relative strength of a given relaxation time τ_i. Assuming a log-normal distribution yields:

$$\frac{\varepsilon}{\varepsilon_s'-\varepsilon_\infty'} = \int_0^\infty \frac{\omega\tau}{1+\omega^2\tau^2}\; \frac{1}{(2\pi)^{1/2}s}\, \exp -1/2\, \frac{\ln\tau/\tau_0}{s}\, d\ln\tau/\tau_0 \qquad (116)$$

where s is the standard deviation of the distribution. This formidable relation comes out to be an errorfunction type of solution for s and has been nicely published in tabular form with instructions as to application by Nowick and Berry.[74]

Analyses of loss peak distribution functions have led to a number of authors concluding that there is a distribution in cation activation energies in glass.[1,75] Several other authors[1,76] have reached the conclusion that the broadening response is due to variations in ion configurations. Thus, if Q is represented as $Q = \Delta F = \Delta H - T\Delta S$, there will be localized variations in the entropy within the glass due to the randomness of the structure.

INTERFACIAL POLARIZATION

Although the atomistic dipolar model may provide a reasonable interpretation of the dielectric behavior of many glasses it does not usually represent a unique interpretation. The frequency response and frequency range of interfacial polarization mechanisms is equivalent to that of the Debye type of model in many cases. Consequently, the possibility of an interfacial interpretation of the dielectric behavior of a glass must always be considered as an alternative to the Debye model. Independent evidence that interfaces are or are not involved in the behavior is required to achieve an unambiguous interpretation of glass behavior.

There are two principal categories of interfacial polarization and both can be important for glasses. The first involves a variation of electrode polarization. The second consists of the dielectric response due to heterogeneities in the glass.

If a sharp compositional gradient is present at one of the electrode-glass interfaces due to electrode polarization, heat treatment, or atmospheric attack, a two layer capacitor results.* As shown in Figure 25, such a two layer capacitor involves a region described by a permittivity, ε_1, a conductivity, σ_1, and a thickness t_1, which is in series with the matrix of ε_2, σ_2, and t_2. Analysis of this model and numerous other series and parallel R-C circuit models of materials have been reported in the literature.[73,77] The frequency dependence of the dielectric parameters of the glass model in Figure 25 are:

$$k'(\omega) = \frac{k_s + k_\infty \omega^2\tau^2}{1 + \omega^2\tau^2} \tag{117}$$

and

$$\sigma(\omega) = \frac{\sigma_s + \sigma_\infty \omega^2\tau^2}{1 + \omega^2\tau^2} \tag{118}$$

and

$$\tan\delta = \frac{\sigma_\infty}{\omega k_\infty} \; \frac{\dfrac{k_\infty \sigma_s}{\kappa_s \sigma_\infty} + \omega^2\tau_\delta^2}{1 + \omega^2\tau_\delta^2} \tag{119}$$

---

*Water films and oxidized electrode layers produce especially strong contributions to interfacial polarization.

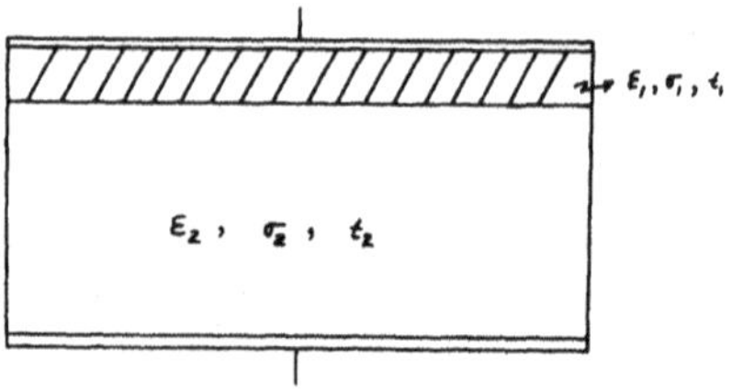

Figure 25. Schematic two-layer model of a glass dielectric.

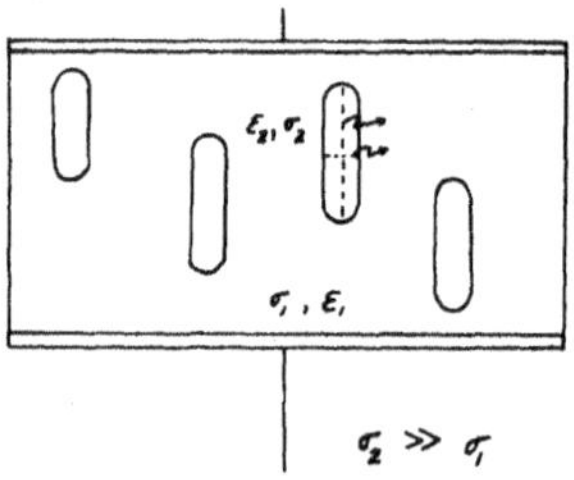

Figure 26. Schematic of a glass dielectric containing high conductivity ellipsoidal heterogeneities of axial ratio a/b.[80]

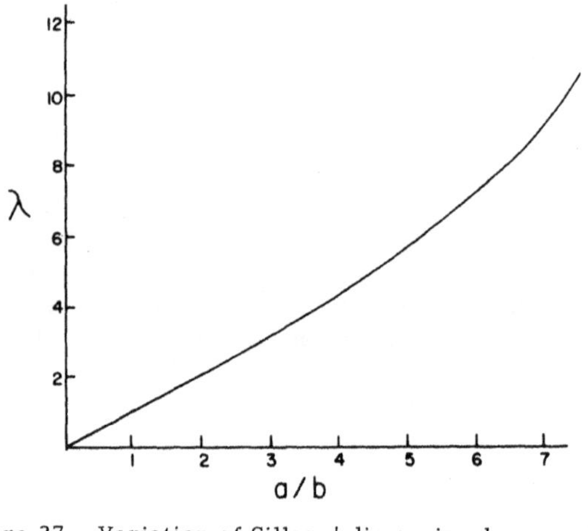

Figure 27. Variation of Sillars' dimensional parameter λ, with axial ratio of ellipsoids, a/b.[80]

Relaxation times and frequency parameters are defined as:

$$\tau = \varepsilon_0 \frac{k_s - k_\infty}{\sigma_\infty - \sigma_s} = \varepsilon_0 \frac{k_1 d_2 + k_2 d_1}{\sigma_1 d_2 + \sigma_2 d_1} \tag{120}$$

$$k_s = \left(\frac{(d_1 + d_2)}{\left(\frac{d_1}{\sigma_1} + \frac{d_2}{\sigma_2}\right)^2} \right) \left(\frac{d_1 k_1}{\sigma_1^2} \right) + \frac{d_2 k_2}{\sigma_2^2} \tag{121}$$

$$\text{and} \quad k_\infty = \frac{d_1 + d_2}{\left(\dfrac{d_1}{k_1} + \dfrac{d_2}{k_2}\right)}; \tag{122}$$

$$\sigma_s = \frac{d_1 + d_2}{\left(\dfrac{d_1}{\sigma_1} + \dfrac{d_2}{\sigma_2}\right)} \tag{123}$$

$$\sigma_\infty = \frac{d_1 + d_2}{\left(\dfrac{d_1}{k_1} + \dfrac{d_2}{k_2}\right)^2}\left(\frac{d_1\sigma_1}{k_1^2} + \frac{d_2\sigma_2}{k_2^2}\right) \tag{124}$$

The frequency response of equations (117), (118), and (119) is equivalent to that shown in Figure 22 for the Debye equations. k' goes through a relaxation decay from k_s at low frequencies to k_∞ at high frequencies described by the relaxation time τ. k" and tan δ likewise go through maxima as a function of frequency. Rearranging equation (119) to obtain the location of the loss angle maximum on a tan δ-log frequency plot yields:

$$f_{m(\tan\delta)} = \frac{\sigma_1 t_2 + \sigma_2 t_1}{2\pi[k_1 t_2 + k_2 t_1]}\left(\frac{k_s}{k_\infty}\right)^{1/2} \tag{125}$$

Equation (125) indicates that a glass dielectric may be responding as a two layer capacitor due to a surface film, for example, and exhibit a loss peak in the same frequency range as Debye peaks if the ratios of σ_1/σ_2 and t_1/t_2 are appropriate. Since σ_1 and σ_2 will also be exponentially temperature dependent, the loss peaks of the two layer glass capacitor will also shift to higher frequencies with increasing temperature. However, the interpretation of such a shift is not straightforward since Q_1 and Q_2 for the two conducting layers may be different. Therefore, the ratio of σ_1/σ_2 may change with temperature as well, which would produce a frequency shift proportional to $\Delta(\sigma_1/\sigma_2)$.

Heterogeneities are another source of polarization in glass. Wagner[78] was the first to extend Maxwell's[79] multilayer dielectric analysis to include dispersed spherical phases. Sillars[80] later generalized heterogeneous dielectrics to include dispersed ellipsoidal phases. An important consequence of the final M-W-S theory is that the dielectric parameters of a glass may be interpreted in terms of a distribution of ellipsoidal particles of a given volume fraction and size, instead of atomistic dipoles as Debye theory requires. Since the frequency location of loss peaks and the frequency and temperature dependence of k' and k" can be equivalent in both the M-W-S and Debye theories independent confirmation of a model for the dielectric behavior of a particular glass series must be obtained.

A schematic of the M-W-S heterogeneous dielectric model is shown in Figure 26. The diagram shows isolated ellipsoidal particles of conductivity σ_2 in a matrix of conductivity σ_1. When σ_2 is greater than σ_1, the phase boundary between the particles and the matrix will serve as a barrier to charge migration and ac charge oscillation and interfacial polarization will occur (Figure 19f). Sillars has shown that for a volume fraction of spheroids, q, of conductivity σ_2 and permittivity ε_2 and axial ratio of a/b, imbedded in a perfect dielectric matrix of conductivity σ_1 and permittivity ε_1, the dielectric parameters of the composite material are:

$$\varepsilon' = \varepsilon_\infty + \frac{\varepsilon_1' N}{1+\omega^2\tau^2} \tag{126}$$

$$\varepsilon'' = \frac{\varepsilon_1' N\omega\tau}{1+\omega^2\tau^2} \tag{127}$$

$$\tan\delta = \frac{\varepsilon_1' N\omega\tau}{\varepsilon_\infty(1+\omega^2\tau^2)+\varepsilon_1' N} \tag{128}$$

where:
$$\tau = \frac{(\varepsilon_1'(\lambda-1)+\varepsilon_2')\ \varepsilon_o}{\sigma_2} \tag{129}$$

and
$$N = q\,\frac{\lambda^2\varepsilon_1'}{\varepsilon_1'(\lambda-1)+\varepsilon_2'} \tag{130}$$

and
$$\varepsilon_\infty = \varepsilon_1'\left(1 + q\,\frac{\lambda(\varepsilon_2'-\varepsilon_1')}{\varepsilon_1'(\lambda-1)+\varepsilon_2'}\right). \tag{131}$$

The relationship between λ and a/b has been evaluated by Sillars and is reproduced in part in Figure 27.

The similarity of the form of the above equation with the Debye equations is readily apparent. However, the location of the M-W-S dielectric relaxation is primarily a function of the axial ratio of the ellipsoidal particles and the conductivity of the dispersed phase. For example.

$$f_{m(\varepsilon'')} = \frac{1}{2\pi\tau} = \frac{1}{2\pi} \frac{\sigma_2}{(k_1'(\lambda-1)+k_2')} \quad (132)$$

If the axial ratio of particles changes during a growth process from spherical to dendritic with an equivalent ellipsoidal ratio of a/b = 7, then the loss peak location would shift downwards by an entire order of magnitude. If the conductivity of the dispersed phase were decreasing as a result of growth, the downward shift would even be more marked. However, elongated growth accompanied by ($a+\Delta\sigma_2$) would tend to be self compensating.

The M-W-S model also shows that the magnitude of both ε' and ε'' are dependent on the volume fraction of the dispersed particles and their axial ratio. Consequently, a change in magnitude of ε' or ε'' cannot be unambiguously interpreted unless other evidence is available to independently estimate axial ratios or the volume fraction. With either information, though, it is possible to obtain the other structural features from the above equations.

COMPOSITION AND STRUCTURAL EFFECTS

The next objective is to attempt to interpret the dielectric behavior of some selected glass compositions in terms of the various theories presented above. Special emphasis will be placed on the relation of structural models to the behavior observed.

A number of dielectric studies of fused SiO_2[81,82] have demonstrated the presence of two dielectric loss peaks at cryogenic temperatures. Recent results[83] have shown that the magnitude of the loss peak at 1.3°K can be related to the H_2O content of the glass (figure 28). The second cryogenic fused silica loss peak occurs in the temperature range of 28-60 °K at audio frequencies and has an activation energy of 1.2 kilocalories per mole. Consequently, the peak

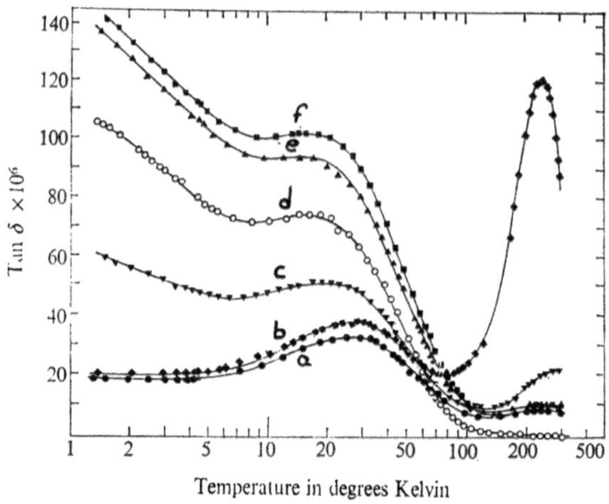

Figure 28. Dielectric loss behavior of a variety of fused silica glasses as a function of temperature.[83]
A.) Spectrosil WF-20ppm OH. B.) Infrasil, 30 ppm OH. C.) Vitreosil OG, 430 ppm OH. D.) Corning 7940, 940 ppm OH. E.) Sample 4, 1160 ppm OH. F.) Spectrosil B, 1190 ppm OH.

shifts to a frequency range of 10^{11}-10^{12} Hz at room temperature, beyond the range of concern for audio or radio circuit applications of fused silica but certainly within the range of interest for microwave devices.

The water based loss peak has been analyzed as an oscillating permanent dipole and can be described by the P_{o2} mechanism shown on the right side of Figure 19c. The higher temperature peak is associated with what Stevels terms "deformational losses." The dipolar mechanism, P_{d2}, that has been suggested for the peak involves the oscillation of the bridging oxygens in nearly colinear Si-O-Si bonds.[84] The model suggests that if the Si-O-Si bond is nearly 180° it will be possible for the oxygen to oscillate between two equivalent positions, as Figure 19c, with only a small energy barrier. No spatial rearrangements or bond-breaking would be involved. This interpretation would seem to be in reasonable agreement with the latest structural interpretation of fused silica advanced by Warren and Mozzi.[85] They concluded that the bond angles could be best described by a skewed Gaussian distribution about 144°. The half breadth of the distribution is nearly 35° and about 10% of the bond angles lie between 175°-180°. It is presumably the oscillation of bonds within the tail of this distribution that produces the "deformational losses."

Perhaps the most detailed atomistic analysis of a dipole relaxation mechanism in glass has been published by Charles.[86] In this publication he attempts a quantitative correlation of dielectric relaxation and dc conductivity in an alkali-silica glass. These are two basic postulations, (1) that near a given non-bridging oxygen ion in a glass there are several equivalent sites for an alkali metal ion, and (2) that the alkali ion will respond to the field in two major types of motion. The alkali ion may oscillate between equivalent sites about the same non-bridging oxygen which leads to ac polarization but no dc conduction. Or, the alkali ion may exchange non-bridging oxygen ions which usually will involve both ac polarization and dc conduction. Charles analyzed these two types of alkali ion motion in detail using a modification of the ionic conductivity and bistable dipole models presented herein and obtained reasonable agreement between theory and observation.

Charles[27] has also examined the influence of microstructure on the dielectric behavior of several alkali-silicate glasses. In this classic experiment he found that a rapid quench of 7.5 mole % and 15 mole % Li_2O-SiO_2 glasses changed a dispersed Li_2O rich glassy phase to a tubular-interconnected phase. The dielectric losses of the rapidly quenched sample were increased by an order of magnitude and the location of the loss peak also shifted by a factor of 10 to higher frequencies. Binary silicate glasses containing higher percentages of Li_2O, up to 40 mole %, did not show appreciable changes in dielectric loss upon changes in quench rate and the microstructure also remained that of a dispersed lithia rich phase in a SiO_2 rich matrix. His interpretation of the change in loss behavior with thermal history of the various glasses is based on Maxwell-Wagner-Sillars interfacial polarization. The conclusion that was reached was that the continuously distributed lithia-rich phases associated with spinodal decomposition gives rise to a large ac loss due to a very high a/b axial ratio in the M-W-S heterogeneous loss model. The nearly spherical dispersed phase, associated with a nucleation and growth phase separation, exhibits a much lower loss due to the low axial ratio in the interfacial polarization model.

An interfacial M-W-S polarization model has also been used by Kinser and Hench [17,18,30,32,33] to interpret the changes in dielectric properties associated with the thermal treatment of several Li_2O-SiO_2 and Na_2O-SiO_2 glasses. They found, as shown in Figure 23, that appreciable losses occur in 30 mole % Li_2O-SiO_2 glasses after a heat treatment of 5 hours at 500°C. Their results also show, Figure 29, that loss peaks are not present for the same samples before they are heat treated in contrast to Charles' data.[86] The source of the dielectric losses in the heat treated glasses has been attributed to the presence of a high conductivity metastable crystal phase of lithium metasilicate. Guinier x-ray diffraction, hot stage electron microscopy, and x-ray small angle scattering were also used to establish the presence of the metastable phase. Analysis of the change in frequency of the dielectric loss peak with thermal treatment employing the M-W-S theory led to the conclusion that the axial ratio of the metasilicate precipitate was decreasing from 100/1 to 5/1 during the nucleation of the equilibrium lithium disilicate phase. A summary of the changes in dielectric loss with thermal treatment is given in Figure 30. Similar results have also been found in 26.4% Li_2O-SiO_2 glasses as well as 33 mole % Li_2O-SiO_2 glasses, which are not phase separated, and 33 mole % Na_2O-SiO_2 glasses. The activation energy for the loss process, Figure 24, was calculated to be 15 kilocalories/mole, equivalent to the activation energy for dc conduction. Extension of the temperature range has shown that the appearance and disappearance of the dielectric loss peaks in the Li_2O-SiO_2 glasses occurs at least as low as 475°C, 24 hours required, and as high as 525°C. Consequently, the presence of dielectric loss peaks in some glass systems may be attributable to annealing treatments that result in precipitation of metastable crystallites which give rise to M-W-S interfacial losses.

Thus, it appears at the present time, based on the case histories cited, that the detailed analysis of dielectric behavior of glasses with well characterized water contents, complete amorphous x-ray structure analysis, transmission electron micrographic evaluation of phase separation, and controlled thermal histories are needed in order to firmly develop a generalized view of the dielectric properties of of glass.

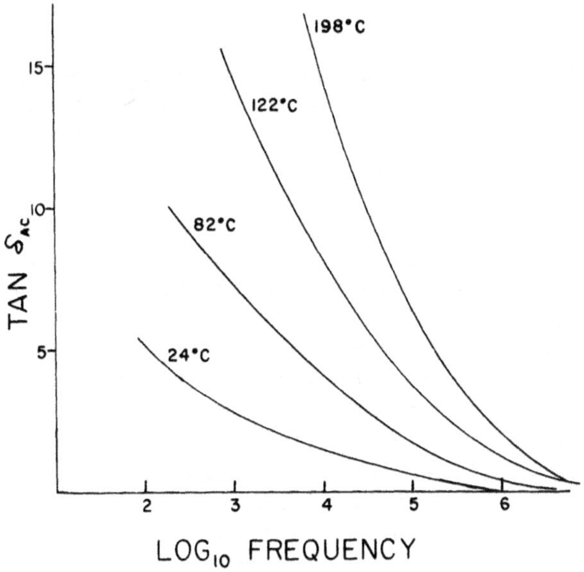

Figure 29. Frequency dependence of the dielectric loss angle of a 30 mole % $Li_2O \cdot SiO_2$ glass as quenched.[30]

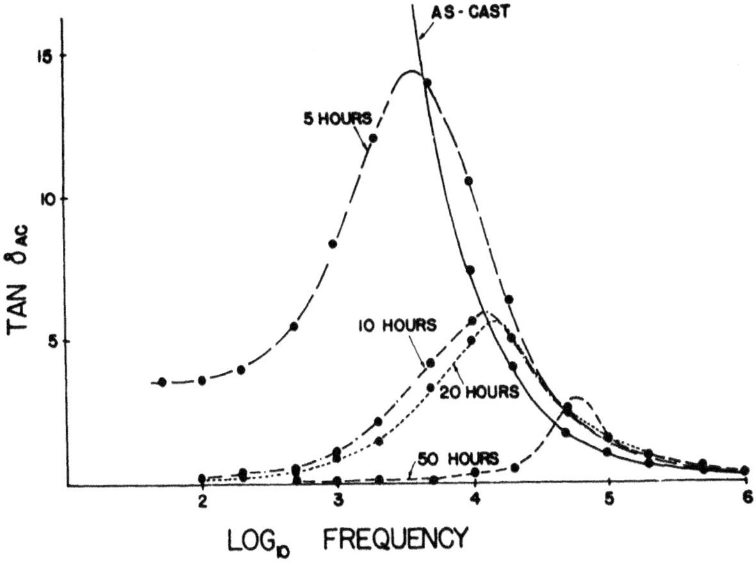

Figure 30. Variation in loss angle of a mole % Li_2O-SiO_2 glass with thermal treatment at 500°C.[18]

SWITCHING BEHAVIOR

In recent years, considerable attention has been given to the ability of amorphous semiconductors (and many crystalline semiconductors) to switch from a state of low conductivity to one of high conductivity. A typical current versus voltage curve is shown in Figure 31.

An explanation of this behavior has been given by Mott.[50] Due to the presence of localized electron states, carriers are injected into the semiconductor by the electrodes. forming a space charge region (Figure 32). This concentration of carriers causes the Fermi level of the semiconductor to drop with respect to the Fermi level of the electrode. Carriers from the electrodes may now be injected into the semiconductor in the high mobility region where the density of semiconductor states is high. This leads to a decreased resistance. When the field is removed, the semiconductor returns to its initial high resistance state. A second type of glass has the ability to remember its low resistance state when the field is removed. This effect has been shown to result from a structural change in a filamtery path between the two electrodes.[87,88] Thus, when a pulse with a slow fall time is applied to the glass, a high conductivity filament is formed due to melting or softening of the glass, and the device is "on." When a pulse with a fast fall time is subsequently applied, the filament is destroyed, the glass returning to its original state, turning the device "off." It appears at this time that the high conductivity filament results partially from cyrstallization of the glass, and partially due to a change in the glass structure resulting from going above the glass transition temperature.[89]

SUMMARY

The foregoing is at best a brief introduction to the extremely broad field of electrical properties of glasses. One unifying principle emerges, however: an applied field results in the charge carriers departing only minutely from zero field equilibrium (with the exception of amorphous semiconductor switching). Both electrons and ions are transported by hopping mechanisms which are determined by the properties and thermodynamic state of the material and not by the strength of the field. Ionic hopping is generally an undesirable characteristic, since it leads to dielectric

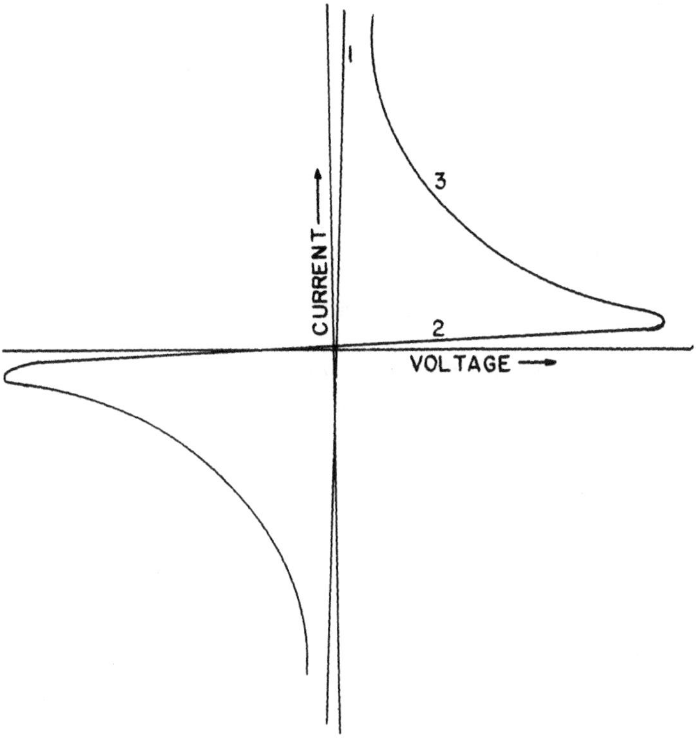

Figure 31. Typical current-voltage characteristics of an amorphous semiconducting switch.

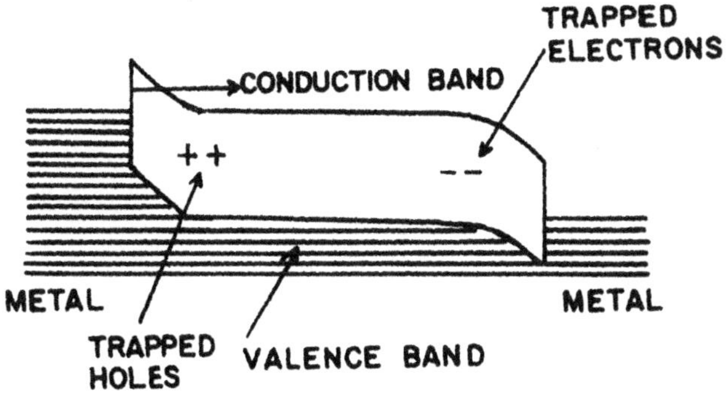

Figure 32. Mott's electrode injection model of reversible switching in an amorphous semiconductor.[50]

loss peaks and power absorption. Heterogeneous glasses whose phases have different dielectric constants and/or conductivities exhibit a similar loss peak due to interfacial polarization which is similar in appearance to a hopping loss peak.

The existence of electronic conduction in semiconducting glasses offers a whole new area of potential applications which are only beginning to be explored.

REFERENCES

1. A. E. Owen, pg. 77 in Progress in Ceramic Science, Vol. 3, Ed. J. Burke, Pergamon Press, N.Y. (1963).
2. J. D. Mackenzie, pg. 126 in Modern Aspects of the Vitreous State, Vol. 3, Ed. J. D. Mackenzie, Butterworths, Washington (1964).
3. A. D. Pearson, Ibid, pg. 29.
4. G. W. Morey, The Properties of Glass 2nd ed., Reinhold Publishing Corp., New York (1954).
5. M. B. Volf, Technical Glasses, Putnam and Sons, Ltd., London (1961).
6. W. Doremus, J. Non-Cryst. Solids, 2 (1970).
7. W. A. Weyl and E. C. Marboe, The Constitution of Glasses Vol. 2, John Wiley & Sons, New York (1964).
8. Electrical Properties of Glass and Glass-Ceramics: A bibliography covering the period 1957-1968, Int. Comm. on Glass, Charleroi, Belgium.
9. G. H. Wannier, Statistical Physics, pp. 441-442, Wiley, New York (1966).
10. E. Warburg, Ann. Physik, 21, 622 (1884).
11. W. D. Kingery, Introduction to Ceramics, John Wiley & Sons, New York (1960).
12. E. Rasch and F. Hinrichsen, Z. Elektrochem, 14,41 (1908).
13. R. L. Myuller, SoV. Phys. Solid State 2, 1213 (1960).
14. D. L. Kinser and L. L. Hench, J. Am. Ceram. Soc., 52, 638 (1969).
15. T. M. Proctor and P. M. Sutton, J. Am. Ceram. Soc., 43, 173 (1960).
16. J. O. Isard, Proc. Instn. Elect. Engrs., 109 B No. 22 p. 440, paper No. 3636, June (1961).
17. D. L. Kinser, "The Early Stages of Crystallization in Alkali-Silicate Glasses," Ph.D. Dissertation, Univ. of Florida, (1968), Diss. Abstracts Univ. of Mich. (1968).
18. D. L. Kinser and L. L. Hench, J. Am. Ceram. Soc., 52,

445 (1968).

19. H. E. Taylor, Trans. Farad. Soc., 52, 873 (1956).
20. H. Moore and R. C. Dilva, J. Soc. Glass Tech., 36, 5T (1952).
21. E. Seddon, E. Tippet, W. Turner, J. Soc. Glass Tech., 16, 450T (1932).
22. A. E. Owen and R. W. Douglas, J. Soc. Glass Tech., 43, 159T (1959).
23. O. V. Mazurin and E. S. Bovisouskii, Sov. Phys.-Tech. Phys., 2 243 (1957).
24. R. J. Charles, J. Am. Ceram. Soc., 48, 432 (1965).
25. B. Lengyel and Z. Boksay, Z. Phys. Chem., 204, 157 (1955).
26. J. O. Isard, J. Soc. Glass Tech., 43, 113T (1959).
27. R. J. Charles, J. Am. Ceram. Soc., 46, 235 (1963).
28. J. T. Littleton and W. L. Wetmore, J. Am. Ceram. Soc., 19, 243 (1936).
29. J. W. Rebbeck, M. J. Mulligan, J. B. Ferguson, ibid, 8, 329 (1925).
30. D. L. Kinser and L. L. Hench, "Electrical Properties of Thermally Treated Alkali-Silicate Glasses", Presented 1968, Am. Ceram. Soc. Meeting, Chicago, Ill.
31. D. L. Kinser and L. L. Hench, J. Materials Science, 3, (1970).
32. L. L. Hench, D. L. Kinser, S. W. Freiman, "The Early Stages of Crystallization of a Li_2O-2 SiO_2 Glass," VIIIth Int. Glass Congress, London, England (1968).
33. L. L. Hench, W. D. Tuohig, D. A. Sanders, "Kinetics of Resorption of a Metastable Precipitate in a Lithia Silica Glass," to be published.
34. S. W. Freiman and L. L. Hench, J. Am. Ceram. Soc., 51, 382 (1968).
35. R. E. Tickle, Phys. Chem. Glasses, 8, 101 (1967).
36. K. Endell and H. Hellbriligge, Glastech. Ber., 20, 277 (1942).
37. J. O. M. Bockris, J. A. Kitchener, S. Sgnatowica, J. W. Tomlinson, Trans. Faraday Soc., 48, 75 (1952).
38. S. Urnes, Glass Ind., 40, 237 (1959).
39. C. Babcock, J. Am. Ceram. Soc., 17, 329 (1934).
40. K. A. Kostanyan, K. S. Saakyan and E. M. Avetisyan, pg. 81 in Electrical Properties and Structure of Glass, Ed. O. V. Mazurin, Consultants Bureau, New York (1965).
41. K. A. Kostanyan, K. S. Saakyan, O. K. Geokchyan, Izv. Akad. Nauk Armyan SSR, 17, 357 (1964).
42. A. E. Martin and G. Derge, Metals Tech, 10, 1 (1943).
43. B. E. Warren, X-Ray Diffraction, Chap. 7, Addison-

Wesley, New York (1969).
44. R. Grigorovici, J. Non-Cryst. Solids, 1, 303 (1969).
45. R. Grigorovici and R. Manaila, Thin Solid Films, 1, 343 (1968).
46. M. V. Coleman and D. J. D. Thomas, Phys. Stat. Solidi, 24, k111 (1967).
47. J. Chang, "Structure of Amorphous Ge Films," M.S. Thesis, Univ. of Florida (1970).
48. N. F. Mott, Adv. Phys., 16, 49 (1967).
49. P. W. Anderson, Phys. Rev., 109, 1492 (1958).
50. N. F. Mott, Festkorperprobleme, 9, 22 (1969).
51. N. F. Mott, J. Non-Cryst. Solids, 1, 1 (1968).
52. H. F. Schaake and L. L. Hench, J. Non-Cryst. Solids, 2, 292 (1970).
53. H. F. Schaake, "Electron Transport Phenomena in Transition Metal Oxide Glasses," M.S. Thesis, Univ. of Florida (1968).
54. I. G. Austin, J. Non-Cryst. Solids, 2, 474 (1970).
55. I. G. Austin and N. F. Mott, Adv. Phys., 18, 41 (1969).
56. A. Miller and E. Abrahams, Phys. Rev., 120, 745 (1960).
57. E. A. Davis and R. F. Shaw, J. Non-Cryst. Solids, 2, 406 (1970).
58. D. Adler, pg. 1 in Solid State Physics Vol. 21, Eds. F. Seitz, D. Turnbull and H. Ehrenreich, Academic Press, New York (1968).
59. J. Appel, Ibid. pg. 193.
60. C. Kittel, Quantum Theory of Solids, Chap. 7, Wiley, New York (1963).
61. T. Holstein, Ann. Phys. (NY), 8, 325 (1959).
62. J. Schnakenburg, Phys. Stat. Solidi, 28, 633 (1968).
63. M. Pollack and T. H. Geballe, Phys. Rev., 122, 1742 (1961).
64. H. F. Schaake in Physics of Electronic Ceramics, Eds. L. L. Hench and D. B. Dove, Plenum Press, New York (in press).
65. A. D. Pearson, pg. 29 in Modern Aspects of the Vitreous State, Vol. 3, Ed. J. D. Mackenzie, Butterworths, Washington (1964).
66. J. D. Mackenzie, Ibid, pg. 126.
67. G. W. Anderson, "Structural Characterization and Optical Transmission Studies of Vanadate Glasses," Ph.D. Thesis, Univ. of Illinois (1969).
68. S. R. Bates and L. L. Hench, to be published.
69. G. H. Shortley and D. E. Williams, Elements of Physics Prentice Hall, (1965).
70. J. M. Stevels, Progress in the Theory of the Physical

Properties of Glass, Elsevier, Amsterdam (1948).
71. L. Azaroff, Electronic Properties of Materials, Addison Wesley, New York (1962).
72. H. Frohlich, Theory of Dielectrics, Oxford Univ. Press, London (1949).
73. V. V. Daniels, Dielectric Relaxation, Academic Press, London (1967).
74. A. S. Nowick, and B. S. Berry, IBM J. Res. Dev., 5, 297 and 312 (1961).
75. M. Gevers and F. K. DuPre, Disc. Faraday Soc. 42N, 47 (1946).
76. A. E. Owen, "Dielectric Properties of Glass," Ph.D. Thesis Univ. of Sheffield, England (1959).
77. L. K. H. vonBeck, pg. 69 in Progress in Dielectrics, Vol. 7, Ed. J. A. Birks, CRC Press (1966).
78. K. W. Wagner, Arch. Electrotech, 2, 371 (1914).
79. J. C. Maxwell, Electricity and Magnetism, Vol. 1, p. 328, Clarendon Press, London (1892).
80. R. W. Sillars, J. Inst. Elec. Engrs., 80, 378 (1937).
81. J. Volger and J. M. Stevels, Phillips Res. Rep. 11, 452 (1956).
82. R. E. Jaeger, J. Am. Ceram. Soc., 51, 57 (1968).
83. S. H. Mable and R. D. McCammon, Phys. Chem. Glasses, 10, 222 (1969).
84. O. L. Anderson and H. E. Bömmel, J. Am. Ceram. Soc., 38, 125 (1955).
85. B. E. Warren and R. Mozzi, "The Structure of Vitreous Silica," submitted to J. Applied Crystallography (1969).
86. R. J. Charles, J. Appl. Phys., 32, 1115 (1961).
87. A. D. Pearson, J. Non-Cryst. Solids, 2, 1 (1970).
88. R. Uttecht, H. Stevenson, C. H. Sie, J. D. Greiner, and K. S. Raghavan, J. Non-Cryst. Solids, 2, 358 (1970).
89. H. Fritzsche and S. R. Ovshinsky, J. Non-Cryst. Solids, 2, 148 (1970).

Glass To Metal Seals

Richard M. Rulon

Alfred University, Department of Chemistry
Alfred, N.Y. 14802

INTRODUCTION

In order to undertake a study of the factors involved in the manufacture of a glass-to-metal seal it is necessary to define carefully the specific type of product to be considered. In general such a seal constitutes an electrical insulation between two or more metallic elements while at the same time forming part of an enclosure or shield, hermetically shielding some electronic component from the surrounding environment.

Seals may be divided into two main groups, the first being composed of two or more metallic conductors insulated from one another and permanently positioned so as to extend through the wall of an all glass envelope, a light bulb or cathode ray tube being typical of this type of seal. The second group consists essentially of a metallic envelope penetrated by a hole or holes into which are positioned wires, insulated from one another and the metallic envelope by a glass component fused and sealed around the wire and into the hole. Extremes of this type might be a transistor and a submarine.

The first type of seal is most generally made by means of a flame-heating of the glass and the use of mechanical presses or forming tools and is known as a "flame seal", while the second type is more often made by means of an indirect heating, using an induction coil to heat the metal components or a furnace, and molds, to heat the entire assembly.

Further classification of seals may be made on the basis of the properties of the glass to be used in the insulating member. In this sense, then, it is desirable to answer the question as to "what is a glass" when used in the manufacture of seals. Although a glass is, per se, non-crystalline, many seals are made with glasses incorporating a relatively large amount of crystal phase either as an additive or as the result of controlled devitrification. Conventional glasses predominate as the insulating member but some excellent seals have been made using glasses containing no silicates at all, and, indeed, some seals use glasses containing no oxides but rather sulfides and related compounds.

"Soft" glass seals are those made with glasses having the working range in the vicinity of 1000°C. and a thermal expansion coefficient in the order of 90 x 10^{-7} in/in per deg. C. Such glasses are generally one of the three types: 1. soda-lime, 2. potash-soda-lead, 3. potash-soda-barium. "Hard" glass seals are those made with glasses having a working range from 1000° C. to 1250°C. and, more importantly, a coefficient of thermal expansion between 30 and 50 x 10^{-7} in/in/deg.C. These glasses are generally of the composition known as borosilicate glasses.

A final method of classifying seals can be made on the basis of the coefficient of thermal expansion of the glass as compared to that of the metal elements, when discussing seals made using a metallic envelope. In those seals where the metal components and the glass have essentially the same expansion, the seal is known as a "matched seal". In some cases the external metal member or "body" has a significantly higher expansion than does the glass insulating member, while the metal terminal sealed into the glass may have still a different expansion. In such seals the glass is under a high permanent compressive strain and the seal is called a "compression" seal.

The following is intended as an outline of the requirements, capabilities and manufacturing procedures for the oven-sealed type of product and will cover hard and soft glass seals, both of compression and matched types. Included in this will be the selection of appropriate metals and glasses, furnace design and requirements, all phases of processing and testing criteria and procedure.

MATERIALS SELECTION AND PREPARATION

In the type of seal to be considered the three structural elements used must be treated separately as well as in their relation to one another to produce a desirable end product.

The innermost metal element or terminal must generally have a fairly high electrical conductivity, either a high rigidity to avoid bending or a high degree of ductility to avoid damage from bending, and a surface which will provide good electrical contact with a mating member. It should also have good durability to withstand the temperature necessary to seal the glass to the metal, resistance to chemical corrosion after sealing and the ability to form a good seal.

Although many terminals are simply shaped, as a piece of square cut wire, there are many instances where special configurations must be formed on the end of the terminal. Among these are small cups, flattened, and pierced ends or threaded ends. For this reason, in many cases, the ability of a metal to be machined or formed is of considerable importance. Some metals such as platinum and related alloys form excellent seals but expense may be prohibitive for most applications. A porous grain structure as may exist in ordinary copper or one which permits long longitudinal faults such as tungsten may result in a seal which fails because of gas leakage through the metal structure of the wire or terminal.

For some special applications the thermal conductivity of a terminal may be important, as may the magnetic characteristics. In all cases the coefficient of thermal expansion of the metal is important in relation to the coefficient of expansion of the glass into which the metal is to be sealed.

The outermost metal element or body must have many of the same properties as the terminal but differs somewhat in that electrical conductivity is seldom of importance. Thermal conductivity may be more significant, cost and machinability are usually more important, the ability to form welds or bonds to other metals may become increasingly important and, because of the larger surface areas involved, corrosion resistance and porosity are more critical in many cases. Thermal durability, coefficient of thermal expansion, magnetic properties and ability to form a good seal are all factors to be considered.

The glass as the insulating, sealing and structural component, is in many respects the most critical material and is, unfortunately, the one least subject to modification by the seal manufacturer in his plant operation. The coefficient of thermal expansion and chemical durability are critical factors, the softening point and working temperature are important in their relationship to the metals used for terminal and body. Three electrical properties must be considered not only at room temperature but as variables related to the projected operating temperature of the finished product. These properties are dielectric strength, dielectric constant and electrical resistivity.

Table I is a list of some metals which have been used in the manufacture of glass to metal seals showing their properties and, by means of the designation B and/or T, whether they are commonly used for body materials, terminals or both.

Table II is a list of some of the more common glasses used for sealing purposes. The numbering system used relates to the Corning Glass Works code. Similar glass materials are often available from a number of other manufacturers.

Table III is a list of a number of metal and glass combinations for both matched and compression seals.

(A) Metal Selection

In general the metal is the material chosen first on the basis of mechanical, electrical or chemical properties with the glass being selected as a compromise to form an adequate seal, the glass chosen on the basis of thermal expansion so as to minimize the probability of gross mechanical failure. It has been fairly well established that the "wetting" of a metal by a molten glass is, at best, poor if the metal is clean and free of oxide but is

Table I. Properties of Metals and Alloys

Metal	Nominal % Composition	Coefficient of Thermal Expansion x 10 °C (25-500°)	Thermal Conductivity $cal/cm^2/cm/sec$	Electrical Resistivity microhm-cm
Tungsten	Nominally pure	46	0.38	5.5
"Kovar"	Ni-29, Co-17, Fe bal.	55	0.05	49
Molybdenum	Nominally pure	55	0.35	4.7
Nickel-iron alloys	Ni-42	78	0.03	66
	Ni-46	92	0.03	46
	Ni-52	98	0.03	43
Platinum	Nominally pure	90	0.17	10
Rhodium	Nominally pure	85	0.20	4.7
"Dumet"	Core. Ni-43, Fe bal. Sheath: OFHC copper	90 radially 65 longitudinally	0.40	4.5
"Sealmet #4"	Ni-42, Cr-6, Fe bal.	110	0.03	34
Titanium alloy	Al-6, V-4, Ti bal.	95	0.01	170
C. R. Steel	Mn-0.4, C-0.1, Trace P. S.	130	0.11	18
"Hastalloy-C"	Ni-54, Mo-17, Cr-15, Fe-5, W-4	130	0.03	130
430 Stainless	Cr-16, C-0.12, Max., Mn-1, Max.	114	0.02	60
"K-Monel"	Ni-66, Cu-29, Al-3	170	0.04	58
304 Stainless	C: 0.08 Max., Mn-2 Max., Si-2 Max., Cr-19, Ni-10	190	0.04	72
316 Stainless	C-0.08 Max., Mn-2 Max., Si-2 Max., Cr-16-18, Ni-10-18, Mo-2-3	170	0.04	74
316 L Stainless	Same as 316 except C-0.03 Max.	170	0.04	74
Copper	OFHC (oxygen free, high conductivity)	170	0.92	1.7

Table II. Properties of Some Common Sealing Glasses.

1	2	3			4		5				6				7	8			9		
		Corrosion Resistance			Thermal Expansion 10^{-7}in/in/°C		Upper Working Temperatures. (Mechanical Considerations Only)				Viscosity Data					Log_{10} of Volume Resistivity			Dielectric Properties at 1Mc and 20°C		
							Annealed		Tempered												
Glass Code	Type	Weath-ering	Water	Acid	0-300°C	Room Temp. Setting Point	Normal Service °C	Extreme Limit °C	Normal Service °C	Extreme Limit °C	Strain Point °C	Annealing Point °C	Softening Point °C	Working Point °C	Density grams per cc	25°C	250°C	350°	Power Factor %	Dielectric Constant	Loss Factor %
0010	Potash Soda Lead	2	2	2	93	100	110	380	---	---	395	435	625	985	2.86	17+	8.9	7.0	.16	6.7	1.
0080	Soda Lime	3	2	2	92	103	110	460	220	250	470	510	695	1005	2.47	12.4	6.4	5.1	.9	7.2	6.5
0120	Potash Soda Lead	2	2	2	89	98	110	380	---	---	395	435.	630	980	3.05	17+	10.1	8.0	.12	6.7	.8
1720	Aluminosilicate	1	1	3	42	52	200	650	400	450	670	715	915	1190	2.52	---	11.5	9.5	.38	7.2	2.7
1990	Potash Soda Lead	3	3	4	124	136	100	310	---	---	330	360	500	755	3.47	---	10.1	7.7	.04	8.3	.33
7040	Borosilicate	3	3	4	48	54	200	430	---	---	450	490	700	1080	2.24	---	9.6	7.8	.20	4.8	1.0
7050	Borosilicate	3	3	4	46	51	200	440	235	235	460	500	705	1025	2.24	16	8.8	7.2	.33	4.9	1.6
7052	Borosilicate	2	2	4	46	53	200	420	210	210	435	480	710	1115	2.28	17	9.2	7.4	.26	4.9	1.3
7056	Borosilicate	2	2	4	51	57	200	460	---	---	470	510	720	1045	2.29	---	10.2	8.3	.27	5.7	1.5
7070	Borosilicate	2	2	2	32	39	230	430	230	230	455	495	---	1070	2.13	17+	11.2	9.1	.06	4.1	.25
7250	Borosilicate	1	2	2	36	38	230	460	260	260	490	540	780	1190	2.24	15	8.2	6.7	.27	4.7	1.3
7720	Borosilicate	2	2	2	36	43	230	460	260	260	485	525	755	1140	2.35	16	8.8	7.2	.27	4.7	1.3
7740	Borosilicate	1	1	1	33	35	230	490	260	290	515	565	820	1245	2.23	15	8.1	6.6	.50	5.1	2.6
7760	Borosilicate	2	2	2	34	37	230	450	250	250	480	525	780	1210	2.23	17	9.4	7.7	.18	4.5	.79
9010	Potash Soda Barium	2	2	2	89	102	110	380	---	---	405	445	650	1010	2.64	---	8.9	7.0	.17	6.3	1.1
9700	Borosilicate	1	1	2	37	39	220	500	---	---	520	565	805	1200	2.26	15	8.0	6.5	---	---	---

Column 5
Normal Service. Freedom from excessive thermal shock is assumed.
Extreme Limits. Glass will be very vulnerable to thermal shock.
Recommendations in this range are based on mechanical considerations only. Tests should be made before adapting final designs.
These data approximate only.

Column 8
Data at 25° extrapolated from high temperature readings and are approximate only.
All data subject to normal manufacturing variations.

Table III. Typical Metal-Glass Seal Combinations.

Body	Glass (Corning Code)	Terminal	Glass Type	Seal Type	Preferred Firing Atmospheres
Copper			Soft	Compression	Neutral to mildly oxidizing
"K Monel"	9010	Steel			
304 Stainless	0080				
316 Stainless		Sealmet #4			
		42% Ni Alloy			
316 L Stainless		46% Ni Alloy			
		52% Ni Alloy			
430 Stainless		Dumet			
		Platinum			
C. R. Steel					
"Hastalloy-C"	7052	Kovar	Hard	Compression	Neutral to mildly oxidizing
	7070	Molybdenum Tungsten			
"Hastalloy-C"	9010	52% Ni Alloy Platinum	Soft	Compression	Neutral to mildly oxidizing
	0080	Rhodium or Platinum-Rhodium Alloy			
"Kovar"	7052 7070 7050	"Kovar"	Hard	Matched	Mildly oxidizing
Titanium	7052	Molybdenum	Hard		Inert gas (prebead terminal)
		"Kovar"		Compression	
52% Ni Alloy	9010	52% Ni-Alloy "Dumet"	Soft	Matched	Mildly oxidizing

tenacious if the metal has an oxide surface to which the glass may adhere. This has led to the argument that in a "matched seal" the glass and metal must be bonded with an oxide bond to avoid leaks, but that the "compression seal" needs no oxide bond since, in effect, the components are "sealed" by physical stress in a manner analogous to the old iron rim on a wooden wheel.

A more careful analysis would show that with "matched seals" the oxidation has been poorly controlled and leaks are occurring for one or more of three reasons: (1) the oxide bond to the metal is poor and a scaling condition exists permitting extensive leaks not between the glass and oxide, but between the oxide and the base metal; (2) the oxide has been partially reduced and reoxidized to form a "sponge", with leakage occurring through the sponge even though the bond is good; (3) the oxide was thin and has been removed completely either by solution in the glass or atmospheric reduction in the furnace. Correspondingly, with the "compression seal" it is often possible to show that there is no real compression stress at the terminal-glass interface by virtue of a peculiarity of the cooling cycle. It is therefore recommended that in almost all seals, care be taken to create an oxide bond which is thin and non-porous in nature, firmly adherent to the metal, easily wet by the glass and fused with enough oxidation potential in the furnace to avoid removal of the oxide.

With a chrome steel, as in the 300 series stainless steels, these conditions can be easily met by using a carefully cleaned steel surface and any atmosphere with either oxygen or water vapor present. In fact, unless the atmosphere is reducing and has a dew point of -60°F or lower the oxide will form with no additional effort when fused at 1700 - 1800°F.

With a metal, such a Kovar or #52 Alloy, the oxide may be formed either (1) as a preoxidation before fusing, say by heating for a short period in air or an oxidizing exothermic gas atmosphere at a temperature of 1700°F or above (care and experience are necessary to control the time so as to avoid a scaling condition.) or (2) by fusing the seal in an oxidizing atmosphere.

Steel and copper or brass may also be given an oxide

coating if handled very carefully but, in general, it is difficult to avoid some scaling, copper is usually protected either by the application of a thin, glassy sodium borate coating as in the case of "Borated" copper or Dumet, while steel and brass are often given a thin nickel plating and then treated in a manner similar to Kovar or #52 Alloy.

The least expensive form of matched seal, one found in many applications uses a terminal of Sylvania #4 alloy or #52 alloy and a body of similar alloys. The glass in this type of seal is most commonly Corning 9010 although, recently #0088 has been used to some extent.

A special matched seal of unusual chemical resistance consists of a terminal made of platinum or platinum-rhodium alloy with a body of Hast alloy - C. These metals, sealed with 9010 glass, are outstandingly resistant to atmospheric corrosion and have been used for some special marine applications. A second group of seals which can fairly well be lumped together under the "matched seal" heading are to be recommended for two main characteristics based, nominally, on (1) lower expansion and, hence, better thermal shock resistance and (2) higher safe operating temperature. In this group the header, or body material, is usually Kovar or its equivalent while terminals of Kovar are generally used, although molybdenum or even tungsten terminals can be used. With a Kovar body and Kovar terminal, 7040, 7050, 7052 and 7065 are excellent sealing glasses, giving an almost perfect match in expansion. All of these glasses are approximately equivalent electrically and have substantially lower dielectric constants then the compression-seal glasses as a group. With reasonable care satisfactory seals can be made using the low expansion "matched seal" metals and such glasses as 7740, 7720, 7250 and 7070. 7720 contains lead oxide and, therefore, to be avoided with carbon molds. Glass #7740 does not "wet" tungsten readily and has as its primary virtue, a considerably higher chemical resistance than most other glasses.

Glass number 7070 is extremely interesting as a glass since, with the exceptions of fused quartz or Vycor, it has the lowest dielectric constant of any commercial glass and can, therefore, be recommended for coaxial connectors. It is not too readily available and requires a long lead-time on purchasing of preforms compared to other glasses such as 9010 and 7052. The 7070 glass makes good seals with Kovar bodies and terminals but should be cooled carefully so as to have the terminal slightly cooler than the body as the glass reaches its strain point.

A point which should be made is that coaxial seals made with a Kovar terminal and a 52 Alloy ring and 7070 glass will give excellent results with the simultaneous sealing of the coaxial ring into a stainless steel body by 9010 glass. In the same manner a coaxial seal of a Kovar terminal and a Kovar ring sealed with 7070 glass and a seal of this ring into a Kovar header with 7052 is also of good quality.

The most common type of compression seal consists of a steel body and #52 alloy terminals with #9010 glass as the sealing component. The seal is inexpensive and generally reliable although the terminals and/or body often need either a nickel plating prior to fusing or a protective plating after fusing to inhibit corrosion. In special cases where higher electrical conductivity is needed terminals may be made of copper cored 52 alloy, Dumet, platinum or, with proper control of the cooling cycle, steel. It is reasonable to use bodies made of 300 series stainless, 400 series stainless, Monel and perhaps copper or brass with a family of glasses such as Corning 0010, 0120, 0080, 9010, 9019, 0088 in so far as thermal expansion is concerned. These will all result in a compression seal and are used with #52 Alloy, #4 Alloy, Dumet or even 1010, 1008 and 1113 steel terminals with care in processing.

Of Corning glasses, 0010 and 0120, like 7720 are high lead glasses and exhibit considerable lead reduction if used with carbon molds or a reducing atmosphere. They should, therefore, be avoided unless some free form mold design can be achieved whereby the carbon does not touch the glass at any time. Glasses 0080 and 9019 are lead free and make excellent mechanical seals but have a much lower (2-3 orders of magnitude) electrical resistivity than do 9010 and 0088 so that if electrical resistance is a major factor it is desirable to avoid these glasses. Some special compression seals have been made using less common combinations such as Hast alloy -C, 7052 Glass and Kovar or Titanium-7052 glass and Kovar. Many such unusual combinations can be used where special requirements must be met.

(B) Metal Preparation

In the machining of metal components it is essential that all burrs and sharp edges be removed from the seal area. Such defects can cause points of stress concentration which then cause mechanical failure in the glass after fusing. In the area of the seal any longitudinal die marks or scratches extending from one side of the seal to the other will cause a leak in the seal since the glass never becomes sufficiently fluid to fill such marks.

Sulfur bearing lubricants or cutting oils should be avoided as should abrasive grit or tumbling compounds of aluminum oxide or silicon carbide. The use of such materials often cause bubbles to form at the glass-metal interface.

The metals should be thoroughly cleaned.

A good general outline for cleaning might be as follows:

(a) Degrease - Hot 140°F trichlorethylene or perchlorethylene is good if done with agitation and clean solutions. Vapor degreasing is also very satisfactory. In all cases care must be taken to insure subsequent removal of the degreasing agent. These materials are quite toxic if the vapors are inhaled and should be used only with good ventilation.

Other solvents such as toluene, petroleium ether, acetone and ether are excellent for degreasing but all are quite flamable as well as being somewhat toxic as vapors. All solvents of the above types are basically skin irritants and prolonged skin contact is to be avoided.

For heavy grease and oils a hot (140 - 160°F) alkaline degreasing will do a good job. These are preferred where the grease is of a saponifiable nature. Since these are all caustic materials, care should be taken to avoid contact with them since severe burns will result.

(b) After degreasing a thorough rinse in warm water is desirable. In the case of the alkaline degreasing this is a "must" before further operation.

(c) Chemical cleaning is done to remove surface contaminents such as oxide and foreign metals on the surface. A general cleaning solution for steel, Kovar and alloys such as #52, #49, #46, and #42 is 5-15% hydrochloric acid heated to 130 - 140°F. An immersion of the part to be cleaned in the acid for 5 - 15 minutes is generally satisfactory, with the time being controlled by the rate of attack on the metal. A mild etch, to a semi-matte surface, is not objectionable except as possibly an appearance defect.

Stainless steels may be cleaned in a similar manner although a modified solution by using an addition of 5% nitric acid will clean faster and should not etch severly.

Another solution recommended for the high chrome stainless steels which have fairly heavy oxide would be 95% nitric acid, 5% hydrofluric acid cold. This may be diluted with water if the reaction is too vigorous. This solution should never be hot before starting work and should not be used with mild steel, it will also attack #52 alloy, etc., quite rapid and is not recommended for these metals. With Monel this step may frequently be eliminated although a 30 second rinse in warm 15% hydrochloric acid may be helpful.

(d) In all cases, after acid cleaning a thorough rinse preferably in running water is necessary. Rinsing should be sufficient to eliminate all residual acids from the metal surface. If the available tap water is high in contaminents it is desirable to further rinse in two steps using deionized or distilled water.

(e) After the water rinse it is necessary to remove the water by successive alcohol rinses to avoid water spotting on the sealing area. A series of three rinses with periodic discarding of the first rinse and down-grading of the second and third with a fresh solvent for the third rinse is good technique.

(f) After removal from the alcohol rinse the parts should be shaken to remove major remaining alcohol and immediately dried either with a hot air dryer or with heat lamps and free air circulation. In no case should the metal surface be handled with the hands or dirty utensils. This is of paramount importance in the seal area. Failure to observe this precaution may result in finger prints and smudges being burned into the exterior surface at the time of sealing and will cause gassing and leaks if the dirt is in the seal area.

Degassing of the metal after cleaning is desirable although not always essential with all metals.

(a) For #52 alloy and related alloys a 20 - 30 minute soak at 1750 - 1800°F in a reducing (either hydrogen or exothermic gas) atmosphere is generally satisfactory if the parts are cooled while still in the inert atmosphere.

In some cases where a preoxidizing is suggested for seal integrity, the parts may be removed from the degassing at a temperature of 1500 - 1600°F and cooled quickly in air with agitation. If this is done care should be taken to insure rapid cooling to avoid the creation of a heavy scaling oxide. A preferred technique would be to cool completely in the reducing atmosphere and then reheat in air or an oxidizing atmosphere to form a thin, adherent oxide film. Experience to establish a heating schedule for a given mass and number of parts is required here.

(b) Stainless steel should be degassed in substantially the same manner except that care must be taken to insure a dew point of at least -40°C or severe oxidation will take place prematurely.

Stainless is seldom preoxidized but if desired this is done by firing to, say, 1800°F for 20 minutes in hydrogen or a 10% hydrogen-nitrogen mix with a dew point of 20°C with cooling in the same atmosphere.

(c) Monel is easily degassed in 15 - 20 minutes at 1800°F but care should be taken to avoid sulfur bearing gases in the atmosphere. Generally if reducing exothermic gas is used, 3 - 6% carbon monoxide content is preferred.

Care should be taken in reprocessing Monel to avoid repeated firing where the atmosphere may alternate from reducing to oxidizing.

A note may be added concerning the use of plated metal parts. Since in any plating there is some question as to the bond strength and adherence of the plate to the bare metal it is desirable to "degas", or "alloy" the plate to the bare metal by heating to, say, 1800°F for 20 - 30 minutes in a reducing atmosphere. If the plating has poor adherence or absorbed gas at the plate-to-metal interface the plating will blister. If this occurs the plating should be rejected, stripped, and redone. To some extent a slower heating say for 10 minutes between room temperature and alloying temperature will allow gases to escape without causing blistering unless the adherence is quite poor.

(C) Glass Selection

Although the softening temperature, chemical properties and thermal expansion of the glass are essential factors to be considered in the formation of a good seal, in most applications the primary function of the glass is to serve as an electrical insulation.

Let us consider, first, the insulation resistance in the general case (Figure 1) of a coaxial configuration involving (1) a metal conductor, (2) a glass insulator and (3) the surrounding metal body.

In actual fact, the electrical resistance of such a unit must be considered to be of two parts, the first a resistor

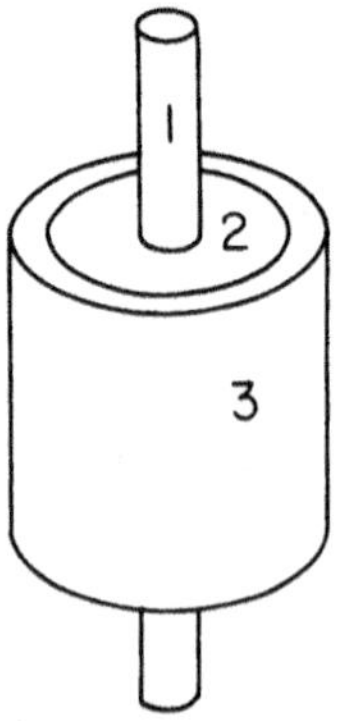

Figure 1. Coaxial seal. 1. metal conductor, 2. glass insulator, 3. surrounding metal body.

consisting of two thin surface layers of glass exposed to air, moisture, etc., and the second consisting of the large volume of bulk of the unit. Calling these respectively R_s and R_v they act in parallel so that the total resistance

$$R_t = \frac{R_s R_v}{R_s + R_v}, \qquad R_t = R_v \text{ when } R_s >> R_v$$

$$R_t = R_s \text{ when } R_v >> R_s \tag{1}$$

If the surface layer is of uniform thickness and resistivity throughout, a value α = ohms/square will define the resistance of a unit area surface with unit length contacts on opposite edges (Figure 2) regardless of dimensions.

On this basis, in order to determine the resistance of a glass surface (Figure 3), such as exists on the coaxial seal, then

$$R_s = \int_{r_1}^{r_2} \frac{\alpha dr}{\int_0^{2n} r d\theta} \tag{2}$$

solving equation 2 for a concentric condition,

$$R_s = \frac{\alpha}{2\pi} \ln \frac{r}{r_1} = 0.366\alpha \log \frac{r_2}{r_1} \tag{3}$$

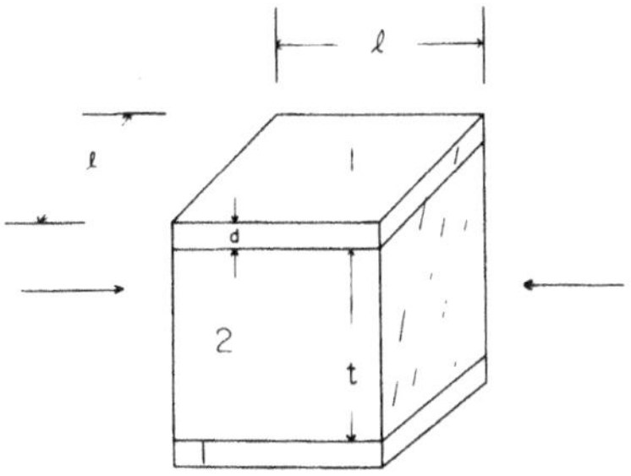

Figure 2. Schematic of glass resistive elements. 1. surface elements, 2. volume element. Arrows point to areas of electrical contact with the elements. If θ is the volume resistivity of the surface, then $\alpha = \frac{\theta \ell}{\ell d} = \frac{\theta}{d}$ = a constant, when d is uniform throughout.

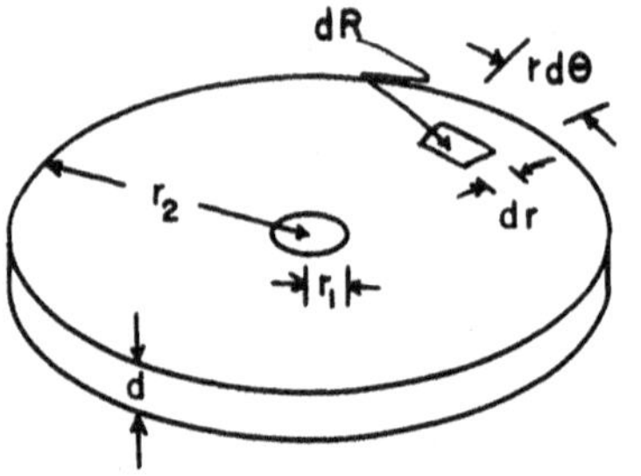

Figure 3. Surface resistive element of the glass insulator of the coaxial seal. r_1 is the radius of the terminal, r_2 is the interior radius of the metal body, dR is a differential element of the surface resistance.

For two such surfaces in parallel,

$$R_s = 0.183\alpha \log \frac{r_2}{r_1} \tag{4}$$

A similar analysis of the volume resistance (Figure 4) for a material with a specific volume resistivity, ϕ = ohm-cm.

$$R_v = \int_{r_1}^{r_2} \frac{\phi dr}{\int_0^{2\pi} rtd\phi} \tag{5}$$

$$R_v = \frac{.366\phi \log}{t} \frac{r_2}{r_1} \quad \text{"t" is in centimeters} \tag{6}$$

since ϕ is generally expressed in ohms-cm, we can convert "t" to inches thus,

$$R_v = \frac{.144}{t} \phi \log \frac{r_2}{r_1} \quad \text{where "t" is in inches} \tag{7}$$

Inserting (4) and (7) into (1) we obtain

$$R_t = \frac{.026\alpha\phi}{.183\alpha t + .144\phi} \log \frac{r_2}{r_1} \text{ or as limits}$$

$$R_t = \frac{.144\phi}{t} \log \frac{r_2}{r_1} \quad \text{when } \alpha t >> \phi \tag{8}$$

$$R_t = .183\alpha \log \frac{r_2}{r_1} \quad \text{when } \phi >> \alpha t$$

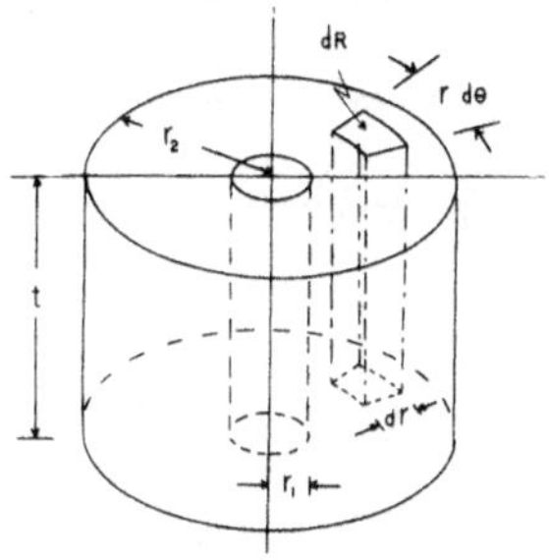

Figure 4. Volume resistive element of the glass insulator of the coaxial seal. t is the length of the glass r_1; r_2 rdθ, dr are as in figure 3. dR here is a differential element of its volume resistance.

The surface resistivity α, is a function of glass composition and humidity. The use of a silicone coating on the glass surface has a marked effect on α since such a film changes the moisture adsorption on the surface. These effects are shown in Figure 5.

The volume resistivity, ϕ, is a function of glass compositions and temperature as shown in Figure 6.

Table IV shows typical values of ϕ, α, R_s, R_v and R_t for Corning 7740 and 0080 glass under various conditions of temperature and humidity with a value of r/r = 5 and t = 0.100 inches.

From reference to Table IV, Figures 1 and 2, it becomes apparent that soda-lime glass is unsatisfactory for

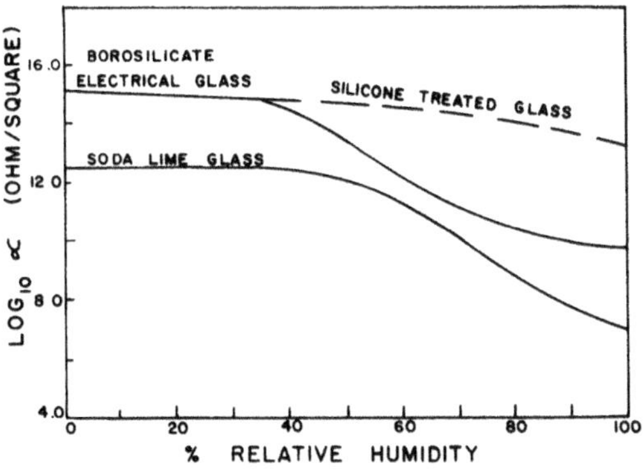

Figure 5. Surface resistivity α of two common glasses as a function of relative humidity. (Courtesy of Corning Glass Works.)

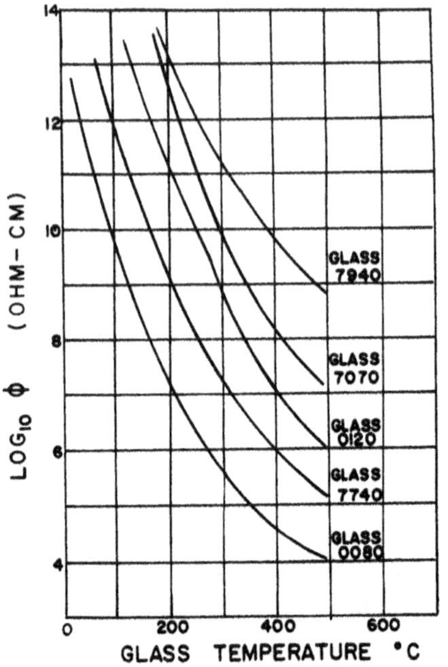

Figure 6. Volume resistant Φ of several commercial glasses as a function of temperature. (Courtesy of Corning Glass Works.)

Table IV. Resistance Properties of 7740 and 0080 Glass

Operating conditions	α (ohm/sq)	Φ (ohm-cm)	R_s	R_v	R_t
	7740 glass (borosilicate glass)				
Rel. Hum. < 40% 20°C	1×10^{15}	1×10^{14}	1.3×10^{14}	1.0×10^{14}	5.6×10^{13}
Rel. Hum. 70% 20°C	1×10^{11}	1×10^{14}	1.3×10^{10}	1.0×10^{14}	1.3×10^{10}
Rel. Hum. < 40% 120°C	1×10^{15}	1×10^{11}	1.3×10^{14}	1.0×10^{11}	1.0×10^{11}
	0080 glass (soda lime glass)				
Rel. Hum. < 40% 20°C	3.5×10^{12}	1.0×10^{12}	4.5×10^{11}	1.0×10^{12}	3.1×10^{11}
Rel. Hum. 70% 20°C	1.0×10^{10}	1.0×10^{12}	1.3×10^{9}	1.0×10^{12}	1.3×10^{9}
Rel. Hum. < 40% 120°C	3.5×10^{12}	1.0×10^{9}	4.5×10^{11}	1.0×10^{9}	1.0×10^{9}

insulation resistances in the order of 10^{12} ohms unless the ratio r_2/r_1 is very large, indeed, even the borosilicate glass is unsatisfactory for use at moderate temperature or humidity.

The use of certain of the lead or barium glasses is to be preferred to the use of soda-lime but caution is necessary to avoid the reductions of the lead oxide or barium oxide during manufacturing since such a reduction will result in the formation of a highly conductive "skin" on the glass surface.

In the event that surface conductivity due to high humidity conditions makes a seal unsatisfactory it is frequently effective to use a silicone coating to prevent the formation of a continuous conductive water film on the glass surface.

While the above figures may be considered reasonable for the glasses given, it should be apparent that reduction to extremely small dimensions, deviations from a true coaxial construction, and manufacturing defects on a given product may cause serious discrepencies in measured values as compared to the calculated values, and that these discrepancies will always be to the detriment of the product.

Another aspect of the electrical insulation-conductivity of a glass which must be considered is the slow electrolysis or ionic migration of the glassy material under long term D.C. potentials or in some cases rectified A.C. potentials created by the nature of the seal itself.

The alkali ions, particularly lithium and sodium, are fairly mobile and can migrate appreciably in a glass. In some cases this migration takes place rapidly enough to cause a considerable change in glass composition in the area of an electrode thereby introducing strain and ultimately causing failure of the glass structure.

The electrical insulation, as calculated, is of primary importance where sensitive electronic equipment using low voltages is concerned but in the case of equipment using high voltage a more important consideration is often simply one of dielectric break down or "arc over". The dielectric breakdown of a seal can again be considered as two separate functions, one the break down of the glass and the second the breakdown of the surrounding air.

Unless severe bubbling takes place in the glass at the time of sealing or excessive contamination is present in the glass (perhaps by solutions of the metal oxide during sealing), the breakdown voltage of most seals will exceed 75-100 volts per mil (.001 inch) of glass thickness as shown by Figure 7.

Air breakdown is a more complicated matter and is influenced not only by the physical space between adjacent metal components but by the sharpness of such metal surfaces at nearest proximity, relative humidity, pressure and specific metal. An approximate value of 30-40 volts per mil is reasonable for most fairly smooth metal surfaces with no point protrusions and "normal" ambient air. In some cases a glass surface which has been exposed to high humidity and which is etched by such exposure, so as to have a high alkali surface which is hygroscopic, will cause a true surface arc over at voltages as low as 10-20 volts per mil.

With high frequency applications such as coaxial cable transmission line connectors the dielectric constant also becomes a factor. Since the loss in electrical signal through such a connector is a function of the different characteristic impedance of adjacent longitudinal sections of the cable, it is desirable to match the impedence of the connector as closely as possible to the cable itself.

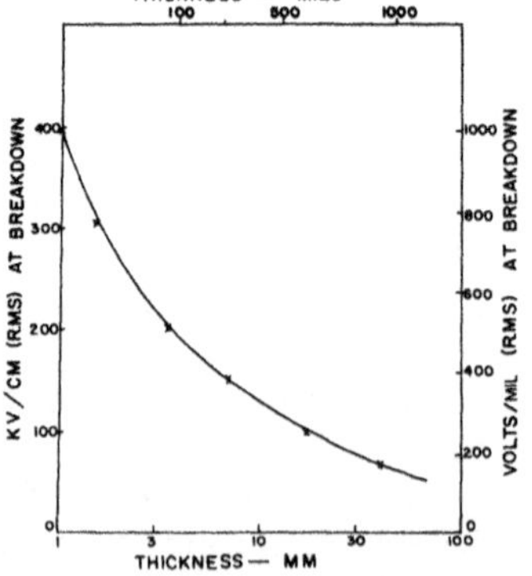

Figure 7. Dielectric breakdown for 7740 glass (60 cycle R. M. S. voltage as a function of glass thickness). (Courtesy of Corning Glass Works.)

Figure 8 gives the relationship between dielectric constant, characteristic impedance and the physical dimensions of a coaxial cable for various dielectrics.

From the graph it would appear that a given impedance in a coaxial cable insulated with polyethlene could be maintained at the glass seal connector simply by changing the body hole/terminal diameters appropriately. However, the abrupt change in dimensions may cause as much signal loss as would the change in impedance with constant dimensional relationship. It is, therefore, almost necessary that a glass such as 7070 be used for seals of this type.

(D) Glass Components

Once the specific glass to be used has been chosen two primary types of glass components are available, either sintered glass preforming or types made from bulk glass. By far the greater number of seals, such as those under consideration, are made using sintered glass compactions or preforms such as Corning Multiform. Almost any glass is available in this form. Many ceramic oxide colors can be added for identification purposes and dimensional tolerances are quite good.

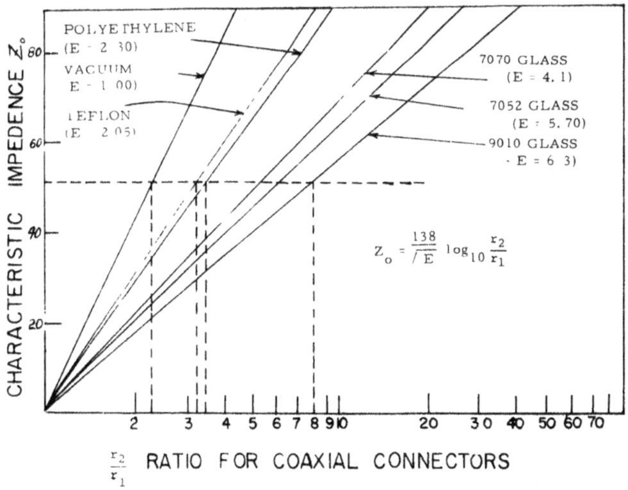

Figure 8. Characteristic impedance of a coaxial connector as a function of dielectric constant E and insulation diameter ratio D/d. (Courtesy of D. G. O'Brien, Inc.). see Fig. 3.

The manufacturing process for such preforms results in the presence of some trapped bubbles in the body of the glass. The presence of these bubbles can create serious problems for the seal manufacturer unless great care is taken to insure that the bubble content is maintained constant by establishing density of the preform glass as compared to the density of the parent glass.

Variations in density of preforms will result in an non-uniform volume of glass after sealing because of bloating of the bubbles during firing. With a low density preform variations in firing time or temperature can cause extreme changes in the quality of the finished glass product.

In addition to preforms,glass components can be obtained in clear glass either by using cut tubing or, more rarely, pressed or punched glass. Obtaining all necessary dimensions in such components may be difficult and the expense is generally much higher than with other preforms.

Since the glass component must be assembled prior to sealing inside of a metal body and with a terminal within it, we must specify dimensions so as to permit reasonable ease of assembly. In the case of a simple tubular glass component diametric tolerances of ±.005 inches are about as good as can be maintained easily. Length variations in the same order are reasonable. The metal components likewise have normal machining variations.

For many small components the dimensional tolerances may be as much as ±3-5% in the nominal dimension. The cumulative effect on volume may be as much as ±10-15%. This, coupled with variations in the metal dimensions, may cause extreme variations in the thickness or "fill" of the glass seal from surface to surface. It is, therefore, fairly common practice in cases of this type to order the glass components by nominal dimensions but specifying the weight accurately to ± 3-5% which is readily maintained.

Glass preforms are usually shipped to the seal manufacturer in sealed containers and have had little or no handling after being heated to sinter the powdered glass. Pressed glass components, likewise, are usually packaged for shipment with a minimum of contact or contamination. These components should rarely, if ever, be given any cleaning prior to assembly and sealing and should never be touched with bare hands or even gloves which may leave lint or powder on the glass surface.

Cut glass tubing may have contaminants such as wax, water borne residue or finger prints on the surface. It is, therefore, often desirable to clean such pieces by means of a careful vapor degrease or washing in such a manner as was used for the metal components.

MANUFACTURE OF MOLDS

In order to accurately mount the terminals and glass in position for fusing onto an external body, it is necessary to make suitable jigs or molds for which the following rules are **important**:

1. Whenever direct contact with the glass during the fusing cycle is important there are only two suitable materials.

(a) Carbon(graphite) - Usually a grade such as L-56 or L-51 from Pure Carbon in St. Marys, Pennsylvania or an equivalent from various other suppliers. (A softer or harder grade may be used but in an oxidizing atmosphere the softer grades erode fairly readily at high temperatures and tend to leave a carbon residue imbedded in the glass.) Such carbon is available in sheet (plates) and rod.

When machining the carbon a good finish is desirable. However, lapping to produce a shiny surface will generally cause increased sticking of the glass with the resultant formation of chips and bubbles in the glass surface.

(b) Boron Nitride - Available from Norton Co., in Worcester and Carborundum Corp., Latrobe, Pennsylvania. An advantage of the boron nitride is that it has a better oxidation resistance than carbon. It is, therefore, recommended when the mold is to be exposed to a long-time, high-temperature, oxidation in an area of critical dimensions.

In addition, boron nitride is electrically insulating in itself and as oxidized. In some cases this may perhaps be a distinct advantage. Much the same requirements as to surface conditions prevail. One disadvantage is that boron nitride is much more expensive than carbon at the present time.

Both carbon and boron nitride should be cleaned and degreased by a short (5 minute or less) heating preferably in a neutral or reducing atmosphere to, say, 1800°F.

2. In general it is best to make the major structural part of the mold of the same metal as the body material of the seal particularly where multi-pin connectors are to be made.

In this respect let us consider a body made of a 300 series stainless steel with two holes 1.000 inches apart on-center into which are to be sealed terminals, concentric with the holes. Upon heating to 1000°C to seal the glass and then cooling to 500°C, approximately the point at which the glass becomes too viscous for free movement, we find that the hole spacing is 1.008 inches in the body while the hole spacing in a carbon mold would hold the terminals at a spacing of 1.002 inches on-center.

We can see that at this point the terminals are no longer concentric with the holes. As the piece continues to cool to room temperature the ends of the terminals held in the carbon mold will return to a 1.000 inch spacing while that portion of the terminals fixed in the glass seal will try to return to a spacing of 0.994 inches.

The effect of this expansion discrepency may cause cracking of the glass, deformation of the terminals and damage to the mold.

A somewhat similar analysis will show that the use of terminals of one metal with a mold of a different material may result in longitudinal displacement of the terminals from the design dimensions unless thermal expansion difference are carefully taken into account.

Two possible mold designs to permit the use of a basically steel mold with carbon in all areas of glass contact are shown in Figure 9.

FURNACE, ATMOSPHERE AND FIRING

The great bulk of seals such as those being considered

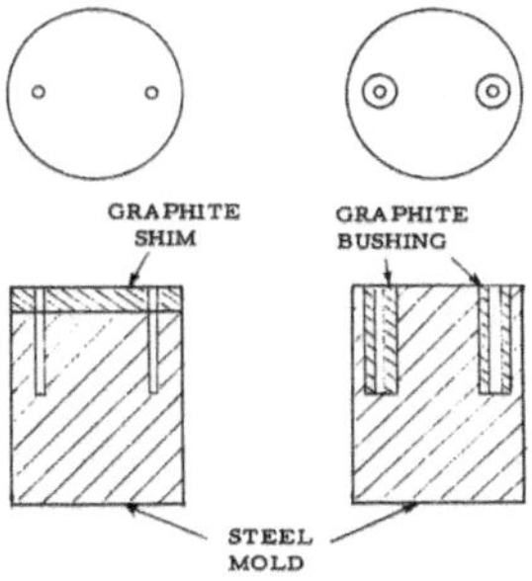

Figure 9. Two possible mold designs using a steel mold and either a carbon shim or carbon bushings. (Courtesy of D. G. O'Brien, Inc.)

are sealed by fusing the glass with the various components assembled in a mold and the assembly being passed through a conveyer furnace or lehr. In order to avoid excessive oxidation of the metal components and of the mold material it is essential that the furnace be of such a design that the atmosphere in the hot zone can be controlled.

Most furnaces are electrically heated, often with two or more heating zones independently controlled, and with a sealed inner jacket into which a preselected atmosphere can be injected. Usually the atmosphere is introduced in three places, the center of the hot zone and also at or near each end of the hot zone so that the controlled atmosphere completely envelopes the mold and seal all during the heating cycle. Furnaces of this type are generally built with the hot zone elevated above the entrance and exit ports so that normal diffusion and convection flow will not allow external air to enter the control zone. Frequently, too, a hanging curtain of fiber glass cloth is used to seal the exit port.

An early type of atmosphere, used in glass sealing was an exothermic gas mixture. This mixture is made by preburning a gas-air mixture, cooling the mixture to condense out all or nearly all of the water and using the remaining combustion products as an atmosphere for sealing chamber. If we assume that the gas being preburned is largely methane then $CH_4 + 2O_2 \rightarrow CO_2 + 2H_2O$ and if we use air as exactly 20% volume of oxygen we can see that a ratio of air to gas of 10:1 is very nearly stochiometric and the gas into the furnace is essentially carbon dioxide plus nitrogen.

If the air to gas ratio is increased slightly there is an excess of oxygen at low partial pressure and the exothermic gas is slightly oxidizing in nature. A reduction of the air-gas ratio below 10:1 results in the formation of an incomplete combustion product, containing some carbon monoxide and perhaps hydrogen at low partial pressures and the atmosphere is a reducing one. In all of these cases there is a partial pressure of water vapor as determined by the degree to which the burned gas mixture is cooled before introduction into the furnace. The presence of the water vapor will make it possible for the atmosphere to be oxidizing to chromium bearing stainless steel and to carbon molds even though it is reducing to iron, nickel or many other metals. (see Figure 10)

To properly use an exothermic gas atmosphere requires great care in the control of the gas to air ratio and in the control of the water content or dew point of the gas as it is introduced into the furnace. The use of Orstat or similar monitoring equipment at least hourly is to be strongly urged.

With the advent of readily available liquid nitrogen, hydrogen and oxygen many installations have gotten away from the use of exothermic gas and have gone to an atmosphere of pure nitrogen or nitrogen with small additions of either oxygen or hydrogen to create neutral, oxidizing or reducing

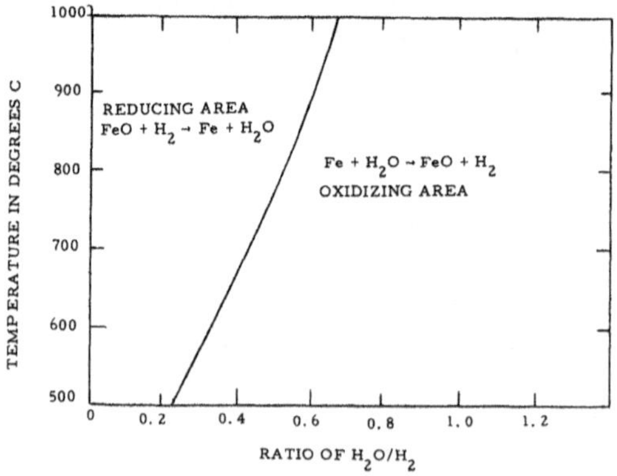

Figure 10. Equilibrium curve for the reaction $H_2O + Fe \rightleftharpoons H_2 + FeO$ as a function of temperature and the ratio H_2O/H_2.

conditions. The use of nitrogen plus 10% air for example may be used to create an oxidizing atmosphere while nitrogen with up to 10% hydrogen will create a non-explosive, reducing atmosphere. By the introduction of controlled amounts of water vapor further variations in control are possible.

Another atmospheric control process is based on the use of ammonia gas either catalytically cracked or burned with air. As in the exothermic gas, careful control of the ammonia to air ratio together with a condensing process can allow the formation of an atmosphere consisting of predominately nitrogen or with controlled small amounts of hydrogen or oxygen in excess of the nitrogen.

In many cases, the use of pure nitrogen separately or in combination with hydrogen or oxygen is the preferred process. Costs are comparable or only slightly higher than for the exothermic gas or ammonia process and the ease of control together with a reduced chance of manufacturing defects makes this process extremely good.

During the sealing operation many manufacturers tend to go to a reducing atmosphere for compression seals so that the metal components leave the furnace with a minimum amount of oxide to be removed before plating or shipping. However, as expressed before, this precludes the possibility of any substantial chemical bond between the metal and the glass. Certainly, in the manufacture of matched seals such a process would be disasterous. It is suggested, therefore, that a normal atmosphere of 0 - 2% oxygen be maintained in a nitrogen atmosphere for most seals. However, in the case of a preoxidized metal system, nitrogen with a partial pressure of hydrogen and water vapor in the ratio of 2:1 may be sufficiently neutral so as to make successful seals.

It is recommended that periodic examination of the glass-metal interface be made by means of a metallographic microscope on all seals as part of the quality control process. In most cases, the presence of 0.0001 - 0.0002 inches of non-porous metal oxide will assist in maintaining seal integrity.

A cursory consideration of the sealing process would suggest that the parts to be sealed are simply heated to some high temperature as dictated by the thermal properties of the glass at which point the glass "melts" and flows so as to form a seal. In general, this is approximately true although by common practice the furnace temperature may be kept nearly constant for many different glasses. 1000°C or 1800°F is used for sealing glasses as different as 9010, 7052 and even 7740. Variations in the speed of the conveyor are more easily and quickly made and will often permit sealing of quite different glasses simply by extending the duration of exposure to the high temperature.

An important factor to consider is that during the sealing operation the assembled components pass through the furnace on a conveyer belt with a rather abrupt rise in temperature as the assembly enters the hot zone of the furnace with a decrease in the rate of temperature rise as the seal assembly approaches the temperature of the furnace. All during the heating period the viscosity of the glass is decreasing and the glass is beginning to flow so as to make contact with both the metal body and the terminal. If the assembly is now maintained at temperature while the assembly passes through the hot zone of the furnace this flow process continues at a maximum rate. As the assembly leaves the hot zone a cooling cycle begins which is essentially the reverse of the heating cycle but it should be remembered that some flow of the glass may take place even during the early or high temperature portion of this cycle.

Depending upon the mass and heat capacity of the mold and seal assembly the shape of the heating and cooling curves may be quite varied. As a result, the sealing process may be as dependent upon these as upon the peak temperature which the seal attains and the time at this temperature.

Since fluidity and hence, flow rate are essentially inverse functions of viscosity it is sometimes desirable to attempt a graphical representation of the relationship between firing time and temperature as in Figure 11. This graph is based on the assumption that all of the sealing operation is done during the soaking period at a given temperature. The graph, as drawn, is only an approximate one and can be used only for low mass seals where temperature rise time and cooling time are extremely short, the glass on which this is based is 9010. Similar graphs can be drawn for other glasses but would usually show nearly

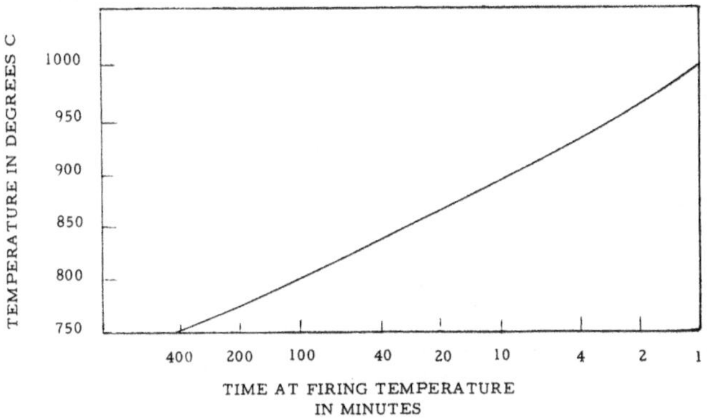

Figure 11. Approximate relationship between fusing temperature and time of fusing.

parallel curves displaced along the time axis.

A somewhat different type of graph can be made by attempting to include both the heating and cooling portions of the firing cycle on a total time-temperature plot where the area under the curve is an integration of the entire sealing cycle. In Figure 12, such a plot is made of the actual measured temperature on a single type of seal fired at two different temperatures for different periods of time. In the two cases the area under the curves are nearly the same and the products, as measured by seal quality, are also equivalent. Other samples of drastically different mass will distort the above curves. However, the total area under the curves will remain almost unchanged. Therefore, seal quality is maintained. In no case should it be considered that such graphs are definitive, but rather, that they are indicative of trends or reasonable approaches to the solution of fairly complex problems.

A typical temperature curve for a seal proceding through a furnace might be as shown in Figure 13. This would be an appropriate profile for the firing of either matched seals or compression seals using a terminal with the same thermal expansion as the sealing glass (9010 in this case).

In one special type of seal, it is desirable to reduce or eliminate the holding time at the annealing temperature. This seal is one in which the coefficient of thermal expansion of a terminal is higher than that of the surrounding

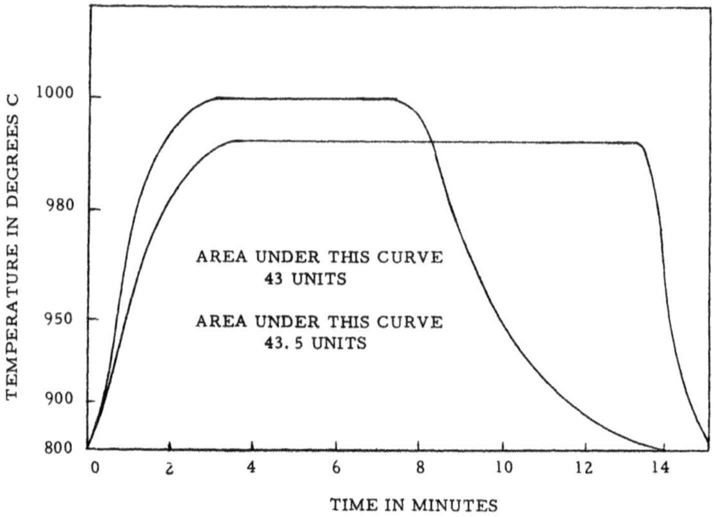

Figure 12. Temperature-time plot of two fusing cycles for a typical seal. Temperature scale is corrected so that equal area fusing cycles will give essentially identical end products. (Courtesy of D. G. O'Brien, Inc.)

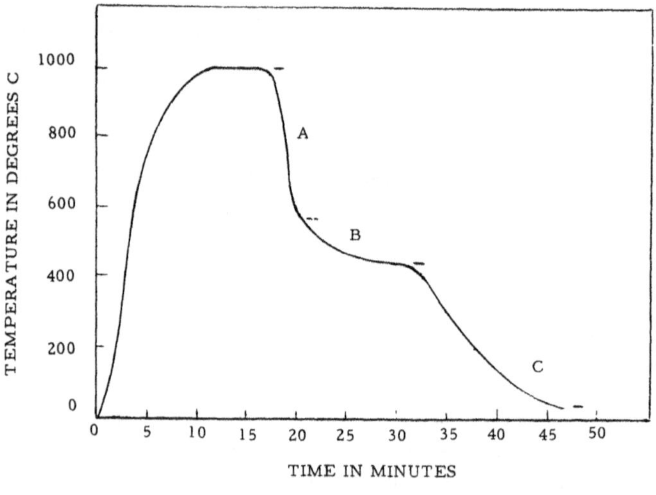

Figure 13. Temperature-time plot for a typical firing cycle. A. Rapid cooling permitted since glass flow will relieve strain quickly. B. Residual permanent strain controlled by cooling rate in this area. C. Rapid cooling here may cause fracture by thermal shock but will not effect permanent residual strain. (Courtesy of D. G. O'Brien, Inc.)

glass. Normally such a condition results in a radial tensile stress existing between the terminal and the glass at their interface. Frequently the tensile stress, as in the case of a steel terminal in 9010 glass, will be such as to cause the seal to leak and even to permit removal of the terminal from the glass by simple pressure with the fingers. It may be possible to overcome this defect by quick cooling of the seal and indeed, in some cases, cooling or radiating fins attached to the terminals may be effective by maintaining a substantial temperature difference between the terminal and the glass immediately adjacent.

Figure 14 is a graph of the thermal expansion curves for 52 alloy, cold rolled steel and 9010 glass. Figure 15 and 16 show the effect created when a metal terminal is fused into a glass and then cooled from the fusing point to room temperature. Assuming that temperature differences may exist during the cooling cycle, the terminal may be either substantially above or below the glass in temperature at the time the glass "sets". The phrase "temperature at which the glass sets" is used here to define a temperature below which glass flow will not relieve any appreciable mechanical strain in the time available as the seal cools. Because of the rather rapid cooling present in the normal manufacturing process, this temperature is generally somewhat above the annealing temperature of the glass reported in the literature. For the purpose of this work a 30°C increase has been used. In Figures 15 and 16 the thermal contraction curves for the

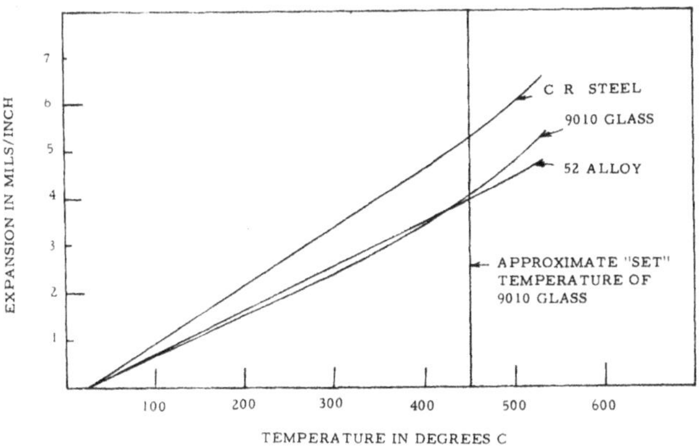

Figure 14. Thermal expansion of 9010 glass and two metals commonly used as terminals with this glass.

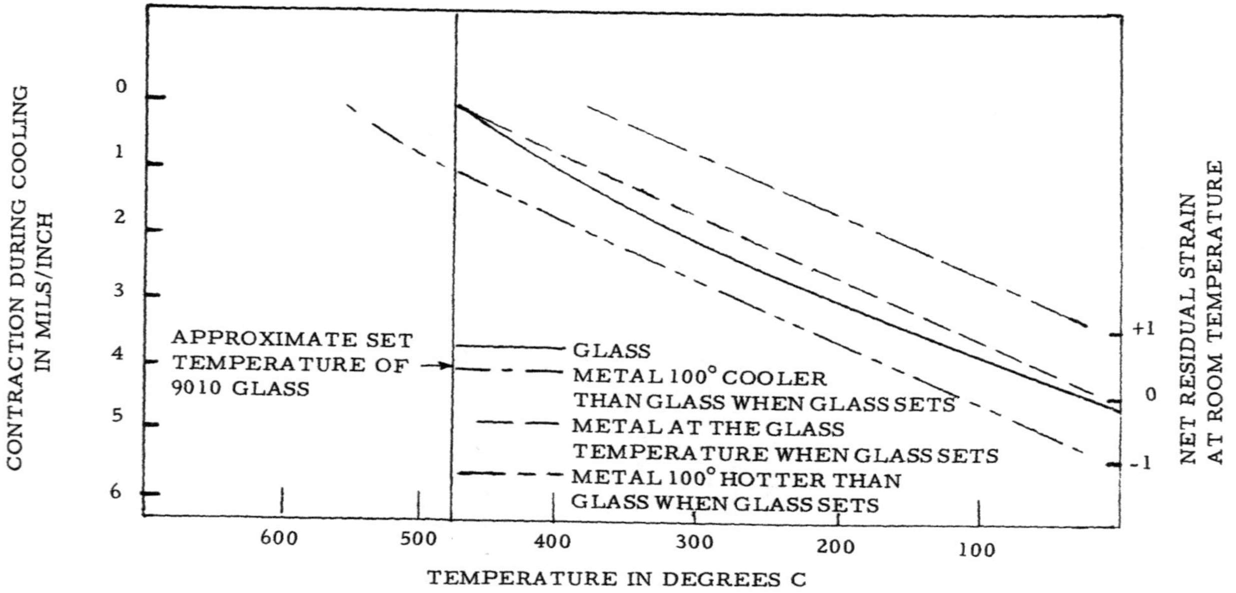

Figure 15. Comparative strains in seals using 52 alloy terminals and 9010 glass. The graph shows a net (+) strain caused by having the metal 100°C cooler than the glass, essentially no strain caused by having the metal temperature equal to the glass temperature, and a net (-) strain caused by having the metal 100°C hotter than the glass when the glass sets. (Courtesy of D. G. O'Brien, Inc.)

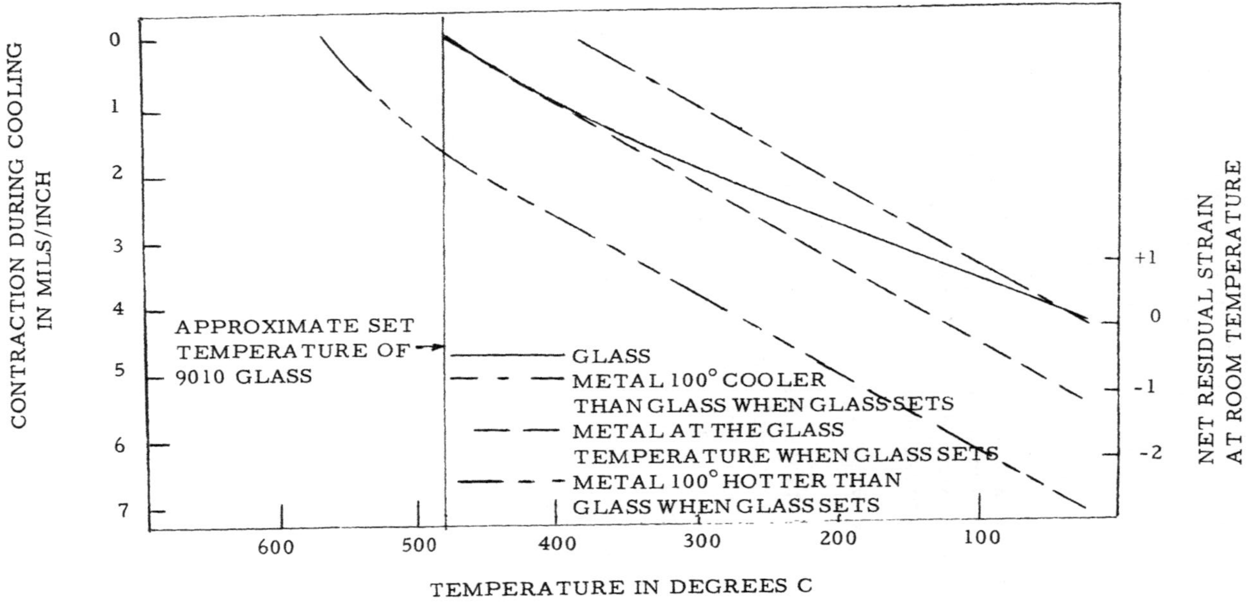

Figure 16. Comparative strains in seals using cold rolled steel terminals and 9010 glass. The graph shows essentially no strain caused by having the metal 100°C cooler than the glass, a net (-) strain caused by having the metal temperature equal to the glass temperature and a very high (-) strain caused by having the metal 100°C hotter than the glass when the glass sets. (Courtesy of D. G. O'Brien, Inc.)

metals are shown in such manner as to indicate the effect on final seal strain for three different cooling processes; (1) the metal temperature below the glass temperature (2) no temperature difference and (3) the metal temperature above the glass temperature. In both figures a net (+) strain indicates that the glass is in circumferential tension at the metal interface and if failure occurs, the glass will exhibit radial cracks. Conversely a net (-) strain indicates that the glass is in radial tension at the metal interface and that if failure occurs it will either be at the interface or the glass will exhibit circumferential cracks around the terminal. In either case if the net strain exceeds approximately 1 mil/inch, failure of the seal becomes likely.

STRESS ANALYSIS AND TYPICAL DEFECTS

While no attempt will be made to give a detailed analysis of the stresses involved, nevertheless, it seems appropriate to illustrate here a few types of seals and possible failures associated with each.

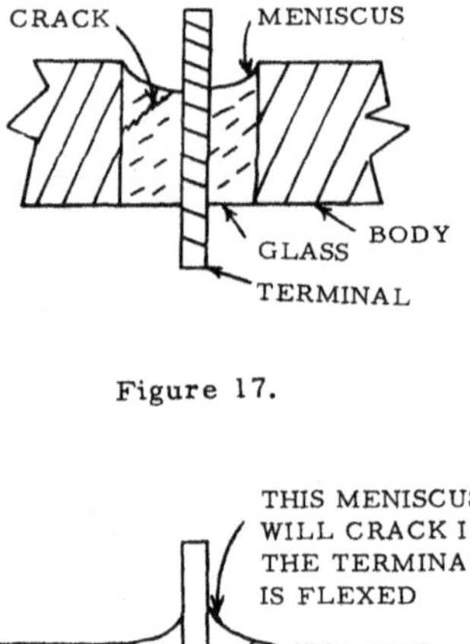

Figure 17.

In Figure 17 a seal is shown in which the metal body has a higher coefficient of thermal expansion than the glass. During cooling a high radial compression stress is created in the glass. Any significant meniscus on the glass, where it is fused to the metal, acts as a cantilever and the glass is frequently cracked in such a manner as to form a ring crack. Often a glass ring can be completely removed from the seal, leaving the seal intact below the ring, but, in other cases, the crack may propagate completely through the seal.

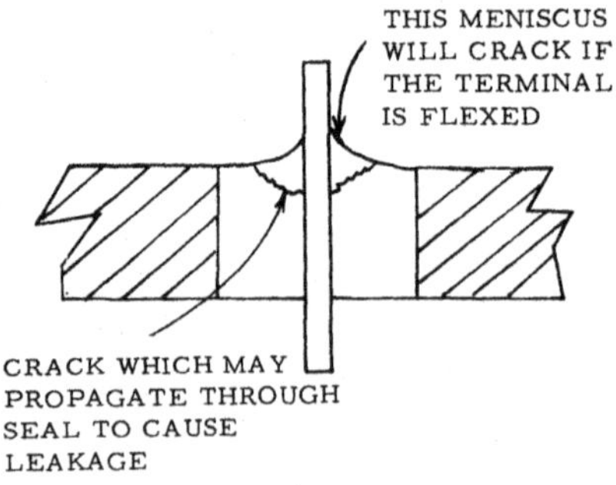

Figure 18.

In Figure 18 a seal is shown in which a high meniscus has been formed around the terminal. In this case any flexure of the terminal will create a crack in the glass around the terminal and seal failure may result. A special case may be made of seals having this configuration in which the terminal has a lower coefficient of thermal expansion than does the glass. As in Figure 17 this meniscus acts as a cantilever and the glass is in a highly strained condition. Again, a ring crack may form completely around the terminal, even without flexure of the terminal.

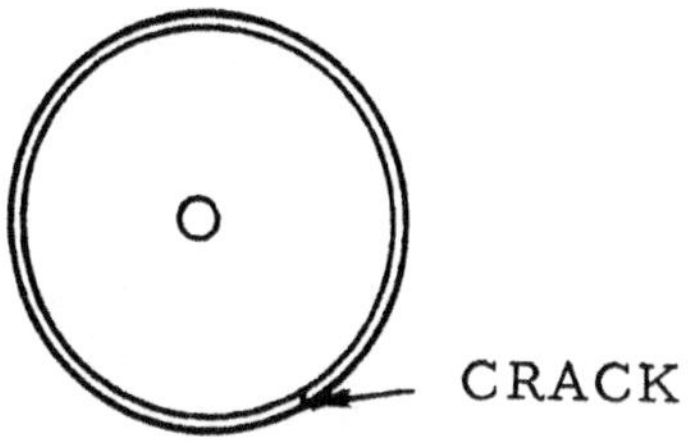

Figure 19.

In Figure 19 a seal is shown in which the outer body has a higher coefficient of thermal expansion than does the glass. If the metal wall thickness is insufficient, the probability is high that the metal body will fail in tension.

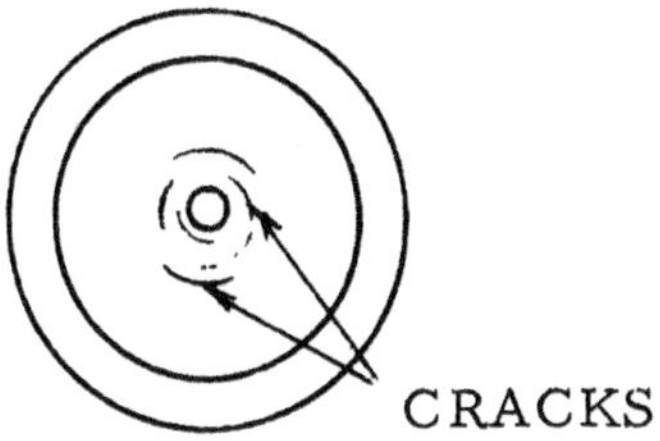

Figure 20.

Figures 20 and 21 show the possible results of differences in coefficient of thermal expansion of the terminal and the surrounding glass. If, as in Figure 20, the terminal is of a higher coefficient than the glass, then the glass is in radial tension, failure may occur in the glass-to-metal bond or by formation of circumferential cracks in the glass around the terminal. In either case leaking of the seal is almost certain.

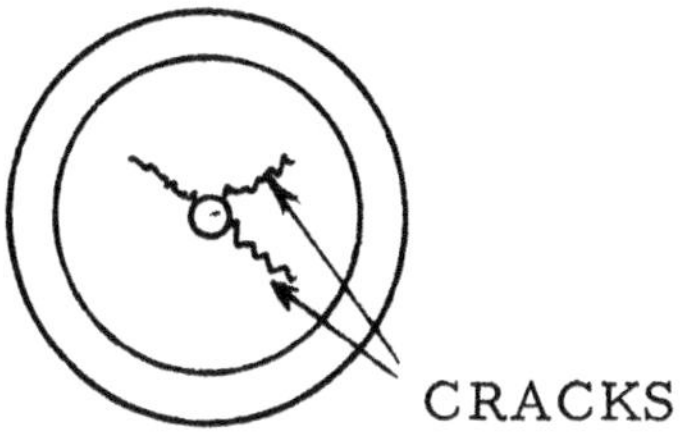

Figure 21.

If, conversely, as in Figure 21, the terminal is of a lower coefficient than the glass then the glass is in radial compression and tangential tension. Radial cracks may easily occur in this case.

A brief analysis of some of the stresses and strain of a hypothetical seal consisting of a solid glass plug in a metal body of simple tubular shape may serve to illustrate some of the techniques applicable to such problems.

The following calculations are based on the assumption that there are no bonding forces between the outer metal tube and the glass plug and that the metal is higher in expansion than is the glass.

If a structure of this configuration has been sealed and is in a strain free condition at the strain point of the glass, upon cooling, differential contraction takes place and would result in different dimensions for the inside metal diameter and the outside glass diameter at room temperature, if free movement were possible.

This difference in dimensions must be reconciled by two factors, 1) a stretching of the metal and 2) a compression of the glass. The stress in the two components may be quite different but the force must be equal and opposite. Tensile stress and displacements are arbitrarily assigned (-) values. (Figure 22)

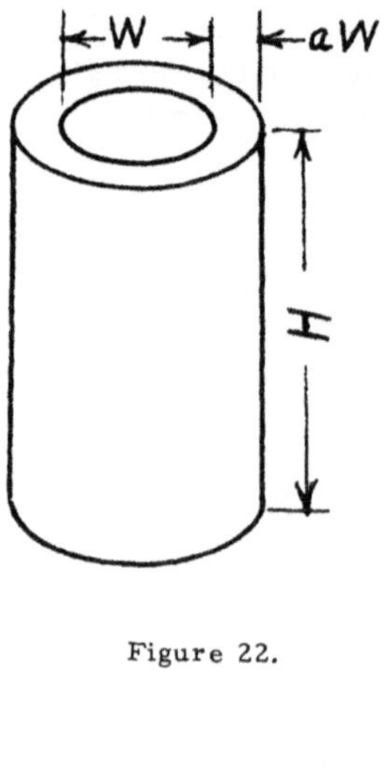

Figure 22.

W_g = glass deformation under stress.

W_m = metal deformation under stress.

E_g = thermal expansion coef. for glass.

E_m = thermal expansion coef. for metal.

ΔT = temperature difference from glass strain point to room temperature.

F_g and F_m = force on glass and metal.

ΔW_g and ΔW_m = dimensional changes in glass and metal.

M_g and M_m = modulus of elasticity of glass and metal.

S_g and S_m = unit stress in glass and metal.

$$\Delta W_g - \Delta W_m = W(E_g - E_m)\ \Delta T = W(\Delta E)\Delta T \tag{1}$$

$$-F_m = +F_g \tag{2}$$

$$\Delta W_g = W(F_g/WH\ M_g) = \frac{F_g}{HM_g} \tag{3}$$

$$\Delta W_m = W(F_m/2aWH\ M_m) \tag{4a}$$

$$\text{if we let } M_m = bM_g,\ \Delta W_m = F_m/2aHbM_g \tag{4b}$$

$$\text{replacing } F_m \text{ by } -F_g \text{ from eq. 2, } = -F_g/2aHbM_g \tag{4c}$$

Substituting 3 and 4c into 1,

$$\frac{F_g}{M_gH} + \frac{F_g}{2abHM_g} = W(\Delta E)\Delta T \tag{5}$$

$$F_g = \frac{2ab\ WHM_g\ (\Delta E)\Delta T}{1 + 2ab} \tag{6}$$

by substituting eq. 6 into eqs. 3 and eq. 4c we get,

$$W_g = \frac{2\ abW(\Delta E)\Delta T}{1 + 2ab} \quad \text{(7a)} \quad \text{and} \quad \Delta W_m = \frac{-W(\Delta E)\Delta T}{1 + 2ab} \quad \text{(7b)}$$

or dividing force by area.

$$S_g = \frac{2ab\ M_g(\Delta E)\Delta T}{1 + 2ab} \quad \text{(8a)} \quad S_m = \frac{-bM_g(\Delta E)\Delta T}{1 + 2ab} \quad \text{(8b)}$$

Dividing 7a into 7b,

$$\frac{\Delta W_m}{\Delta W_g} = -\frac{1}{2ab} \tag{9}$$

Similiarly 8a into 8b gives

$$\frac{S_m}{S_g} = \frac{-1}{2a} \tag{10}$$

To illustrate the magnitude of S_m, S_g, ΔW_m and ΔW_g let us choose as a metal a stainless steel for which $M_m = 6 \times 10^7$ lbs/in^2, $E_m = 170 \times 10^{-7}$ in/in/deg. C and 9010 glass, $M_g = 1 \times 10^7$ lbs/in^2 and $E_g = 90 \times 10^{-7}$ in/in/deg C.

Then $b = 6$, $\Delta E = 80 \times 10^{-7}$ in/in/deg.C. If also $\Delta T = 475^\circ C - 25^\circ C = 450^\circ C$, we can thus calculate values as follows

$$\Delta W_g = \frac{2 \times a \times 6 \times 80 \times 10^{-7} \times 450W}{1 + 12a} = \frac{4.3 \times 10^{-2} aW}{1 + 12a} \text{ inches} \tag{11}$$

$$\Delta W_m = \frac{3.6 \times 10^{-3} W}{1 + 12a} \text{ inches} \tag{12}$$

$$S_g = \frac{4.3 \times 10^5 a}{1 + 12a} \text{ lbs/in}^2 \tag{13}$$

$$S_m = \frac{2.16 \times 10^5}{1 + 12a} \text{ lbs/in}^2 \tag{14}$$

Thus by equation 14 we see that the stress in the metal for a given glass and metal combination has an upper limit and that the actual stress is a function of the wall thickness of the metal compared to the diameter of the glass as represented by the term "a" in the equation.

S_m can be given limiting values based on either the ultimate strength of the metal or on the elastic limit of the metal and value of "a" determined for each of these.

If $S_m = 1.5 \times 10^5$ lbs/in^2 for the ultimate strength then a = 0.036 which means that for a 1.00 inch diameter glass core, a tublar wall of less than .036 will fail (split) in tension in a seal as described. If $S_m = 2 \times 10^4$ lbs/in^2 is used as the upper limit of elasticity then a = 0.81 which means that for any value of wall thickness less than 0.81 around a 1.00 inch glass core, the tubular metal element will be permanently stretched during the cooling cycle. Ideally this is a minimum value for "a".

One particularly significant result of this type of calculation may be observed by using $S_m = 3 \times 10^4$ lbs/in^2 as the normal "strength" of the steel which is defined in terms of permanent deformation to the extent of 0.2%.

In this case a = 0.52W, but more importantly the metal tube will have been permanently stretched 0.002 inches on a glass diameter of 1.00 inches. It should be noted here that for those cases where S_m is greater than the elastic limit of the metal, all values calculated on the basis of the preceding equations are in error because the permanent distortion of the metal ΔW_m is greater, ΔW_g is less, and S_m and S_g are less than theoretical.

The significance of these high stress values and the permanent distortion of the metal lies primarily in the effect on the seal when it is reheated in service. For example, in the case being discussed here the differential expansion using a 1.00 inch diameter glass plug is 0.0036 inches. If now the "a" value of 0.52 is chosen and the steel is stretched to the extent of 0.2% or 0.002 inches, then the actual difference in "free" size is only 0.0016 inches. Upon reheating in operation a temperature rise of 200°C (to 225°C) will result in complete elimination of any compression stress. Any further heating will, in effect, open a gap between the steel and the glass causing a leak. This again points out the desirability of an oxide bond even on compression seals, particularly where high tensile stresses and permanent deformation of the metal body can occur.

LEAK CRITERIA AND TESTING

In any glass to metal seal all of the mechanical and electrical parameters are significant criteria of seal quality and usefulness but, without question, the one single determination of the usefulness of the seal is: "Is it a seal?"

Ideally, a seal should permit no migration of any gas, air or material from one side of the seal to the other. In point of fact, this ideal is rarely, if ever, achieved and indeed, theoretically, cannot be if we speak in terms of extremely long periods of time. The definition of a good seal should, more realistically then, be defined on the basis of permitting only an insignificant migration through the seal during the useful life of the device in which the seal is to be used.

In general for applications such as light bulbs or vacuum tubes the permissible leak rate is so low as to be unmeasurable except by the device itself. For many solid state devices and certain pieces of electronic hardware, although a "perfect" seal is desirable, something less than perfect is usable.

A Helium Mass Spectrometer is the primary means of evaluating the magnitude of a leak (or leak rate) in a conventional glass to metal seal. In this test the seal to be tested is mounted on a manifold attached to a mass spectrometer specifically adjusted to detect Helium. The manifold is exhausted to a high vacuum and the exterior side of the seal thus enclosed in a plastic bag or envelope containing pure Helium at atmosphere pressure. If a leak exists in the seal, the magnitude of the leak can be determined and correlated with the spectrometer current.

The leak rate, as measured, is reported in either micron-cu ft/hr or cc- atm/sec. for air. A micron-cubic foot of air is one cubic foot of air at one micron pressure and a cc-atm being one cubic centimeter of air at one atmosphere pressure, a leak rate of 1 micron-cu ft/hour being equal to 1.04×10^{-5} cc-atm/sec. The maximum acceptable leak in most cases would be in the order of 1×10^{-8} cc atm/sec and in many cases even lower leak rates are specified as the upper limit. For the purpose of understanding the magnitude of this leak rate it should be noted that 1×10^{-8} cc-atm/sec is equivalent to a quantity of gas of 4.5×10^{-13} moles/sec, 1.4×10^{-5} moles/year or 4.2×10^{-4} grams of air/year.

This method of leak rate measurement is always made with a pressure gradient across the seal of essentially 1 atmosphere. If a higher pressure gradient is to be encountered in service operation,the maximum permissable leak rate,as tested,must be adjusted.

The mass spectrometer method of testing seals is somewhat slow, laborous and hence expensive where large number of seals are to be tested. A secondary test for leak rate is "Spot Check" dye-penetrant and developer made by Magnuflux Corp.

In this process it is necessary to start with the test area completely free of moisture (seals should be tested immediately after cooling from the fusing operation or after drying for 1 - 2 hours at 110-125°C preferably under vacuum). The dye penetrant, an oil-based dye dissolved in a hydrocarbon vehicle, is painted onto the test area, care being taken to insure that all surfaces where a leak may exist are covered with the penetrant. The developer, essentially a whitish powder dispersed in a solvent, is painted onto the other side of the seal and allowed to dry. Avoid an excess of the developer since a thin film will give more easily observed results than a thick heavy coating. After application of penetrant and developer the seal is set aside for observation. Depending upon the size of any existing leaks a spot or spots of red penetrant will begin to show through on the developer after a period of time. Normally about 2 hours observation will be sufficient to show any single leak in the order of 10^{-8} cc-atm/sec. Observation in the order of 12-24 hours should show any single leak in the order of 10^{-10} cc-atm/sec.

In this evaluation of leaks shown by the dye penetrant it is necessary to recognize the fact that, with a multiple terminal connector, there may be a multiplicity of leaks in the range of 10^{-9} - 10^{-10}, cc-atm/sec. Failure to observe the seal for a sufficient period of time will not show these to be present and yet the summation of them will exceed the 10^{-8} value as measured on a mass spectrometer. The Spot Check test is obviously a good production control test but should not replace the mass spectrometer as a primary reference test.

A special group of seals requires an entirely different evaluation than simple leak testing as outlined above. This group of seals includes numerous types used for marine applications, among which would be connectors for marine cables and for hull penetration connectors in submarine equipment.

In the case of a deep submergence vessel, such as the Trieste, it is not sufficient to say that the seal shall not leak air at a pressure differential of 1 atmosphere. A seal to be used on electronic gear connected through the hull of a vessel submerged to a depth of nearly 6 miles in the ocean must also be able to insure against water leaks at a pressure of at least 20,000 Psi or preferably 30,000 Psi for a safety margin.

Such seals are frequently designed with a fairly long glass sealing length, a taper in the metal hole so that the glass is driven into the taper by external pressure, and is generally of a compression type seal to further strengthen the glass.

Testing such seals for leaks is usually composed of two parts: the first being as a standard seal test and the second being similar to the dye penetrant method but with the seal assembled into a pressure manifold so that high test pressures can be maintained for several hours. The high pressure check would normally specify no dye leaks at say 30,000 Psi pressure in 2 - 6 hours. On a production basis failure of even 1-2% of a series of such seals may indicate poor design.

SUMMARY

It has been shown that a great number of different types of glass-to-metal seals can be made with various types of metals and glasses. The electrical properties of the seal are determined by the type of glass chosen while the choice of metal largely determines the chemical and mechanical properties of the seal.

In the case of seals made with differences in expansion coefficient between the different components stress may be very high. Structural design must be such as to reduce some of these effects or, alternatively, make certain that the stress is compressional on the glass to avoid failure.

Chemical processing of the metal is extremely important and the presence of a metal oxide to which the glass may bond is imperative in most cases.

Specific glass and metal combinations have been discussed, with general manufacturing methods and properties of finished seals. Criteria and testing of seals for hermiticity has been explained in some detail.

REFERENCES

1. Kohl, W. H., Materials and Techniques for Electron Tubes, Reinhold (1960).
2. Shand, E. B., Glass Engineering Handbook, McGraw-Hill (1958).
3. Morey, G. W., Properties of Glass, Reinhold (1938).
4. Corning Glass Works, "Engineering with Glass".
5. Corning Glass Works, "Properties of Selected Commercial Glasses" (B-83).
6. Wilbur B. Driver Co., "Glass-To-Metal Sealing Alloys".
7. Smithells, C. J., Metals Reference Book, Plenum (1967).
8. Amer. Soc. for Metals, Metals Handbook.
9. Partridge, J. H., "Glass-To-Metal Seals", The Society of Glass Technology, Elmfield, Sheffield, England (1949).
10. Dartnell, R. C., Fairbanks, H. V., and Toehler, W. A., "Investigation of the Adherence of Glass to Metals and Alloys", J. Am. Ceram. Soc., 34, 357-360 (Nov. 1951).
11. Dietzel, A., "Reactions and Bonding Between Glass and Metal on Melting", in German, Glastech. Ber., 24, 263-268 (Nov. 1951).
12. Fulrath, R. M., Mitoff, S. P., and Pask, J. A., "Fundamentals of Glass-To-Metal Bonding: III, Temperature and Pressure Dependence of Wettability of Metals by Glass", J. Am. Ceram. Soc., 40, 269-274 (1957).
13. Weiss, W., "Wetting Properties and Mechanical Strength of Glass-To-Metal Seals", in German, Glastech. Ber., 29, 386-392 (Oct. 1956).
14. Hagy, H. E. and Ritland, H. N., "Viscous Flow in Glass-To-Metal Seals", J. Am. Ceram. Soc., 40, 58-62 (Feb. 1957).
15. Monack, A. J., "Glass-To-Metal Seals in Electronic Components and Applications", Elec. Mfg., 39, 96-101, 162, 164, 166, 168, 170, 172, 174, 176, 178, 180 (Feb. 1947).
16. Redston, G. D. and Stanworth, J. E., "Glass-To-Metal Seals", J. Soc. Glass Technol., 29, 48-76T (132) (1945).
17. Dusing, W., "Glass-To-Kovar Seals for High-Vacuum Tubes", Telefunken Zeitung, 26, 111-120 (Mar. 1953).
18. Oldfield, L. F., "Sealing Glass to Metal", The Times Review of Industry, The Times Publishing Co., Ltd., London (Sept. 1958).

19. Rawson, H., "The Theory of Stresses in Two-Component Glass-To-Metal Tube Seals", J. Sci. Instr., 26, 25-27 (Jan. 1949).
20. Adam, H., "The Theoretical Foundations of Glass Compression Seals and Their Practical Consequences", Trans. Soc. Glass Technol., 38, 285-296 (1954).
21. A.S.T.M. Specification F204-50, "Test for Surface Flaws in Tungsten Seal Rod and Wire".
22. Chilcot, A. L., "Leakage of Water Through Tungsten Seals", Electronic Eng., 17, 693 (Sept. 1945).
23. Davis, M., "Glass-To-Metal Seals of Strong Adherence", Services Electronics Res. Lab. Tech. J., 4, 1-11 (Feb. 1954).
24. Stanworth, Rawson H., and Knowles, M., "Surface Treatment of Molybdenum and Tungsten for Glass Sealing", U. S. Patent 2,836,935 (June 3, 1958).
25. Rawson, H. and Denton, E. P., "The Glass Sealing Properties of Titanium and Zirconium", Brit. J. Appl. Phys., 5, 352-353 (Oct. 1954).
26. Fink, C. G., "Copper-Sheathed Sealing Wire", U. S. Patent 1,498,908 (June 24, 1924).
27. Eldred, B. E., "Copper-Sheathed Wire for Glass Seals", U. S. Patent 1,140,134/135/136 (May 18, 1915).
28. Kingston, W. E., "Low-Expansion Alloys for Glass-To-Metal Seals", Trans. Am. Soc. Metals, 30, 47-67 (Mar. 1942).
29. Monack, A. J., "Method of Sealing Glass to Iron", U. S. Patent 2,345,278 (Mar. 28, 1944).
30. Hidnert, P., "Thermal Expansion of Heat-Resisting Alloys: Ni-Cr, Fe-Cr, and Ni-Cr-Fe Alloys", J. Res. Bur. Standards, 7, 1031-1066 (RP388) (Dec. 1931).
31. ASTM Standard Specification for 17-percent Chromium-Iron Alloy for Sealing to Glass (S256-53).
32. Hull, A. W., Burger, E. E., and Navias, L., "Glass-To-Metal Seals II", J. Appl. Phys., 12, 698-707 (Sept. 1941).
33. Redston, G. D., "Iron-Nickel-Cobalt Alloy for Sealing to Glass", J. Sci. Instr., 23, 53-57 (Mar. 1946).
34. "Sealing Glass to Kovar", Bull. 145, Stupakoff Ceramic and Manufacturing Co., Latrobe, Pa.; Carborundum Co., Refractories Div. (1945).
35. Mairs, K. H., "Fe-Ni-Co Alloys for Glass-To-Metal Seals", J. Metals, 40, 460-464 (May 1952).
36. Pask, J. A., "New Techniques in Glass-To-Metal Sealing", Proc. IRE, 36, 286-289 (Feb. 1948).

SUBJECT INDEX

AUTHOR INDEX

GPSR Compliance

The European Union's (EU) General Product Safety Regulation (GPSR) is a set of rules that requires consumer products to be safe and our obligations to ensure this.

If you have any concerns about our products, you can contact us on ProductSafety@springernature.com

In case Publisher is established outside the EU, the EU authorized representative is:

Springer Nature Customer Service Center GmbH
Europaplatz 3
69115 Heidelberg, Germany

Batch number: 10761588

Printed by Printforce, the Netherlands